PALAEOECOLOGY OF AFRICA

VOLUME 18

Van Zinderen Bakker collecting pollen of *Senecio brassica* in the afroalpine vegetation on Mt Kenya above Sirimon track.

PALAEOECOLOGY OF AFRICA

AND THE SURROUNDING ISLANDS

VOLUME 18

Edited by
J.A.COETZEE (D.Sc.)
Department of Botany, University of Orange Free State, Bloemfontein

A.A.BALKEMA / ROTTERDAM / BROOKFIELD / 1987

Published by

A.A.Balkema, P.O.Box 1675, 3000 BR Rotterdam, Netherlands

A.A.Balkema Publishers, Old Post Road, Brookfield, VT 05036, USA

ISBN 90 6191 689 5

ISSN 0168-6208

Printed in the Netherlands

CONTENTS

FOREWORD

This volume marks a change in the editorship of *Palaeoecology of Africa* as E.M. van Zinderen Bakker who fulfilled this task during twenty years has relinquished his responsibility. It is very gratifying to be able to announce that Professor K.Heine from the University of Regensburg, West Germany, has willingly consented to take on the task of Editor-in-Chief from Volume 19 onwards. We are sure that under his guidance this publication will continue to receive the same support as it did in the past.

The series was started in 1966 with the compilation in Volume 1 of the eight Reports on *Palynology in Africa* written by Van Zinderen Bakker covering the years 1950-1963. Since then *Palaeoecology of Africa* has become an independent international medium for palaeoenvironmental studies in Africa. The flow of manuscripts which reaches us indicates the need for a publication of this type. I wish to thank the authors and the publisher for their continuous support.

This volume is dedicated to our former Editor-in-Chief as a token of friendship of his many colleagues. At the express wish of our octogenarian, this publication is not a 'Festschrift' but an ordinary volume although several colleagues have contributed to it with the intention of extending their good wishes to an old friend.

J.A.Coetzee
Acting Editor

Bloemfontein
1 May 1987

DEDICATION TO E.M. VAN ZINDEREN BAKKER

This volume is dedicated to Eduard Meine van Zinderen Bakker Sr., the founder and editor of *Palaeoecology of Africa.* He has been its mentor for twenty years and under his aegis the journal has developed into a well-known and widely circulated publication.

As he has reached his 80th birthday on 15 April 1987 he has decided to relinquish his editorship and it will, therefore, be very appropriate to give a short summary of his career in this 18th volume.

Our editor spent forty years of his life in the Netherlands and forty years in South Africa. Since childhood he has had a profound love for all living things and his great reverence for life has been his mainstay in all his wanderings.

Born in Friesland and educated at the Municipal University of Amsterdam he devoted the first half of his life to education, nature conservation and plant ecology in Holland. As one of the founders of the youth hostel movement in the Netherlands he often took his pupils of the Royal College ('Koninklijke HBS') and Grammar School in Apeldoorn on hikes to bring them into direct contact with the marvels of nature. In these years he was actively engaged in ecological studies of the marshes, fens and lakes of western Holland on which he published two books.

During the tragic years of the war he continued his work as far as possible and at real risk of his life was engaged in the resistance movement. After the liberation of his country the Netherlands Government invited him to take part in the political screening of the national education system. He also worked in close contact with the headquarters of the Canadian Army in the Netherlands for the well-being of the soldiers.

After this very active and productive period in his life he started to look towards other horizons. Since boyhood he had been an avid reader of books on faraway countries, eventually concentrating on studying the geology and biology especially of Africa, the dark continent which fascinated him most of all and which he longed to visit. Early in 1947 his dreams were realised and he emigrated with his wife and two sons to South Africa where he became engaged in teaching botany at the University of the Orange Free State in Bloemfontein. His very rich background, culture, sharp wit and sense of humour which he brought into the lecture rooms enriched the lives of his students, many of whom have become his friends.

In his new surroundings he soon set out to apply the method of pollen analysis to the Quaternary problems of Africa. In order to improve his knowledge of palynology he paid extended visits to the fathers of pollen analysis, H.Godwin, F.Firbas, G.Erdtman and J.Iversen. After initial palynological studies in South Africa and Lesotho the scarcity of suitable material made it necessary for him to move northwards to Angola, Tanzania, Kenya and Ethiopia. During these long safaris with his sons he intensely enjoyed primitive camp life and the beauty of Africa. He was fortunate in being able to make deep borings in the swamps and crater lakes on the high East African mountains, where preliminary work by O.Hedberg and M.E.S.Morrison had already shown the potentials of pollen analytical studies. I had the much appreciated privilege to carry out the analyses of most of Van Zinderen Bakker's valuable material which he unfortunately had no time to work on. The results indicated that considerable changes in temperature which accompanied the last glacial maximum in Africa were coeval with similar changes in Europe. Many others extended this research independently and showed the enormous consequence these climatic variations had on the geology and biogeography of Africa. The results of this research were in close agreement with parallel studies done by T.van der Hammen and his colleagues in the high Andean mountains.

In connection with the pollen analytical results in Africa, Van Zinderen Bakker was later chiefly engaged in studies of the reconstruction of palaeoenvironments in the Sahara, East- and southern Africa and especially in the history of the Namib desert.

The work on the East African mountains showed Van Zinderen Bakker the basic significance of the temperature factor and consequently of the energy budget of the earth for understanding climate. He stressed the importance of his parameter as a decisive factor in palaeoclimate chronology. This principle aroused his interest in the Antarctic region. After trying for many years to visit the sub-Antarctic he was eventually able to equip an expedition in 1965 to study the geology and biology of the islands Marion and Prince Edward. The incredible energy that went into the organisation of the expedition and the planning of the research was typical of the way in which Van Zinderen Bakker had always carried out his work. The results of this long expedition were published in co-operation with thirty-five specialists in a voluminous monograph. Under his leadership scientists from various countries were able to do research on a wide range of subjects such as pollen analysis, glacial geology, limnology, mineral cycling, bioenergetics, etc.

This stimulating programme brought Van Zinderen Bakker into close contact with SCAR, the Scientific Committee on Antarctic Research. In this international body he took an active part especially as chairman of the Group of Specialists on Late Cenozoic Studies. In this capacity his horizon was considerably enlarged through close co-operation with scientists of different disciplines working in the sixth, white continent and the surrounding oceans. His pollen work on Marion Island, confirmed by related disciplines, proved to him that large temperature changes had indeed a considerable influence on the entire globe. In order to emphasise the important role the history of the cold Antarctic continent played and still plays on the evolution of the climate of the earth three international symposia were organised by Van Zinderen Bakker on behalf of SCAR in Cambridge (UK), Canberra and Birmingham (UK), the proceedings of which were published in book form.

In the meanwhile Van Zinderen Bakker had been very active in yet another field, namely aquatic ecology. From his pre-war days in Europe he had a strong inclination towards limnology and while expanding his research programme as head of the Botany Department of the University he visited Holland, Germany and England to become acquainted with the newest techniques in this field of study. This enabled him to establish a flourishing School of Limnological Studies at Bloemfontein. All his research activities were at a later stage combined in the Institute for Environmental Sciences of which he became the first director.

Van Zinderen Bakker has published ninety articles in scientific journals, wrote six books and edited fifteen volumes of *Palaeoecology of Africa,* and three other books.

The research efforts by Van Zinderen Bakker were widely recognised by many scientific bodies. His University, which strongly supported his research during forty years, conferred on him the degree DSc. honoris causa. The Council for Scientific and Industrial Research granted him a unit for palynological research and gave him considerable financial support for decades. He was elected Fellow of the Royal Society of South Africa and made Honorary Member of the Deutsche Quartärvereinigung, the Southern African Society for Quaternary Research and the South African Association of Botanists. He serves on the editorial board of several international scientific journals and was recipient of a number of gold medals for researth in the Antarctic, biology (2x), botany and for furthering cultural relations with the Netherlands (2x). After having been Honorary Consul of the Netherlands for twenty years he was appointed Officer in the Netherlands Order of Oranje-Nassau.

Eduard van Zinderen Bakker will always be known for his dynamic personality, high principles, vision and intense scientific interest in the mysteries of the earth. For these attributes he has been deeply respected by his co-workers and colleagues all over the world. It is small wonder that many of his inspired students today occupy high posts in biological research. His long and distinguished career bears witness to an unusually versatile scholar and universal scientist so rare in the age of increasing specialisation. True to his nature he will not yet be leaving the scientific arena but will continue with his research and we wish him every success for the future.

J.A.Coetzee

N.B. After going to press we received the news that Prof. van Zinderen Bakker has been elected honorary member of INQUA at the Ottawa Conference in July, 1987.

OBITUARY G.THANIKAIMONI (1938-1986)

We deeply regret to announce the sad demise of Dr G.Thanikaimoni, Head of the Palynology Department of the French Institute, Pondichery, who fell victim to hijackers' terrorism at Karachi on 6th September 1986.

Dr G.Thanikaimoni had acquired international reputation as a versatile palynologist. While on his way to the USA to attend the Second International Conference of Palaeo-Oceanography at Woodshole, he was snatched away from our midst. He was to present at this conference the results of a regional programme on the mangroves of Asia and Pacific and a state-of-the-art report on marine palynology. He was also preparing a manual of palynology of mangroves at the request of UNESCO.

Ganapathy Thanikaimoni was born in Madras city on the 1st January 1938. A Master of Science (botany) of the University of Madras, he was in 1961 awarded the Fyson Prize of the Presidency College, Madras, for his proficiency in plant taxonomy. He joined the French Institute, Pondichéry, in 1962, where he pursued his scientific career.

He could develop his talents to the maximum, thanks to the influence of the eminent scientists, H.Gaussen, B.G.L.Swamy, M.Van Campo, Y.De Ferre, P.Legris and Ph.Guinet, in whose contact he came in the earlier phase of his career.

In 1970, he received his Doctor of Science (Doctorat d'Etat) from the University of Montpellier for his doctoral dissertation on the 'Palynology and Systematics of the Palmae', prepared under the guidance of Professor Van Campo. This work has been noted for his original ideas on the contribution of pollen morphology to the taxonomy and phylogeny of the large family Palmae. In the framework of evolutionary systematics, he investigated the palynotaxonomy of several predominantly tropical families, Araceae, Arecaceae, Eriocaulaceae, Lentibulariaceae, Menispermaceae, Aponogetonaceae and Sonneratiaceae and also provided a putative biogeographic evolution of Arecaceae and Menispermaceae.

Many young botanists received their training under him. Among them are Dr G.Vasanthy, noted for her contribution to fundamental palynology, and

Dr Y.N.Seetharam, noted for his Ph.D. thesis on the 'Palynology and Systematics of Clusiaceae'.

As a Quaternary palynologist, Thanikaimoni's contributions to the understanding of the vegetational history of the hills of southern India with Dr F. Blasco, and of the evolution of mangroves with Dr C.Caratini and Dr C.Tissot are significant. He also investigated the pollen spectra of the archaeological sites of Borobudur (Indonesia), Laang Spean (Kampuchea) and Satingpra (Thailand) reconstructing their past environmental histories.

As a pollen morphologist, Thanikaimoni maintained an active collaboration with many specialists. He was a member of the International Working Group on Terminology and worked on primitive angiosperms with Dr Le Thomas. He was invited by the Linnean Society of London in 1983 to present the synthesis of his findings on the form and function of pollen apertural types.

Within the framework of the Indo-French cooperation, he organised with Dr C.Caratini, a workshop of the experts of Tertiary pollen of India and Africa, at the French Institute, Pondichéry, in April 1983. He was envisaging enlarging the scope of this study to an international level.

Thanikaimoni built up a reference collection of pollen and spores of over 20 000 extant species and assembled a vast literature on palynology. Both of these have been frequently consulted by scientists from India and abroad. He served on the Scientific Advisory Board of the Birbal Sahni Institute of Palaeobotany, Lucknow, and was also a fellow of Linnean Society of London.

Thanikaimoni's monumental *Bibliographic index on the morphology of the pollen grains of angiosperms* (five volumes) which he began in 1972 to provide a ready reference for the use of research workers, is a unique contribution to palynology. It includes 9 224 genera covering 5 618 publications. In view of the lacuna of such a work for the lower groups of plants he had prepared under the sponsorship of Dr P.H.Raven, Director of the Missouri Botanical Gardens, a *Bibliographic index on the morphology of the spores of Bryophytes and Pteridophytes* according to the same model. Dr G.Thanikaimoni indeed brought his department to the forefront of world palynology.

He leaves behind his wife, a son, a daughter and a large circle of friends and fellow scientists who had benefitted from his expert advice.

LIST OF PUBLICATIONS G.THANIKAIMONI (1938-1986)

1965 Contribution to the pollen morphology of Eriocaulaceae. *Pollen Spores* 7 (2): 181-191, 2 pl.

1965 *Laurentia longiflora* (Linn.) Endl. in Pondicherry. *J. Bombay nat. Hist. Soc.* 62 (2): 323-324.

1966 Contribution à l'étude palynologique des palmiers. *Inst. fr. Pondichéry, trav. sec. sci. tech.* 5 (2): 1-92.

1966 Pollen morphology of the genus *Utricularia. Pollen Spores* 8 (2): 265-284, 5 pl.

1966 (with D.M.A.Jayaweera). Pollen morphology of Sonneratiaceae. *Inst. fr. Pondichéry, trav. sec. sci. tech.* 5 (3): 1-12, 3 pl.

1968 Morphologie des pollens des Ménispermacées. *Inst. fr. Pondichéry, trav. sec. sci. tech.* 5 (4): 1-57, 16 pl. h.t.

1968 Palynology in vegetation studies. 21st International Geographical Congress. In *Proc. Symp. Phytogeography and Vegetation Mapping:* 1-4.

1969 Esquisse palynologique des Aracées. *Inst. fr. Pondichéry, trav. sec. sci. tech.* 5 (5): 1-31, 20 pl. h.t.

1970 Les Palmiers: palynologie et systématique. Thèse Doct. sci. nat. Montpellier. Arch. orig. Centre Document. CNRS, Paris, no. 4173, 221 p., fig. schémas, 22 pl. h.t.

1970 Pollen morphology, classification and phylogeny of Palmae. *Adansonia* 10 (3): 347-365, 2 pl.

1970 (with W.C.Elsik). *Bomarea lyncina* Herb. (Amaryllidaceae) and *Auriculiidites* Elsik. *Pollen Spores* 12 (2): 177-180.

1972 Index bibliographique sur la morphologie des pollens d'Angiospermes. *Inst. fr. Pondichéry, trav. sec. sci. tech.* 12 (1): 1-337.

1972 (with P.Legris). Palynology in Pondicherry. *Jour. Paly.* 8: 156-162.

1972 Pollen morphologie. In J.Bogner (ed.), Revision der Arophyteae (Araceae). *Bot. Jahrb. syst.* 92 (1): 1-63.

1972 (with G.Vasanthy). Sarraceniaceae: palynology and systematics. *Pollen Spores* 14 (2): 143-155, 2 pl.

1973 (with J.Pias). Signification de la présence de pollens de Chénopodiacées dans les croûtes et encroûtements calcaires du Quaternaire récent en Afghanistan. *C.R. Acad. Sci. Paris* 276D: 2509-2512.

1973 Index bibliographique sur la morphologie des pollens d'Angiospermes. Supplément 1. *Inst. fr. Pondichéry, trav. sec. sci. tech.* 12 (2): 1-164.

1973 (with C.Caratini & F.Blasco). Relation between the pollen spectra and the vegetation of a South Indian mangrove. *Pollen Spores* 15: 281-292.

1974 Palmae. In: Pollen et Spores d'Afrique tropicale. Association des Palynologues de Langue Française. *Travaux et Documents de Géographie Tropicale* 16: 210, 211, 214-217, 220, 221. CEGET, Talence.

1974 (with C.Caratini). Palmae. In: Pollen et Spores d'Afrique tropicale. Association des Palynologues de Langue Française. *Travaux et Documents de Géographie Tropicale* 16: 212, 213. CEGET, Talence.

1974 (with C.Caratini & B.Lugardon). Palmae. In: Pollen et Spores d'Afrique tropicale. Association des Palynologues de Langue Française. *Travaux et Documents de Géographie Tropicale* 16: 218-219. CEGET, Talence.

1974 (with F.Blasco). Late quaternary vegetational history of southern region. In K.R. Surange, R.N.Lakhanpal & D.D.Bharatwaj (eds), *Aspects and appraisal of Indian palaeobotany*: 632-643. BSIP, Lucknow.

1975 (with F.Blasco, C.Caratini & S.Chanda). Main characteristics of Indian mangroves. In *Proc. Int. Symp. Biol. Management Mangroves (Hawaii)* 1: 71-87.

1976 Index bibliographique sur la morphologie des pollens d'Angiospermes. Supplément 2. *Inst. fr. Pondichéry, trav. sec. sci. tech.* 13: 1-386.

1976 Palynological literature in French. *Geophytology* 6 (1): 112-121.

1977 L'analyse pollinique des débris archéologiques du Borobudur. *EFEO Mém. Archéol.* 12: 69-72.

1977 Pollen analysis. In R.Mourer, Laang Spean and the prehistory of Cambodia. *Mod. Quaternary Res. SE Asia* 3: 29-56.

1977 Pollen morphological terms: Proposed definitions 1. In *Proc. IVth Int. Palynol. Conf. (Lucknow)*, Abstracts: 180.

1978 Principal works on the pollen morphology of the Compositae. In V.H.Heywood, J.B.Harborne & B.L.Turner (eds.), *The biology and chemistry of the Compositae.* Ch. 8: Pollen morphology in the Compositae and in morphologically related families. Appendix 1: 249-265.

1978 Pollen morphological terms: Proposed definitions 1. In *Proc. IVth Int. Palynol. Conf. (Lucknow)* 1: 228-239, 2 pl. h.t.

1978 (with S.Chanda & B.Lugardon). On the ultrastructure of pollen aperture in Calectasia R.Br. (Xanthorrhoeaceae). *Pollen Spores* 20 (3): 351-365.

1979 (with F.Roland-Heydacker). Pollen morphology of primitive angiosperms: some neglected aspects. In *Proc. IVth Int. Palynol. Conf. (Lucknow)* 1: 542-545, 1 pl. h.t.

1979 (with G.Vibichanaraw). Morphological index to Van Campo's African Pollen Atlas. *Bull. IFAN Dakar* 41, Ser. A (2): 286-299.

1980 Mangrove formations in Indian Peninsula. In M.Nagaraj & C.P.Malik (eds.), *Current trends in botanical research*: 233-235.

1980 Quatrième index bibliographique sur la morphologie des pollens d'Angiosperms. *Inst. fr. Pondichéry, trav. sec. sci. tech.* 17: 1-337.

1980 Taxonomic index to pollen morphological literature. In *Proc. 5th Int. Palynol. Conf. (Cambridge),* Abstracts: 384.

1981 (with C.Caratini & C.Tissot). Mangroves of India: Palynological study and recent history of the vegetation. In *Proc. IVth Int. Palynol. Conf. (Lucknow)* 3: 49-59.

1981 Index palynologius. In *Proc. IVth Int. Palynol. Conf. (Lucknow)* 3: 344-348.

1983 Palynological investigation on the Borobudur monument. *Bull. EFEO (Paris)* 72: 237-250, 27 pl. h.t.

1983 (with I.K.Ferguson, J.Dransfield & F.C.Page). Notes on the pollen morphology of *Pinanga* with special reference to *P.aristata* and *P.pilosa* (Palmae. Arecoideae). *Grana* 22: 65-72.

1984 (with collaboration of F.Roland, I.K.Ferguson, M.T.Cerceau & L.Derouet). Menispermacées: Palynologie et systématique. *Inst. fr. Pondichéry, trav. sec. sci. tech.* 18: 1-135, 67 pl. h.t.

1984 (with C.Caratini, S.Nilsson & E.Grafstrom). Omniaperturate Euphorbiaceae pollen with striate spines. *Bull. Jard. Bot. Nat. Belg.* 54 (1/2): 105-125, 11 pl.

1984 Palynology and systematics of Menispermaceae. In *Proc. 6th Int. Palynol. Conf. (Calgary),* Abstracts: 163.

1984 (with C.Caratimi, B.S.Venkatachala, C.G.K.Ramanujam & R.K.Kar). Selected Tertiary Angiosperm pollen from India and their relationship with African Tertiary pollen. In *Proc. 6th Int. Palynol. Conf. (Calgary),* Abstracts: 163.

1984 (with C.Caratini, B.S.Venkatachala, C.G.K.Ramanujam & R.K.Kar). Pollens d'Angiospermes du Tertiaire de l'Inde et leurs relations avec les pollens du Tertiaire d'Afrique. (With English version.) *Inst. fr. Pondichéry, trav. sec. sci. tech.* 19: 1-93, 72 pl., 1 carte.

1985 Aponogetonaceae: Palynology and phylogeny. *Bibliotheca Botanica* 137: 11-14.

1985 Contribution de l'Institut Français de Pondichéry. In *Proc. 9th Symposium Association des Palynologues de Langue Française (Montpellier).*

1985 Variation de l'aperture des Annonacées: quelques tendances nouvelles. In *Proc. 9th Symposium Association des Palynologues de Langue Française (Montpellier).*

986 Pollen apertures: form and function. In S.Blackmore & I.K.Ferguson (eds.), *Pollen and spores, form and function. Linnean Society Symp.* series 12: 119-136.

1986 Studies on mangroves by French Institute, Pondicherry. In *Proc. Natl. Symp. Estuarine Biology (Parangipettai).*

1986 Cinquième index bibliographique sur la morphologie des pollens d'Angiospermes. *Inst. fr. Pondichéry, trav. sec. sci. tech.* (in press).

1986 Evolution of Menispermaceae. *Canadian J. Bot.* (in press).

OBITUARY OLIVER DAVIES (1905-1986)

Oliver Davies died tragically on 26 August 1986, mindlessly slain by intruders into his Pietermaritzburg home, where he had lived for the greater part of nearly forty years.

He was born on 7 May 1905. After his junior schooling, in which he excelled, he attended Rugby, the famous English public school, where he obtained an exceptional knowledge of Greek, Latin, French and German. After completion of his schooling he proceeded to a Master of Arts degree and thereafter, a Doctorate, his field of scholarship and research being classics. After a research scholarship in Athens, he pursued his classical studies yet further at the University of Istanbul.

His particular field of interest at that time was ancient mining, particularly in Britain, his early authoritative book on this subject having recently been re-published.

He arrived in South Africa in 1948 to take up the chair of Classics at the University of Natal in Pietermaritzburg, which post he held until 1951. During this time he developed a keen interest in local archaeology, particularly as related to river terrace and raised beach gravels.

Oliver Davies left Natal at the end of 1951 to take up an archaeological post at the University of the Gold Coast, as it then was, at Legon near Accra. Having found it to be partially underlain by an iron age archaeological site, he proceeded to excavate the greater portion of the central town gardens of Accra as part of his local research.

He returned to Natal on his 'retirement' at the end of 1966.

One of the first tasks in which he involved himself was the resurrection of the until then defunct Natal Branch of the Archaeological Society of South Africa. He did this to such good effect that the Branch continues to function strongly at the present day.

Having been appointed Honorary Keeper of Antiquities at the Natal Museum in 1968, he used the Museum as a base for all his later work, a situation of which he was very appreciative. In the years ollowing, he was involved in numerous excavations of caves, shelters and open sites of various ages, the results of which were published in many papers and articles.

Oliver Davies was also much interested in sea-level and climatic changes as related to archaeology. This led him to undertake, over a period of some years, field trips and investigations covering virtually the entire length of the South African coastline. This resulted in the publication of an extensive series of papers dealing with the various constituent portions of the South African coastline in respect of evidence for local and overall sea-level and climatic changes, depositional history, related palaeontology, and archaeology.

In recognition of his contribution to archaeology in Natal, the University of Natal awarded Oliver Davies the honorary degree of Doctor of Literature in 1979. In the same year for the same reason he was made an Honorary Research Associate in Archaeology by the Natal Museum.

Oliver Davies was a founding member of the South African Society for Quaternary Research (SASQUA) in 1969, holding successively the offices of Secretary in that Society from 1973 to 1975, Vice President from 1975 to 1977 and President from 1977 to 1979. For many years he served on the African Shorelines Sub-commission of INQUA.

Oliver Davies also played an important part as Convenor of the Tertiary-Quaternary Working Group of the South African Committee for Stratigraphy whose work culminated in 1980 with the publication of the new Stratigraphy of South Africa.

A confirmed bachelor all his days, Oliver Davies was midly eccentric in his dress and life-style. Who (if you were ever in the field with him) could possibly forget his appearance in those ragged puttees that seemed to have served as a source of sustenance for quite a few generations of various voracious anthropods, or his formerly pale blue towelling soft hat of uncertain shape that at various times was used for a wide variety of purposes. Also, a ride in his vintage Land Rover with Oliver at the helm was a never to be forgotten experience.

Who, also, ever having had correspondence with Oliver, can ever forget that microscopic illegible calligraphy of his, usually written on the back of some previously used scrap paper, and received in a second-hand envelope. Occasionally, if one was lucky, one would receive correspondence, or a draft of a paper, unmistakeably typed by him on his ancient typewriter. A carry-over of his English public school days, Oliver characteristically addressed and referred to most people by their surnames only.

Oliver Davies was an inveterate traveller and an attender of conferences, having travelled in Europe as recently as April and May of this year. He always enjoyed returning to Greece, and Macedonia in particular. By no means averse to a few beakers of his 'blushful Hippocrene', possibly as a result of his connection with Greece, Oliver thoroughly enjoyed social occasions. One of the more notable of these, for which unbelievably for him, he appeared in a dark suit, was the formal celebration last year of his eightieth birthday. This was held in that other home of his, in which he was a longtime habitué, the Victoria Club of colonial atmosphere and fame in Pietermaritzburg.

I was fortunate in accompanying Oliver on an extended field trip through coastal Natal and the eastern and southern Cape coasts last year to see some of the exposures and features about which he had written. Armed with his rough bush-cut cudgel-walkingstick, he was still capable of walking several kilometres along beaches to an exposure without undue effort, and this in his eightieth year. I found it remarkable how he could return with unswerving accuracy to a small exposure he had last visited about twenty years ago, in spite of marked changes in the local environment in the interim. His mind continued completely lucid, his dry wit and reminiscences about such famous archaeological figures of the past as the Abbé Breuil and Van Riet Lowe, livening many an hour.

The fine scholar and inimitable character that was Oliver Davies will be missed by all whose pleasure it was to know him. Fortunately, his memory will live on in the wealth of publications he produced, particularly in the latter years, during his long and full life.

Rodney Maud

A UNIQUE RECORD OF THE PAST

J.DESMOND CLARK
Department of Anthropology, University of California, Berkeley, USA

The seventeen volumes of *Palaeoecology of Africa* constitute a unique record of the progress of palynological, palaeo-climatic, palaeo-environmental and prehistoric archaeological research in the African continent. These volumes show the development and expansion of this research from small beginnings until it now includes reports of virtually every major project in these fields being undertaken in the continent today. If anyone were to attempt today to compile a history of palaeo-environmental research in Africa, it would necessitate a lengthy and extensive library search for literature, often in hard to find publications. In fact, a work of this kind would be magnum opus. This is precisely what Professor Van Zinderen Bakker, with the help of his long-time colleague and friend, Dr J.A.Coetzee and his publisher A.A.Balkema, has given us in this most valuable record of the progress of the discipline. Our debt to him is a very great one for carrying out this task which could not easily be repeated. In these volumes he has kept us all informed of what is being done, where and by whom, and has thereby not only stimulated a number of new field projects but has provided the basis for dialogue between investigators from one end of the continent to the other. This is an outstanding achievement covering, as it does, some thirty-five years.

The first serious introduction to palynology in Africa for me, and probably for most archaeologists working in the southern part of the continent, was an article by Professor Van Zinderen Bakker who had recently joined the Botany School of the University of the Orange Free State at Bloemfontein to develop palynology research. The article 'Archaeology and palynology' in an early number of the *South African Archaeological Bulletin* (Van Zinderen Bakker 1951) described palynology as 'South Africa's latest archaeological weapon' and was an eloquent appeal for cooperation from all parts of the sub-continent. It spoke of the methods, limitations and prospects of what had already been achieved in other countries, especially in north-western Europe. It certainly brought it home to the prehistorians that there was, indeed, a gold mine of information about the palaeo-habitats of the past populations we were studying if the right kind of deposits could be found that contained pollen. It was especially encouraging to learn from the early volumes of *Palaeoecology of Africa* that such deposits existed in South

Africa and in tropical regions besides, since the 1950s was a time when archaeologists were beginning to look for evidences of the behavioural traits implicit in the artifacts, bones and other data at the buried occupation sites they were excavating. This swing to trying to reconstruct prehistoric behaviour followed the development of the radiocarbon method of dating by Willard Libby and, consequently, the construction of a reliable chronology. To archaeologists working in Africa the vital importance of palaeo-climatic and palaeo-ecological data and the use of these to reconstruct the preferred habitats of early humans became readily apparent and, as the *Palaeoecology* volumes show, quickly became the basis for trying to understand cultural variability and change during the Quaternary.

I think the first application of pollen studies to a stratified archaeological sequence in the continent was carried out from the early 1950s by Professor Van Zinderen Bakker in association with the late Dr A.C.Hoffman, at the Florisbad mound spring in the Orange Free State, where, some years before, had been found the partial cranium of *Homo helmei*. His report and pollen diagram was delivered at the Third Pan-African Congress on Prehistory held in Livingstone in 1955 (Van Zinderen Bakker 1957) and clearly demonstrated the very different ecological conditions that pertained between the earlier and later spring deposits with their associated artifacts and fauna. From then on, botanists began to make increasingly more significant contributions to palaeo-ecological understanding of archaeological sites throughout the continent.

It was also in the early to mid 1950s that I began work at the Kalambo Falls prehistoric site at the south end of the Lake Tanganyika section of the Rift Valley, where waterlogging of the earlier sediments in a small basin had preserved large quantities of macro- and micro-plant remains in Earlier and Middle Stone Age contexts. Professor Van Zinderen Bakker was invited to collaborate and carried out a study tour of south-central and east Africa that provided comparative data for his interpretation of the later Pleistocene temperature curve dictated by vegetational changes in the Kalambo Falls sediments and comparison with that for Western Europe. This was first published in 1964 (Clark & Van Zinderen Bakker 1964) as well as in the *Eighth Report on Palynology* (Van Zinderen Bakker 1966) and many related fields in the closing section of the first volume covering the years 1962 and 1963. This fruitful collaboration continued, side by side with investigations in Northeast Angola, sponsored in the 1960s by Diamang, the diamond company operating the large and deep, open-cast exposures through the redeposited Kalahari sands and related sediments. Here again, important evidence for considerably lowered temperatures, by 5-6°C, was found in pollens associated with later Pleistocene fluvial sediments in the river valleys, showing that grassland had, to a great extent, replaced forest during the time of the Last Glacial in that part of the Zaïre Basin.

By this time, the *Palaeoecology* volumes had become the main place where reports, usually but not always short, on recent field and laboratory research in Africa were presented. The scope of the publication expanded now to include palaeo-botanical and temperature work in Antarctica and the Islands. Research in the African continent itself is now usually presented by regions or under relevant disciplines and sub-disciplines. Thus, geology, geomorphology, palynology, palaeontology, historical biogeography and archaeology of a particular region will be covered. Of especial significance is the amount of research reported for the Sahara and its climatic system and the geomorphology and archaeology of the desert. In Ghana, the Lake Bosuntwi core can be expected to provide the first reliable understanding of West African temperature and vegetational changes for most or all of the Quaternary. In Kenya and Ethiopia, the emphasis has been on the morphology and hydrology of lake basins, on glaciations and on geochronology based on the chemical composition of extruded volcanics. In South Africa, interest now centres on the coastal areas, on the palaeo-environments of the Plio-Pleistocene high veld cave breccias and on the changing climates during and immediately after the Last Interglacial when archaic forms of *Homo sapiens* were being replaced by anatomically modern humans. Long-range correlation of stratified sequences is being greatly helped by the evidence from deep sea cores in the oceans surrounding Africa and the climatic interpretations as these can be made from the changing nature of their organic, mineralogical and chemical composition. It is now possible to construct – as the volumes show – reliable models for palaeo-environments in broad regions of the continent based on the accumulated research of the past thirty-five or more years, for example, for the Sahara, the Namib, eastern Ethiopia, for North Africa in comparison with the Middle East, and even for changing vegetation patterns as a whole in the continent south of the Sahara.

These palaeo-climatic and environmental data and the constructs they permit are the essential baseline for all attempts by archaeologists to understand human behaviour, whether it be in the later Pliocene or the late Holocene. Without this knowledge, the role of climate in contributing to the biological and cultural changes shown by the archaeological record cannot be properly understood. In particular is this so for the earlier stages of hominid evolution where long-range changes in climate and palaeo-ecology are likely to have been the major factor in initiating the dietary changes and attendant behavioural adjustments these required. The *Palaeoecology of Africa* volumes constitute the major reference work for anyone investigating the past history of plant and animal populations in the continent.

These painstakingly compiled volumes with their many bibliographies and reference lists of radiocarbon dates, etc., we owe primarily to the long devotion to researching and disseminating knowledge of African palaeo-ecology by Professor Van Zinderen Bakker. This has indeed been a major and quite unsurpassed work and in honouring you, all of us send to you, Eduard, on this the occasion of your 80th birthday, our sincerest thanks and very best wishes for continued work and success in the future.

Berkeley
November 27th, 1986

J. Desmond Clark

REFERENCES

Clark, J.D. & E.M.van Zinderen Bakker 1962. Pleistocene climates and cultures in Northeastern Angola. *Nature* 196: 639-642.

Clark, J.D. & E.M.van Zinderen Bakker 1964. Prehistoric culture and Pleistocene vegetation at the Kalambo Falls, Northern Rhodesia. *Nature* 201: 971-975.

Van Zinderen Bakker, E.M. 1951. Archaeology and palynology. *South Afr. Archaeol. Bull.* 6 (23): 80-87.

Van Zinderen Bakker, E.M. 1957. A pollen analytical investigation of the Florisbad Deposits (South Africa). In J.D.Clark & S.Cole (eds.), *Third Pan-African Congress on Prehistory (Livingstone, 1955)*: 56-67. London: Chatto & Windus.

Van Zinderen Bakker, E.M. 1966. Palynology: Southern Africa. Chapter 11 (196-198) in *Eighth Report on Palynology and many related fields covering the years 1962 and 1963* (E.M.van Zinderen Bakker). Cape Town: Balkema.

CLIMATOLOGY

EAST AFRICAN RAINS OF 1961/62 AND THE ABRUPT CHANGE OF THE WHITE NILE DISCHARGE

H.FLOHN
Universität Bonn, F.R.Germany

ABSTRACT

The excessive rainy season 1961/2 over East Africa – causing a discontinuous rise of Lake Victoria and the discharge of the White Nile lasting at least twenty years – coincides with an even larger anomaly of sea surface temperature, surface winds and convective cloudiness at the western equatorial Indian Ocean. Together this anomaly extended over at least 6 x 10^6 km^2 and lasted more than four months. In many respects, it resembles the ENSO events in the equatorial Pacific; in contrast to it it is characterized by a lack of surface westerlies. Eighty to ninety percent of the area-averaged rainfall (in the order of 16-24 mm/d) is contributed by large-scale convergence of water vapour transport; the liberation of latent heat by rainfall during November 1961 surpasses, over areas of more than 10^5 km^2, the solar constant.

INTRODUCTION

One of the most promising ways to understand palaeoclimatology – at least for the (geologically short) period in which the earth's topography remained essentially the same – is to look at the vagaries of the contemporaneous weather and to compare extreme anomalies of to-day with past climatic anomalies. During the three cold winters of 1939-1942, the author was impressed by the fact that the quite unusual anomalies of the atmospheric circulation up to an altitude of 16 km (100 hPa) demonstrated a situation which should have led, if maintained over a sufficiently long time, to a glaciation of European mountains and also to a pluvial in northernmost Africa. At this time it was generally thought that both were simultaneous; during the last twenty years of the evolution of the climatic history of Africa (to which E.M.van Zinderen Bakker has contributed so many detailed studies and thoughtful reviews (Van Zinderen Bakker & Mercer 1986)) we had learned that the true climatic history was not as simple.

One of the most impressive sources of this history is given by the river

Nile. Its history from the late Miocene until now (Williams & Faure 1980, Said 1981) shows that its long course from Lake Victoria to the Mediterranean is only a geologically young evolution. Climatic studies (Hassan 1981, Riehl et al. 1979) suggested that the abrupt changes of its discharge, measured at the cataracts during the Middle Kingdom (Bell 1975) and at the Roda Nilometer at Cairo (Hassan 1981, Shahin 1985) since more than 1 200 years, are due to changes of the discharge of the White Nile. This coincides with a recent anomaly: the sudden rise of the Lake Victoria level (and consequently White Nile runoff at Jinja) beginning in fall 1961 was maintained – quite surprisingly – with minor fluctuations at this higher level, until (at least) 1982. This is the more amazing as after the catastrophic rains in whole Eastern Africa during late 1961 and early 1962 the rainfall (except perhaps at the islands and some coasts of Lake Victoria) returned to normal.

In this chapter it is intended to describe in some detail the extreme rainy season 1961/2 in East Africa (including southern Somalia and the Seychelles), together with some important simultaneous data from the Indian Ocean given by Reverdin et al. (1986). Together these data show similarities to the well-known ENSO events in the Pacific (Rasmusson & Carpenter 1982, Behrend 1984). In earlier papers (Flohn 1983, Flohn & Burkhardt 1985) the results of the diurnal variations of the thermally heated circulation system of Lake Victoria have been described, together with their key role for its water budget. This local circulation with a diameter of about 250 km is only a small side effect within the large-scale anomaly of the rainy season 1961/2 extending not only over a considerable part of the African continent, but also over the Indian Ocean. Its diameter is in the order of 5 000 km, its duration five (or perhaps six) months, quite similar to the ENSO event of 1982/3 in the Pacific (Rasmusson & Wallace 1983).

The recent evolution of climatology as a geophysical science has become the more fascinating, since the occurrence of 'abrupt' changes of 'climatic crises' (Flohn 1986, Rognon & Curie 1983, Yamamoto et al. 1985) – e.g. during the last glaciation and at the transition between Late Glacial and Holocene – presents evidence for 'chaotic' behaviour, for an instability of the deterministic, non-linear climatic system. It is this evolution, which motivated the author to a quite simple investigation within the limits of classical (descriptive) climatology: such studies are needed when a possible change of classical paradigmata ('natura non facit saltum') appears at the horizon (Flohn 1986, Rognon & Curie 1983).

EAST AFRICAN RAINFALL, AUGUST 1961 – JANUARY 1962

Fortunately comprehensive annual and monthly climatic data have been published during this time by the East African Meteorological Department in Nairobi for Kenya, Uganda and Tanzania. The rainfall data are arranged by 1 x 1° fields; they are given, together with their 'normals', referring to the average of all available observations for this month at each station. No distinction is made to the variable length of the record; in areas with a dense station network we have mostly used records covering at least ten years. Because of this strong inhomogeneity of the 'normals', only area-averaged data for 1 x 1° fields for an approximated spatial coverage have been used; in arid areas some fields lack observations.

Recently some rainfall data from southern Somalia became available (Flohn, in press): five stations south of latitude 5°N, extending the data from northern Kenya to the coast.

During a visit to Nairobi during August 1964 the data of this remarkable rainfall period have been discussed with the local meteorologists. Due to the failure of the visual pilot-balloon ascents at days with heavy cloudiness, our knowledge of the low-tropospheric wind field was very unsatisfactory. Radiosonde ascents (Nairobi, Entebbe) were too far apart (and also incomplete); a comparative survey of synoptic maps did not lead to convincing results.

The most striking result derived from monthly data was the long duration of above-normal rainfall (up to six months) and the large extension of positive anomalies, covering the whole of East Africa (with southern Somalia about 2.1 million km^2) and extending even into adjacent countries. For selected catchments monthly amounts and anomalies (in %) are given in Table 1. In Central Tanzania (Rufiji River) rainfall during February 1962 reached 133% of normal, the average for September-February was 1 074 mm = 185%. In the Lake Rukwa catchment above-normal rainfall extended even into March; here the average for September-February reached 1 061 mm = 161%.

In the equatorial part of East Africa – where (as in Somalia) a relatively dry period during northern summer interrupts the rainy seasons (long rains March-May, short rains October-December), the peak of the 1961/2 rains was centered during the 'short' rainy season – in absolute values (Table 1) during November 1961. To investigate the space-time structure of this 'event', this month was selected for a study based on daily observations. For East Africa about 900 station series were available. Here we selected some areas of particular interest (semi-arid northern Kenya) and areas with a dense station network: the highlands of Central Kenya and of northeastern Tanzania; in the latter we selected ten or fifteen of the longest records in each 1 x 1° field. For comparison, the islands of Zanzibar and Pemba were chosen, together with the Seychelles (about 1 100 km off the coast near Mombasa).

The central Kenya highlands (latitude 1°N-2°S, 35-38°E) cover nine adjacent 1 x 1° fields with together 110 000 km^2; only the SW corner (about 8 000 km^2) is not represented. The highlands and low-level farmlands in NE Tanzania (and adjacent Kenya) cover seven 1 x 1° fields (with some adjacent corners) between latitude 3 and 7°S and longitude 36-39°E. The station density at the islands of Zanzibar and Pemba (c. 1 000 km^2) and at the Seychelles (together 400 km^2) is greater than at the continents.

In Central Kenya one hundred fairly equally distributed station records have been selected. The average monthly rainfall at these stations was

Table 1. Rainfall anomalies in selected catchment areas, August 1961 – January 1962

	Rainfall in mm						Sum	Percent of normal						Sum
	A	S	O	N	D	J	A-J	A	S	O	N	D	J	A-J
Northern Kenya	143	43	205	365	119	54	929	176	103	443	496	242	241	295
Athi River, Mombasa	35	120	216	480	189	115	1 155	106	404	352	404	230	280	318
Tana River	33	111	441	557	120	81	1 343	131	427	499	325	126	230	301
Rift Valley (whole)	120	58	133	357	192	100	960	147	109	284	560	233	217	264
Lake Victoria (whole)	153	136	180	364	180	100	971	128	136	202	397	225	191	181
Kyogo-Victoria	173	165	211	287	104	53	993	123	140	200	351	186	153	187
Seychelles[3]	300	402	264	417	279	273	1 383	337	330	156	228	86	98	261
Pangani River	1	69	201	362	172	172	[2]976	–	297	452	457	193	298	[2]333
Daressalaam Hinter L.	1	62	269	338	287	172	1 128	–	222	484	443	295	160	309
Rufiji River	1	7	54	163	333	286	843	–	97	444	359	218	151	207
Lake Rukwa	1	5	18	216	284	306	829	–	128	106	300	158	152	182

1. August values below normal.
2. September-January; September-February (see text).
3. Sum and percentage of normal for August-November (12 stations).

480.4 mm or 446% of the normal (107.8 mm). The average rain frequency was 24.9 days, so that on each day about 83% of the total area received rainfall; only on three days less than 50% of the area was affected, while on sixteen days more than 90% of the area received rainfall. The average amount per station (R) was on these sixteen (respectively three) days 22.7 (respectively 10.8) mm. Equally distributed over the whole area (R*), the average daily rainfall of the month was 16.0 mm; on the sixteen (respectively three) days just mentioned it was 21.6 (4.4) mm/d. It should be mentioned that an area-averaged rainfall of 16 mm/d releases a latent energy of 455 W/m^2 i.e. one-third more than the average amount of the solar radiation received during 24 hours at the upper limit of the atmosphere (340 W/m^2).

Table 2A contains similar monthly averages for all areas, together with the anomaly of rainfall in percent. Table 2B gives results from selected days with the highest percentage of stations receiving rain. If the spatial distribution of stations would be homogeneous (which it is not), this parameter could also be defined as the percentage of area with rain at a particular day.

For comparison, some data for southern Germany are given in the Appendix.

Attempting to derive some results on the mechanism of rainfall during

Table 2. Statistical rainfall parameters, November 1961

A. Whole month	n	np	R	R*	$\overline{RR}$
Northern Kenya (NK)	13	54%	17.8	9.5 mm/d	592%
Central Kenya (CK)	100	83	18.6	16.2 mm/d	446
NE Tanzania (NET)	70	63	18.1	11.3 mm/d	407
Lake Victoria (LV)	27	73	19.2	14.1 mm/d	362
Zanzibar + Pemba (ZP)	20	71	23.0	16.4 mm/d	343
Seychelles (S)	10	90	14.4	12.9 mm/d	209

B. Selected days	NK	CK	NET	LV	ZP	S	
Number of days	5	16	8	11	13	16	
% of stations with rain above	85%	90%	85%	85%	90%	100%	
Average percentage of s.r.	89%	95%	91%	93%	96%	100%	
Average intensity per s.r.	31.2	24.2	24.3	23.5	28.9	16.2	mm/d
Maximum intensity per s.r.	36.4	31.9	48.8	30.2	52.6	35.6[1]	mm/d
In November	10	8	7	10	9	25	

1. Only at 90% s.r.

n = total number of stations used.
np = percentage of stations with rainfall at a given day.
R = average daily rainfall at stations with rain (s.r.).
R* = average daily rainfall for the whole area.
$\overline{RR}$ = average monthly rainfall in % of 'normal' (see text).

such a macro-scale tropical anomaly, this can be done based on the water budget equation of an atmospheric column:

$$ET - R + \partial W/\partial t = \mathrm{divh}\,(\vec{vW})$$

Here ET denotes evapotranspiration, R = rainfall, W = water vapour content of the atmospheric column (precipitable water), $v\vec{W}$ the vertically integrated water vapour transport vector within the column, all values expressed in mm water column.

Above 850 hPa (which is nearly the average height of the stations) W is equivalent to a precipitable water of about 24 mm (= g/cm^2) (Peixoto & Oort 1983). Assuming saturation, W of equatorial air above 850 hPa can be estimated to be 42 mm. The daily evaporation ET in the tropics is about 5 mm/d over Lake Victoria, oceans or rain forest; under the nearly permanent clouds (Nairobi 86%, Nakuru 91%) during this month the average daily evaporation could not have been much more than 1-2 mm/d, on the sixteen rainiest days < 1 mm/d.

Rainfall R is given in Table 2A for the whole month, in 2B for the rainiest days. For a longer rainfall period, $\partial W/\partial t$ cannot have been much more than about 50% of W; this gives for the whole month a change of 12 mm or 0.4 mm/d, for the sixteen rainiest days (taken as one single period) a change of 21 mm or 1.3 mm/d. This yields for the whole month for Central Kenya:

$$\mathrm{divh}\,(vW) = 1.5 - 16.2 + 0.4 = -14.3 \text{ mm/d}$$

or for the sixteen rainiest days:

$$\mathrm{divh}\,(vW) = 1.0 - 24.2 + 1.3 = -21.9 \text{ mm/d}$$

Numerical results for the other areas are fairly similar. The essential result is that during the peak of the macro-scale anomaly about 90% of the rainwater must have been advected by large-scale horizontal convergence of the water vapour transport. At the scale of this event, this means an even larger entrainment area: as we shall see, a true macro-scale event.

In reality, the area of very high rainfall over East Africa was much larger (Table 1). The semi-deserts of northern Kenya (about 250 000 km^2) received – taking the above-mentioned thirteen stations as representative – 285 mm of rain instead (normally) 42.6 mm, i.e. 669%.

The daily rainfall amounts at individual stations were not excessive: very few scattered values surpassed 150 mm, none 200 mm. Observers had the impression of a rather uniform rainfall over vast areas, with few interruptions.

While no data could be obtained from adjacent Ethiopia, the available (five) stations from southern Somalia received about 400% of the normal.

To demonstrate the spatial pattern, several maps based on 1 x 1° field averages were constructed. But the gaps without observations and the small representativeness of fields containing only one or two stations did not permit convincing results. A further inspection of the patterns showed a great role of topography, thus excluding a possible use of a more sophisticated

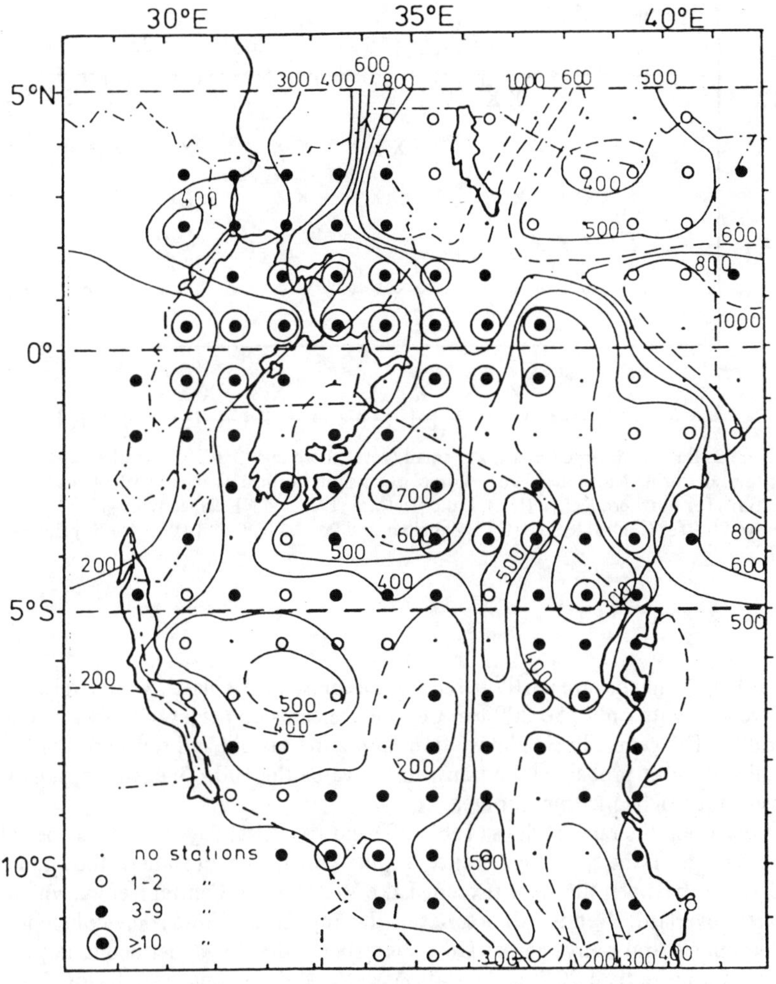

Figure 1. Rainfall above East Africa in November 1961, estimated in percent of long-year average for 1°x1° rectangles.

mathematical approach to obtain smoothed patterns. Here we give only one example of a subjectively smoothed pattern, taking into account the number of stations and the length of available records (Fig. 1). During November 1961, the greatest rainfall surplus occurred in the semi-arid areas of North and East Kenya, well above 1 000%. Around the Rift Valley area along longitude 35°E the anomaly reached 500-700%; similar areas in eastern and

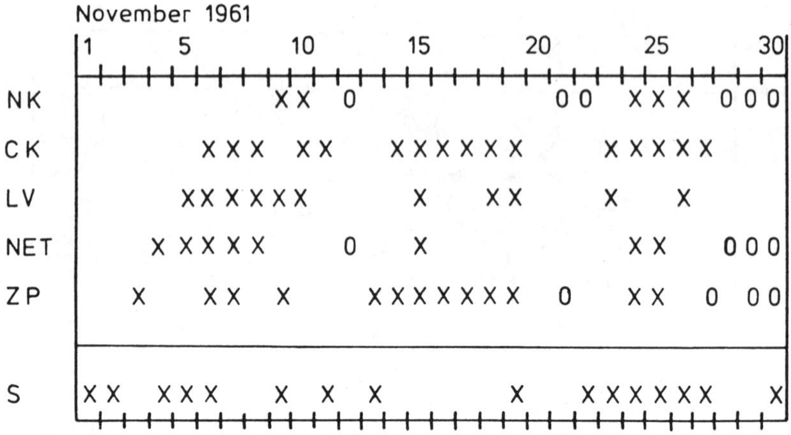

Figure 2. Days with large (crosses) or small (zeros, here less than 20%) areal extension of rainfall during November 1961. 'Large' extensions defined as above 85% of the stations for northern Kenya (NK), Lake Victoria (LV) and NE Tanganyika (NET), above 90% for Central Kenya (CK), Zanzibar and Pemba (ZP) and 100% for Seychelles (S).

southern Tanzania are partly based on unrepresentative data. A relatively large area with only 150-200% in Central Tanzania coincides with local semi-aridity. The generally small deviations along the eastern and southern border indicate (most probably) the transition towards the 100% line, which is, however, not reached within our area.

Selecting the rainiest days (Table 2B) and the driest days in our six special areas (Fig. 2), the spatial coherence is quite convincing, in spite of the large distances between the areas (except Lake Victoria and Central Kenya, which partly overlap). Even the time series of the Seychelles shows some coherence. Here, in Central Kenya and at Lake Victoria no 'dry' days (defined as days with not more than 20% of the stations with rain!) are observed during this month.

Regarding the evolution of the anomaly, it should be mentioned that during July 1961 (i.e. at the peak of the Southern Hemisphere dry season) an unusually strong rainy period happened in SE Tanganyika (latitude 4-9°S, 36-40°E, 83 stations) with 488% of normal rainfall or 9.8 mm per rainy day (46% of all days). It was even stronger in the isolated highlands around Mbeya (c. 9°S, 33°E, eight stations) with 621% of normal, 16.3 mm per rainy day (68% of all days). During this period even Lake Victoria was reached: at the Sesse Islands in NW 361 respectively 277 mm (about 250% of normal) fell on twelve (thirteen) days, similarly at Bukoba 325% on ten days. It is, however, not clear, in which way these seasonally unusual precursors could play any role in the later evolution which started in these regions not before September.

SIMULTANEOUS ANOMALIES AT THE EQUATORIAL INDIAN OCEAN

In a recent paper, Reverdin et al. (1986) have shown that at the equatorial Indian Ocean the normal mode of the atmospheric circulation is subject to large interannual variability. One of the most remarkable anomalies happened during October-December 1961, i.e. simultaneous with this East African rainfall event. This gives an opportunity to interpret the 1961/2 event as another example of macro-scale events at a time-scale of several months occurring over equatorial oceans and the nearby continents.

Investigations on the interannual variability at the Indian Ocean (Cadet & Diehl 1984) indicated, during November, variance maxima of cloudiness along the equator between longitude 50 and 70°E, those of the zonal wind component and sea surface temperature at latitude 5°S between longitude 70 and 90°E. The greatest variance of the zonal wind at 1°S, 84°E – along one of the most frequent shipping routes – occurs during (northern) fall.

Rainfall at the (very moist) western coast of Sumatra (Padang) was extremely low – only about 30% of normal – during October-December 1961. At the same time the spatial variance of zonal wind and cloudiness along the equator – indicating great differences in an E-W direction – reached an extreme (and unique) maximum, much higher than every other value of the 23-year period (Reverdin et al. 1986: Figs. 6-8). A cloudiness maximum at longitude 60-70°E contrasted with a minimum near 90°E. Reverdin et al. (1986) compared this observational evidence with a heat-induced circulation, forced by latent heating through convective rainfall above an area with unusually high sea surface temperature. Their Figures 12-13 indicated a sea surface temperature maximum along the equator between longitude 60 and 70°E with a deviation from normal of more than 2°C, together with a maximum of (probably convective) cloudiness. These features are balanced by a similarly impressive minimum at longitude 85-90°E, i.e. 1 000 km off the Sumatra coast. This large-scale circulation system (which cannot yet be mapped) provides the convergence of water-vapour transport into East Africa as mentioned above.

The Indian Ocean wind pattern (Reverdin et al. 1986: Fig. 14) shows for November 1961 a broad sinuous convergence zone along the equator instead of the usual broad band of westerlies straddling the equator, with two convergence zones on both sides. While the sea surface temperature is above 28°C in a broad zone extending from the Bengal Gulf to the Somali coast, a large area of high cloudiness (6-7 Oktas) covers most of the equatorial ocean west of longitude 70°E, while a low cloudiness area (3-4 Oktas) lies in its eastern part between latitude 0 and 15°S. Reverdin et al. (1986) conclude that in

1961 the Indian Ocean played an active role in the maintenance of the anomalies, which show also some similarities to the ENSO features at the Central Pacific.

Taking these features together with the 228% rainfall at the Seychelles, it is easily recognized that the apparently very large rainfall anomaly in East Africa is nothing but the western margin of a much greater anomaly, resembling the 1982/3 ENSO event in the equatorial Pacific. This excessive rainfall anomaly centered in October-December 1961 extended, between about latitude 5°N and 7°S, along the equator between longitude 30 and 70°E, i.e. over about 1 300 x 4 500 km, with an area estimated at about 6 million km^2. The comparable ENSO event – much better documented by satellite data measuring the area-averaged radiation temperature (in the equatorial region a reliable parameter for height) of cloud tops (Heddinghaus & Krueger 1981) and the precipitable water (Prabhakara et al. 1985) – extended, e.g. during December 1982, over about 1 500 x 5 000 km and was surrounded at three sides by a much larger area of subsidence indicated by warm (low-level) clouds and low precipitable water. The whole Pacific macro-system had a zonal (meridional) diameter of at least 110 (respectively 25) degrees, i.e. of more than 30 million km^2.

In the 1961 Indian Ocean/East Africa event the whole area cannot now be reconstructed, but our estimates of the role of the convergence of the (vertically integrated) water vapour transport indicate a similar magnitude; together with a not too dissimilar dynamics. During 1984 another macro-scale event of this type occurred in the Atlantic (Philander 1986). Macro-systems of this size and this persistence over several months up to half a year had never been recognized in high latitudes, but apparently can occur above equatorial latitudes and are not limited to the Pacific.

LAKE VICTORIA LEVEL AND WHITE NILE DISCHARGE UNTIL 1982

In earlier papers (Flohn 1983, Flohn & Burkhardt 1985) the author described the sudden rise of Lake Victoria between late 1961 and early 1964; the runoff of the White Nile at Jinja reached in January 1962 184%, January 1964 260% of the average during 1948-60. Further downwards, regional rainfall caused earlier relative peak values during April/May 1963. In contrast to earlier (but smaller) peak values in 1917 (measured) and probably also around 1893 and 1878 (estimated from significant correlations with the Aswan low season discharge), the lake level and the runoff data of the White Nile remained high for another two decades. If one compares the ten-year average 1962-71 with the thirteen-year average 1948-60, the ratio is at the Kagera river (flowing from W into Lake Victoria) 1.74, at Jinja 2.18, at Lake Kyoga 2.55, at Panyango (L.Albert) even 2.98 (?) and at Mongalla, about 830 km downstream of Jinja, still 2.21. Even after the great water losses in the swamp region and after including the waters of Bahr-el-Ghazal and Sobat, at Malakal, this ratio is still high (1.44). Here the rise of the discharge in the years 1962-74 is rather slow and the peak (relative and absolute) is not reached before December 1964 - February 1965. Below Khartoum, this signal cannot be recognized clearly, due to the overwhelming contribution from the Blue Nile and to the increasing consumption and seasonal manipulation by man.

These results were derived from the very valuable data published by Shahin (1985). More recent data from Jinja have been given by Kite (1981); lake levels until 1980 and Mongalla runoff data until 1982 have been available through correspondence.

At Mongalla the runoff ratio between the 50-year period 1912-61 and 1962-82 is 1.94; the difference between both averages (Fig. 3) is greater than the sum of the standard deviations of both subseries. In spite of the high evaporation of Lake Kyoga and Lake Albert (in both cases estimated to be higher than rainfall at the lakes) the rainfall at the catchment causes an increase of runoff between Jinja and Mongalla by 33%. In contrast to the small trend of Lake Victoria level after 1964, the Mongalla runoff shows a negative trend, statistically significant at $>$ 99%. If this trend can be extrapolated – which can be assumed if the rainfall of the catchment downstream of Jinja remains normal – the runoff value of the 50-year 'normal' period will be reached shortly after 2000.

For comparison Table 3 gives some statistical parameters of the Mongalla runoff for the subperiods 1912-61 and 1962-82 and for the whole period (71 years); see also Figure 3.

SUMMARY AND DISCUSSION

The 1961/2 event of widespread and long-lasting rainfall in East Africa was responsible for a conspicuous rise of Lake Victoria and a jump-like increase of the White Nile discharge by a factor $\sim$ 2 lasting until 1982 – later data being not available. The maintenance of these (unprecedented) rains needed a strong contribution of horizontal convergence of water vapour transport estimated to be 80-90%.

Comparative simultaneous data of surface winds, sea surface temperature and cloudiness at the western equatorial Indian Ocean – as presented by Reverdin et al. (1986) – indicate, together with rainfall data from the Seychelles, the existence of an even larger area of excessive rainfall extending, on both sides of the equator, up to longitude 70°E. The whole area covers at least an area of 12° latitude x 40° longitude or nearly 6×10^6 km^2. Looking to the high sea surface temperature data and the strong convection lasting for four-six months over such a vast area, this event resembles in several aspects the ENSO events in the central equatorial Pacific. An essential difference is demonstrated by the surface winds: instead of the normal westerlies

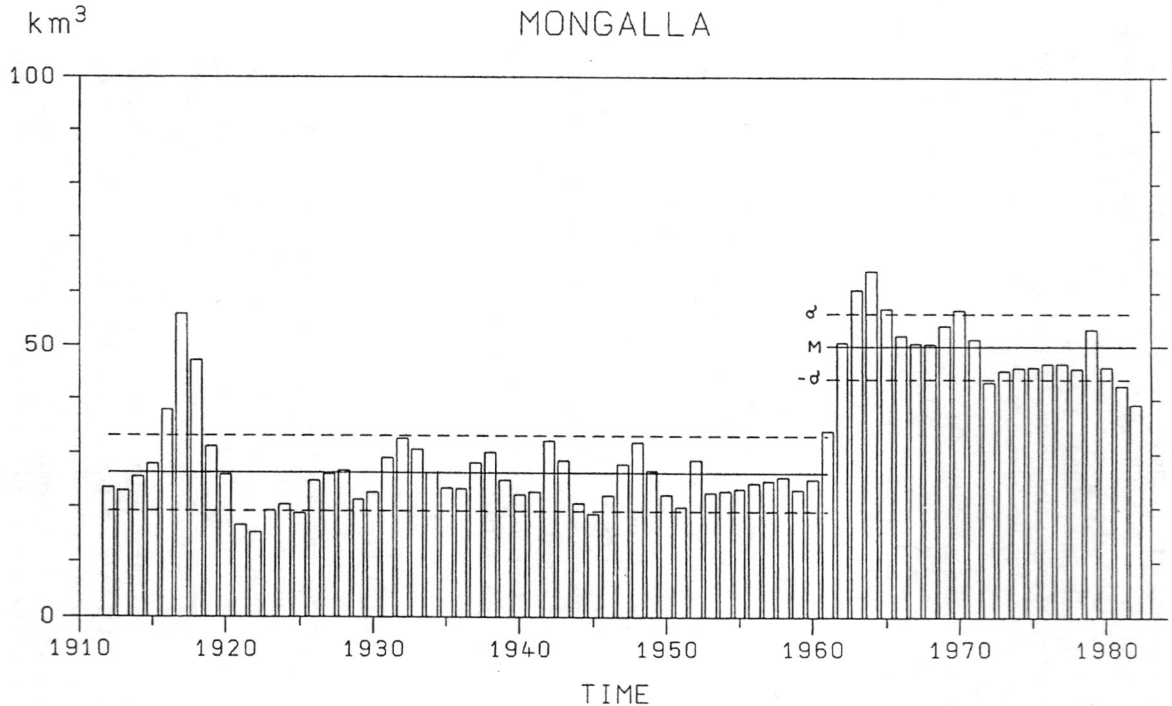

Figure 3. Runoff of the White Nile (1912-1982) at Mongalla (5.2°N, 31.9°E, c. 850 km downstream of Lake Victoria); M = average, σ = standard deviation in km^3/a. Note that the peak of 1916/7 remained only short-lived. The Lake Victoria level remained high until the last available data (1980); the downward trend after 1963 (at Mongalla $-0.83\,km^3/a$) is much weaker.

Table 3. Annual variation of the White Nile runoff at Mongalla (a = 1912-61, b = 1962-82, c = 1912-82)

		J	F	M	A	M	J	J	A	S	O	N	D	Year
Average	a	189	162	172	178	220	212	235	274	269	263	231	211	2617×10^7 m^3
	b	396	351	370	368	408	395	421	468	467	473	445	432	4996×10^7 m^3
	c	250	218	231	234	276	266	290	332	328	325	294	276	3320×10^7 m^3
Median	a	176	152	162	174	218	204	226	265	255	234	210	190	2489×10^7 m^3
	b	386	357	363	358	394	386	419	459	462	473	445	432	5040×10^7 m^3
Coefficient	a	32	32	32	26	28	30	27	26	33	36	32	32	27%
of variation	b	12	14	13	15	18	16	15	15	18	17	12	11	12%
	c	44	46	45	43	39	40	37	34	39	41	41	43	39%
Maximum	a	485	408	441	411	438	491	499	542	643	735	530	485	5580×10^7 m^3
	b	526	453	478	479	605	564	572	632	686	734	576	527	6400×10^7 m^3
Minimum	a	98	80	85	91	119	116	124	156	145	152	129	108	1526×10^7 m^3
	b	309	239	292	277	282	289	322	345	357	373	373	346	3911×10^7 m^3

encircled by two intertropical convergence zones at both flanks (which also characterize the peaks of the Pacific ENSO events) one observes a single broad convergence zone formed by weak easterlies both at its northern and southern flank.

In addition to the fairly frequent ENSO cases at the Pacific (including Indonesia), very large-scale climatic anomalies of about the same scale (both in time and space) seem to occur also at other oceans. In the Atlantic, Philander (1986) and other authors have described such an event during 1984. Some data from an earlier study (Henning & Flohn 1980) on the fluctuation of the heat fluxes at the air-sea interface in the Atlantic apparently suggest a similar (?) anomaly during 1963, partly simultaneous with a large-scale circulation anomaly over the northern Atlantic and Europe during early 1963. Here further studies are needed, since during 1962/63 a remarkable – hitherto unexplained – cooling occurred in the tropospheric temperature average over the polar cap (Flohn 1986: Fig. 6). Our knowledge of such large-scale teleconnections – as suggested by Sir Gilbert Walker, Jerome Namias and other scientists – is still insufficient. The ENSO event of 1982/3 (Rasmusson & Wallace 1983, Flohn 1986) has indicated, that such anomalous giant areas of enhanced convective rainfall, extended over areas of the order 10 million km^2 and persisting for four-six months are – due to the enormous amount of latent heat liberated by the rains – able to disturb the atmospheric circulation of large sections of the globe in a hitherto unknown manner.

It should not be forgotten that the ice-covered polar regions at both hemispheres cover some 7% of the global area, while the tropics cover about 30-40%; the latitude belt between 10°S and 10°N alone covers 17%. Recent evidence shows the irregular occurrence of giant convective macro-cells with a life-time of half a year or so – rather frequent in the Pacific section, rather rare in the Indian and Atlantic section of the globe. Due to the highly effective liberation of latent heat by rainfall, these events create anomalous patterns of the atmospheric circulation of a hitherto unsuspected scale. Their role for climatic change (and for long-range prediction) deserves attention also over the other equatorial regions.

APPENDIX

For comparison, some representative data of area-averaged rainfall are given for southern Germany (Ungewitter 1968), i.e. for an area of about 120 000 km^2 in latitude 48-50°N. These data are based on a twelve-year sample for the whole year, with bracketed data referring to the summer tertial (May-August) alone. On 12.8 (10.6) % of all days all synoptic stations (n = 40) receive precipitation, with an average daily amount of 7.8 (8.9) mm. On 5% of all days the area-averaged rainfall is greater than 10 mm/d, in less than 0.2% (three cases) greater than 20 mm/d. For northern Germany (n = 37) with few mountains the figures are somewhat smaller.

ACKNOWLEDGEMENTS

The author acknowledges the valuable assistance of Mr M.Shahin (Delft, Netherlands, formerly Cairo), in providing unpublished runoff data from the Nile catchment, as well as of Mr Macodras (Nairobi) for Lake Victoria data and long rainfall series from Kenya. Furthermore the help of Ms Alice Kapala with figures and the diligent work of Ms I.Hayes at the word processor is gratefully acknowledged.

REFERENCES

Behrend, H. 1984. Teleconnections of tropical rainfall anomalies and the southern oscillation. *Bonner Meteor. Abhandl.* 31: 1-50.

Bell, B. 1975. Climate and the history of Egypt. *Am. J. Archeol.* 79: 224-269.

Cadet, C.L. & B.C.Diehl 1984. Interannual variability of surface fields over the Indian Ocean during recent decades. *Monthly Weather Rev.* 112: 1921-1935.

East African Meteorological Department 1962. *Summary of rainfall in Kenya and Seychelles (Tanganyika and Zanzibar, Uganda) for the year 1961. The daily rainfall of East Africa during November 1961.*

Flohn, H. 1983. Der Katastrophenregen 1961/2 und die Wasserbilanz des Viktoria-See-Gebietes. *Wiss, Ber. Meteor. Inst. Karlsruhe* 4: 17-34.

Flohn, H. 1986. Singular events and catastrophes now and in climatic history. *Naturwissenschaften* 73: 136-149.

Flohn, H. (in press). *Rainfall teleconnections in northern and eastern Africa.*

Flohn, H. & Th.Burkhardt 1985. Nile runoff at Aswan and Lake Victoria: a case of discontinuous climate time series. *Zeitschr. Gletscherkde. Glazialgeol.* 21: 125-130.

Hassan, F.A. 1981. Historical Nile floods and their implications for climatic change. *Science* 212: 1142-1145.

Heddinghaus, T.R. & A.F.Krueger 1981. Annual and interannual variations in outgoing long-wave radiation over the tropics. *Monthly Weather Rev.* 109: 1208-1218.

Henning, D. & H.Flohn 1980. Some aspects of evaporation and sensible heat flux in the tropical Atlantic. *Contrib. Atmos. Phys.* 53: 430-441.

Kite, G.M. 1981. Recent changes in level of Lake Victoria. *Hydrol. Sci. Bull.* 23: 233-243.

Prabhakara, C., D.A.Short & B.E.Vollmer 1985. El Niño and atmospheric water vapour: Observations from Nimbus 7 SMMR. *J. Clim. Appl. Meteor.* 24: 1311-1324.

Peixoto, J.P. & A.H.Oort 1983. The atmospheric branch of the hydrological cycle and climate. In A.Street-Perrott et al. (eds.), *Variations in the Global Warter Budget*: 5-65. Dordrecht: Reidel.

Philander, S.G.H. 1986. Unusual conditions in the tropical Atlantic Ocean in 1984. *Nature* 322: 236-238.

Rasmusson, E.M. & T.Carpenter 1982. Variations in tropical sea surface temperature and surface wind fields associated with the Southern Oscillation/El Niño. *Monthly Weather Rev.* 110: 354-384.

Rasmusson, E.M. & J.M.Wallace 1983. Meteorological aspects of the El Niño/Southern Oscillation. *Science* 222: 1195-1202.

Reverdin, G., D.L.Cadet & D.Gutzler 1986. Interannual displacements of convection and surface circulation over the equatorial Indian Ocean. *Quart. J. Roy. Meteor. Soc.* 112: 43-67.

Riehl, H., M.el-Bakry & J.Meitin 1979. Nile river discharge. *Monthly Weather Rev.* 107: 1546-1553.

Rognon, P. & M.Curie 1983. Les crises climatiques de courte durée (quelques années à quelques siècles) et leur enregistrement dans la sédimentation continentale. In A. Ghazi (ed.), *Palaeoclimatic research and models*: 114-123. Dordrecht: Reidel.
Said, R. 1981. *The geological evolution of the River Nile.* Berlin: Springer.
Shahin, M. 1985. *Hydrology of the Nile basin.* Amsterdam: Elsevier.
Van Zinderen Bakker, E.M. & J.H.Mercer 1986. Major late Cainozoic climatic events and palaeoenvironmental changes in Africa viewed in a world-wide context. *Palaeogeogr., Palaeoclim., Palaeoecol.* 56: 217-235.
Williams, M.A.J. & H.Faure 1980. *The Sahara and the Nile.* Rotterdam: Balkema.
Yamamoto, R., T.Iwashima & S.N.Kazadi 1985. Climatic jump: a hypothesis in climate diagnosis. *J. Meteor. Soc. Japan* 63: 1157-1160.

ANTARCTICA

THE ANTARCTIC ICE SHEET DURING THE LATE NEOGENE

J.H.MERCER
Institute of Polar Studies, Ohio State University, Columbus, Ohio, USA

ABSTRACT

Until recently the ice sheet in East Antarctica was believed to have changed little since first covering the continent in the late Middle Miocene. Recent discoveries of marine diatoms apparently of Pliocene age, and of plant remains, including wood of a *Nothofagus* species, associated with glacial sediments of the Sirius Formation at high elevations in the Transantarctic Mountains, have dramatically changed this picture. Some workers have inferred open-marine conditions in what are now subglacial basins on the East Antarctic craton throughout most of the Pliocene, ending when dropping temperatures allowed the ice sheet to occupy these basins, partly submerging the Transantarctic Mountains where it deposited the diatom-bearing Sirius Formation in late Pliocene to early Pleistocene time. Others point to the scarcity of evidence for prolonged warmth in the late Pliocene and favour restricting open-marine conditions in East Antarctica to the well-documented early Pliocene warm interval about 5-4 Ma, believing the species diachroneity of 1-2 million years accounts for the apparent age of the youngest diatom assemblage. However, evidence that long-distance ice-rafting, which implies the production of tabular bergs from a cold ice sheet, continued during the early Pliocene conflicts with the presence of open-marine embayments and the growth of vegetation, including tree species, at latitude 85°S, which imply that ice was temperate at sea level. Such conditions may have prevailed during a warm interval centred ~ 8 Ma, immediately before the well-documented terminal Miocene refrigeration and glacial build-up and before the start of long-distance ice-rafting. This chronology is supported by the characteristics of the *Nothofagus* wood, which closely resembles that of one species in southern South America and one in Tasmania. Reasons are given for believing that the antarctic *Nothofagus* species was not introduced from South America during the Pliocene, but was an antarctic survival, probably related to the Tasmanian species and part of an endemic plant assemblage whose hardiest constituents were extinguished by the terminal Miocene refrigeration. A major obstacle for this hypothesis is that it requires a 4-5 million-year diachroneity of diatom assemblages.

BACKGROUND

Until the last few years, a consistent picture of the history of the Antarctic ice cover was emerging from studies of the distribution in time and space of ice rafted detritus and the stable isotopic composition of foraminiferal tests in ocean cores: first appearance of calving glaciers at the end of the Oligocene ~ 26 Ma, followed by build-up of the ice sheet in East Antarctica between 16.5 and 13 Ma, and formation of the marine ice sheet in West Antarctica 7-9 Ma (Rutford et al. 1972, Shackleton & Kennett 1975, Ciesielski & Weaver 1983). The ice sheet in East Antarctica was thought to have changed little after an initial period of instability, especially 14.8-14 Ma (Woodruff et al. 1981). On the other hand the ice sheet in West Antarctica, being a marine ice sheet based below sea level and thus dependent on the presence of fringing ice shelves which cannot form or persist except as cold glaciers below the pressure melting point, was believed to have been much more vulnerable to climatic variations and had probably disintegrated repeatedly since it first formed, most recently, perhaps, during the last interglacial (Mercer 1968). Matthews & Poore (1980), however, disputed this reconstruction of Antarctic glacial history; by interpreting the oxygen isotopic curves differently, they concluded that a major ice sheet had formed in East Antarctica as early as the Eocene-Oligocene transition ~ 38 Ma. Others have since come to similar conclusions. Despite these differences of opinion there was general agreement that an ice sheet had covered the East Antarctic craton continuously since the late Middle Miocene. Although Ciesielski & Weaver (1974) interpreted changes in the distribution of silicoflagellates as showing that ~ 4.25-4 Ma, surface waters at latitude 63°S in the southwest Pacific Ocean were 8-10°C warmer than they are today, they concluded that the ice sheet in East Antarctica would have suffered only marginal recession at that time.

SIRIUS FORMATION: MARINE DIATOMS AND PLANT REMAINS

Two developments in the 1980s have called for radical reassessment of the late Neogene history of the Antarctic Ice Sheet. Both concern the fossil content of the Sirius Formation, which consists of widespread, isolated outcrops of glacial sediments deposited beneath and at the margins of wet-based ice, disconformably covering pre-Tertiary rocks in the Trans-antarctic Mountains. In 1983, sedimentary clasts containing marine diatoms were found in till of the Sirius Formation at latitude 86°S and 2 400 m elevation in the Reedy Glacier area. The diatoms have been identified as of Lower Oligocene, Upper

Oligocene to Lower Miocene, Middle Miocene and Pliocene age; Pliocene species dominate the assemblage (Harwood, in press).

Webb et al. (1984) argued convincingly against an in situ origin, or transport from West Antarctica, and concluded that the diatomaceous sedimentary clasts had been carried by East Antarctic ice from former marine basins on the East Antarctic craton, now the Wilkes and Pensacola subglacial basins. They inferred extensive deglaciation of East Antarctica throughout most of the Pliocene, probably brought about by warmth-induced removal of ice shelves and marine-based ice; when temperatures fell again to levels at which ice shelves and marine-based ice could survive – that is, water temperature below -1.5°C and summer air temperature below 0°C – the diminished ice sheet covering the highlands of East Antarctica was able to expand and re-occupy the marine basins, picking up marine sediments which, mixed with other glacial debris, were deposited in the Sirius Formation on the Transantarctic Mountains. The youngest diatoms in the Sirius Formation are mid- to late Pliocene in age, according to the biostratigraphy developed in high middle latitudes by McCollum (1975), Weaver & Gombos (1981) and Ciesielski (1983). The Sirius Formation is of course, younger than the youngest diatoms.

In 1985 plant remains were found at several horizons in sediments of the Sirius Formation at Oliver Bluffs near the northern end of the Dominion Range massif, adjacent to the Beardmore Glacier at latitude 85°S and 1 800 m elevation (Webb & Harwood, in press). The plant remains include small stems of wood (the largest 10 cm by 2 cm), leaves and pollen grains which, according to Askin & Markgraf (in press), include those of *Nothofagus* of the *fusca* type. The wood has been identified as *Nothofagus*; it most closely resembles that of the South American species *N.betuloides*, but is also very similar to that of the Tasmanian species *N.gunnii* (Carlquist, in preparation). The small size of the stems suggests the *krummholz* form, growing above the tree line near the altitudinal limit of the species. This implies that summer temperatures at the site were about 4.5°C, which is the estimated temperature at the upper limit of *N.antarctica* scrub at 700 m in Tierra del Fuego, using an atmospheric lapse rate of 0.65°C per 100 m (Mercer 1978 and in press, McQueen 1976). Mercer (1985) inferred that the open-marine Wilkes and Pensacola basins were re-occupied by ice after temperatures had fallen enough for ice shelves and marine-based ice to form, but the summer warmth shown by the plant remains conflicts with this reconstruction of events, implying that the ice sheet in East Antarctica re-occupied these basins when it was temperate at sea level and temperatures were too high for the presence of ice shelves and marine-based ice. It could have done this only if the basins had risen above sea level, allowing the ice to advance across dry (and presumably vegetated) land. This conclusion receives some support from Drewry's (1983) map of the isostatically adjusted bedrock surface of Antarctica after complete deglaciation, according to which marine waters would extend only to latitude 77°30'S in the Wilkes basin, and to 86°S in the Pensacola basin; in between for a distance of 1 600 km behind the

Transantarctic Mountains from the Skelton to the Reedy glaciers, the surface would be above sea level, much of it more than 500 m above. Therefore, the floors of the Wilkes and Pensacola basins, which were at least 100 m below sea level when the diatoms were living (Harwood, in press), have risen several hundred meters relative to sea level since the last marine incursion.

AGE OF THE SIRIUS FORMATION

The Sirius Formation was deposited by wet-based ice at a time when the Antarctic climate was much warmer than it is today – warm enough, at least on occasion, for shrubby vegetation to grow at latitude 85°S – and after a comparably warm interval when open-marine conditions prevailed on parts of the East Antarctic craton. There is no reason to believe that these were not consecutive parts of the same warm interval. Any age assignment for the Sirius Formation must take into account this evidence for prolonged warmth.

Webb & Harwood (in press), by applying to high southern latitudes the diatom stratigraphy developed from sub-Antarctic ocean cores, conclude that open-marine conditions prevailed on the East Antarctic craton during most of the Pliocene; they believe that the Sirius Formation itself, being younger than the youngest diatoms, is late Pliocene to early Pleistocene in age. Support for the ocean core based diatom chronostratigraphy comes from Cockburn Island, Antarctic Peninsula, where the pecten conglomerate, for which Harwood (1986) had previously proposed an age of 2.8 2 MY based on associated marine diatoms has been K/Ar-dated at less than 3.65 MY (Webb & Andreasen, in press). Hennig (1911) noted that some of the bryozoan species need a minimal water temperature of ~ 10°C, which is ~ 12 °C higher than today's. However, Johnson & Nigrini (1985) find that even in the tropics, biostratigraphic non-synchroneity may be more than 1 MY, and they suspect that it could be even greater across oceanographic and climatic gradients. Harwood (1986) suspects that diatom biostratigraphy from the Southern Ocean may not apply directly to Antarctic interior and coastal sites; he believes that much research is needed to understand species migration and diachronous first and last appearance data across latitudes and oceanic fronts in the Southern Ocean and Antarctic intracratonic basins.

The chief problem with accepting a late Pliocene or early Pleistocene age for deposition of the Sirius Formation is the scarcity of evidence for high-latitude warmth at that time, commensurate with the growth of woody vegetation, including tree species, at latitude 85°S (Mercer 1985).

Late Pliocene climates

Although Kennett & Vella (1975) infer a prolonged warm interval ~ 3.4-2.6 Ma at DSDP Site 284 in the Tasman Sea, Hodell et al. (1986) find evidence for repeated glacial-interglacial (or cold-warm) events 4.1-2 Ma at

DSDP Site 588 on the Lord Howe Rise east of Brisbane, and Ciesielski & Grinstead (1986) conclude that at DSDP Site 514 at latitude 46°S in the South Atlantic, long-term cooling began between 3.97 and 3.86 Ma. They believe that a hiatus at this site from 3.86 to 3.18 Ma resulted from the expansion of sea ice and ice shelves in the Antarctic region; as supporting evidence they note that the middle of this interval, a major glaciation at latitude 49° 30′S in southern Argentina has been dated by K/Ar at 3.6 Ma (Mercer 1983). In the Northern Hemisphere, ice-rafting began in Baffin Bay at DSDP Site 645 (latitude 68°N) about 3.5 Ma, and at the same time, surface waters of the Labrador Sea (DSDP Site 646) became permanently colder (Arthur, Srivastava et al. 1986). In the eastern North Atlantic, waters at DSDP Site 548 southwest of England were warmer than they are today 3.4 Ma, and then cooled 3.4-3.2 Ma (Loubere & Moss 1986). Repeated K/Ar-dated glaciations began in Iceland about 3.15 Ma (McDougall & Wensink 1966, Saemundsson & Noll 1974), and the middle latitude ice sheets first formed about 2.5 Ma (Shackleton, Backman et al. 1984). On the other hand some biological evidence, unsupported by K/Ar dating, supports late Pliocene warmth in high northern latitudes; in northeast Greenland the remains of a forest similar to that now growing in Labrador 2 500 km to the south has been assigned to the late Pliocene or Pleistocene on the evidence of the associated molluscan fauna (Funder et al. 1985), and on the north coast of Alaska, biological evidence suggests that sea ice was scarce or absent some time after 3 Ma and before 2.4 Ma (Carter et al. 1986).

Early Pliocene climates

In this early Pliocene, much evidence points to a warm interval, accompanied by marine transgression and decrease in $\delta^{18}O$ levels, lasting three-quarters to one million years, following a time of cooling, ice build-up and a drop in sea level at the end of the Miocene ~ 6.2 to 5 Ma (Ciesielski & Weaver 1974, Keany 1978, Vail & Hardenbol 1979, Mercer & Sutter 1982, Kennett 1986, Kennett & von der Borch 1986, Hodell et al. 1986). Mercer (1985) and Ciesielski & Grinstead (1986) suggested an early Pliocene age for the open-marine conditions in the Wilkes and Pensacola basins. However, although the early Pliocene seems a more likely time than the late Pliocene for Antarctic warmth, temperatures then were probably still too low for deglaciation of the Wilkes and Pensacola basins, and for the growth of woody vegetation at latitude 85°S, which implies that summer temperatures were above freezing point and glaciers were temperate throughout coastal Antarctica. At these temperatures glaciers such as the Beardmore would have ended on reaching tidewater, calving icebergs of Greenland type, orders of magnitude smaller than the largest tabular bergs. Ice shelves and floating glacier tongues and, therefore, tabular bergs would have been absent, so that the northern limit of IRD (ice-rafted detritus) would have been far to the south of its present position.

Available evidence strongly suggests that icebergs have been reaching to about their present northern limit in middle latitudes without major interruption, though in varying quantities, since 7-6.5 Ma (short interruptions, comparable in duration to a Pleistocene interglacial, would be hard to detect). In the South Atlantic, which has the best record, the northern limit of IRD moved back and forth during the latest Miocene and Pliocene, but it never retreated far to the south. At DSDP Site 513 at latitude 47°35′S, for instance, a hiatus spans the interval 3.86-3.16 Ma, but otherwise IRD was deposited continuously, though at varying rates, after its first appearance ~ 6.9 Ma (Bornhold 1983). At DSDP Site 328 at latitude 49°49′S north of South Georgia, deposition of dropstones, which are better evidence for iceberg-rafting than are quartz grains, began in the late Miocene and continued without interruption thereafter, except for hiatuses at the Miocene-Pliocene and Pliocene-Pleistocene transitions (Barker, Dalziel et al. 1976, Plafker et al. 1976). At DSDP Site 266 at latitude 56°24′S in the Indian Ocean south of Australia, ice-rafting began slightly after 6.5 Ma, continued at low levels during the Gilbert chron 5.2-3.4 Ma, and decreased during the Gauss chron 3.4-2.45 Ma. However, its detailed history at this site is uncertain because of hiatuses and poor core recovery (Piper & Brisco 1975).

Other factors that weigh against acceptance of a Pliocene age for the Sirius Formation are: 1. the present high elevations of the diatom-bearing till outcrops, in relation to the calculated uplift history of the Trans-antarctic Mountains; and 2. evidence that the Sirius Formation in places covers previously unglaciated surfaces.

High elevation of Sirius Formation outcrops

The highest known outcrop of the Sirius Formation is at 4 114 m on Mt McKellar in the Beardmore Glacier area (Prentice et al. in press). In the Dominion Range, also in the Beardmore Glacier area, it ranges between 1 800 and 2 300 m. At all these sites the glacial sediments must have been uplifted many hundreds of meters since deposition. The outcrop where plant remains are interbedded with glacial sediments is now at 1 800 m elevation, but must have been within ~ 500 m of sea level when the vegetation was growing, otherwise temperatures at sea level would have been too high for the existence of an ice sheet. This implies uplift of at least 1 300 m. The uplift rate of the Transantarctic Mountains during the Neogene, based on observations in Victoria Land, is generally believed not to have exceeded 100 m per million years (Fitzgerald & Gleadow 1985, Smith & Drewry 1984), which would imply an age of ~ 13 MA for the plant remains, compared to the age of less than 3.5 MA inferred from the diatom content of the Sirius Formation. However, faulting of the Sirius Formation in the Dominion Range, with displacement of several hundred meters, is evidence for major tectonism since the vegetation was growing (Webb et al., in press), implying that in this part of the Transantarctic Mountains, either uplift has greatly exceeded 100 m per million years during the late Neogene, or the accepted age ranges of the diatoms do not apply in these high latitudes.

Sirius Formation covers previously unglaciated surfaces

On Tillite Ridge adjacent to the Reedy Glacier, diatom-bearing till of the Sirius Formation overlies rock debris produced by weathering and solifluction (Mercer 1968). On Orr Peak in the Miller Range, Nimrod Glacier area, sediments of the Sirius Formation cover basement gneiss oxidized and fragmented to a depth of 40 m (Webb et al., in press). This implies that these sites had not been glaciated during the preceding several million years, which seems unlikely if the Sirius Formation was Pliocene in age, in view of the evidence for eustatic sea level drop and, by inference, ice-sheet build-up, at the end of the Miocene (e.g. Kennett & von der Borch 1986, Mercer & Sutter 1982).

If a Pliocene age for the Sirius Formation is ruled out, it must be older, but the further its inferred age is pushed back before 3-4 Ma, the less credible becomes the required species diachroneity of the youngest diatoms that it contains. A latest Miocene age is ruled out by the abundant evidence for cold, glacial build-up, and long-distance ice-rafting at that time. However, immediately preceding this terminal Miocene event, there is evidence for a two million year-long warm interval centered bout 8 Ma.

Late Miocene warm interval

For the past 6.5 million years – that is, since the well-documented latest Miocene cooling, glacial build-up and eustatic sea-level drop – uninterrupted ice-rafting to middle latitudes strongly suggests that temperatures have been too low for open-marine conditions to have prevailed on the East Antarctic craton, or shrubby vegetation to have grown at latitude 85°S. If so, the warm interval must have occurred before 6.5 Ma. Kennett (1986) finds evidence on the Lord Howe Rise east of Brisbane for relative warmth 9-6.5 Ma, between a cool interval 11-9 Ma and the terminal Miocene cold interval starting about 6.2 Ma. From changes in nannoflora distribution, Haq (1980) infers a warm interval 9-7.4 Ma in the North Atlantic and 9-8 Ma in the South Atlantic; in the North Atlantic this was a particularly warm interval, when a tropical assemblage penetrated as far north as the Bay of Biscay. At DSDP Site 357 in the South Atlantic, Boersma & Shackleton (1977) noted that both the planktic and benthic $\delta^{18}O$ records show a warm (or low ice-volume) peak ~ 8 Ma, followed by a cold (or high ice-volume) peak ~ 6 Ma; the range is 0.6% for benthics and 1% for planktics. In the northeast Pacific Wolfe * Poore (1982) find a marked Late Miocene warming centered ~ 8 Ma, which is also detected in the land flora of the adjacent North American coasts.

DISCUSSION

The marine diatoms embedded in, and the plant remains interbedded with, the glacial sediments of the Sirius Formation, are conclusive evidence for a prolonged final warm interval when Antarctic temperatures were much higher than they are today. Problems arise when trying to assign an age-span for this warm interval as there are well-founded objections to all the choices. Three possibilities are considered: late Pliocene to early Pleistocene (after 3.5 Ma); early Pliocene (~ 5-4 Ma); and Late Miocene (~ 9-7 Ma). Evidence can be marshalled for and against each, as follows, ignoring for the moment the implications of Carlquist's conclusions after examining the internal structure of the wood from the Beardmore Glacier.

Late Pliocene-early Pleistocene after ~ 3.5 Ma

Evidence for: It is consistent with the palaeomagnetically-dated diatom stratigraphy for the Southern Ocean, to which the diatom assemblage on Cockburn Island, northern Antarctic Peninsula, conforms.

Evidence against: Evidence for high-latitude warmth at that time is scarce, particularly for the prolonged and exceptional level of warmth that the field evidence demands. It requires an uplift rate for the Transantarctic Mountains many times greater than has been inferred from other evidence. The absence of evidence for earlier incursions of the ice sheet in East Antarctica onto the Transantarctic Mountains is hard to explain, in view of the evidence for the presence of a major ice sheet at the end of the Miocene and probably also about 3.6 Ma. The northern extent of ice-rafting was similar to today's.

Early Pliocene ~ 5-4 Ma

Evidence for: Evidence for prolonged warmth is abundant.

Evidence against: It would require species diachroneity of 1-2 million years. There is no evidence for major restriction in the northern extent of ice-rafting. It would require an uplift rate of the Transantarctic Mountains much greater than has been inferred from other evidence. The absence of evidence for an earlier incursion of the ice sheet in East Antarctica onto the Transantarctic Mountains at the end of the Miocene is hard to explain.

Late Miocene ~ 9-7 Ma

Evidence for: It was the last interval without long-distance ice-rafting. Uplift rates for the Transantarctic Mountains are greater than anticipated, but acceptable. It is reasonable to suppose that the ice sheet first reached full-bodied size at the end of the Miocene, covering previously unglaciated surfaces in the Transantarctic Mountains.

Evidence against: Major diachroneity – about 5 million years – of diatom assemblages is required.

Unless some radiometric – preferably isotopic – age control is obtained for this final interval of Antarctic warmth, it may be hard to demonstrate unequivocally just when it occurred. However, the Antarctic plant remains may perhaps provide a valuable clue, if they can be shown to be either from extinct endemic species, or from an extant species now living outside Antarctica. If the warm interval occurred during the Pliocene, the vegetation could not have survived the cold at the end of the Miocene, and must have been reintroduced; if, on the other hand, the warm interval preceded this terminal Miocene cold, the plants could be the surviving remnant of a long-established Antarctic plant cover, in which case they should show marked endemism.

The age of the Antarctic Nothofagus wood

Carlquist (in preparation) has determined that the wood found in Antarctica at latitude 85°S is *Nothofagus.* Its internal structure closely resembles that of only two other species: *N.gunnii* of Tasmania, and *N.betuloides* of southern South America. It is significantly different to that of the other two southern South American species, *N.pumilio* and *N.antarctica.* This has important implications for the ancestry of the Antarctic *Nothofagus,* for the age of the wood specimens and, therefore, for the age of the Sirius Formation, in which the wood-bearing horizons occur.

N.gunnii is a hardy deciduous species which, in its native Tasmania, grows under extremes of cold and wind exposure at elevations above 1 300 m, where it forms thickets 1-2.5 m high in rocky areas near or above treeline (Beadle 1981). This is the sort of environment that probably prevailed adjacent to the Beardmore Glacier when *Nothofagus* shrubs were growing there. The evergreen *N.betuloides,* on the other hand, grows in the wettest and most isothermal parts of southern South America, chiefly on the western side of the cordillera where annual precipitation ranges from about 1 000 mm to more than 7 000 mm. Such a climate must differ markedly from that in which the Antarctic *Nothofagus* grew. Precipitation on the coasts of an open-marine Ross embayment would certainly have been higher than it is today, but not as high as in the present *N.betuloides* forests, where the orographic effect of the Andean cordillera augments the frontal precipitation from a stream of eastward-travelling low pressure systems. Furthermore, at latitude 85°S, where the sun is nearly six months continuously above the horizon and six months below it, the climate is not likely to have been isothermal.

East of the *N.betuloides* forests in southern South America, the deciduous *N.pumilio* begins to dominate where rainfall drops below 600 mm annually. However, the hardiest South American species is the deciduous *N.antarctica,* which forms a belt of prostrate shrubs extending as much as 100 m above the treeline of both the *N.betuloides* forest on the west and the *N.pumilio* forest on the east (McQueen 1976). In Tierra del Fuego, where summer sea-level temperature averages about 9°C, *N.antarctica* scrub extends up to about 700 m elevation (McQueen 1976); summer temperature there can average no more than about 4.5°C (Mercer 1978). Thus *N.antarctica,* not *N.betuloides,* is the hardy South American counterpart of the Tasmanian *N.gunnii*; both occupy habitats similar to that inferred for the shrubby *Nothofagus* that grew adjacent to the Beardmore Glacier. Carlquist (in preparation), however, has determined that the wood from the Beardmore Glacier is not *N.antarctica,* and *N.betuloides,* whose wood it most closely resembles, is the least suited of the three southern South American species to harsh conditions. It is, therefore, hard to believe that *N.betuloides,* rather than *N.antarctica,* would have gained a foothold in Antarctica if that continent was re-colonized during the Pliocene, after an interval of severe cold at the end of the Miocene that must have exterminated tree species. Evidence for major glaciation in southern Argentina at the end of the Miocene (Mercer & Sutter 1982) implies that Antarctica was then at least as cold as it is today.

If the Antarctic *Nothofagus* species is not a Pliocene immigrant from South America, what is it and what is its history? *N.gunnii* has similar wood structure and is tolerant of cold, harsh conditions, but there is no obvious factor that might have favoured recolonization of Antarctica by this species from a distance of 3 000 km instead of by a South American species from 900 km away. Carlquist (in preparation) suggests that the wood may not be that of either *N.gunnii* or *N.betuloides,* but is that of an extinct species that presumably is related to one of them. This indeed seems to be the most probable answer. *N.gunnii* is the more likely relative because its habitat resembles more closely that inferred for the Antarctic species. If so, the two species probably diverged during the Palaeogene, before Tasmania had separated too far from Antarctica.

If this reasoning is valid, it has important implications for the age of the Antarctic wood, because tree species could not have survived the cold interval at the end of the Miocene ~ 6.5-5 Ma. Therefore it must predate it, representing one of the last survivors of a long-established plant assemblage destined for extinction before the start of the Pliocene. This implies that at least some of the Sirius Formation, with which the wood-bearing horizons are interbedded, also predates the latest Miocene.

ACKNOWLEDGEMENTS

Supported by National Science Foundation grants DPP-8117889A01 and DPP-8420622A01, awarded to the Ohio State University. Institute of Polar Studies contribution no. 579.

REFERENCES

Arthur, M., S.Srivastava et al. 1986. High-latitude palaeoceanography. *Nature* 320: 17-18.

Askin, R.A. & V.Markgraf (in press). Palynomorphs from the Sirius Formation, Dominion Range Antarctica. *Antarctic J. of the US.*

Barker, P., I.W.D.Dalziel et al. 1976. Site 328, DSDP 36. In P.F.Baker, I.W.D.Dalziel et al. (eds.), *Initial reports of the Deep-Sea Drilling Project* Vol. 36: 87-141. Washington DC: US Government Printing Office.

Beadle, N.C.W. 1981. *The vegetation of Australia.* Stuttgart: Fischer. 690 pp.

Boersma, A. & N.Shackleton 1977. Tertiary oxygen and carbon isotope stratigraphy, Site 357 (Mid-latitude South Atlantic). In P.R.Supko, K.Perch-Nielsen et al. (eds.), *Initial Reports of the Deep-Sea Drilling Project* Vol. 39: 911-924. Washington DC: US Government Printing Office.

Bornhold, B.D. 1983. Ice-rafted debris in sediments from Leg 71, southwest Atlantic Ocean. In W.J.Ludwig & V.A.Krasheninikov (eds.), *Initial Reports of the Deep-Sea Drilling Project* Vol. 71: 307-316. Washington DC: Government Printing Office.

Carlquist, S. (in preparation). Pliocene *Nothofagus* wood from the Transantarctic Mountains. *Aliso* (Journal of the Rancho Santa Ana Botanic Garden).

Carter, D.L., J.Brigham-Grette, L.Marincovich, V.L.Pease & J.W.Hillhouse 1986. Late Cenozoic Arctic Ocean sea ice and terrestrial paleoclimate. *Geology* 14: 675-678.

Giesielski, P.F. 1983. The Neogene diatom stratigraphy of DSDP, Leg 71. In W.J.Ludwig & V.A.Krasheninikov (eds.), *Initial Reports of the Deep-sea Drilling Project* Vol. 71: 635-665. Washington DC: Government Printing Office.

Ciesielski, P.F. & G.P.Grinstead 1986. Pliocene variations in the position of the Antarctic Convergence in the southest Atlantic. *Paleoceanography* 1: 197-232.

Ciesielski, P.F., M.T.Ledbetter & B.B.Ellwood 1982. The development of Antarctic glaciation and the Neogene paleoenvironment of the Maurice Ewing Bank. *Marine Geology* 46: 1-51.

Ciesielski, P.F. & F.M.Weaver 1974. Early Pliocene temperature changes in the Antarctic seas. *Geology* 2: 511-515.

Ciesielski, P.F. & F.M.Weaver 1983. Neogene and Quaternary paleoenvironmental history of Deep-Sea Drilling Project Leg 71 sediments, southwest Atlantic Ocean. In W.J. Ludwig & V.A.Krasheninikov (eds.), *Initial reports of the Deep-Sea Drilling Project* Vol. 71: 461-477. Washington DC: Government Printing Office.

Drewry, D.J. 1983. Isostatically adjusted bedrock surface of Antarctica. In *Antarctica: Glaciological and Geophysical Folio* Sheet 6. Cambridge: Scott Polar Research Institute.

Fitzgerald, P.G. & A.J.W.Gleadow 1985. Uplift history of the Transantarctic Mountains, Victoria Land, Antarctica. *Sixth Gondwana Symposium*: Abstract. Columbus: Ohio State University.

Funder, F., N.Abrahamsen, O.Bennike & R.W.Feyling-Hansen 1985. Forested Arctic: evidence from North Greenland. *Geology* 13: 542-546.

Haq, B.U. 1980. Biogeographic history of Miocene calcareous nannoplankton and paleoceanography of the Atlantic Ocean. *Micropaleontology* 26: 414-443.

Harwood, D.M. 1986. *Diatom biostratigraphy and paleoecology with a Cenozoic history of Antarctic ice sheets,* unpublished Ph.D. dissertation, Ohio State University.

Harwood, D.M. (in press). Recycled siliceous microfossils from the Sirius Formation. *Antarctic J. of the US.*

Hennig, A. 1911. Le conglomérat pléistocène à pecten de l'Ile Cockburn. *Wissenschaftliche Ergebnisse der Schwedischen Südpolar-expedition 1901-1903,* Band 3 (10: 73 pp.

Hodell, D.A., K.M.Elmstrom & J.P.Kennett 1986. Latest Miocene benthic $\delta^{18}O$ changes, global ice volume, sea level and the 'Messinian salinity crisis'. *Nature* 320: 411-414.

Johnson, D.A. & C.A.Nigrini 1985. Synchronous and time-transgressive Neogene radiolarian datum levels in the equatorial Indian and Pacific oceans. *Marine Micropaleontology* 9: 489-523.

Keany, J. 1978. Paleoclimatic trends in Early and Middle Pliocene deep-sea sediments of the Antarctic. *Marine Micropaleontology* 3: 35-49.

Kennett, J.P. 1986. Miocene to early Pliocene oxygen and carbon isotope stratigraphy in the southwest Pacific, Deep Sea Drilling Project Leg 90. In J.P.Kennett, C.C.von der Borch et al. (eds.), *Initial reports of the Deep-Sea Drilling Project* Vol. 90: 1383-1411. Washington DC: Government Printing Office.

Kennett, J.P. & P.Vella 1975. Late Cenozoic planktonic foraminifera and paleoceanography at DSDP Site 284 in the cool subtropical South Pacific. In J.P.Kennett & R.E.Houtz (eds.), *Initial report of the Deep-Sea Drilling Project* Vol. 29: 769-782. Washington DC: Government Printing Office.

Kennett, J.P. & C.C.von der Borch 1986. Southwest Pacific Cenozoic paleoceanography. In J.P.Kennett, C.C.von der Borch et al. (eds.), *Initial reports of the Deep-Sea Drilling Project* Vol. 90: 1493-1517. Washington DC: Government Printing Office.

Loubere, P. & K.Moss 1986. Late Pliocene climatic change and the onset of Northern Hemisphere glaciation as recorded in the northeast Atlantic Ocean. *Geological Society of America Bull.* 97: 818-828.

Matthews, R.K. & R.Z.Poore 1980. Tertiary $\delta^{18}O$ record and glacioeustatic sea level fluctuations. *Geology* 8: 500-514.

McCollum, D.W. 1975. Diatom stratigraphy of the Southern Ocean. In D.Hayes & L. Frakes (eds.), *Initial reports of the Deep-Sea Drilling Project* Vol 28: 515-571. Washington DC: Government Printing Office.

McDougall, I & H.Wensink 1966. Paleomagnetism and geochronology of the Pliocene-Pleistocene lavas in Iceland. *Earth & Planetary Science Letters* 1: 232-236.

McQueen, D.R. 1976. The ecology of *Nothofagus* and associated vegetation in South America. *Tuatara* 22: 38-68.

Mercer, J.H. 1968. Antarctic ice and Sangamon sea level. In *International Association of Scientific Hydrology, General Assembly of Berne*, Publ. 79: 217-225.

Mercer, J.H. 1978. Glacial development and temperature trends in the Antarctic and in South America. In E.M.van Zinderen Bakker (ed.), *Antarctic glacial history and world palaeoenvironments*: 73-93. Rotterdam: Balkema.

Mercer, J.H. 1983. Cenozoic glaciation in the Southern Hemisphere. *Ann. Rev. Earth and Planetary Sciences* 11: 99-132.

Mercer, J.H. 1985. When did open-marine conditions last prevail in the Wilkes and Pensacola basins, East Antarctica, and when was the Sirius Formation emplaced? *S.Afr. J. Sci.* 81: 243-245.

Mercer, J.H. (in press). Southernmost Chile: a modern analog of the southern shores of the Ross embayment during Pliocene warm intervals. *Antarctic J. of the US.*

Mercer, J.H. & J.F.Sutter 1982. Late Miocene – earliest Pliocene glaciation in southern Argentina: implications for global ice-sheet history. *Palaeogeogr., Palaeoclim., Palaeoecol.* 38: 185-206.

Piper, D.J.W. & C.D.Brisco 1975. Deep-water continental margin sedimentation. DSDP Leg 28, Antarctica. In D.Hayes & L.Frakes (eds.), *Initial reports of the Deep-Sea Drilling Project* Vol. 28: 727-755. Washington DC: Government Printing Office.

Pläfker, P., S.Bartsch-Winkler & A.T.Ovenshine 1976. Paleoglacial implications of coarse detritus in DSDP Leg 36 cores. In P.F.Barker, I.W.D.Dalziel et al. (eds.), *Initial reports of the Deep-Sea Drilling Project* Vol. 36: 857-864. Washington DC: Government Printing Office.

Prentice, M.L., G.H.Denton, T.V.Lowell, H.C.Conway & L.E.Heusser (in press). Pre-late Quaternary glaciation of the Beardmore Glacier region, Antarctica. *Antarctic J. of the US.*

Rutford, R.H., C.Craddock, C.M.White & R.L.Armstrong 1972. Tertiary glaciation in the Jones Mountains. In R.J.Adie (ed.), *Antarctic geology and geophysics*: 239-250. Oslo: Universitetsforlaget.

Saemundsson, K. & H.Noll 1974. K/Ar ages of rocks from Husafell, western Iceland, and the development of the Husafell central volcano. *Jökull* 24: 99-132.

Shackleton, N.J., J.Backman et al. 1984. Oxygen isotope calibration of the onset of ice-rafting and history of glaciation in the North Atlantic region. *Nature* 307: 620-623.

Shackleton, N.J. & J.P.Kennett 1975. Paleotemperature history of the Cenozoic and the initiation of Antarctic glaciation: oxygen and carbon isotope analyses in DSDP sites 277, 279 and 281. In J.P.Kennett & R.E.Houtz (eds.), *Initial reports of the Deep-Sea Drilling Project* Vol. 29. Washington DC: Government Printing Office.

Smith, A.G. & D.J.Drewry 1984. Delayed phase change due to hot asthenosphere causes Transantarctic uplift? *Nature* 309: 536-538.

Vail, P.R. & J.Hardenbol 1979. Sea-level changes during the Tertiary. *Oceanus* 22: 71-79.

Weaver, F.M. & A.M.Gombos 1981. Southern high-latitude diatom biostratigraphy. In *DSDP: a decade of progress*, SEPM Special Publication 32: 445-470.

Webb, P.N. & J.E.Andreasen (in press). K-Ar dating of volcanic material associated with the Pliocene Pecten Conglomerate (Cockburn Island) and Scallop Hill formation (McMurdo Sound). *Antarctic J. of the US.*

Webb, P.N. & D.M.Harwood (in press). The terrestrial flora of the Sirius Formation: its significance in interpreting the late Cenozoic glacial history. *Antarctic J. of the US.*

Webb, P.N., D.M.Harwood, B.C.McKelvey, J.H.Mercer & L.D.Stott 1984. Cenozoic marine sedimentation and ice-volume variations on the East Antarctic craton. *Geology* 12: 287-291.

Webb, P.N., D.M.Harwood, B.C.McKelvey, M.C.G.Mabin & J.H.Mercer (in press). Late Cenozoic tectonic and glacial history of the Transantarctic Mountains. *Antarctic J. of the US.*

Wolfe, J.A. & R.Z.Poore 1982. Tertiary marine and nonmarine climatic trends. In *Climate in earth history*: 154-158. Washington DC: National Academy Press.

Woodruff, F., S.M.Savin & R.G.Douglas 1981. Miocene stable isotope record: a detailed deep Pacific Ocean study and its paleoclimatic implications. *Science* 212: 665-668.

OCEANOGRAPHY

PLEISTOCENE CLIMATES IN CHINA AND EUROPE COMPARED TO OXYGEN ISOTOPE RECORD

G.J.KUKLA
Lamont-Doherty Geological Observatory, Columbia University, Palisades, N.Y., USA

The best palaeoclimatic records of the last 2.5 million years come from oceanic sediments and from sequences of alternating loess and soils in Europe and Asia. In deep-sea sediments the ratio of oxygen isotope O^{18} to O^{16} in foraminifera is controlled predominantly by the volume of glacier ice, the primary measure of the Plio-Pleistocene global climate state. Deep-sea sediments are dated by ^{14}C, palaeomagnetic reversals, and by tuning the climatic signal to the astronomic chronology (Hays et al. 1976, Imbrie et al. 1984).

Based on the oxygen isotope stratigraphy deep-sea sediments subdivided into stages of alternating relatively high and relatively low $O^{18/16}$ ratios (Emiliani 1955) with the boundaries drawn at the levels of fastest change. Even numbered units represent cold intervals and odd numbered are relatively warm (Fig. 1). Boundaries of pronounced isotopic maxima following pronounced minima are called 'terminations' by Broecker & Van Donk (1970), and are numbered by Roman numerals. A 'glacial cycle' is labeled by capital letters. The last two completed glacial cycles (B and C) lasted 116 k.y. each. There are eight glacial cycles in Brunhes and ten above Jaramillo.

The most complete and detailed records of Pleistocene climates on land come from loess. The most studied deposits are those in Central Europe and China. Fossils and the loess soils document harsh continental climate. The intervening soils mark relatively mild and wet conditions. Grey podzolic soils with associated snails mark the presence of hardwood forests in Central Europe and by definition, an interglacial climate (Fairbridge 1972). Sharp boundaries between glacial loess and overlying interglacial soils are called marklines delimiting a glacial cycle and labeled by capital letters in order of increasing age (Kukla 1961). Marklines and glacial cycles in loess are the time equivalent of equally designated terminations and glacial cycles of the oxygen isotope record.

Many loess sections in Central Europe are fills on top of river terraces so that the episodes of river downcutting and terrace aggradation can be related to the loess sequence. On slopes several interstadial soils in each glacial cycle can usually be identified. In contrast, in the more common subhorizontal platforms of Europe and China, each glacial cycle is represented only by a single polygenetic soil and an overlying single unit of loess. The strongly weathered mid-Brunhes soils of cycles F and G, and the exceptionally thick loess units of cycles B, C, F and K are useful markerbeds in Central Europe (Fig. 1). A major erosional episode accompanied by deep river downcutting and a change in the hydrographic network occurred in cycle F of mid-Brunhes age (Kukla 1977).

Loess units in China are prefixed by L and soils by S and numbered in order of increasing age (Fig. 2). Four major stratigraphic units are distinguished: older loess is relatively compact and rich in carbonate concretions. The thick paleosol S 5 and the exceptionally thick and coarse grained loess units L 1, L 2, L 5, L 6, L 9 and L 15 are the prominent marker beds. Unconformities were found at the base of the Lower and Upper Lishi loess (Liu 1985).

The loess sequences in Europe and China closely parallel the deep-sea stratigraphy. All three records mark a pronounced deterioration of climate at the Gauss/Matuyama boundary (about 2.5 Ma), indicating of more intense glaciation starting at about 0.9 Ma and strong glacials in cycles B (oxygen stage 2 and Xifeng loess L 1), C (6 and L 2), F (12 and L 5), H (16 and L 6) and K (22 and L 9). Greater amplitude of the oxygen isotope glacial cycles since about 1.3 Ma in the North Atlantic cores (Fig. 1) corresponds approximately to the boundary between Wucheng and Lower Lishi in China. Two episodes of increased rate of crustal uplift were prominent in Europe: one in the mid-Brunhes at about 450 Ka and the second in the upper Matuyama at around 850 Ka. They are not present in the deep-sea record. Both were followed by major faunal dispersions (Repenning 1984).

There is a discrepancy between the number of glaciations recorded in the last approximately one million years from the ocean floor and from the loess sequences and the four glacials and three interglacials recognized from the same interval in the classical Pleistocene subdivisions in Europe and North America (Fig. 3). The discrepancy is due to the different concepts of an interglacial and to the uneven intensity of the Middle Pleistocene glaciations. Classical Pleistocene subdivisions are based on climatostratigraphic criteria. An interglacial is defined as a time with global climate comparable to the present one and recognized in Europe in lake beds with pollen of deciduous forests (Fairbridge 1972). In the Alpine system, the interglacials are represented by steps separating subsequent terraces, which were incorrectly interpreted as being of climatic origin.

The clue to the discrepancy between the few glacials in classical systems and many glacial cycles in the deepsea and loess stratigraphy are the 'skipped' glacials during which the ice margins did not reach far enough to leave recognizable geomorphic evidence. Cycles D and G (oxygen stages 8 and 14) are these types. While the peak oxygen isotope ratios in the SPECMAP series of Imbrie et al. (1984) are 1.81 for stage 2, 1.58 for 6, 1.12 for 10 and 1.50 for 12, the peak value of stage 8 is 0.89 and that of stage 14 only 0.79. Such low values point to the global ice volume being lower by at least 20% compared to stage 2. Since most of the fluctuation of Pleistocene ice was due to the

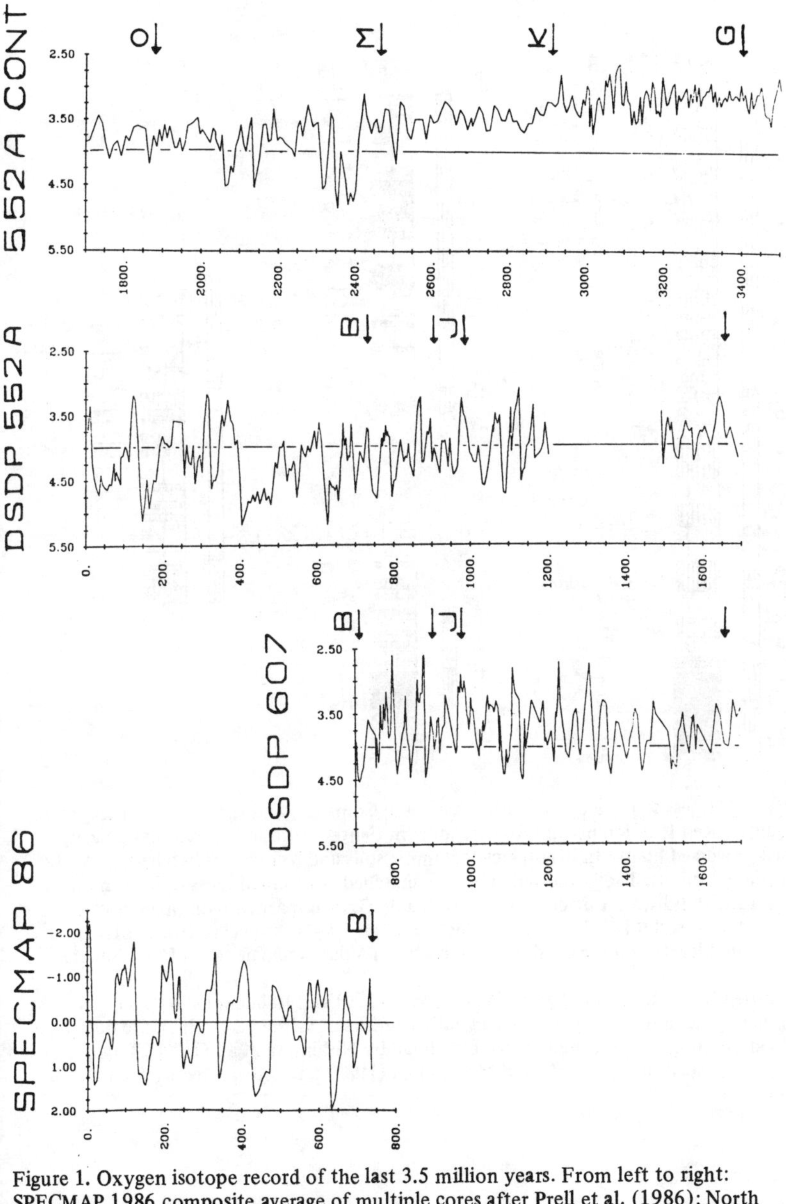

Figure 1. Oxygen isotope record of the last 3.5 million years. From left to right: SPECMAP 1986 composite average of multiple cores after Prell et al. (1986); North Atlantic Deep-sea Drilling Project (DSDP) Site 607 from Ruddiman et al. (in print); DSDP 552 A after Shackleton et al. (1984 and unpublished).
Vertical scale shows time in thousands of years before present, horizontal scale the oxygen ratio. Arrows: Palaeomagnetic reversals as determined in the cores; B: base of Brunhes; J: base of Jaramillo; O: base of Olduvai; M: base of Matuyama; G: base of Gauss; K: top of Kaena.

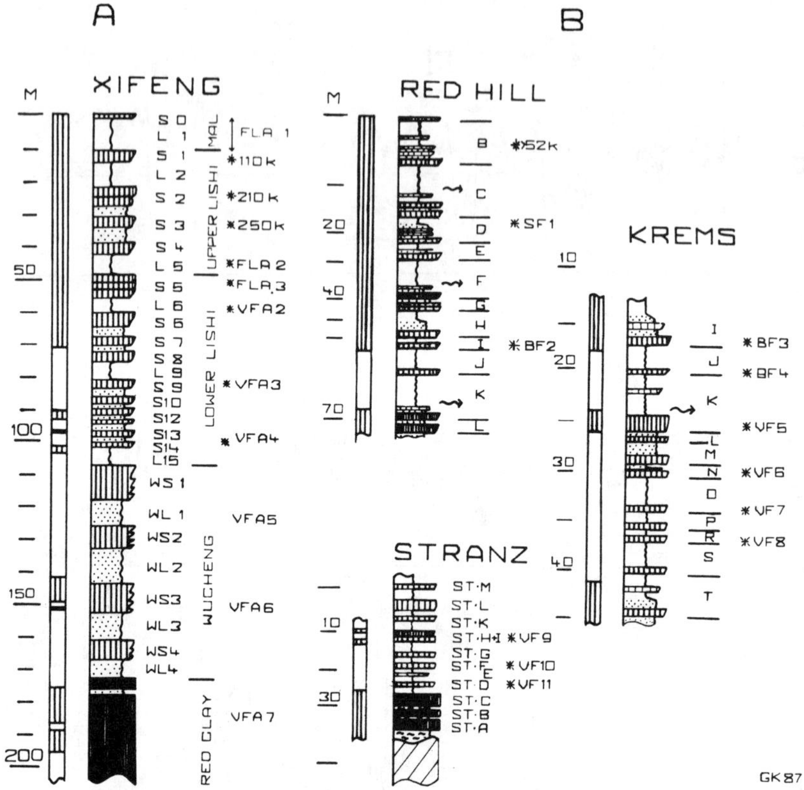

Figure 2. Loess sequence at localities Xifeng in China (A) and the combined record at localities Red Hill, Krems and Stranzendorf in Central Europe (B). Normal polarity black, reversed blank. In lithological columns: soils and soil groups hatched, strongly weathered rubefied soils and red clays crosshatched, weathered loess or loess interspersed with hillwash dotted, other loess blank. Gravels marked with open circles. S 0 – WS 4: soil units; L 1 – WL 4: loess units; B-U: glacial cycles. Horizontal wavelines equal levels of deep erosion. Vertebrates of Villafranchian age: VF, of Biharian age: BF.
Modified from: X.M.Liu et al. (1985), Kukla (1977) and Rabeder (1981). Thermoluminescence dates in Xifeng marked with asterisks in S 1 and S 2 which are determined on samples from the locality Luochuan by Nishimura et al. (1984). Polarity of the Xifeng section measured by X.M.Liu et al. (1985) and interpreted by Kukla et al. (in press).

Laurentide ice sheet, it is highly unlikely that the continental ice margins during oxygen stages 8 and 14 could have reached far enough south to leave recognizable moraine tracks. Supporting the oceanic record, the loess of cycle D indicates a climate less severe than in cycles B, C or F and the loess of cycle G is very thin in Europe and completely missing in China.

Thus the oxygen isotope record and the loess sequences in Europe and China both infer relatively long intervals during which the continental ice sheets probably did not reach the middle latitudes of the northern hemisphere. The first episode occurred during oxygen isotope stages 7, 8 and 9, lasting from about 340 ka to 185 ka. The second covers stages 13, 14 and 15 and lasted from about 620 Ka to 480 Ka. The dates are after the model of Imbrie et al. (1984).

The classical Fenoscandian glacial stages and those of the Alpine system were directly correlated with the loess sequences and through the latter linked with the deep-sea record (Kukla 1977). The moraines with sharp relief, originally charted in Europe as Weichsel (Fig. 3), correlate not only with the O^{18} stage 2 but also with stage 6 (Warthe advance of Woldstedt (1954)). Likewise, the Würm terrace in the Alps includes sediments of the O^{18} stages 1 through 6 (Kukla 1977). The Ehringsdorf travertine in Germany, the famous site of the Eemian interglacial flora (Woldstedt 1954), correlates at least in part with stage 7 (Blackwell & Schwarcz 1986), while sediments of the Eemian marine transgression in northern Germany and Denmark, which cuts through the Warthe moraines overlain by Weichsel drift, correlate with oxygen substage 5e (Knudsen 1984). The stratigraphic position of the type Eem interglacial deposits in Holland is still unclear. The Holstein was correlated with the O^{18} stage 11 because of its relation to the overlying loess (Kukla 1977) and because of radiometric dating (Sarnthein et al. 1986). The ice-free interval during oxygen stages 13 through 15 is coeval with the exceptionally strong polygenetic soils of cycles F and G in Central Europe, with the Ferreto on Alpine terraces, and also with the polygenetic marker soil S 5 in China. Two glaciations in northwestern Europe, the type Saale and the Late Elster, separated by deposits of the Holstein interglacial, are older than the Ehringsdorf interglacial (oxygen stage 7), and probably coeval with or younger than the erosional episode of cycle F (Kukla 1977). Moraines of Late Elster are known only from subsurface, frequently located below the present sea level, while the deposits of type Elster (s.s.) are positioned high above the present river valleys. This may indicate that the Elster (s.s.) is older than the crustal shift of cycle F.

Tentative correlation of Elster and Saale with the deep-sea stratigraphy is not without problems. Five glacial cycles (16, 14, 12, 10 and 8) fall within the critical interval. Two of them, 14 and 8, are unlikely to have developed enough ice for a far-reaching southward advance. Frontal moraines of Saale s.s. and Elster s.s. reached further south than Weichsel or Warthe. Only O^{18} stages 16 and 12 but not 10 are comparable in amplitude with 6 and 2. However, because Holstein correlates with the O^{18} stage 11 and Ehringsdorf with

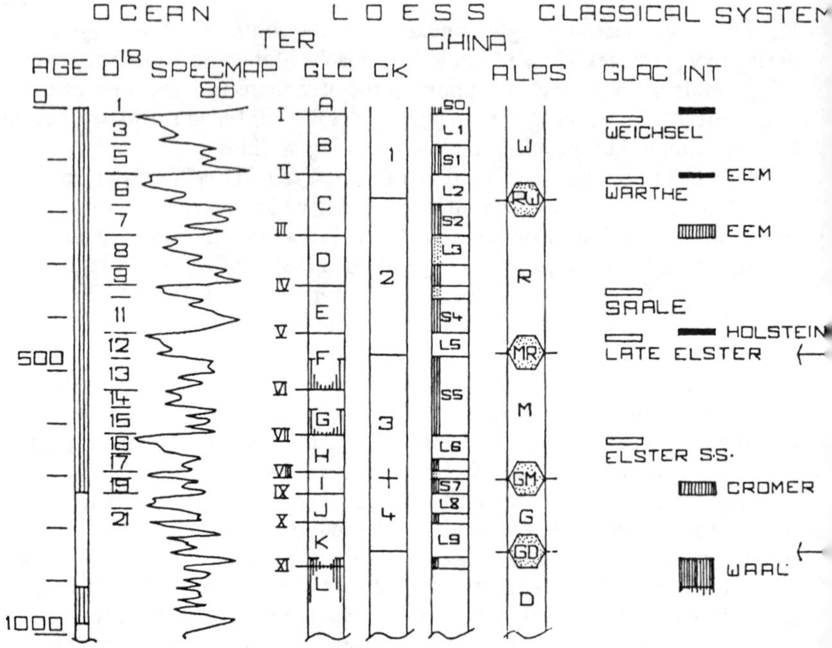

Figure 3. Correlation of deep-sea and loess records with the Alpine and North European classical subdivisions. Oxygen isotope stages (O^{18}) and ages (in thousands of years) of SPECMAP 1984 isotopic record after Imbrie et al. (1984). Extension to 1 Ma from DSDP 552 A after Shackleton et al. (1984). Terminations (TER) and glacial cycles (GLC) marked with the same capital letters for deep-sea and Central European loess after Kukla (1977). CK: shows stratigraphic position of terraces at Red Hill. Soils and loess units: same as in Figure 2.
Chronostratigraphic position of the type units (full boxes) or radiometrically dated units (open boxes) of the Alpine (ALPS) and North European classical stages modified from Kukla (1977). F: faunal dispersal vents; W: Würm; R: Riss, M: Mindel; G: Gunz; and D: Donau.

7, the type Saale can only belong to a relatively weak oxygen stage 10. If so, a specific crustal configuration may have produced a situation in which a relatively moderate global ice volume reached relatively deep into the middle latitudes. Alternatively the described chronostratigraphic model may be wrong.

Palaeoclimatic reconstruction below oxygen stage 16 is even more unclear. Relatively few deep-sea cores, analysed for O^{18} ratio, cover this interval and the records have few features in common. At least two episodes of high ice volume occurred between Jaramillo and the Brunhes, namely in oxygen stages 20 and 22. The warm stages 17 and 19 of the oxygen isotope stratigraphy are poorly expressed. This may indicate a time of relatively large global ice volume or a locally colder ocean surface at the sites of the ocean cores. The relatively low eustatic sea level reported by Hearty et al. (1986) in the Mediterranean would be in agreement with the former alternative. However, in contrast to the deep-sea record, the terrestrial soils and fauna of the interglacial correlative with stage 19, point to a fully expressed interglacial climate with Hippopotamus and monkeys in Central Europe (Maglio 1975).

A drastic change in the composition of terrestrial faunas took place in Europe between Jaramillo and Brunhes (Repenning 1984). As the rearrangement of land bridges and forest belts is a likely cause of major faunal migrations and because Late Matuyama orogenic movements are known from several locations around the globe (cf. Zuchiewicz 1984), the increased severity of glaciations younger than 850 ka may be due to the shift of the atmospheric and oceanic circulation caused by the uplift of young mountain belts and by the rearrangement of the sea floor (Ruddiman et al., in print).

The cold stage 22 was preceded in Europe by a lengthy episode of temperate climate (Waalian in northwest Europe, Krems soils KR 7, 8 and 9 in Austria).

Few records with sufficient climatic and time stratigraphic resolution are at hand to draw more than only a tentative picture of climate fluctuations prior to 1 Ma. In contrast to the Middle and Late Pleistocene, only mountain glaciers, not continental ice sheets, left traces in the northern middle latitudes. The only indication of a continental size glacier is known from a borehole in Iowa where a diamicton overlain by a silt with 2 MY old volcanic glass was interpreted as glacial till (Boellstorff 1978). However, the layer is composed of locally derived detritus resting on the bedrock and the glass shards in the silt may be reworked so that the glacigenic origin and age of the diamicton remains questionable. The existence of a continental ice sheet in North America between 2.2 and 2.5 million years is however theoretically possible since the North Atlantic DSDP borehole 552 A (Fig. 1) shows a pronounced oxygen isotope anomaly at this time with an amplitude comparable to the Late Pleistocene (Fig. 1). The anomaly is accompanied by the occurrence of ice rafted detritus indicative of the formation of active glaciers in the high latitudes (Shackleton et al. 1984), by pronounced cooling of the North Atlantic (Loubere & Moss 1986), by cold fauna invasion of the Mexican Gulf and of the Mediterranean Sea (Arias et al. 1980, Kukla et al. 1979) and by appearance of colder flora in Western Europe (Zagwijn 1975). The onset of loess deposition at the end of the Gauss and in the lowermost Matuyama shows a shift in Europe and China toward markedly drier and colder conditions (Liu 1985). During the Olduvai and Jaramillo the climate was relatively warm.

Exceptionally thick loess units which we interpret as indicative of strong glaciations and/or mountain uplift events are located: a. about half way between the Olduvai and Jaramillo, b. at the top and c. below the bottom of the Olduvai zone, and d. near the base of Matuyama (cf. Fig. 2). The boundaries of the basal loess in both Europe and China point to the oscillatory nature of the environmental shift from the warm and wet Pliocene climates into the much colder and drier Pleistocene regimes. Since major epeirogenetic movements were noted at this time in the Alpine belt (Zuchiewicz 1984, Arias et al. 1980), it is possible that the onset of the Plio/Pleistocene cooling was due to the uplift of young mountain belts in the northern hemisphere and its impact on surface albedo, hydrography and atmospheric and oceanic circulation (Emiliani & Geiss 1959).

ACKNOWLEDGEMENTS

Preparation of Figure 1 provided by J.Gavin. Comments by R.Fairbanks, D.Kent, J.Gavin and H.Johnson are gratefully acknowledged.

REFERENCES

Arias, C., A.Azzrolli, G.Bigazzi & F.Bonadonna 1980. Magnetostratigraphy and Pliocene-Pleistocene boundary in Italy. *Quat. Res.* 13: 65-74.

Blackwell, B. & H.P.Schwarcz 1986. U-series analyses of the lower travertine at Ehringsdorf, DDR. *Quat. Res.* 25: 215-222.

Boellstorff, J. 1978. North American Pleistocene stages reconsidered in light of probable Pliocene-Pleistocene continental glaciations. *Science* 202: 305-307.

Broecker, W.S. & J.van Donk 1970. Insolation changes, ice volumes, and the O^{18} record in deep-sea cores. *Rev. of Geophysics and Space Physics* 8: 169-198.

Emiliani, C. 1955. Pleistocene temperatures. *J. Geol.* 63: 538-578.

Emiliani, C. & J.Geiss 1959. On glaciations and their causes. *Geol. Rundschau* 46: 576-601.

Fairbridge, R.W. 1972. Climatology of a glacial cycle. *Quat. Res.* 2: 283-302.

Hays, J.D., J.Imbrie & N.J.Schackleton 1976. Variations in the earth's orbit: pacemaker of the Ice Ages. *Science* 194: 1121-1132.

Hearty, P.J., G.H.Miller, C.E.Searns & B.J.Szabo 1986. Aminostratigraphy of Quaternary shorelines in the Mediterranean Basin. *Geol. Soc. Am. Bull.* 97: 850-858.

Imbrie, J., J.D.Hays, D.G.Martinson, A.McIntyre, A.C.Mix, J.J.Morley, N.G.Pisias, W.L. Prell & N.J.Shackleton 1984. The orbital theory of Pleistocene climate: support from a revised chronology of the marine delta ^{18}O record. In A.L.Berger et al. (eds.), *Milankovitch and climate,* Part I: 269-305. Boston: Reidel.

Knudsen, K.L. 1984. Foraminiferal stratigraphy in a marine Eemian-Weichselian sequence at Apholm, North Jutland. *Bull. Geol. Soc. Denmark* 32: 169-180.

Kukla, G.J. 1961. Quaternary sedimentation cycle – Survey of Czechoslovak Quaternary – Czwartorzed Europy srodkovej i wschodniej. In *Proc. INQUA VIth Intern. Congr.* (Instytut Geologiczny, Prace) 34: 145-154.

Kukla, G.J. 1977. Pleistocene land-sea correlations. *Earth-Science Reviews* 13: 307-374.

Kukla, G.J., B.P.Collins & M.L.Bender 1979. Radiometric age of the Arctica Islandica Boundary in Italy: 2 M.Y. *Ann. Geol. Pays Hellen.* Tome hors serie 1979, fasc. II: 699-709.

Kukla, G.J., F.Heller, X.M.Liu & T.S.Liu (in press). *Pleistocene climates in China dated by magnetic susceptibility.*

Liu, T.S. 1985. *Loess and the environment.* Beijing: Ocean China Press. 251 pp.

Liu, X.M., T.S.Liu, M.Y.Cheng, C.Liu & T.C.Xu 1985. The primary study on magneto-

stratigraphy of a loess profile in Xifeng area, Gansu Province. In *Proc. Intern. Symp. Loess Res.* (October 5-16, Xian), 5 pp.
Loubere, P. & K.Moss 1986. Late Pliocene climatic change and the onset of Northern Hemisphere glaciation as recorded in the northeast Atlantic Ocean. *Geol. Soc. Amer. Bull.* 97: 818-828.
Maglio, V.J. 1975. Pleistocene faunal evolution in Africa and Eurasia. In K.W.Butzer & G.L.Isaac (eds.), *After the Australopithecines*: 419-476. The Hague: Mouton.
Nishimura, S., J.V.Miao & S.Sasajima 1984. Thermoluminescence dating of palaeosol in the Louchuan loess sequence. In S.Sasajima & W.Yongyan (eds.), *Recent research of loess in China*: 69-78. Kyoto: Kyoto Univ.
Rabeder, G. 1981. Die Arvicoliden (Rodentia, Mammalia) aus dem Pliozän und dem alteren Pleistocän von Niederösterreich. *Beiträge Paläont. Osterr.* 8: 1-373.
Repenning, C.A. 1984. Quaternary rodent biochronology and its correlation with climatic and magnetic statigraphies. In W.C.Mahaney (ed.), *Correlation of Quaternary chronologies.* Norwich (England): Geo. Books.
Ruddiman, W.F., M.Raymo & A.McIntyre (in press). *Matuyama 41,000 year cycles: North Atlantic ocean and Northern Hemisphere ice sheets.*
Sarnthein, M., H.E.Stremme & A.Mangini 1986. The Holstein Interglaciation: time-stratigraphic position and correlation to stable-isotope stratigraphy of deep-sea sediments. *Quat. Res.* 6: 283-298.
Shackleton, N.J., J.Backman, H.Zimmerman, D.V.Kent, M.A.Hall, D.G.Roberts, D. Schnitker, J.G.Baldauf, A.Desprairies, A.Homrighausen, P.Huddlestun, J.B.Keene, A.J.Kaltenack, K.A.O.Krumsiek, A.C.Morton, J.W.Murray & J.Westberg-Smith 1984. Oxygen isotope calibration of the onset of ice rafting and history of glaciation in the North Atlantic region. *Nature* 307: 620-623.
Woldstedt, P. 1954. Saaleeiszeit, Warthestadium und Weichselseiszeit im Norddeutschland. *Eiszeitalter und Gegenwart* 415: 34-48.
Zagwijn, W.H. 1975. Variations in climate as shown by pollen analysis, especially in the Lower Pleistocene of Europe. In A.E.Wright & F.Mosely (eds.), *Ice ages: ancient and modern. Geological J.* spec. issue 6: 137-152. Liverpool: Seel House Press.
Zuchiewicz, W. 1984. The Late Neogene-Quaternary tectonic mobility of the Polish West Carpanthians. A case study of the Dunajec Drainage Basin. *Ann. Soc. Geol. Poloniae* 54: 133-189.

SURVEY OF PALYNOLOGICALLY ANALYSED DEEP-SEA CORES IN THE NORTHEAST ATLANTIC OFF NORTHWEST AFRICA

H.HOOGHIEMSTRA
Institute of Palynology and Quaternary Sciences, University of Göttingen, F.R.Germany

INTRODUCTION

Pollen analysis of deep-sea cores from the northeast Atlantic has experienced a fast development during the last decade. In the study of deep-sea sediments on a global scale, however, pollen studies play a very modest part. Recent results in marine palynology argue for a better use of pollen as an information-carrier in deep-sea research.

The objective of this chapter is to provide a survey of available marine pollen records in the northwest African area. Several deep-sea cores are still available for sampling. Continuous high-resolution pollen records of well-dated deep-sea cores proved to be relevant sources of continuous records of palaeoclimatological and palaeoecological variables, which play an essential part in climatological model studies of the past. In addition, isopollen maps of relevant pollen types of selected time-slices of the past are helpful in assigning the distribution of the vegetation zones on the northwest African continent.

STRATIGRAPHY

The zonation of the recently analysed marine pollen records has been brought into line with the oxygen isotope chronology as much as possible. Each pollen zone has been numbered according to the oxygen isotope stage, which is represented by this pollen zone.

Oxygen isotope stages are subdivided by figures (e.g. 5.1 to 5.5; see Imbrie et al. 1984), whereas pollen zones are subdivided by letters (e.g. 5A to 5E; in some earlier published palynological publicatons: 5a to 5e). In this way evidence of oxygen isotope records and pollen records can easily be compared.

Table 1. Survey of palynologically analysed deep-sea cores off northwest Africa

(1)	(2)	(3)	(4)	(5)	(6)
8057B	37°41′	10°05′	2811	11-290	R_p/R_b index
M 15669-1	34°53′	07°49′	2030	0.5-479.5	d^{18}/O
M 80-17B	33°37′	09°25′	3016	50-180	^{14}C
M 16004-1	29°59′	10°39′	1512	12-440	d^{18}O
M 15627-3	29°10′	12°05′	1024	0-864	d^{18}O
M 123-09-3	26°50′	15°07′	2760	20-750	d^{18}O
M 123-92-1	25°10′	16°51′	2575	10-960	d^{18}O/forams
M 123-10-4	23°30′	17°43′	3080	10-556	d^{18}O/forams
M 16017-2	21°15′	17°48′	800	20-332	d^{18}O/^{14}C
KS 10	21°	20°05′	3775	10-620	Forams
KL 10	21°	20°05′	3775	1-395	Forams
KL 11	21°	19°10′	3263	10-560	Forams
KL 15	21°	18°40′	2552	1-780	Forams
KL 14	21°	18°10′	1885	10-320	Forams
ODP 658	20°45′	18°35′	2662	Upper 300 m	d^{18}O/pal.mag./m.f.
M 123-29-6	19°22′	19°56′	3320	75-840	d^{18}O/forams
M 13289-2	18°05′	18°01′	2490	96.5-153	d^{18}O/^{14}C
ODP 659	18°05′	21°02′	3070	Upper 102 m	d^{18}O/pal.mag./m.f.
KL 17	16°25′	20°45′	3871	1-300	Forams
M 12347-2	15°50′	17°51′	2576	90-550	Forams
M 12345-5	15°29′	17°22′	945	5-475	Forams/d^{18}O
M 12344-6	15°26′	17°21′	625	60-195	Forams
M 16402-2	14°28′	20°33′	4234	59.5-119	d^{18}O
K 6	14°15′	21°25′	4450	15-295	Forams
M 13239-1	13°53′	18°19′	3156	94.5-172.5	d^{18}O
				233-301.5	d^{18}O
DSDP 367	12°29′	20°02′	4748	201-1744	Forams
M 13209-2	12°29′	20°03′	4713	51-130	C_{org}/carb.
V 132-18-1	12°02′	18°00′	3285	0-570	Forams
ODP 661	09°27′	19°23′	4006	Upper 105 m	d^{18}O/pal.mag./m.f.
M 16416-2	09°52′	19°24′	4336	201.5-409.5	Carb.
M 16415-2	09°34′	19°06′	3841	62.5-235	d^{18}O
				2-59	d^{18}O
				239-748	d^{18}O
M 30 (KL 184)	07°42′	14°24′	2330	5-766	–

(1): Core number.
M: Meteor.
V: Valdivia.
DSDP 367: Deep Sea Drilling Project, Leg 41: site 367.
ODP 658: Ocean Drilling Program, Leg 108: site 658, Hole 658A.
ODP 659: Leg 108: site 659, Hole 659A.
ODP 661: Leg 108: site 661, Hole 661A.
(2): Latitude (°N).
(3): Longitude (°W).
(4): Depth of borehole
(5): Core interval analysed (cm).

	(8)	(9)	(10)
- 20	28	Yes	7
-122	44	Yes	7
- 15	13	No	1
-140	48	Yes	7
-250	66	Yes	7
-130	37	No	1
-145	40	No	1
-130	25	No	1
- 22	33	Yes	4
-150	23	Yes	3
-100	16	Yes	3
- 90	21	Yes	3
-?60	28	Yes	3
-100	20	Yes	3
.5 Ma	191	Yes	In prep.
-c.225	39	No	1
- 22	5	Yes	6
.3 Ma	55	Yes	In prep.
- ?	11	Yes	3
- 22.5	5	Yes	8
- 22.5	13	Yes	8
- 14	2	Yes	8
- 26	6	Yes	6
- ?	15	Yes	3
- 22.5	5	Yes	6
- 19.5	3	Yes	6
-200?	21	Yes	2
- 24	5	Yes	6
- 65?	29	No	1
Ma	59	Yes	In prep.
-150	29	Yes	5
-140	43	Yes	5
- 65	7	Yes	In prep.
-480	52	Yes	In prep.
- ?	34	No	1

(6): Dating method.
R_p/R_b index: Planktonic/benthic foraminifera index.
$d^{18}O$: Oxygen isotope stratigraphy.
^{14}C: Carbon-14 dating.
forams: Foraminifera stratigraphy.
pal.mag.: Palaeomagnetism.
m.f.: Micro-fossil stratigraphy.
carb.: Carbonate stratigraphy.
(7): Approximate time interval represented (X 1 000 years BP).
Ma: Million years.
(8): Number of samples analysed.
(9): Pollen concentration data available.
(10): Reference number.

REFERENCES CONCERNING TABLE 1

1. Agwu, C.O.C. 1979. Vegetations- und klimageschichtliche Untersuchung an marinen Sedimenten vor der westafrikanischen Küste. Dissertation, Georg-August-Universität, Göttingen. 119 pp.
   Agwu, C.O.C. & H.-J.Beug 1982. Palynological studies of marine sediments off the West African coast. *'Meteor' Forschungs-Ergebnisse* C 36: 1-30.
   Agwu, C.O.C. & H.-J.Beug 1984. Palynologische Untersuchungen an marinen Sedimenten vor der westafrikanischen Küste. *Palaeoecol. of Africa* 16: 37-52.
2. Bonnefille, R., M.Rossignol-Strick & G.Riollet 1982. Organic matter and palynology of DSDP site 367 Pliocene-Pleistocene cores off West Africa. *Oceanologica Acta* 5 (1): 97-104.
3. Caratini, C., J.Bellet & C.Tissot 1979. Etude microscopique de la matière organique: palynologie et palynofacies. In M.Arnould & R.Pellet (eds.), *Orgon III: Mauritanie, Sénégal, îles du Cap-Vert*: 215-247. Paris: CNRS.
4. Hooghiemstra, H. (in press). Changes of major wind belts and vegetation zones in NW Africa 20,000-5000 yr B.P. as deduced from a marine pollen record near Cap Blanc. *Review of Palaeobotany and Palynology.*
5. Hooghiemstra, H. & C.O.C.Agwu (in press). *Changes in the vegetation and trade winds in west equatorial Africa 140,000-70,000 yr B.P. as deduced from two marine pollen records.*
6. Hooghiemstra, H., A.Bechler & H.-J.Beug (in press). *Isopollen maps for 18,000 yr N.P. of the Atlantic offshore of northwest Africa: evidence for palaeo-wind circulation.*
7. Hooghiemstra, H., H.Stalling & C.O.C.Agwu (in prep.). *Changes in trade transport and the vegetation in northern NW Africa during the last 140,000 years: evidence from four new marine pollen records.*
8. Rossignol-Strick, M. & D.Duzer 1979. Late Quaternary pollen and dinoflagellate cysts in marine cores off West Africa. *'Meteor' Forschungs-Ergebnisse* C 30: 1-14.
   Rossignol-Strick, M. & D.Duzer 1979. West African vegetation and climate since 22,500 BP from deep-sea cores palynology. *Pollen et Spores* 21: 105-134.

OTHER REFERENCES

Imbrie, J., J.D.Hays, D.G.Martinson, A.McIntyre, A.C.Mix, J.J.Morley, N.G.Pisias, W.L. Prell & N.J.Shackleton 1984. The orbital theory of Pleistocene climate: support from a revised chronology of the marine d^{18}O record. In A.L.Berger et al. (eds.), *Milankovitch and climate*, Part 1: 269-305. Dordrecht: Reidel.

Ericson, D.B. & G.Wollin 1968. Pleistocene climates and chronology in deep-sea sediments. *Science* 162: 1227-1234.

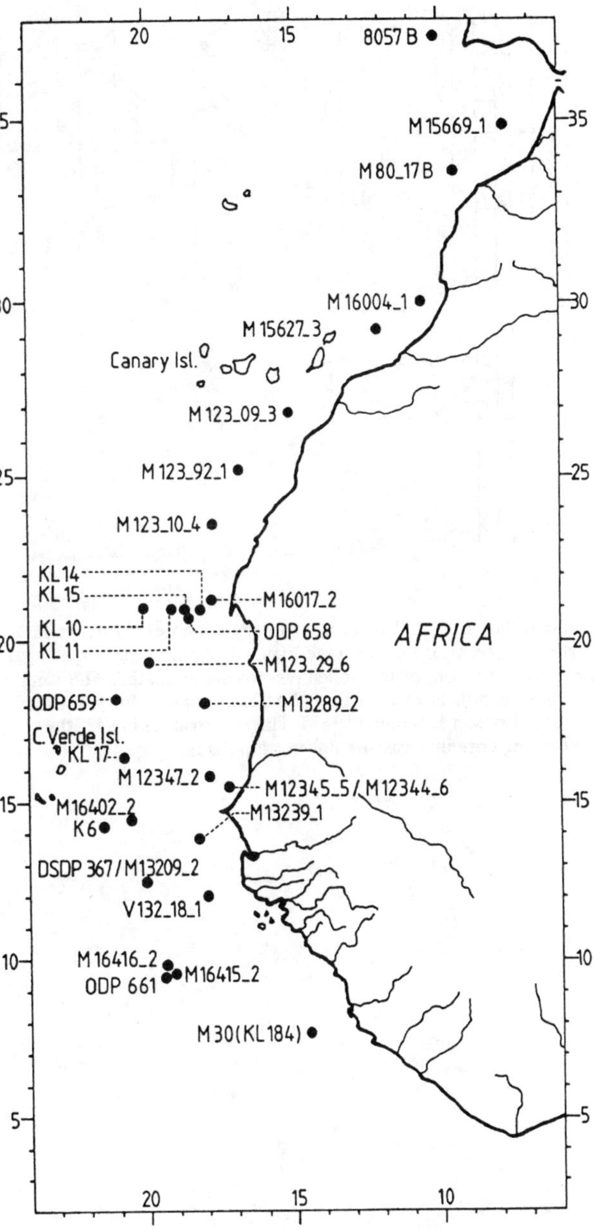

Figure 1. Map with locations of palynologically analysed deep-sea cores in the northeast Atlantic off northwest Africa

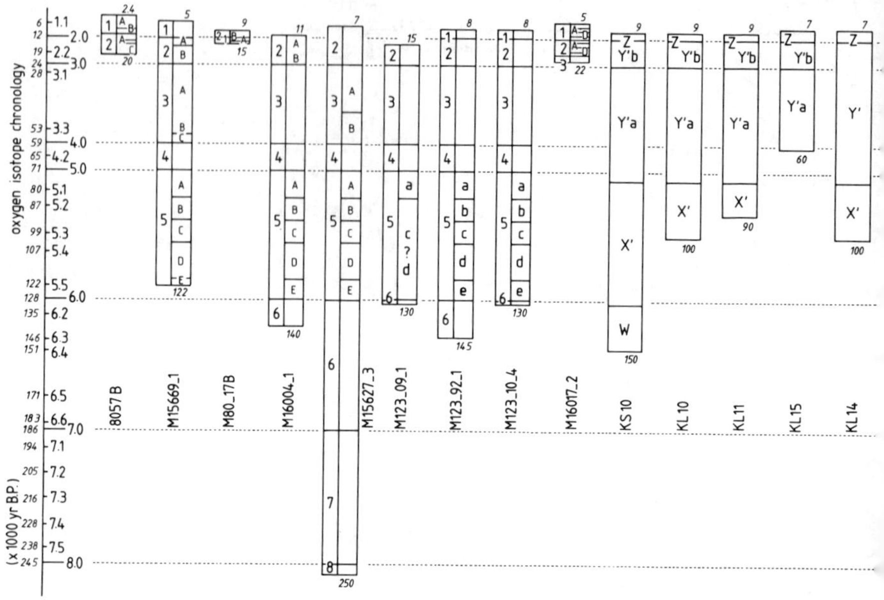

Figure 2. Stratigraphy and correlation of pollen records of deep-sea cores off northwest Africa, arranged from north to south. For locations see Figure 1. Approximate ages (X 1000 years BP) of the top and bottom of the pollen records are indicated. The zonation of earlier published pollen records has been adopted without change. Foraminiferal zones (Z, Y, X, W and V) after Ericson & Wollin (1968). The time scale is linear: the time intervals and the concerning core intervals are not in proportion.

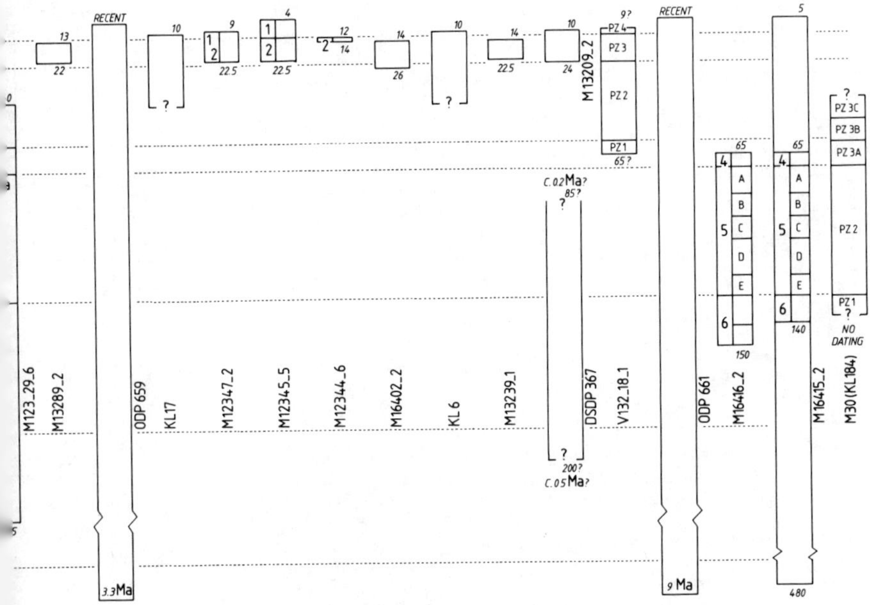
RECENT
3.3 Ma
M123_29_6
M13289_2
ODP 659
KL17
M12347_2
M12345_5
M12344_6
M16402_2
KL 6
M13239_1
DSDP 367
V132_18_1
M13209_2
PZ 4
PZ 3
PZ 2
PZ 1
C.02 Ma?
C.05 Ma?
RECENT
9 Ma
ODP 661
M16416_2
M16415_2
M30 (KL184)
PZ 3C
PZ 3B
PZ 3A
NO DATING

HISTORY OF THE BENGUELA CURRENT OFF SW AFRICA (DSDP SITES 362 and 532)

L.DIESTER-HAAS
Geographisches Institut der Universität Mannheim, Abt. für Geologie, F.R.Germany

ABSTRACT

A coarse fraction analysis of Miocene-Pliocene-Quaternary sediments of Site 532 and 362 from the Eastern Walvis Ridge off SW Africa has been performed. Radiolaria and diatom contents show short-term and long-term cyclic variations. They can be related to carbonate dissolution variations, expressed by benthos/plankton foraminiferal ratios. Carbonate dissolution is strong in glacial, and weak in interglacial periods.

In the Late Miocene (10-6 MY) opal maxima are related to carbonate dissolution maxima, that means to glacial periods. In the interval 5.2 MY to the present the opposite relation is observed: opal maxima correspond to carbonate dissolution minima, thus to interglacial periods.

The following history of the Benguela Current is reconstructed from these findings. During interglacial periods of the 10-6 MY interval the Benguela Current turned from the SE-NW direction to the west within the Cape Basin. During glacial periods of this 10-6 MY interval the current system shifted northward. The Benguela Current then turned to the west over the Walvis Ridge and transported opal from the near-coastal upwelling zone to the Walvis Ridge Terrace.

During the interval 6-5.2 MY the Benguela Current shifted southward and northward over the Walvis Ridge and finally, at 5.2 MY, the present and Late Quaternary situation has been attained. The Benguela Current shifted equatorwards in glacial periods and polewards in interglacial periods. Thus the westward-turning which is responsible for the opal supply from the shelf upwelling area to the continental slope, occurred over the Walvis Ridge in interglacial periods and in the Angola Basin in glacial periods.

The long-term cyclic changes in opal content can be related either to an on/offshore displacement of the Current or to variations in the velocity of the Benguela Current. The Late Pliocene/Early Quaternary opal maximum can be related to a northward shift of the polar front in the South Atlantic.

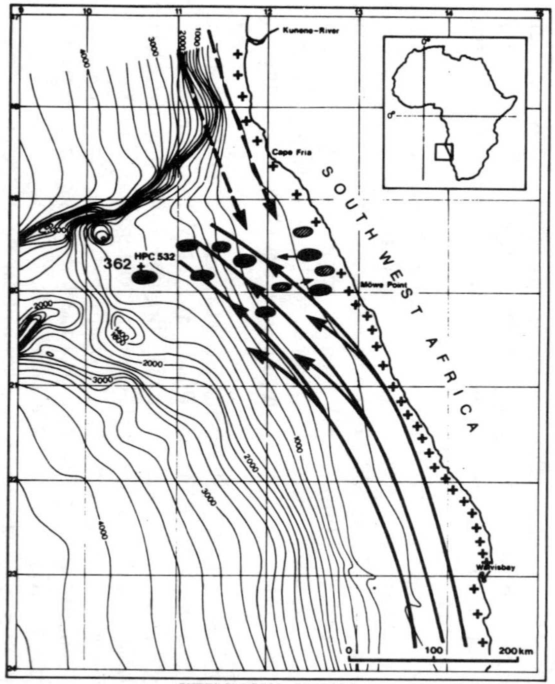

INTERGLACIAL PERIODS

Benguela Current
tropical water
eddies of tropical water drifting landward
eddies of upwelled water drifting seaward
upwelling

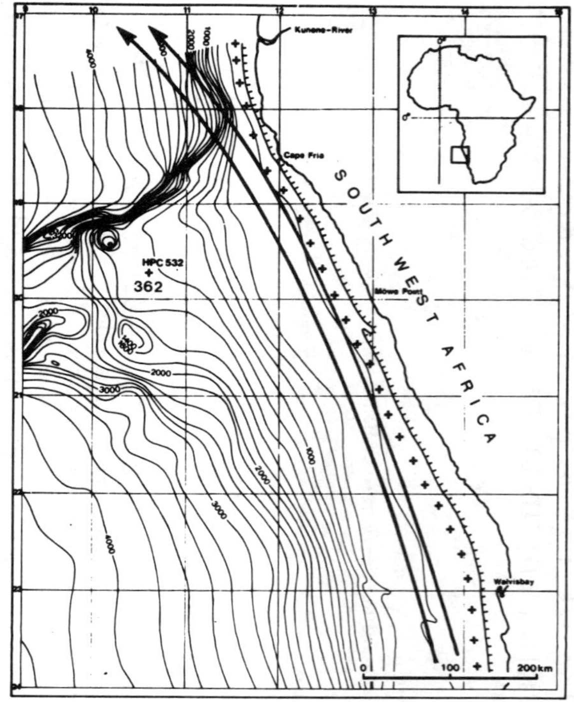

GLACIAL PERIODS

Benguela Current
glacial coast line
Upwelling

INTRODUCTION

The history of the Benguela Current off SW Africa is an interesting problem for terrestrial research in SW Africa (biological and palaeoclimatological questions) as well as for palaeoceanographic reconstructions in the S-Atlantic. Since the publication of the results of Siesser (1980) it has generally been accepted that the Benguela Current started at 10 MY on the Walvis Ridge. In the present investigation variations in the position of the Benguela Current are reported for the interval 10 MY to the present.

The sediments used for the reconstruction of the Benguela Current are from DSDP Sites 362 (Leg 40) and 532 (Leg 75), eastern Walvis Ridge (362: 19°45.45'S, 10°31.95'E, 1 325 m water depth; 532: 19°44.61'S, 10°31.13'E, 1 341 m water depth) (Fig. 1) (Bolli, Ryan et al. 1978, Hay, Sibuet et al. 1984).

Situated about 180 km offshore, the sites are outside the direct influence of coastal upwelling, which is centered on the inner shelf and shelf edge (Hart & Currie 1960, Summerhayes 1983). The Benguela Current, which flows parallel to the coast off SW Africa, leaves the SE-NW direction at about 20-23°S and turns to the west (Martin 1981). The Walvis Ridge Terrace is thus an area of mixing of water masses: warm tropical surface water meets the cold Benguela Current and eddies of cold, upwelled water from the near-coastal area are drifted seawards by the Benguela Current (Fig. 1). These up-welled water masses contain opal skeletons which are moved to the west (Schuette & Schrader 1981, Hagen et al. 1981).

This actual oceanographic situation represents the key to the reconstruction of the Benguela Current in the past. Sediments from the last about 200.000 years revealed (Diester-Haass 1985) that in warm, interglacial periods the amount of opal skeletons of radiolaria and diatoms is high, whereas in glacial periods opal skeletons are rare in the sediments from Site 532 on the Walvis Ridge Terrace. The explanation for this difference can be derived from the present oceanographic variations (Fig. 1): in cold, glacial periods the Benguela Current flowed parallel to the coast until about 17°S, without turning westward over the Walvis Ridge (Van Zinderen Bakker 1976). The westward-turning occurred only in the Angola Basin, where influence of upwelling during the last glacial period has been observed (Jansen et al. 1984). During Late Quaternary warm, interglacial periods the same oceanographic situation as today leads to a supply of opal skeletons in the surface water by the westward-turning Benguela Current from the shelf/shelf edge to the Walvis Ridge Terrace.

Thus the amount of opal skeletons on the Walvis Ridge is an indicator of the westward-turning Benguela Current. In this investigation the variations in opal content in Site 532/362 sediments are described and related to the position of the Benguela Current in glacial and interglacial periods.

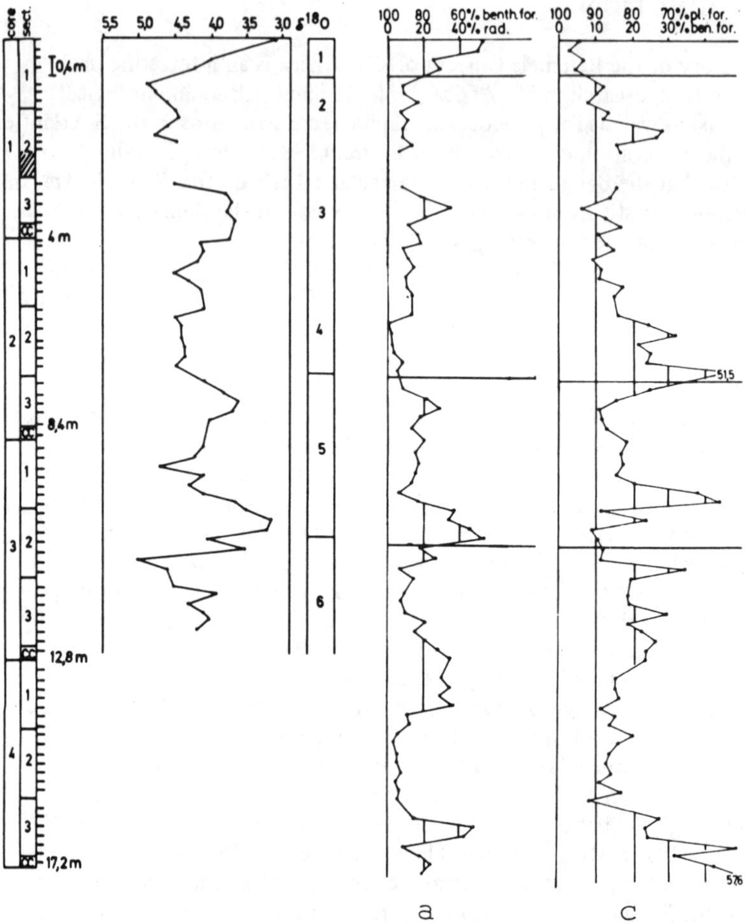

Figure 2. Site 532, core 1-4. Left side: core, section and meters below surface; small horizontal lines mean samples. Oxygen isotope curve (after N.Shackleton) and oxygen isotope stages 1-6.
a. Radiolarian/benthos foraminiferal ratio, calculated as (radiolaria/radiolaria + benthos foraminifera) X 100.
b. Benthos/plankton foraminiferal ratio, calculated as (benthos for./benthos for.+ plankton for.) X 100.

MATERIAL AND METHODS

The sediments of Site 362/532 are pelagic biogenous oozes, partly siliceous, partly calcareous. A coarse fraction analysis has been performed. The sediments have been washed and sieved and the 40-63 and > 63 μm fractions were obtained. The > 63 μm fraction has been dry sieved into the following subfractions: 63-125, 125-250, 250-500, 500-1 000, > 1 000 μm. The weight of the individual fractions has been determined. In each fraction about 800 grains (if present) have been counted. About 25 biogenous, terrigenous and authigenic components were distinguished. The composition of the 40-63 and of the > 63 μm fraction was calculated following the method of Sarnthein (1971).

The opal content in the coarse fraction only forms the smallest part of the total biogenous opal content in the sediments. But the > 40 μm sized opal and the total biogenous opal content vary in a parallel way (cf. Figs 2-4 a and b with Hay, Sibuet et al. 1984, their Fig. 4). Normally all opal is dissolved in the water column and at the sediment/water interface. Only when the opal supply is sufficient to saturate the interstitial water with respect to opal, a preservation of opal is possible (Johnson 1974, Berger 1976). Thus opal supply and opal preservation are intimately linked.

The opal content in percent values of the coarse fraction or of the total sediment is not a reliable parameter for opal supply/preservation variations, because it is subject to dilution by other components. Especially the amount of planktonic foraminifera shows strong variations because of considerable carbonate dissolution (Diester-Haass et al., in press; Diester-Haass & Rothe, in press; Diester-Haass, in press). Thus radiolarian and diatom percentages, as the main opal producers, increase, where planktonic foraminifera are strongly dissolved. Ratios have been used in order to avoid these problems.

The radiolarian/diatom ratio (calculated as (radiolaria/radiolaria + diatoms) X 100) is a sensitive indicator of opal supply and preservation. The higher the diatom supply, the better the preservation of radiolaria and diatoms (Goll & Bjorklund 1974) and the lower the radiolarian/diatom ratio. The radiolarian supply/preservation can best be evaluated by means of the radiolarian/benthos foraminiferal ratio (calculated as (radiolaria/radiolaria + benthonic foraminifera) X 100), since benthonic foraminifera are supposed to be three times less dissolved than planktonic ones (Arrhenius 1952). So the variations in this ratio are assumed to indicate variations in supply/preservation of radiolaria.

For the stratigraphic framework of this investigation the models of Proto-Decima et al. (1978) and Steinmetz et al. (1984) have been followed (Figs. 2-4).

Oxygen isotope data are not available for the total sediment sequence. Only core 1-4 of Site 532 and parts of core 17 of Site 532 B have been analyzed by means of oxygen isotopes (Diester-Haass 1985, Diester-Haass et al., in press). In these examples it has been found that during glacial periods there was strong carbonate dissolution in this rather shallow water depth of

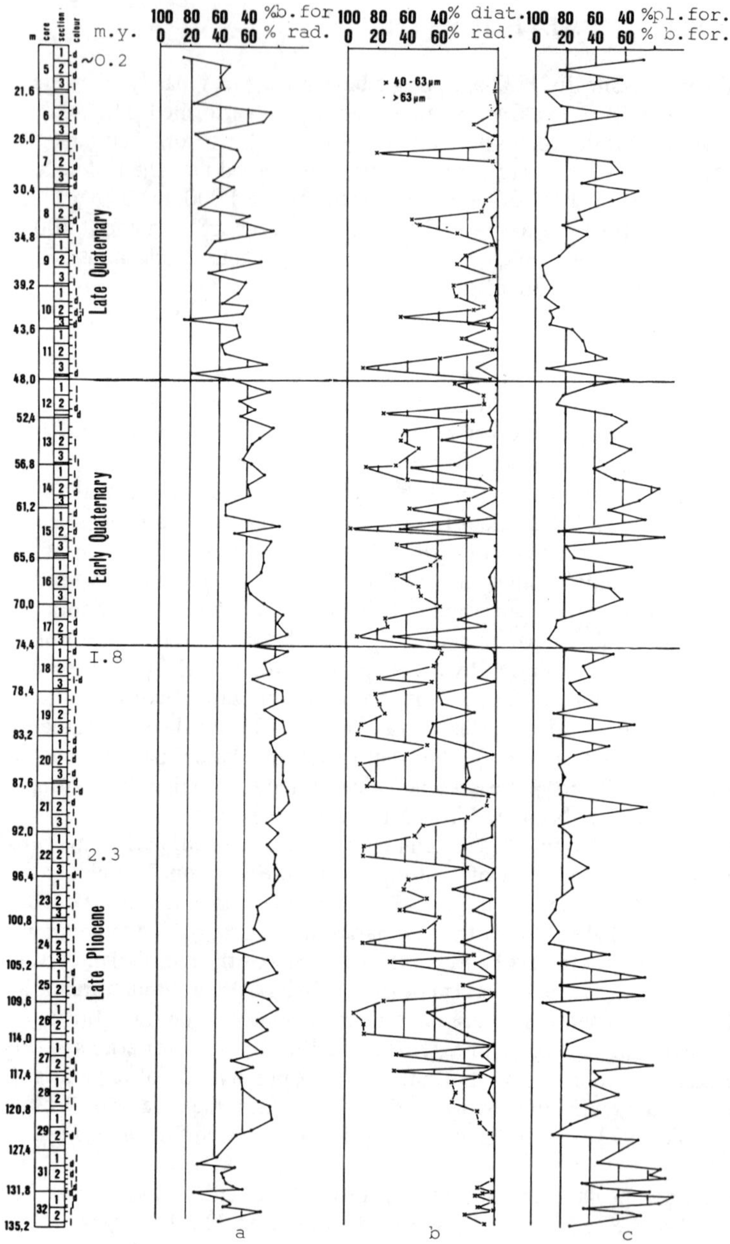

m
core
section
colour
m.y.
~0.2
I.8
2.3
Late Quaternary
Early Quaternary
Late Pliocene
100 80 60 40 %b.for
0 20 40 60 % rad.
100 80 60 40% diat.
0 20 40 60% rad.
100 80 60 40 %pl.for.
0 20 40 60 % b.for.
× 40 - 63 μm
· >63 μm
a
b
c

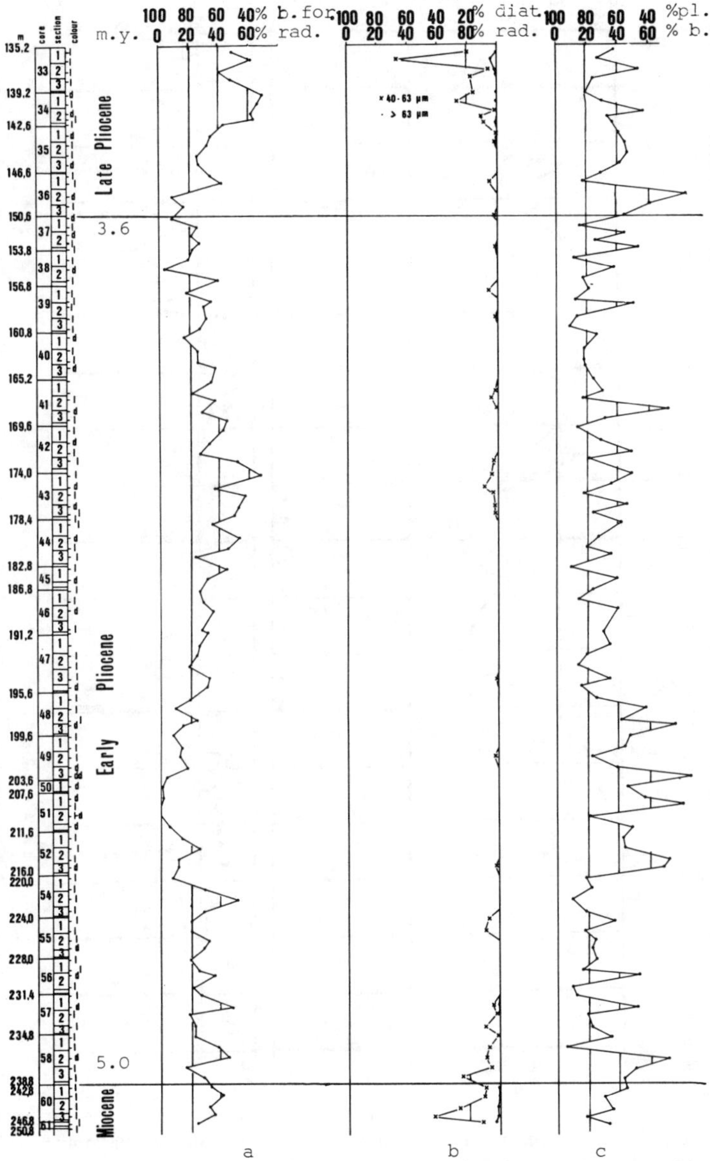

Figure 3. Site 532, core 5-61. Left side: meters below surface, core, section. Small horizontal lines mean sample. 1 = light sediment; d = dark sediment. Stratigraphy after Steinmetz et al. (1984); absolute ages im MY.

a. Radiolarian/benthos foraminiferal ratio, calculated as in Figure 2.

b. Radiolarian/diatom ratio, calculated as (radiolaria/radiolaria + diatom) X 100.

c. Benthos/plankton foraminiferal ratio, calculated as in Figure 2.

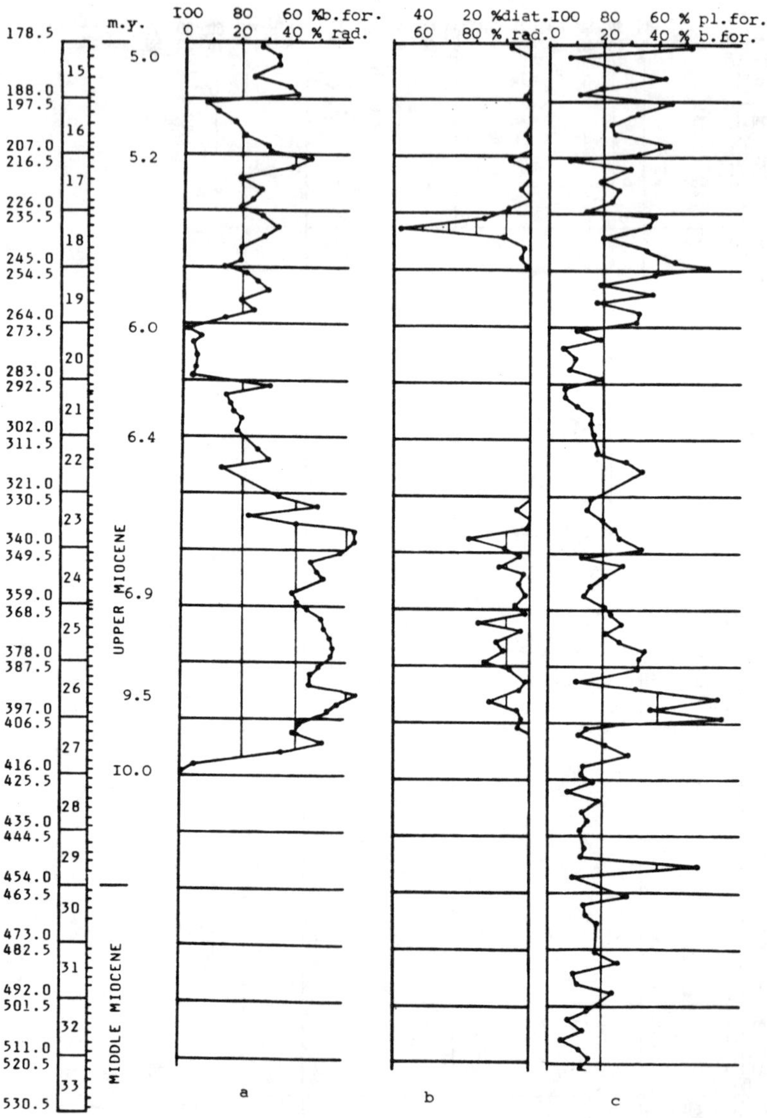

Figure 4. Site 362. Left side: meters below surface, core, section. Site 362 has only been spot-cored, thus there is a non-cored interval of about 10 m between the cores. Small horizontal lines mean sample. Stratigraphy after Proto-Decima et al. (1978); absolute ages.

a. Radiolarian/benthos foraminiferal ratio, calculated as in Figure 2.

b. Radiolarian/diatom ratio, calculated as in Figure 3.

c. Benthos/plankton foraminiferal ratio, calculated as in Figure 2.

about 1 300 m, whereas in interglacial periods carbonate preservation was good. This observation is based on the fragmentation of planktonic foraminifera and on the benthos/plankton foraminiferal ratios, which are good indicators of carbonate dissolution (Thunell 1976) and which both increase considerably in glacial periods. The explanation for this finding proposed by Diester-Haass et al. (in press) is the following: during glacial periods, when sea-level was lowered, supply of organic matter from the shelf to the slope was increased by a factor of 1.5-2 compared to interglacial periods. This organic matter led to increased carbonate dissolution, because it was bacterially decomposed and CO_2 was generated (Berger 1970, Emerson & Bender 1981).

The examples from oxygen isotope stages 1-6 and from the Early Quaternary, where carbonate dissolution increases (decreases) in glacial (interglacial) periods, allowed to generalize this finding and to conclude that sediment sections with strong carbonate dissolution have been deposited during periods of glacially lowered sea-level, sediments with weak carbonate dissolution have been deposited during interglacial periods.

RESULTS

On Figures 2-4 the results on the opal content variations are shown: the radiolarian/benthos foraminiferal ratios (a) and the radiolarian/diatom ratios (b). The benthos/plankton foraminiferal ratios as a measure of $CaCO_3$ dissolution are plotted in order to show the relation between opal content and climatic variation (c). The scales are different in the three figures, because sample distances (20 cm in the Late Quaternary, HPC 532, core 1-4; 70 cm, Quaternary/Pliocene, HPC 532; 150 cm, Miocene of Site 362) necessitate different scales for presentation of results.

In the sediments older than 10 MY no biogenous opal has been found in fractions > 40 μm. At about 10 MY (core 27 of Site 362), > 40 μm sized opal appears for the first time and increases in a short interval to high values. The ratios of radiolaria versus benthos foraminifera show two kinds of variations: small-term cyclic variations, which are superimposed on large-scale cyclic variations. The most prominent large-scale events are a maximum after the onset of opal supply (about 10-6.5 MY), a minimum at about 6.4-6 MY, two minima in the Early Pliocene, an increase in the Late Pliocene until very high ratios in the latest Late Pliocene and Early Quaternary, a decrease in the Late Quaternary (Figs. 2-4a). Diatoms are absent in sections with ratiolarian/benthos foraminiferal ratio minima and they increase parallel to an increase in the latter ratio. Highest diatom amounts, that means lowest radiolarian/diatom ratios are in the Late Pliocene/Early Quaternary (Fig. 3b). The radiolarian/diatom ratio shows – besides the large-scale cyclic variations – small scale cyclic variations, which also parallel the radiolarian/benthos foraminiferal ratio cycles. In the Late Pliocene the radiolarian/benthos foraminiferal ratios are very high with nearly no small-scale variations, whereas the radiolarian/diatom ratios show small-scale cycles.

The benthos/plankton foraminiferal ratio has been plotted in Figures 2-4c. The normal value of this ratio in a water depth of 1 300 m is 3-10% (pers. comm. M.Bremner) and it might increase up to 15-20% with increasing fertility in surface waters in arid areas (Diester-Haass 1978). Values higher than 20% can be due either to lateral supply of benthos foraminifera from shallow water or to dissolution of planktonic foraminifera. An analysis of species of benthonic foraminifera did not suggest any lateral supply (no shelf species present; pers. comm. B.Zobel and D.Schnitker). It can therefore be concluded that the strongly varying benthos/plankton foraminiferal ratios can be explained by the dissolution of planktonic foraminifera.

In the period prior to 10 MY values are small (< 20%); at about 10 MY there is a rapid increase up to 60%. The general trend in carbonate dissolution shows large-scale variations with minima in the Late Miocene at about 6-6.4 MY, in the early Early and late Early Pliocene, in the late Late Pliocene and Early Quaternary. Superimposed on these large-scale cycles are small-scale cyclic changes with considerable variations in the degree of carbonate dissolution within short intervals.

When comparing maxima and minima of carbonate dissolution and of radiolarian/diatom ratios resp. radiolarian/benthos foraminiferal ratios one finds the following correlation: in the Late Miocene, 10-6 MY interval, every increase in carbonate dissolution is correlated to an increase in radiolaria compared to benthos foraminifera and in diatoms compared to radiolaria. However, for the period 6-5.2 MY this correlation does not exist. Starting at 5.2 MY an opposite correlation is noted: all dissolution maxima are opal minima and all dissolution minima are opal maxima. A few irregularities occur at the bottom of the Pliocene. In the Late Pliocene/Early Quaternary the radiolarian/benthos foraminiferal ratios are high without considerable small-scale cyclicity. But the radiolarian/diatom ratio cycles exactly correlate to benthos/plankton foraminiferal ratio cycles: each increase in the latter is accompanied by a decrease in diatoms.

In the Late Quaternary cores 10 to 7 there is no correlation between radiolarian/benthos foraminiferal ratios and benthos/plankton foraminiferal ratios. But radiolarian/diatom ratios correlate with dissolution variations. This interval needs a more detailed analysis by means of closely-spaced samples.

In the latest Quaternary oxygen isotope stages 1-6 diatoms are absent, but radiolarian/benthos foraminiferal ratios correspond to the benthos/plankton foraminiferal ratios with radiolarian maxima occurring in periods with low dissolution (Fig. 2).

DISCUSSION

The carbonate dissolution curve (benthos/plankton foraminiferal ratios) shows strong variations and suggests long-term and short-term variations in sea-level. The opal content is clearly related to the carbonate dissolution/sea-level variation curve. Since sea-level variations in the Neogene are mainly the consequence of variations in the polar ice-caps (Vail et al. 1977), it is concluded that fluctuations in the opal content are of a glacial-interglacial nature. Glacial events are known to occur since the Oligocene in Antarctica (Poore & Matthews 1984) and since about 2.6 MY in Arctica (Shackleton et al. 1984).

Taking into account the present-day oceanographic situation, the variations in diatom and radiolarian content in the coarse fraction can be explained either by variations in the velocity or in the position of the Benguela Current. Sancetta (1984) also assumes a changing intensity and/or on- and offshore movements of the Benguela Current as a reason for the varying diatom content in Site 532. A supply of opal skeletons from the near-coastal upwelling area without transport by the Benguela Current does not seem possible, because the Sites are situated about 180 km offshore.

The velocity of the Benguela Current might have changed in a glacial-interglacial cyclicity. Since the publication of the CLIMAP results (1976) it is generally agreed that during glacial periods the velocity of the main surface currents in the ocean increased compared to interglacial ones. But this cannot be the reason for our opal content variations, because in the Late Miocene we find opal content maxima in glacial periods (= strongest carbonate dissolution), whereas in the Pliocene and Quaternary opal maxima occur in interglacial periods, when the Benguela Current velocity is supposed to be weaker.

Thus variations in the position or path of the Benguela Current seem to be a more realistic possibility to explain the cyclic variations in opal content. The shifting of the main axis of the Benguela Current in the Angola Basin, which has been described for the last glacial period (see Introduction), is assumed to have occurred during all glacial periods of the Quaternary and Pliocene. With this glacial northward migration of the Current, the area of westward-turning of the current also shifts northward, so that the eddies of upwelled, opal-loaden water are transported mainly into the Angola Basin, whereas on the Walvis Ridge Terrace the opal supply is reduced (Fig. 5).

In the Late Miocene, after the onset of the Benguela Current influence on the Walvis Ridge, we find the opposite relation: higher opal content in glacial periods than in interglacial ones. Prior to 10 MY the Benguela Current did not reach the Walvis Ridge at all. At about 10 MY, within a 150 cm thick sediment unit, deposited in about 40 000 years (Diester-Haass & Rothe, in press), its northward migration is reflected, leading to the rapid increase in opal content of core 27 of Site 362. During the first glacial interval after the long warm Middle Miocene period the Benguela Current influence is evident on the Walvis Ridge. During the following warmer interval, its influence becomes weaker. These findings led to the reconstruction of the Benguela Current as presented in Figure 5: in the period prior to 10 MY and in the interglacial periods of the interval 10-6 MY the Benguela Current flowed along

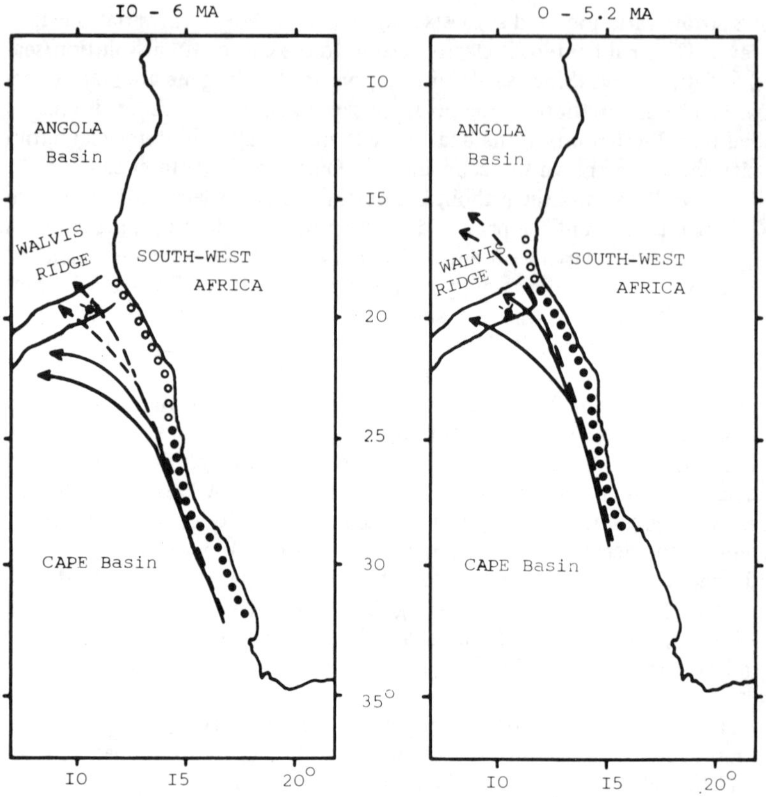

Figure 5. Position of the Benguela Current in the interval 10-6 MY and 5.2 MY to the present, during glacial (– – –) and interglacial (——) periods. Near-coastal upwelling in glacial (○○○) and interglacial (●●●) periods. ✱ position of Site 362. Glacial/interglacial shifting of shore-line has not been shown.

the SW African coast and the main axis turned to the west within the Cape Basin. During glacial periods of this interval 10-6 MY the Benguela Current shifted northwards and turned to the west over the Walvis Ridge, transporting more opal to Site 362.

During the period 6-5.2 MY the Current shifted northward and southward over the Walvis Ridge and finally at 5.2 MY the present and Late Quaternary conditions have been attained.

In the Late Pliocene/Early Quaternary opal contents are extremely high (Fig. 3 a, b and Hay, Sibuet et al. 1984). Radiolarian amounts are constantly high, whereas diatom/radiolarian ratios show the glacial/interglacial cyclicity with higher interglacial diatom values. This opal maximum can be explained either by a change in the path or in the velocity of the westward-turning Benguela Current. The first possibility implies an onshore-offshore shifting of the Current and suggests that in the Late Quaternary as well as in the early Pliocene Site 532 lay at the boundary of the westward-turning Benguela Current, whereas in the latest Pliocene and early Quaternary it was situated in its center. The second possibility may have produced changes in the amount of opal supplied to the west. The Late Pliocene increase in opal content seems to have occurred simultaneously with an important northward advance of the polar front in the South-Atlantic (Wise et al. 1985), which might have produced the increase in velocity, thus in opal supply, of the Benguela Current.

CONCLUSIONS

1. In Late Miocene to recent sediments from the Walvis Ridge Terrace off SW Africa (DSDP Site 362 and 532) cyclic short-term and long-term variations in opal skeleton amounts have been found. They can be related to benthos/plankton foraminiferal ratios, which are an expression of carbonate dissolution. Carbonate dissolution is strongest during glacial periods, weak during interglacial periods (Fig. 6).

2. In the interval 10-6 MY each opal maximum occurs in a glacial period, each opal minimum in an interglacial period. In the interval 5.2 MY to the present the opposite relationship is found: highest (lowest) opal content in interglacial (glacial) periods. The 6-5.2 MY periods shows a mixture of both types of correlation.

3. The Late Quaternary oceanographic situation has existed since 5.2 MY: the Benguela Current turned to the west over the Walvis Ridge in interglacial periods and transported opal skeletons from the coastal upwelling area to the Walvis Ridge Terrace. The Current shifted as far north as the Angola Basin in glacial periods without turning to the west over the Walvis Ridge. That is why opal contents from glacial sediments are much smaller on the Walvis Ridge.

In the very beginning of the Benguela Current influence on the Walvis Ridge at 10 MY, the main axis of the Current turned to the west in the Cape Basin in interglacial periods, leaving a weaker opal content than in glacial periods of the interval 10-6 MY, when the Current shifted equatorwards and turned to the west over the Walvis Ridge.

4. The very high opal contents of the Late Pliocene/Early Quaternary are probably due either to a change in the path of the central axis of the Benguela Current (sea-ward displacement) or to an increase in velocity of the Benguela Current, which can be related to an important northward shift of the polar front in the South-Atlantic.

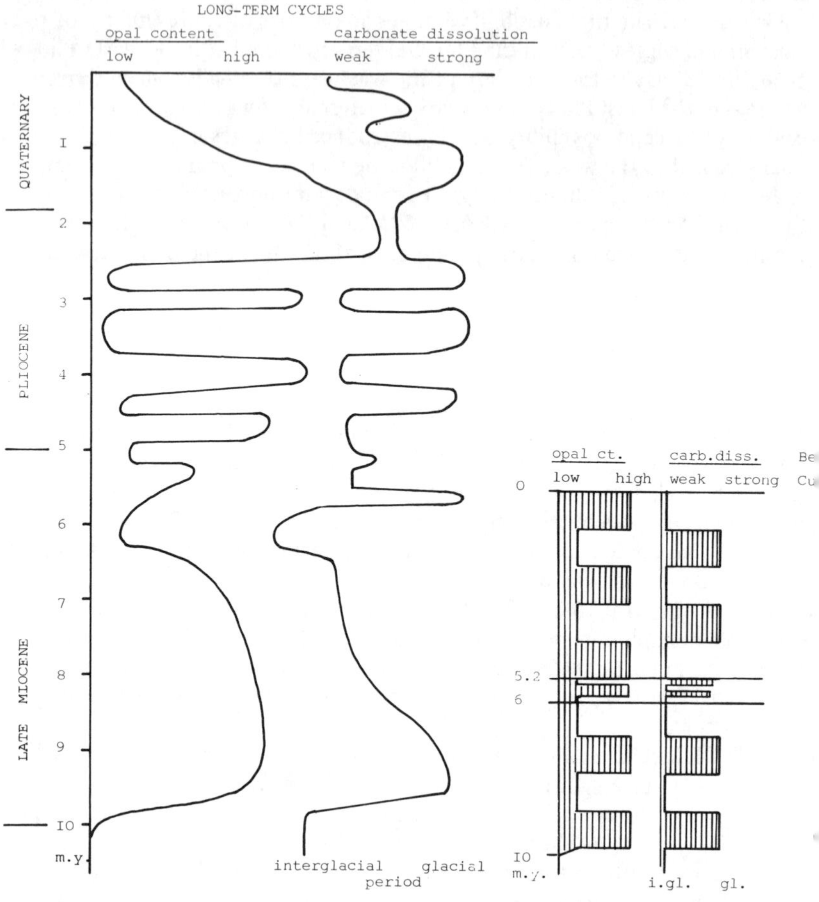

a. b.

Figure 6. Schematic summary.
a. Long-term cycles in opal content (of coarse fraction) and in carbonate dissolution. Strong carbonate dissolution occurs in glacial periods, weak dissolution in interglacial periods.
b. Schematic graph, showing relation between opal content (ct.) and carbonate dissolution/Climate in the small-scale cycles, which are superposed on the long-term cycles, shown in Figure 6a, for the interval 10-6 MY, 6-5.2 MY and 5.2 MY to the present.
i.gl. = interglacial period; gl. = glacial period.
Column on the right: area where Benguela Current turns to the west:
W.R. = Benguela Current turns to the west on the Walvis Ridge.
S. = Benguela Current turns to the west south of the Walvis Ridge, in the Cape Basin.
N. = Benguela Current turns to the west north of the Walvis Ridge, in the Angola Basin.

ACKNOWLEDGEMENTS

The samples for this investigation have been provided by DSDP/ODP through the assistance of the US National Science Foundation.

I thank several colleagues for data, advice and discussions: M.Bremner, P. Meyers, P.Rothe, N.Shackleton, D.Schnitker, C.P.Summerhayes, B.Zobel. I gratefully acknowledge the laboratory work done by M.Killinger and A.Resch. I thank the German Research Society (DFG) for financial support.

REFERENCES

Arrhenius, G. 1952. Sediment cores from the East Pacific. In *Report of the Swedish Deep-Sea Expedition 1947-48,* 5: 1-228.

Berger, W.H. 1970. Planktonic foraminifera: selective solution and the lysocline. *Marine Geology* 8: 111-138.

Berger, W.H. 1976. Biogenous deep-sea sediments. In J.P.Riley & R.Chester (eds.), *Treatise on chemical oceanography* 5: 265-388. New York: Academic Press.

Bolli, H.M., W.B.F.Ryan et al. (eds.) 1978. *Initial Reports of the Deep-Sea Drilling Project,* vol. 40. Washington DC: US Govt. Print. Off. 1079 pp.

CLIMAP Project Members 1976. The surface of the ice-age earth. *Science* 191: 1131-1137.

Diester-Haass, L. 1978. Sediments as indicators of upwelling. In R.Boje & M.Tomczak (eds.), *Upwelling Ecosystems*: 261-281. Berlin: Springer Verlag.

Diester-Haass, L. 1985. Late Quaternary sedimentation on the eastern Walvis Ridge, SE Atlantic (HPC 532 and four piston cores). *Marine Geology* 65: 145-189.

Diester-Haass, L. (in press). Sea-level changes, carbonate dissolution and history of the Benguela Current in the Oligocene-Miocene off SW Africa (DSDP Site 362. Leg 40). *Marine Geology.*

Diester-Haass, L. & P.Rothe (in press). Plio-Pleistocene sedimentation on the Walvis Ridge, SE Atlantic (DSDP Leg 75, Site 532): influence of surface currents carbonate dissolution and climate. *Marine Geology.*

Diester-Haass, L., P.A.Meyers & P.Rothe (in press). Light-dark cycles in opal-rich sediments near the Plio-Pleistocene boundary, DSDP Site 532, Walvis Ridge Continental Terrace. *Marine Geology.*

Emerson, S. & M.Bender 1981. Carbon fluxes at the sediment-water interface of the deep-sea: calcium carbonate preservation. *J. Mar. Research* 39: 139-162.

Goll, R.M. & K.R.Bjorklund 1974. Radiolaria in surface sediments of the south Atlantic. *Micropal.* 20/1: 38-75.

Hagen, E., R.Schemainda, N.Michelsen, L.Postel, S.Schulz & M.Below 1981. Zur küstensenkrechten Struktur des Kaltwasserauftriebs vor der Küste Namibias. *Geod. Geophys. Veröff.,* Reihe IV, Heft 36: 99 pp.

Hart, T.J. & R.T.Currie 1960. The Benguela Current. *Discovery Rep.* 31: 123-298.

Hay, W.W., J.C.Sibuet et al. (eds.) 1984. *Initial Reports of the Deep-Sea Drilling Project,* vol. 75. Washington DC, US Govt. Print. Off.: 1303 pp.

Jansen, J.H.F., T.C.F.van Weering, R.Gieles & J.van Iperen 1984. Middle and Late Quaternary oceanography and climatology of the Zaire-Congo fan and the adjacent eastern Angola basin. *Netherlands J. Sea Res.* 17 (2-4): 201-209.

Johnson, T.C. 1974. The dissolution of siliceous microfossils in surface sediments of the eastern tropical Pacific. *Deep-Sea Research* 21: 851-864.

Martin, R.A. 1981. Benthonic foraminifera from the Orange-Lüderitz shelf, Southern African continental margin. *Bull. Mar. Geosci. Unit,* No. 11, Dept. of Geol., Univ. Cape Town: 82 pp.

Poore, R.Z. & R.K.Matthews 1984. Oxygen isotope ranking of Late Eocene and Oligocene planktonic foraminifers: implications for Oligocene sea-surface temperatures and global ice-volume. *Mar. Micropal.* 9: 111-134.

Proto-Decima, F., F.Medizza & L.Todesco 1978. Southeastern Atlantic Leg 40 calcareous nannofossils. In H.M.Bolli, W.B.F.Ryan et al. (eds.), *Initial Rep. of the Deep-Sea Drilling Project* 40: 571-634. Washington DC: US Govt. Print. Off.

Sancetta, C. 1984. Diatoms from Leg 75, Deep-Sea Drilling Project. In W.W.Hay, J.C. Sibuet et al. (eds.), *Initial Rep. of the Deep-Sea Drilling Project* 75: 755-762. Washington DC: US Govt. Print. Off.

Sarnthein, M. 1971. Oberflächensedimente im Persischen Golf und Golf von Oman. II. Quantitative Komponentenanalyse der Grobfraktion. – *'Meteor' Forsch. Ergebn.* C, 5: 1-113.

Schuette, G. & H.J.Schrader 1981. Diatom taphocoenoses in the coastal upwelling area off South West Africa. *Mar. Micropal.* 6: 131-155.

Shackleton, N.J., J.Backman et al. 1984. Oxygen isotope calibration of the onset of ice rafting and history of glaciation in the north Atlantic region. *Nature* 307, no. 5952: 620-623.

Siesser, W.G. 1980. Late Miocene origin of the Benguela upwelling system off northern Namibia. *Science* 208: 283-285.

Steinmetz, J.C., E.J.Barron, A.Boersma, B.Keating, C.McNulty, C.Sancetta & H.Stradner 1984. Summary of biostratigraphy and magnetostratigraphy of Deep-Sea Drilling Project Leg 75. In W.W.Hay, J.C.Sibuet et al. (eds.), *Initial Reports of the Deep-Sea Drilling Project* 75: 449-458. Washington DC: US Govt. Print. Off.

Summerhayes, C.P. 1983. Sedimentation of organic matter in upwelling regimes. In J.Thiede & E.Suess (eds.), *Coastal upwellings. Its sediment record.* NATO Conf. Ser., IV, 10 b: 29-72. New York: Plenum Press.

Thunell, R.C. 1976. Calcium carbonate dissolution history in Late Quaternary deep-sea sediments, Western Gulf of Mexico. *Quat. Res.* 6: 281-297.

Vail, P.R., R.M.Mitchum & S.Thompson 1977. Seismic stratigraphy and global changes of sea-level, part 4: global cycles of relative changes of sea-level. In C.E.Payton (ed), *Seismic stratigraphy-applications to hydrocarbon exploration. AAPG Memoir* 26: 83-97.

Van Zinderen Bakker, E.M. 1976. The evolution of Late Quaternary paleoclimates of southern Africa. *Palaeoecology of Africa* 9: 160-202.

Wise, S.W., A.M.Gombas & J.P.Muza 1985. Cenozoic evolution of polar water masses, southwest Atlantic Ocean. In K.Hsü & H.J.Weissert (eds.), *South Atlantic paleoceanography*: 283-324. Cambridge.

DESERT PALAEOENVIRONMENTS

ZUM ALTER JUNGQUARTÄRER SEESPIEGELSCHWANKUNGEN IN DER MITTLEREN KALAHARI, SÜDLICHES AFRIKA

K.HEINE
Institut für Geographie, Universität Regensburg, FR Germany

ABSTRACT

The age of late Quaternary lake-level fluctuations in the Middle Kalahari, southern Africa.

The late Quaternary lake-level fluctuations in the Middle Kalahari (Lake Ngami, Mababe Depression, Makgadikgadi Basin) are discussed. The most valuable survey of the geomorphology, younger Kalahari Bed stratigraphy, sedimentology, and palaeontology published in 1904 by Passarge is used as basic reference. Additional data from field-work, laboratory analyses and radiocarbon age determinations supplement Passarge's observations. Furthermore, all available radiocarbon dates of the region are used to support the final conclusions of this review. All results tend to indicate that we have to distinguish between periods with brackish water bodies, fresh water lakes, and dry pans or basins. During the Late Quaternary brackish environments were typical for the pans apart from a few episodes. The Holocene is characterized by greater aridification (dessication of lakes) with only a few minor lake-level fluctuations.

The vast Ngami-Makgadikgadi area did not show synchronous fluctuations in lake-level; the Gidikwe Ride west of the Makgadikgadi basin seems to be a geomorphic feature separating the old Makgadikgadi pans that existed at least since the Middle Pleistocene from the Ngami-Boteti area. The latter was flooded with brackish lakes up to the 936 m level before 30 000 years BP and 25 000-c. 14 000/12 000 years BP; during these periods the Makgadikgadi pans were not occupied by huge brackish water bodies. Fresh water lakes existed c. 30 000-25 000 yrs BP in the Ngami-Mababe-Boteti area as well as in the Makgadikgadi basin; because of lack of field evidence we cannot decide yet whether both areas were inundated by one hunge lake or by two lakes with different lake levels. Other fresh water episodes occurred around 20 000 to 19 000 yrs BP and c. 12 000 yrs BP in the western Makgadikgadi basin.

The Ngami-Mababe-Boteti area southeast of the Okavango Delta derived much of its water from the local and the Angolan catchment; therefore fluctuating brackish high-level lakes could even exist during the late glacial maximum when most of the Makgadikgadi basin had dried up.

The lacustrine landforms have attracted wide ranging palaeoenvironmental interpretations. The results may settle the controversy concerning the palaeoenvironmental reconstructions of the Middle Kalahari since the concepts of very humid as well as of very arid Lake Quaternary climates must be questioned. The Middle Kalahari experienced semi-arid climates throughout the Lake Quaternary with only gradual modifications to more humidity or more aridity.

EINLEITUNG

In jüngster Zeit ist wiederholt versucht worden, jungquartäre Klima- und Landschaftsrekonstruktionen für die Kalahari zu erstellen (Heine 1982, Van Zinderen Bakker 1982, Cooke & Verstappen 1984, Deacon, Lancaster & Scott 1984). Besonders schwierig erwies sich dabei, Einigkeit über die Ausdeutung der Paläoenvironment-Indikatoren zu erzielen. Allein eine vorsichtige Interpretation und Zusammenschau möglichst vieler Paläoenvironment-Indikatoren kann zu richtungsweisenden Ergebnissen führen. Beispielhaft sind die Arbeiten von E.M.van Zinderen Bakker Sr, der sich stets bemühte, die Ergebnisse aller mit der Paläoökologie befassten Disziplinen in seine Betrachtungen einzubeziehen; seine zusammenschauenden Darstellungen zum Klima- und Landschaftswandel im südlichen Afrika sind daher immer von besonderem Wert gewesen. Meinem Freund und Kollegen Eduard M.van Zinderen Bakker sei dieser Beitrag zum 80. Geburtstag gewidmet.

Nachfolgend sollen die jungquartären Seespiegelschwankungen in der Mittleren Kalahari (Fig. 1) erneut vor dem Hintergrund der Angaben Passarges (1904) und eigener Gelände- und Laboranalysen diskutiert werden.

FORSCHUNGSSTAND

Für die Kalahari, dem zentralen Gebiet des ariden Inneren Südafrikas, spielt die Datierung einer innerwürmzeitlichen Feuchtphase eine grosse Rolle. Zwei Modelle werden diskutiert: 1. Pluviale Verhältnisse existierten vor > 30 000 bis c. 19 000 BP, d.h. vor dem letzten Hochglazial, und 2. das letzte Hochglazial, besonders zwischen 16 000 und 13 000 BP, war humid.

In der neueren Literatur nennt Grove (1969) zuerst als eindrucksvollsten Beweis früherer pluvialer Bedingungen die Strandlinie (Gidikwe Ridge) eines ausgedehnten Palaeo-Makarikari-Sees; zum Alter wird ausgeführt 'that it must have pre-dated a Middle Stone Age site described by Bond & Summers (1954), which lies in the deltaic area of the Nata River' (Grove 1969: 203). Beobachtungen über hygrische Klimaschwankungen in der Kalahari sind bereits von Passarge (1904), der auch die ältere Literatur zu diesem Problem auswertet, erwähnt worden. Street & Grove (1976) berichten aufgrund eines ^{14}C-Alters

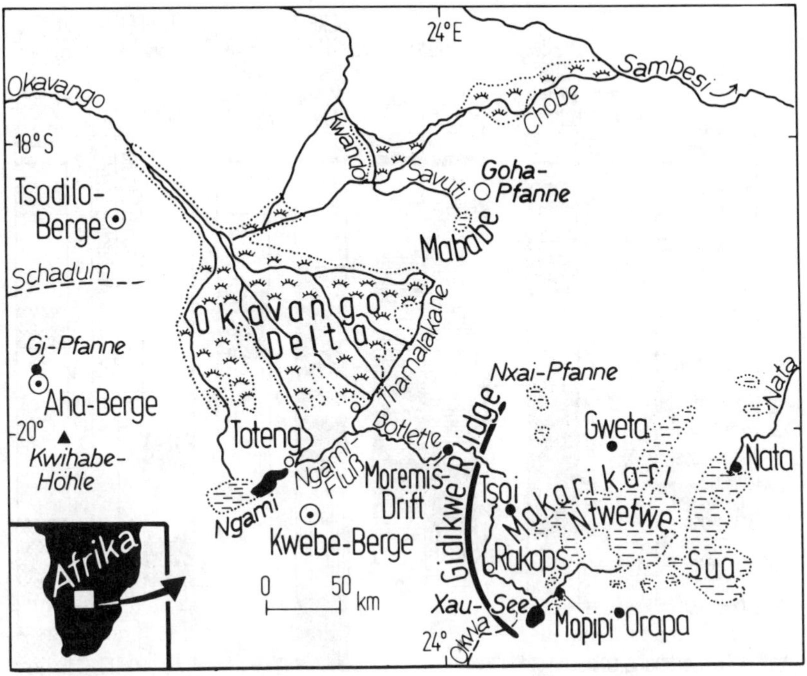

Figur 1. Ubersichtskarte

von einem pluvialen See im Makarikari-Becken um 18 000 BP. Diese Vorstellung eines letztkaltzeitlichen Pluvialsees widerspricht den teilweise theoretischen Uberlegungen von Heine (1977, vgl. dort die Kartenbeilage). Aufgrund von ^{14}C-datierten Höhlensintern und Calcretes in West-Ngamiland postuliert Cooke (1975: 430) ebenfalls, dass 'one major wet episode appears to have been contemporaneous with the Würm maximum glacial advance in higher latitudes'. Diese Aussage wird bei Grey & Cooke (1977), Cooke & Verhagen (1977), Lancaster (1979a) und Cooke (1979, 1980, 1984) bekräftigt. Demgegenüber ergeben sich aufgrund der Untersuchungen von Heine (1978a und b, 1979, 1982) pluviale Bedingungen von über 30 000 bis c. 19 000 BP. Zwischen den Vorstellungen von Cooke (und Mitarbeitern) und Heine konnte bisher keine Annäherung erzielt werden, obgleich Street & Grove (1979), Lancaster (1979b, 1981) und Deacon, Lancaster & Scott (1984) die Erkenntnisse von Heine ihren Paläoklimarekonstruktionen zu Grunde legen. In jüngsten Arbeiten werden für das Makarikari-Becken von Cooke & Verstappen (1984), Cooke (1984) und Shaw (1986) neben der hochglazialen Pluvialphase (c. 17 000-12 000 BP) folgende Zeiten mit hohem Seespiegel für das Jung-

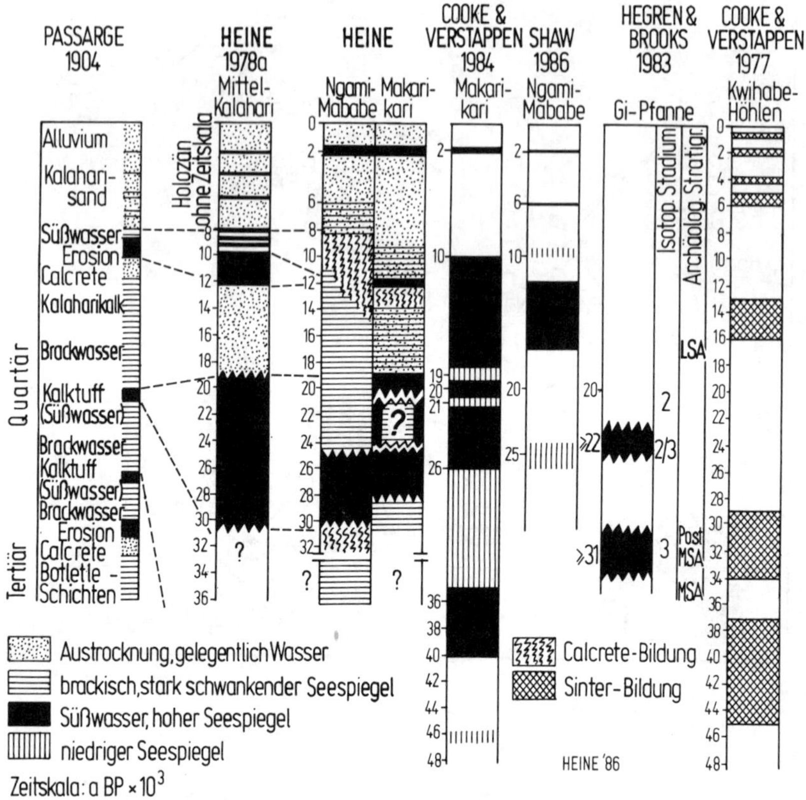

Figur 2. Zusammenstellung der Paläoenvironments der Mittel-Kalahari nach verschiedenen Autoren

quartär genannt (Fig. 2): zwischen 40 000-35 000 BP und zwischen 26 000 und 10 000 BP (mit niedrigem Seespiegel um 21 000 und 19 000 BP). Damit wird allgemein akzeptiert, dass im Mittelwürm zwischen c. 40 000 und 19 000 BP hohe Seespiegel im Makarikari-Becken auftraten; die verschiedenen Phasen mit niedrigem Seespiegel zwischen 35 000 und 26 000 BP, um 21 000 BP und um 19 000 BP im Makarikari-Becken (Cooke 1984) und um 25 000 BP im Ngami-Gebiet (Shaw 1986) geben Hinweise auf eine unterschiedliche Ausprägung der pluvialen Verhältnisse. Unklar bleibt in der Literatur nach wie vor die Beurteilung der hochglazialen Phase zwischen c. 19 000 und 12 000 BP.

Im folgenden sollen die Befunde ausgewählter Autoren, die einerseits zur Rekonstruktion hochglazialer niedriger Seespiegel und andererseits zur Annahme hochglazialer hoher Seespiegel geführt haben, näher betrachtet und kritisch diskutiert werden.

BEFUNDE

Die Beobachtungen von Passarge

Aufgrund umfangreicher Untersuchungen kann Passarge bereits im Jahre 1904 eine äusserst detaillierte Beschreibung der Mittleren Kalahari vorlegen. Die von Passarge (1904) mitgeteilten Befunde legt Heine (1978a und b, 1979, 1982) seinen Untersuchungen zugrunde. Die Diskussion über die jungquartäre Klimaentwicklung der Mittel-Kalahari hat gezeigt, dass wichtige Beobachtungen Passarge (1904) unberücksichtigt bleiben. Daher müssen zuerst die für die Diskussion relevanten Ergebnisse aus dem Werk von Passarge (1904) vorgestellt werden.

Passarge (1904) gibt folgende Stratigraphie der Deckschichten für das Ngami-Okavango-Makarikari-Gebiet an (Tabl. 1): Im Liegenden befinden sich die Botletle-Schichten; diese bestehen aus ausserordentlich mannigfaltigen Gesteinen, die überraschend schnell wechseln können. Die Botletle-Schichten zeigen in Abhängigkeit von den Oberflächenformen des Grundgebirges eine ungleichmässige Lagerung; im Okavango-Becken sind sie besonders mächtig ausgebildet, und nur lokal greifen sie auf die Plateaugebiete über. Die Botletle-Schichten weisen in älteren Ablagerungen eingekieselte Chalzedonsandsteine auf, die einer ersten Periode der Chalzedonlösung während einer trockenen Klimaphase zugeschrieben werden. Im mittleren Teil der Botletle-Schichten finden sich kalkige Ablagerungen, Pfannensedimente oder Seebeckensedimente mit Brackwasser-Diatomeenassoziationen; diese Schichten werden einer ersten Phase der Kalklösung unter feuchteren Klimabedingungen zugeordnet. Harte buckelige Oberflächen mit Klakrinden weisen auf die zweite Periode der Chalzedonlösungen unter ariden Klimaverhältnissen hin.

Die Grenze zwischen den Botletle-Schichten und dem hangenden Kalaharikalk ist scharf; es gibt vielerorts Hinweise auf starke Abtragung und Zerstörung der Botletle-Schichten vor der Akkumulation des Kalaharikalks, der in den liegenden Partien sehr häufig Schutt zeigt. Der Kalaharikalk ist faziell unterschiedlich ausgebildet: Sinterkalke, mürbe Kalksandsteine, Pfannenkalktuffe, Salzmergel, Salzpelite und Salzlager werden beschrieben. Die harten Sinterkalke enthalten viel Quarzsand und weisen Sinterstrukturen und lamellige Rinden sowie teilweise Brekzien aus Botletle-Schichten und Anstehendem auf. Die Sinterkalke enthalten selten Gastropoden (*Physa, Succinea*), Diatomeen fehlen meist. Die mürben Kalksandsteine, ein mit feinem Kalkschlamm verkitteter Sand mit oft zelliger Struktur und feinen Kalkröhren (ehemalige Wurzelgänge), sind oft kochsalzhaltig und von härteren kompakten Bänken durchzogen, die nach oben hin zunehmen. Molluskenschalen sind im Kalksandstein häufig, und zwar sind es Schalen rezenter Gastropoden von Land- und Sumpfschnecken der heutigen Süsswassersümpfe und -flüsse und Salzpfannen; nur bei Lilokwalo wurde *Corbicula* gefunden. Brackwasserdiatomeen sind häufig; es sind Formen von meist an Schilf und Pflanzen festsitzenden Arten. Die Periode der Kalaharikalk-Bildung war eine Zeit energischer Kalkablagerung.

Tabelle 1. Stratigraphie (schematisch) der Deckschichten im Ngami-Okavango-Makarikari-Ge

Formation	Fazies		
	Flusstäler von Ngami und Botletle	Ngami- und Makarikari-Pfannen	Klein Pfann
Alluvium	Weisser Flusssand, z.T. auch Schlamm	Beckenschlamm, auch Salze, Salzmergel u.a.	
Kalahari-sand	Unreine Sande, kalkfrei, grau bis weiss	Unreine Sande, kalkfrei, grau bis weiss	
	Im Liegenden: Kalkhaltige Sande mit Kalkknollen, Geröllen, Süss- und Brackwasserdiatomeen		
	Scharfe Grenze	Fluviale Prozesse	
Kalahari-kalk	Im Hangenden: z.T. Dolomitisierung, Verkieselung, Auslaugung der Salze, Sinterkruste mit Brackwasserdiatomeen Murbe Kalksandsteine mit Brackwasserdiatomeen, stellenweise Kalktuffe mit Gastropoden (Süsswasser), Salzpelit, Salzlager u.a. Im Liegenden: Kalksandsteine, Kalkmergel, Brekzien aus Schutt, Geröllen (z.T. Botletle-Schichten), starke Erosion		Kalk z.T. merg u.a. Pfan sedi
	Scharfe Grenze, Zertrümmerung der Botletle-Schichten durch E		
Botletle-Schichten	Harte buckelige Kalkrinen Kalkige Sedimente und Seebeckensedimente mit Brackwasserdiatomeen Pfannensandstein, fossilfrei, verkieselt Eingekieselte Chalzedonsandsteine		

ımengestellt nach den Angaben bei Passarge (1904)

Plateaus	Paläoökologische Interpretation		Bildung von Brackwasser-seen	Bildung von Süsswasser-seen
Decksande		semi-arid, mit geringen hygrischen Schwankungen		
Unreine Sande, kalkfrei, rot-grau-weiss, teilweise auch mit Kalk-knollen und Ge-röllen, bes. im-Liegen. Im Liegen-den z.T. mit Schotterlage				
fe Grenze	feuchter			
Sinterkalk [= Calcrete] Kalksandstein [= Calcrete] Erosion, Schutt, Gerölle	semi-arid 2. Periode der Kalklösungen: feuchter	semi-arid, mit feuchteren Abschnitten		
	feuchter			
Rinden [= Calcrete] Mannigfaltige Ge-steine, bes. an. Hängen Brekzien	2. Periode der Chalzedonlösungen: semi-arid 1. Periode der Kalklösungen: etwas feuchter 1. Periode der Chalzedonlösun-gen: semi-arid	semi-arid		

Im Hangenden treten der Kalaharisand und der Decksand auf. Der Kalaharisand beginnt am Botletle mit einer kalkreichen Schicht im Liegenden, die eine gemischte Daitomeenflora führt; diese enthält die Brackwasserformen des Kalaharikalks zusammen mit den Süsswasserdiatomeen der heutigen Sümpfe und Flüsse. *Spiculae* und *Spongillen* sind auch häufig. Die Kalaharisande bestehen in der Regel – mit Ausnahme der kalkreichen liegenden Partien – aus abgerundeten bis eckigen Quarzkörnern; Quarzsplitter sind reichlich vertreten, ebenso Fragmente anderer Gesteine, dazu Feldspäte, Epidot, Chalzedon und Glimmer. Diese recht unreinen Sande sind immer kalkfrei, wenn sie rot oder rötlich gefärbt sind, können aber als weisse bis hellgraue Sande auch kalkhaltig wein. Als Decksand wird ein über anstehendem Gestein auskeilender Kalaharisand bezeichnet.

Die Beobachtungen von Passarge (1904) habe ich in Figur 3 dargestellt; sie enthält die von Passarge (1904) ausführlich beschriebenen Profile aus dem Ngami-Botletle-Makarikari-Gebiet und geomorphologische Beobachtungen. Im einzelnen lassen sich folgende Grundzüge der jüngeren geomorphologischen Entwicklung nachzeichnen: Das Ngami-See-Gebiet wird im Süden von einer markanten Stufe, die zum Ngami-Plateau überleitet, begrenzt. Im Westen des Ngami-Sees bilden zwei Wälle aus Chalzedonsandstein die Begrenzung; die ihnen aufliegenden Sande sind reich an Chalzedonknollen und Kalkgeröllen. Das Ngami-Becken ist mit Kalkmergel und Schlamm gefüllt; die Kalkmergel werden im Norden – dorthin steigt die Ngami-Alluvialebene sanft an – von Flusssanden bedeckt, die teilweise recht mächtig sein können.

Der Ngami-Fluss, der ohne sichtbares Gefälle eine Verbindung zwischen Ngami-See und Tamalakane schafft, ist in Schichten des Kalaharikalks eingeschnitten, die von Chalzedonsandstein (Botletle-Schichten) unterlagert und vom Kalaharisand überlagert werden. Der Kalaharikalk zeigt im Hangenden eine harte Sinterschicht und eine scharfe Grenze gegen die hangenden Sande. Bei Mori Mossetla sind in die mürben Kalksandsteine des Kalaharikalks fossilreiche Kalktuffe eingeschaltet.

Der Botletle weist im ersten Abschnitt ein deutliches Gefälle in Richtung Makarikari-Becken auf. Es folgt ein Flussstück mit kanalartigem Lauf. Schliesslich durchbricht der Botletle obernalb von Meno a Kwena in einem Cañonartigen Tal recht mächtige Sandablagerungen und bei Meno a Kwena auch die hangenden Kalaharikalkschichten.

Die geologisch-sedimentologische Ausbildung der Schichten zwischen Meno a Kwena und Mochoto Drift wechselt nur wenig. Allein bei Meno a Kwena befindet sich über der harten Kalaharikalk-Kruste ein 2 m mächtiger diatomeenreicher, lagig ausgebildeter Kalksandstein, der westlich von Meno a Kwena schnell auskeilt. Der harte Oberflächenkalk im Hangenden des Kalaharikalks verliert sich entlang des Botletle-Laufs von Namessán nach Süden. Nahe Rakopo's Drift sind die jüngeren Beckensedimente, ein weisser, feiner, mürber Kalkschlamm, weit verbreitet.

Die Ntschokutsa-Pfanne zeigt ebenfalls salzreiche Kalkmergel; der Süd- und

Ostrand der Pfanne wird aus Kalksandsteinen der Botletle-Schichten und Gesteinen des Grundgebirges aufgebaut.

Zwischen dem Ngami-See und den Makarikari-Pfannen erstreckt sich das Hainafeld, das von Passarge (1904) in vier Zonen gegliedert wird (Fig. 3). Das Sandfeld, im Westen von den Kwebe Bergen begrenzt, reicht bis an die Haina Hügel heran; Pfannen sind äusserst selten; die Oberfläche ist sehr schwach gewellt. Unter dem Kalaharisand, der von östlichen Winden umgelagert wurde, fehlen Kalkkrusten; es stehen Grauwacken an. Im Hainafeld sind Pfannen zahlreich, die mit Kalksand und Kalkton ausgekleidet sind. Weder im Sandfeld noch im Hainafeld wird unter dem Kalaharisand, der hier rötlich auf den Wällen und grauweiss und kalkhaltig in den sanften Senken ausgebildet ist, eine Kalkkruste über dem Grauwacken-Grundgebirge angetroffen. Die Oberfläche des Hainafeldes ist deutlich stärker gewellt als diejenige des Sandfelds.

Ein markanter Unterschied besteht zwischen dem Hainafeld und dem Grasland im Osten. Deutlich tritt die Grenze zwischen beiden Landschaften hervor: ein sandiges Gebiet, das sich etwa in NW-SE-Richtung erstreckt. Der Untergrund des Graslands wird von Kalaharikalk aufgebaut; es sind Pfannensedimente, die auch als Gerölle und Scherben den hangenden Sanden beigemischt sind. Die Sande sind grau und kalkreich. Das Grasland ist eine wirkliche Ebene, deren Oberfläche ganz leicht gewellt ist. In den Niederungen befindet sich schwarzes humoses Bodenmaterial. Ein Sandstreifen durchzieht das Grasland in NW-SE-Richtung. Fast NNW-SSE verläuft die östliche Begrenzung des Graslands. Wiederholt zeigen sich Sandwälle, die von ehemaligen Flusstälern voneinander getrennt werden. Dieser Bereich der Sandwälle wird abrupt von der Ebene der Makarikari-Pfannen im Osten abgelöst. Hier ist der Untergrund sehr kalkhaltig.

Quartärmorphologische Deutung der Befunde von Passarge

Die Beobachtungen und Ergebnisse der Untersuchungen von Passarge (1904) können einerseits geomorphologisch und andererseits stratigraphisch-sedimentologisch gedeutet werden.

Die sehr detaillierten Angaben zu den geomorphologischen Verhältnissen der Mittleren Kalahari durch Passarge (1904) in Verbindung mit der Interpretation neuerer Karten und Satellitenbilder habe ich zur Vorbereitung meiner Geländearbeiten im Jahre 1977 benutzt. Folgende Grundzüge der quartären Oberflächenformen der Mittleren Kalahari konnten dabei herausgearbeitet werden: Sowohl im Makarikari-Gebiet als auch im Ngami-Becken lassen sich verschiedene pluviale Seebildungsphasen belegen. Die grösseren pluvialen Seen müssen älter als die kleineren sein. Das Alter dieser pluvialen Seen war im Jahre 1977 unbekannt.

Für einen älteren grösseren pluvialen Brackwassersee im Makarikari-Gebiet lassen sich mehrere Beobachtungen anführen. Im Westen des Sandwalles, der von Passarge (1904) noch nicht als durchlaufender Uferwall, sondern als Flussuferwälle verschiedener Gewässer erkannt, der von Grove (1969) als

Westufer eines Sees identifiziert und der in jüngster Zeit von Cooke & Verstappen (1984) genauer beschrieben wurde (Gidikwe Ridge), befindet sich die 'wirkliche Ebene' (Passarge 1904: 231) des Graslands. Es ist ein ehemaliger Seeboden, wie die morphologischen und bodenkundlichen Beschreibungen von Passarge (1904) zweifelsfrei belegen. Dieser ehemalige Pluvialsee wird im Westen durch Ufersandwälle begrenzt. Seine Oberfläche zeigt mitunter 'sehr flache Terrainwellen' (Passarge 194: 231). Der Sand ist hellgrau, rote Sande fehlen; im Westen wird der Sand mächtiger, die eigentliche Fläche des Graslands ist nur mit einer dünnen Sandschicht, die Knollen mürben Kalksandsteins enthält, bedeckt. In den sogenannten 'Brackpans' kommt der mürbe Kalksandstein des Untergrundes in ausgedehnten Flächen an die Oberfläche. Bemerkenswert ist ferner das Auftreten schwärzlichen sandig-humosen Bodens in einigen flachen Niederungen; er ähnelt den schwärzlichen Schlammböden im Gebiet des Okavangosystems.

Für einen jüngeren kleineren Pluvialsee im Makarikari-Gebiet sind die von Passarge (1904) beschriebenen Verhältnisse beiderseits des Botletle zwischen Meno a Kwena und Rakopo's Drift Zeugen. Hier ist das Land noch ebener und der Boden noch kalkreicher als im Grasland (S. 232). Im Westen dieser ebenen Flächen beginnt ganz plötzlich eine Waldzone auf Sand, der noch grau ist, aber auch leicht rötlich werden kann; diese Sandzone, der Gidikwe Ridge, trennt somit einen jüngeren Pluvial-Seeboden von einem älteren.

Der Gidikwe Ridge selbst muss aufgrund der Angaben von Passarge (1904) älter sein als die diatomeenreichen, lagigen weichen Kalksandsteine, die etwa 2 m mächtig bei Meno a Kwena aufgeschlossen sind, denn diese Schichten lassen sich nicht Botletle-aufwärts verfolgen; im Gegenteil, die Sande des Gidikwe Ridge nehmen von Meno a Kwena nach Westen hin zu, während die Kalaharikalk-Schichten abnehmen. Zwischen Matabele Drift und dem Zusammenfluss von Ngami-Fluss und Tamalakane treten die Kalksandsteine des Kalaharikalks am Botletle-Ufer ganz zurück und der Untergrund wird aus Botletle-Schichten aufgebaut.

Die Ausdehnung des durch den Gidikwe Ridge im Westen eingeschlossenen kleineren jüngeren Pluvialsees wird durch die mürben Kalksandsteine des Kalaharikalks repräsentiert (vgl. die Profile in Figur 3). Es sind in der Regel mürbe sandreiche Kalke mit zelliger Struktur und Wurzelröhren; Brackwasserdiatomeen sind charakteristisch für die Ablagerungen. Gastropoden kommen in verschiedenen Profilen vor. Reine Süsswassermollusken sind nur im Profil Lilokwalo beobachtet worden. Oft enthalten die zelligen Kalke auch Beimengungen von Kochsalz. Die Versinterung und teilweise auch Verkieselung nimmt gewöhnlich in den obersten Partien des Kalaharikalks zu. Paläoökologisch können die Ablagerungen der Kalaharikalk-Schichten östlich des Gidikwe Ridge nicht einem ausgedehnten pluvialen Süsswassersee zugeschrieben werden. Schon Passarge (1904) sieht in ihnen Pfannensedimente, die sich durch ihren Fossilinhalt (S. 783) und ihre sedimentologischen Eigenschaften im wesentlichen als Brackwasserablagerungen zu erkennen geben. Bei den Landschnecken zeigt sich eine Ubereinstimmung mit den semi-ariden Savannen im Süden, speziell Südwestafrikas, bei den Süsswasserconchylien dagegen eine grössere Ubereinstimmung mit der trophisch-afrikanischen Fauna (*Ampullaria, Lanistes, Vivipara* und *Mutela*) (Passarge 1904: 758 f). Wiederholte Uberflutungen und nachfolgende Austrocknung des Makarikari-Gebiets sind damit belegt. Während dieser Zeit muss auch die letzte Ausformung der Sandwälle des Gidikwe Ridge erfolgt sein, d.h. zur Zeit der Brackwasser-Kalksandstein-Sedimentation haben Winde aus östlichen Richtungen einerseits das fluvial eingeschwemmte Sandmaterial teilweise umgelagert und zu Sandwällen akkumuliert und andererseits Dünensandmaterial eingeweht; die von Passarge (1904) beschriebenen Sande des Gidikwe Ridge deuten diese Vorgänge an.

Besonders interessant ist das von Passarge (1904: 207 f., 217) beschriebene Profil von Meno a Kwena, das über den Brackwasser-Kalksandsteinen mit hartem Sinterbank-Abschluss einen 2 m mächtigen Horizont aus weichem Kalksandstein aufweist, dessen hangende Schichten sandärmer, kalkreicher und dichter erscheinen (harte Oberflächenbank); dieser konkordant auflagernden Schicht verleihen feine abschilfernde Lagen eine besondere Struktur. Dieser Kalktuff ist sehr reich an ziemlich gut erhaltenen Diatomeen (*Anomoeoneis sculpta*, Ehr., selten; *Mastogloia lanceolata*, Thwaites, selten; *Epithemia Argus* (Ehr.), Kütz., sehr häufig; *Campylodiscus Clypeus*, Ehr., sehr häufig; *Campylodiscus bicostatus*, W.Sm.) (Passarge 1904: 764 f.). Die oberste harte Bank des Kalksandsteins im Liegenden dieses Kalktuffs ist ebenfalls reich an Diatomeen, die aber weniger gut erhalten sind (*Rhopalodia gibberula* (Kütz.), O.Müller; *Epithemia Argus*, Ehr., sehr häufig; *Campylodiscus Clypeus*, Ehr., sehr häufig) (Passarge 1904: 765).

Ohne scharfe Grenze entwickelt sich im Profil Meno a Kwena der hangende Kalaharisand aus dem Kalktuff; die untersten Partien des Sandes sind kalkhaltig und reich an zersetzter vegetabilischer Substanz und Diatomeen (Passarge 1904: 208). Passarge (1904: 731) gibt folgende Beschreibung: 'Kalkreicher graubrauner, mässig fester Sand. Mikroskopisch: Sand mit Hülle von Kalkpatkörnern, zahlreiche Diatomeen und Spiculae. Nach Auflösen des Kalks wird schmutzig braune flottierende organische Substanz sichtbar, reichlich; Aussehen wie der Tierkot der Seekreide. Flockenhülle um die Quarzkörner. Die Diatomeen sind nach H.Reichelt Süss- und Brackwasserformen. Der Sand gleicht dem Kalk-Schlamm-Sand der norddeutschen Seen.' Dieser liegende Kalaharisand von Meno a Kwena, welcher von allen untersuchten Proben am reichsten an Arten ist, enthält eine Mischung von Süsswasserdiatomeen des Ngami-Sees mit den Brackwasserdiatomeen der Kalaharikalke und Tuffe (*Phytolitharien*, Ehr.) und Kieselnadeln von Schwämmen wurden von Reichelt 44 Arten von Diatomeen bestimmt (Passarge 1904: 763 f.). Die liegenden Partien des Kalaharisandes bei Meno a Kwena (Kalk-Schlamm-Sand) geben Hinweise auf Zufuhr grösserer Süsswassermengen. Da westlich von Mochoto Drift ein markanter Hangknick die Sandwälle des Gidikwe Ridge gegen die Ebenen der Makarikari-Region begrenzt, wird dieser von mir als Ufersaum eines post-Kalaharikalk-zeitlichen Sees gedeutet und mit dne liegenden Kalk-Schlamm-Sand-Ablagerungen bei Meno a Kwena korreliert. Die von Passarge

(1904) für das Ngami-See-Gebiet beschriebenen Verhältnisse lassen ebenfalls eine Rekonstruktion ehemaliger grösserer Seen zu. Die im Westen des Ngami-Sees gelegenen Sandwälle trennen einerseits die Brackwasser-Ablagerungen der Matanya-Pfannen von dem westlich angrenzenden, mit rotem Kalaharisand bedeckten Gebiet und andererseits trennt der östliche Ufersandwall die Matanya-Pfannen von den Ngami-Pfannen. Im Norden verschwinden die Uferwälle in fluvialen Sandakkumulationen, die von dem sich dort verzweigenden Tauche-Fluss abgelagert wurden, und zwar nach der maximalen Seeausdehnung. Die jüngsten Sedimente im Ngami-See bestehen aus dunklem, eingetrocknetem Schlamm, sind frei von Kalk und sehr leicht und reich an Diatomeen, die ausschliesslich (wie auch die jüngsten Ngami-Fluss-Sedimente) Süsswasserarten enthalten. Am Ngami-Fluss gibt das Profil von Mori Mossetla (vgl. Fig. 3) eindeutige Hinweise auf verschiedene Phasen mit Süsswasserseen; den etwa 8 m mächtigen mürben Kalksandsteinen sind bis 0,30 m dicke Bänke gelblichen mürben Kalktuffs, der an Gastropoden sehr reich ist, eingelagert.

Nach Passarge (1904) sind die Ablagerungen des Kalaharikalks zwischen dem Zusammenfluss von Ngami-Fluss und Tamalakane und Matabele Drift nur sehr geringmächtig oder gar nicht mehr anzutreffen. Demnach könnte man mit Passarge (1904) an zwei getrennte Becken – Ngami und Makarikari – denken, in denen wiederholt Brackwasserseen und zeitweise auch Süsswasserseen gebildet wurden.

Die Gebiete des Ngami-Plateaus, des Sandfelds und Hainafelds zwischem dem Ngami-Becken und den Makarikari-Pfannen waren nicht mehr von Brackwasser-/Süsswasserseen erfasst; die Kalaharisande liegen dort unmittelbar auf dem Grundgebirge. Auch weist das Fisch-Vley c. 7 km südlich des Ngami-Sees keine Kalaharikalk-Kalksandsteine auf, die auf eine Ausdehnung des Ngami-Pluvialsees nach Süden deuten. Zwischen Toting und den Kwebe Bergen beschreibt Passarge (1904: 570 f.) Wellentäler; es handelt sich um sehr alte Längsdünensysteme, die auch vom Schadum-Fluss genannt werden und die nur ausserhalb der quartären Okavangosedimente auftreten. Auch das Fehlen der Kalkkrusten (Kalaharikalk) unter den Sandfeld-Sanden, und über dem liegenden Grundgebirge belegt das hohe Alter (prä-Kalkkrustenzeitlich) der 'Wellental'-Oberflächenformen. Allerdings bemerkt Passarge (1904) wiederholt, dass der Kalaharisand vielerorts vom Wasser transportiert und abgelagert worden ist, der Wind trug anschliessend zu seiner Verbreitung bei, so auch im nördlichsten Hainafeld unmittelbar südlich des Botletle.

Die Beobachtungen Passarges (1904) geben eindeutige Hinweise auf (1) mindestens zwei unterschiedlich alte und unterschiedlich ausgedehnte Seen, die (2) aufgrund der stratigraphischen und paläontologisch-sedimentologischen Befunde verschiedene Brackwasser- sowie Süsswasser-Phasen verkörpern. (3) Die reinen Süsswasser-Sedimente sind recht geringmächtig im Vergleich zu den Brackwasser-Sedimenten. Die Brackwasser-Sedimente enthalten stets viel Sand, ein Hinweis auf den starken äolischen und/oder fluvialen Eintrag dieser Sande in die Ngami- und Makarikari-Pfannenbereiche. Die Süsswasser-Kalktuffe sind Quarzsand-arm. (4) Die letzte Phase mit Süsswassersee-Bildung belegen die liegenden Kalaharisand-Sedimente, die als sandiger Kalkschlamm ausgebildet sein können. Passarge (1904) stellt alle Kalaharikalk-Bildungen in das Quartär, ohne eine genauere Altersandsprache geben zu können. Die von Passarge (1904) als 'harte Sinterkalke' bezeichneten Kalaharikalke sind grösstenteils als Calcretes anzusprechen; Passarge (1904) vergleicht die Kalaharikalke der Inselgebirge mit Kalkkrusten der Halbwüsten und vermutet, dass ihre Bildung in Gebirgen durch Einwehung (Staubstürme) beträchtlicher Kalkmengen, die dann bei Zunahme der Niederschlagsmengen zu Kalkabsätzen geführt haben konnten, erklärt werden kann (Passarge 1904: 198).

Eigene Beobachtungen

Die zuvor dargelegten Beobachtungen Passarges (1904) konnte ich meinen Arbeiten in der Mittelkalahari zu Grunde legen. Für mich stellte sich daher vordringlich die Aufgabe, die von Passarge (1904) beschriebenen Formen und Sedimente mit unseren heutigen paläoökologischen Vorstellungen des Quartärs zu vergleichen und vor allem chronostratigraphisch zu erfassen, wobei erste Anhaltspunkte einer absoluten Altersbestimmung der von Passarge (1904) genannten Sedimente und Fossilien im Mittelpunkt standen.

Im folgenden werden die von mir im Jahre 1977 aufgenommenen Profile vorgestellt (Fig. 4).

Am Nordrand der Ngami-Alluvialebene, innerhalb des deutlich ausgebildeten inneren Ufersandwallsystems, befindet sich in etwa 930 m Höhe eine Abfolge aus Kalaharisand über Beckensedimenten. Die liegenden Schichten werden von einem mürben, nur leicht verfestigten Kalksandstein gebildet; die in dem hellen Kalkschlamm eingebetteten Sandkörner sind teils gut gerundet, teils gerundet und zeigen in der Mehrheit helle Verwitterungsrinden, die auf wieder aufgearbeitete ältere Dünensande hinweisen. Das Sediment enthält in einer recht einheitlichen Grundmasse aus Sand und Kalkschlamm Wurzelröhren, die mit jüngerem Sand und Kalkausscheidungen ausgefüllt sind. Die Körner der Wurzelröhrensandfüllungen zeigen keine Verwitterungsrinden Vereinzelt sind auch Fragmente umgelagerter, älterer Calcretes zu beobachten. Zwischen 60 und 65% des Materials besteht aus Kalk, von dem ein ^{14}C-Alter vorliegt: 14 620 ± 90 (Hv 8381). Der hangende Kalaharisand gliedert sich in den schwach kalkhaltigen (3-5% $CaCO_3$) liegenden und den kalkfreien humosen grauen hangenden Sand; es sind vorwiegend gut gerundete helle, vielfach glasklare Körner, die ihr Maximum in der Fraktion 0,2-0,1 mm ∅ haben (un 50%); dies spricht für äolische Um- und Ablagerung des Materials, weniger für fluvialen Transport. Das ^{14}C-Alter gibt nur näherungsweise Auskunft über die Zeit der Sedimentbildung, da sowohl kleinere Fragmente älterer Calcretes als auch die Füllungen der Wurzelröhren den ermittelten Wert beeinflusst haben können.

Das Profil vom Ngami-Nordrand lässt sich mit den Profilen von Passarge (1904; vgl. Fig. 3) korrelieren. Die 'Flusssande' Passarges (1904) sind äolisch umgelagerte fluviale Sedimente; die 'Kalkmergel' Passarges (1904) sind

Beckensedimente aus Sand und Kalkschlamm, die als Brackwassersedimente bis gegen Ende der letzten Kalkzeit (c. 15 000-12 000 BP) sedimentiert wurden. Sie enthalten Sandkörner, deren Herkunft vorwiegend im verwitterten Grundgebirge und in einer Abtragung der Botletle-Schichten vermutet wird.

C. 3 km nordwestlich des Ngami-Flusses (Lake River) in 940 m Höhe befindet sich über einem Beckensediment aus Kalkschlamm und Sand ebenfalls der Kalaharisand (Fig. 4). Der liegende mürbe Kalksandstein i.S. Passarges (1904) ist leicht verfestigt, lässt sich jedoch zwischen den Fingern zerreiben. Im Dünnschliff erkennt man eine Ummantelung der Sandkörner mit Kalk; Verwitterungsränder sind seltener als im Beckensediment-Sand vom Ngami-See. Die ^{14}C-Analyse ergab ein Alter von 14 300 ± 190 BP (Hv 8386). Das Profil lässt sich mit den obersten Partien des Profils von Mori Mossetla (Passarge 1904; vgl. Fig. 3) korrelieren. Auch hier wird die Sedimentation des Kalksandsteins der Kalaharikalk-Formation Passarges (1904) durch eine schwache Verfestigung im Hangenden (beginnende Calcrete-Bildung) abgeschlossen. Dieser Vorgang der Calcrete-Bildung erfolgte gegen Ende der letzten Kaltzeit vor rund 15 000-13 000 BP. Die obersten Beckensedimente des Brackwassersees wurden demnach vor der schwachen Calcrete-Bildung, d.h. im letzten 'Hochglazial', sedimentiert.

In 945 m Höhe befindet sich ein Profil am unteren Thamalakane (Fig. 4). Das Liegende bildet ein zu einer Kalkkruste (Calcrete) verbackener Sand mit verkieselten Pflanzenresten; der Sand weist häufig die charakteristischen Verwitterungsrinden an den Sandkörnern auf, die deren Herkunft aus älteren abgetragenen Sedimenten erkennen lassen. Die Oberfläche der Calcrete ist wellig-wulstig und trägt einen 3-10 mm dicken Uberzug aus hartem Kalksinter. Das ^{14}C-Alter der obersten Schicht der Calcrete (Mischprobe aus sandreicher Kalkkruste und Kalksinterhaut) beträgt 8 720 ± 95 BP (Hv 8383). Das Alter an sich ist nur bedingt aussagefähig, setzt es sich doch aus dem Alter der Calcrete (Spätglazial) und dem Alter des Kalksinter-Uberzugs (Holozän) zusammen. Hinweise auf limnische Sedimente wurden in diesem Profil nicht gefunden.

3 km östlich der Strassenbrücke bei Maun über den Thamalakane zeigt ein Profil in 940 m Höhe im Hangenden c. 1 m mächtige graue schluffige Sande. Darunter liegen helle weissgraue, sehr lockere, fein abschilfernde, geschichtete Beckensedimente aus Kalkschlamm und Sand; die hangenden Partien dieser c. 1 m dicken Lage haben ein ^{14}C-Alter von 9 390 ± 80 BP (Hv 8378). Darunter folgt eine harte Calcrete-Bank mit einer auffällig wulstig-buckeligen Oberfläche; sie ist zu 98% aus $CaCO_3$ zusammengesetzt und enthält feinste Haarwurzelröhren; Sandkörner sind in ihr nicht zu finden. Im Schliff ist in der feinen Matrix eine schwach wolkige Struktur zu erkennen, die der 'Zellenstruktur' Passarges entspricht. Das ^{14}C-Alter ist 31 750 ± 500 BP (Hv 8387). Unter dieser Kalkkruste, die nach unten weicher wird, liegt ein Seekreide-Horizont von 0,2-0,3 m Mächtigkeit; dieser ist ebenfalls völlig frei von Sandkörnern, enthält etwa 87% $CaCO_3$ und geht zum Liegenden über in mürbe kalkig-tonige geschichtete Beckensedimente. Die Seekreide weist feine Haar-

wurzelröhren auf und hat ein ^{14}C-Alter von 27 050 ± 450 BP (Hv 8379). Eine Korrelierung dieses Profils mit dem Profil Passarges (1904) bei Mori Mossetla zeigt wesentliche Ubereinstimmungen in den hangenden Schichten. Die von Passarge (1904) beschriebenen fossilreichen 'Kalktuffe', die in die mürben Kalksandsteine eingelagert sind (vgl. Fig. 3), treten als Seekreide und harte Calcrete-Bank hervor. Die Spuren der Haarwurzelröhren belegen, dass die Ablagerungen in rund 940 m Höhe in flachem Wasser erfolgten. Da die ^{14}C-Alter von Calcrete-Bildungen oft nur orientierende Altersangaben liefern, zumal der 'Reservoir-Effekt' (Geyh 1983: 31 f., 93) oft zu hohe Alter bedingt, wird angenommen, dass das Calcrete-Alter von 31 750 ± 500 BP zu hoch ist, während das Alter der Seekreide eher dem wahren ^{14}C-Alter der Sedimentbildung entspricht. Auf jeden Fall belegen die Altersangaben, dass um 25 000-30 000 BP ein ausgedehnter Süsswassersee im Bereich von Ngami-Fluss und Thamalakane existierte, dessen Seespiegel bis fast 940 m Höhe reichte. Brackwassersedimente im Liegenden und Hangenden dieser Süsswasserablagerungen bezeugen, dass einerseits vor mehr als 30 000 BP und andererseits zwischen c. 25 000 BP und dem frühen Holozän vorwiegend Uberflutungen und Austrocknung die Entstehung der Brackwassersedimente zur Folge hatte. Die Brackwasserablagerungen weisen im Profil bei Mori Mossetla (Fig. 3) wiederholt 'Kalktuff'-Lagen auf (Passarge 1904); dies darf als Hinweis gewertet werden, dass bereits vor > 30 000 BP verschiedene Süsswasserseen bestanden haben, deren Sedimente jedoch in meinen Profilen nicht mehr erfasst werden (Fig. 4).

Die Profile nördlich des Botletle-Flusses (Fig. 4) ergänzen die zuvor beschriebenen Beobachtungen. Der Seekreide-Horizont ist durchgehend zu verfolgen. Während die Seekreide und die hangenden Brackwassersedimente der Profile in 940 m Höhe deutlich die Spuren von Wurzelröhren zeigen, fehlen diese in den Seekreiden der Profile in 920 m Höhe. Dies darf als Hinweis auf die Seespiegelhöhe zur Zeit der Seekreidebildung gedeutet werden; in 940 m Höhe war Pflanzenwachstum möglich, nicht mehr jedoch wegen der c. 20 m mächtigen Wasserbedeckung in 920 m Höhe. Nach der Bildung der Seekreide wurden Brackwassersedimente akkumuliert, die mit einer Calcrete-Bildung im Hangenden abschliessen. Der Calcrete kann plattig, knollenartig oder auch als harte Kruste ausgebildet sein. Bedeckt werden die Brackwasser-Kalkmergel stets von grauen schluffig-tonigen Sanden. Verschiedene ^{14}C-Alter liegen aus den Profilen vor. Für die Seekreide ergeben sich Alter von 30 250 ± 520 BP (Hv 8382), 24 330 ± 270 BP (Hv 8364) und 20 835 ± 355 BP (Hv 8365). Das jüngste, zuletzt genannte Alter erscheint recht jung, erklärt sich aber aus den Umstand, dass in der Seekreide mitunter frische (rezente) Wurzelreste nachzuweisen sind. Das Alter der Calcrete von 13 275 ± 110 BP (Hv 8380) fügt sich in die Gruppe der oben erwähnten Calcrete-Alter ein. Auch das Alter von 13 275 ± 110 BP ist eine orientierende Altersangabe, denn die Probe setzt sich aus Kalkbildungen mit vereinzelten Sandkörnern zusammen; im Schliff sind mehrere Phasen der Kalkausscheidung zu erkennen, die jedoch einem relativ begrenzten Zeitabschnitt des Spätglacials angehören.

Die Profile nördlich des Botletle-Flusses, am unteren Thamalakane und Ngami-Fluss, dokumentieren aufgrund der Sedimente und ^{14}C-Alter, dass während des Jungquartärs zwischen c. 35 000 BP und dem Holozän vorwiegend mürbe Kalksandsteine bzw. Kalkmergel gebildet wurden, die als Brackwassersedimente auf ständige Wechsel von Uberflutungen und Trockenfallen des Gebietes hinweisen; nur die Seekreide-Lage repräsentiert einen Süsswassersee mit bis zu 20 m Tiefe in der Periode zwischen c. 30 000 und 25 000 BP. Besonders auffällig ist, dass die Seekreide keine fluvialen oder/ und äolischen Sandkörner enthält.

Innerhalb der durch den Gidikwe-Uferwall umschlossenen Makarikari-Pfannen liegt das Profil '910 m' (Fig. 4). Hier schliessen sandige geschichtete Beckensedimente mit einer Silcrete im Hangenden ab. Die Kalksandsteine sind reich an Mollusken: Muscheln aus der *Corbicula fluminalis*-Gruppe, *Melanoides (Melanoides tuberculata)* und *Unioniden* belegen eine tropische Süsswasserfauna (vgl. Passarge 1904; Brown 1978). Ein ^{14}C-Alter ergab 19 170 ± 660 BP (Hv 8366). Landschnecken, zumeist *Succineiden* und *Xerocerastus* sp., die im oberen Abschnitt gesammelt wurden, haben ein Alter von 11 920 $^{+1\,630}_{-1\,270}$ BP (Hv 8367). Die Beckensedimente werden im Hangenden durch eine Silcrete-Bildung abgeschlossen, die im Schliff recht gut gerundete Sandkörner erkennen lässt, die von $CaCO_3$-Ausscheidungen umkleidet werden, während SiO_2 die Zwischenräume ausfüllt. Die Silcrete wird von einer dünnen Lage aus Kalksinter überzogen, der in Vertiefungen der buckeligen Oberfläche auch einige cm mächtig werden kann. Das ^{14}C-Alter dieser Kalksinter ist 4 025 ± 110 BP (Hv 8689). C. 0,5 m mächtige graue Sande bedecken die harte Silcrete/Kalksinterbank.

Eine unmittelbare Korrelierung zu den Profilen Passarges (1904) kann nicht vorgenommen werden. Bemerkenswert ist jedoch, dass die jüngsten Beckensedimente Süsswassermuscheln enthalten, die auf einen Süsswassersee und/oder perennierendes Gewässer vor c. 19 000 BP weisen. Hochglaziale Seebildungen lassen sich in dem Profil nicht belegen. Inwieweit die Landschnecken mit einem Alter von c. 12 000 BP für feuchtere Verhältnisse sprechen – wie ich früher angenommen habe (Heine 1978a und b) –, kann nicht eindeutig beurteilt werden; es besteht die Möglichkeit, dass die Landschnecken vor der Bildung der Silcrete von der Oberfläche her in die Beckensedimente eingedrungen sind; in diesem Fall würden die Landschnecken-führenden Sedimente nur bedingt paläoökologisch interpretierbar sein.

C. 5 km westlich Gweta fehlen in einem Profil in 915 m Höhe die Beckensedimente. Uber 1,3 m grauer schluffreicher Sand bedeckt einen mächtigen Calcrete-Horizont, der sich auf einem fest verbackenen Schutt gebildet hat. Vermutlich stehen hier Brekzien der Botletle-Schichten i.S. Passarges (1904) an. Das Profil lässt erkennen, dass innerhalb des Makarikari-Beckens die verschiedenen Pfannen durch niedrige Gesteinsschwellen getrennt werden.

Nahe der Nata-Flussmündung in die Sua-Pfanne (~ 930 m Höhe) ist in 1-2 m Tiefe ein braun-grauer Kalkschlamm aufgeschlossen, der vereinzelte

Gerölle enthält. Charakteristisch sind weisse, gut gerundete Gerölle aus Diatomeen; dieses Material hat einen Kalkanteil von unter 5%, während das schluffige, jedoch sandfreie Beckensediment c. 40-50% $CaCO_3$ aufweist. Unter dem Mikroskop fallen wiederholt Neubildungen von feinen Calcit-Aggraten auf. Das ^{14}C-Alter einer Probe aus dem seekreideartigen Kalkschlamm ergab 23 750 ± 250 BP (Hv 8688). Uber diesem Sediment liegt ein Horizont, der durch seinen Geröllanteil hervortritt; vorwiegend dunkle Gerölle aus dem Gebiet östlich des Nata-Flusses und Gerölle aus Diatomeen befinden sich in einer kalkreichen (13% $CaCO_3$), tonig-schluffig-sandigen Matrix. Die Fraktion 0,65-0,2 mm Ø besteht hauptsächlich aus fluvial transportierten Quarzkörnern, während die Fraktionen 0,2-0,1 und 0,1-0,063 mm Ø (15 kzw. 6%) nur einen geringen Quarzanteil zeigen. In der Siltfraktion bildet sich ein zweites Korngrössenmaximum aus den Calcit-Neubildungen. Der Tonanteil erreicht 20% Diese Ablagerungen sind äusserst fossilreich; die Fauna eines Süsswassersees mit *Melanoides* (aber nicht die typische tuberculata), *Corbicula, Unioden* und *Bulinus* der tropicus-Gruppe gesellen sich zu den Strandgeröllen. Zwei ^{14}C-Alter der Mollusken ergaben 25 910 $^{+1\,210}_{-1\,000}$ BP (Hv 8371) und 27 350 ± 550 BP (Hv 8370). Ein Diatomeengeröll hat ein ^{14}C-Alter von 2 220 ± 220 (Hv 8373). Das Hangende bilden grau-schwarze Sande, die aufgrund ihrer Zurundung und Mattierung sowie ihres Korngrössenspektrums als charakteristische fluviale Sedimente angesprochen werden können, die äolisch umgelagert wurden. Die Decksande enthalten c. 1,5% $CaCO_3$.

Die Deutung der Befunde des Profils östlich der Nata-Flussmündung (Fig. 4) belegt einen Süsswassersee um 28 000-25 000 BP im Bereich der nördlichen Sua-Pfanne. Die ^{14}C-Alter der Muscheln sind dafür eindeutige Belege. Auch das Alter der liegenden Beckensedimente (23 750 ± 250 BP) ergänzt diese Befunde; die Laboranalysen ergaben, dass jüngere Calcit-Aggregate neu gebildet wurden; dies dokumentiert auch die geringe Aufkalkung der äusserst porösen Diatomeen-Gerölle und das ermittelte Alter von 2 220 ± 220 BP. Daher wird angenommen, dass die liegenden Beckensedimente ein um einige Jahrtausende zu junges Alter ergeben haben. Gleichzeitig dokumentieren die Beobachtungen und ^{14}C-Daten, dass auch in diesem Gebiet die grösste Ausdehnung eines Süsswassersees um 28 000-25 000 BP war, während danach (Hochglazial) kein Süsswassersee, aber auch keine Brackwasserseestände das 930 m-Niveau mehr erreichten. Erwähnt sei schliesslich noch der Hinweis auf etwas feuchtere Verhältnisse um 2 200 BP, die sich aus den Calcit-Mineralneubildungen ableiten lassen.

Die jungquartären pluvialen Phasen der Mittelkalahari aufgrund der Beobachtungen von Passarge und eigener Befunde

Grundlage für die Rekonstruktion der jungquartären pluvialen Phasen sind die chronostratigraphischen Beziehungen der Sedimente. Ergänzend wird die Lage der Sedimente in Beziehung zu den geomorphologischen Formen interpretiert. So werden beispielsweise die hangenden Süsswasser-Sedimente innerhalb bestimmter Strandwälle als Zeugen der jeweils jüngsten pluvialen Süsswasserseen angesprochen; die Ausdehnung der Süsswasserseen ergibt sich dann aus der Verbreitung der Süsswasserablagerungen bzw. aus den Verlauf der Uferwälle, die diese Sedimente umschliessen. Die Uferwälle bzw. Strandwälle können in ihrer Anlage älter oder auch jünger als die Süsswasserseen sein, denn die Strandwälle wurden während der Brackwasserphasen gebildet, die den Süsswasserphasen vorausgingen bzw. nachfolgten. Genetisch dürfen die Strand- bzw. Uferwälle nicht mit den Süsswassersee-Sedimenten verknüpft werden. Die Aussagen über die Ausdehnung der Süsswasserseen sind daher teilweise hypothetisch.

Pluviale Süsswasserseen sind für das Botletle-Gebiet westlich Matabele Drift zwischen c. 30 000 und 25 000 BP und zwischen 28 000 und 25 000 BP für die nördliche Sua-Pfanne nachgewiesen worden (s.o.). Unsicher bleibt vorerst die Rekonstruktion des Seespiegels für diese innerwürmzeitlichen Pluvialseen. Sowohl die Profile in 940 m Höhe am Botletle als auch das Profil in c. 930 m Höhe bei Nata liegen im Bereich der Uferzonen ehemaliger Seen. Die Differenz von c. 10 m kann eine Erklärung in dem Umstand finden, dass das gesamte Gebiet des Okavango-Beckens und der Makarikari-Pfannen stark tektonisch gestört ist. SW-NE-streichende Verwerfungen haben zu einer Verstellung einzelner Schollen geführt; bereits Passarge (1904: 214 f.) vermutet eine Schwelle zwischen dem Thamalakane und Moremis Drift am Botletle aufgrund der dort auskeilenden Kalaharikalke. Reeves (1974; vgl. auch Breyer 1982, Jones 1982, Mallick et al. 1981) belegt Störungen anhand von Gravitätsmessungen. Tektonische Aktivität lässt sich im Okavango-Graben seit dem Präkambrium bis in jüngsten Zeiten nachweisen (Dingle et al. 1983: 293), weshalb hier nicht völlig ausgeschlossen werden kann, dass infolge tektonischer Verstellungen die Höhen von 940 m und 930 m ein und demselben Seespiegel eines innerwürmzeitlichen Süsswassersees zuzuordnen sind.

Nach 25 000 BP müssen die Süsswasserseespiegel gefallen sein. Bis c. 19 000 BP können aufgrund der Befunde sich ständig verkleinernde Süsswasserseen existiert haben, die zuletzt im Makarikari-Becken nur noch bis c. 910 m Höhe reichten. Spätestens nach 19 000 BP wurden nur noch Brackwassersedimente gebildet; gleichzeitig erfolgte eine Sandanwehung am Gidikwe Ridge. Das Hochglazial weist keine Süsswassersedimente und Faunen auf. Damit erscheint mir der Beweis erbracht, dass im Ngami- und Makarikari-Gebiet keine hochglazialen Süsswasserseen existierten. Im Botletle-Gebiet bedarf die Mächtigkeit und die absolute Höhenlage der Brackwasserablagerungen der Periode zwischen c. 19 000 und c. 12 000 BP einer Erklärung. Passarge (1904) spricht von einer Periode der Kalklösungen und ordnet diese Zeit einer feuchteren Klimaphase im Vergleich zu heute zu. Demnach müssen wir annehmen, dass das Hochglazial zwar keine Süsswasserseen hervorbrachte, jedoch wiederholt ausgedehnte Brackwasserseen kannte. Der mitunter relativ hohe Anteil an roten, äolisch in die hochglazialen Brackwassergebiete eingetragenen Sande, die den fossilen roten Dünen entstammen, bezeugt einerseits hohe Windgeschwindigkeiten mit Sandtransport und andererseits starke

Evaporation. Im Spätglazial (c. 14 000-12 000 BP) endet die Brackwassersedimentation und eine schwache Clacrete-Bildung setzt in den höher gelegenen Bereichen ein, während die am tiefsten gelegenen Sedimente vermutlich infolge immer noch auftretender Uberflutungen keine Calcrete-Krusten oder Kalksinterüberzüge aufweisen.

Besonders wichtig sind die Beobachtungen von Passarge (1904), dass die liegenden Partien des Kalaharisandes in den Profilen östlich des Gidikwe Ridge (Fig. 3) kalkhaltig sind und sowohl Assoziationen von Brackwasser- als auch Süsswasserdiatomeen führen. Hierin sehe ich mit Passarge (1904: 220) einen Hinweis auf Brackwasserbecken, die von Süsswasserflüssen überflutet wurden, Ereignisse, die nach Beendigung der Brackwasser-Sedimentbildung und nach bzw. während der spätglacialen Calcrete-Bildung auftraten und weite Bereiche der Makarikari-Pfannen innerhalb des Gidikwe Ridge-Strandwallsystems erfassten. Vielleicht ist der Hinweis auf feuchtere Verhältnisse um 12 000 BP in dem Profil '910 m' (Fig. 4) mit diesen Süsswasserereignissen in Zusammenhang zu bringen. Zur Bildung flächen hafter Süsswassersedimente (z.B. Seekreide) kam es nicht mehr. Daher dürfen die Indizien für das (oder die) Süsswasserereignis(se) des Spätglazials und/oder frühen Holozäns paläoklimatisch nicht überbewertet werden.

Das Holozän selbst kann anhand der limnischen Sedimente und geomorphologischen Erscheinungen nicht wesentlich gegliedert werden. Allein um 2 200 BP ergeben sich Hinweise auf geringe geoökologische Veränderungen infolge einer leichten Verstärkung der Humidität.

Der Vollständigkeit halber sei angemerkt, dass aus den Beschreibungen von Passarge (1904) in Verbindung mit meinen Datierungen hervorgeht, dass vermutlich im Ngami-Gebiet vor über 30 000 BP weitere Süsswassersee-Phasen bestanden haben und dass andererseits die Sedimente für grosse Abschnitte des Mittelwürms gleiche oder ähnliche Brackwasserseen belegen, wie sie für das Hochglazial charakteristisch waren. Somit sind die Phasen der Süsswasserseen 'Ausnahmesituationen' während der letztkaltzeitlichen Entwicklung. Das Holozän unterscheidet sich von der letzten Kaltzeit dadurch, dass weder grössere, länger andauernde Süsswassersee-Phasen noch Brackwassersee-Phasen existierten. Diese Befunde veranlassten Passarge (1904) von einer grossen Pluvialzeit und deren Abklingen bis zur Gegenwart (= Austrocknung des Inneren Südafrika) zu sprechen.

Argumente für hochglaziale pluviale Seespiegel?

Gegen die Vorstellung, dass die maximalen letztkaltzeitlichen Seenbildungen vor dem letzten Hochglazial auftraten, wandte sich Cooke (1979: 107). Dabei bezog sich Cooke (1979) auf eine Anzahl von ^{14}C-Daten von Höhlensintern und Calcretes in Tälern der Kwihabe Berge im nordwestlichen Ngamiland etwa 300 km westlich des Makarikari-Beckens (Cooke & Verhagen 1977, Grey & Cooke 1977). Diese Daten sollen nach Cooke (1979) feuchte Bedingungen zwischen 16 000 und 13 000 BP anzeigen. Auch wird von Cooke (1979) bezweifelt, dass der ältere See ausgedehnter war als der jüngere; dies wird nicht stratigraphisch, sondern geomorphologisch belegt: Die Uferbereiche des kleineren Sees sind stärker zerschnitten und damit älter als die weniger zerstörten Uferformen des grösseren Sees. Inzwischen hat Cooke (1984: 276 f.) ebenfalls den höheren Seespiegel (945 m) als den älteren erkannt, den niedrigeren (920 m) als den jüngeren, ungeachtet seiner früheren morphologischen Schlussfolgerung (Cooke 1979).

Problematisch bleiben jedoch die Datierungen der verschiedenen pluvialen Phasen bei einem Vergleich zwischen meinen Ergebnissen und denen von Cooke (zuletzt 1984).

Die von Cooke (1984) publizierte Chronologie der Makarikari-Seespiegelschwankungen basiert auf folgenden Annahmen: 1. Wenn Calcretes im Makarikari-Becken in tiefen Lagen gebildet werden, muss der Seespiegel entsprechend niedrig gewesen sein; 2. Calcretes der Strandterrassen wurden ± in einer Höhenlage gebildet, in der der Seespiegel lag; und 3. Calcretes im Liegenden von Strandwällen und Dünen sind älter als die sie bedeckenden Sedimente. Damit wird die Chronologie der Seespiegelschwankungen nicht stratigraphisch und in Verbindung mit einer paläontologischen sowie sedimentologischen Bearbeitung der Sedimente ermittelt, sondern rein morphologisch, nämlich aufgrund der Höhenlage der Calcretes. In Figur 5 sind die Daten Cookes (1984) erfasst. Da die Altersbestimmung von Calcretes mittels ^{14}C immer nur orientierende Angaben liefern kann (Geyh 1983, Netterberg 1978, Blümel 1982), können die absoluten Werte von Cooke (1984) nur sehr eingeschränkt Verwendung finden. Zum anderen können die ermittelten Calcrete-Alter von über 25 000 BP bis > 46 000 BP wohl kaum in der Weise interpretiert werden, dass das jeweilige Alter auch das Jahrtausend der Calcrete-entstehung bei entsprechend niedrigen Seespiegeln ausweist (vgl. Arakel 1986). Schon Passarge (1904: 505 ff.) beschreibt beispielsweise aus der Pfanne von Ntschokutsa (= Rysana Pan bei Cooke (1984: 274)) verschiedene Kalksandsteine und Sinterkalke; wenn Cooke (1984) eine Calcrete vom Ostrand dieser Pfanne in 911 m Höhe auf $46\,200\,^{+1\,700}_{-1\,400}$ BP datiert, daraus niedrige Seespiegel um 46 000 BP und ein subhumides bis semi-arides Klima ableitet, dann zeigt sich die Fragwürdigkeit dieses methodischen Vorgehens.

Allein die ^{14}C-Alter der Proben (Cooke & Verstappen 1984), die nicht aus Calcrete bestehen, führen zu ergänzenden Aussagen. Südlich von Tsoi wurden weicher Calcrete (= mürber Kalksandstein = Brackwassersediment i.S. Passarges) auf 19 680 ± 100 BP und 19 420 ± 160 BP und ein Torf auf 19 160 ± 250 BP datiert (Cooke (1984) gibt unter denselben Labornummern leicht abweichende Alter an). Eindeutig belegt die Torfprobe einen niedrigen Seespiegel um 19 600 BP, was den eigenen Befunden entspricht: Austrocknung eines Süsswassersees um c. 19 000 BP. Gleichzeitig dokumentieren die Angaben von Cooke (1984) dass nach c. 19 000 BP keine Brackwassersedimente mehr über dem Torf bzw. den 'Soft Calcretes' sedimentiert wurden. Somit finden sich hier Hinweise auf eine gegenüber dem Botletle-Gebiet differenzierte Entwicklung. Bei Tsoi bedecken fluviale Deltasedimente und äolische Sande die

Torfe und 'Soft Calcretes' (Cooke & Verstappen 1984). Hochglaziale Brackwassersedimente sind nicht ausgebildet; das Hochglazial scheint durch die äolischen Sande repräsentiert zu werden.

Bei Zhumo werden 'Soft Calcretes' eines früheren Flussufers auf 3 130 ± 50 BP, bei Toromoja Knochen einer Terrasse auf 1 590 ± 70 BP und Muscheln einer Terrasse der Gwi-Pfanne auf 1 710 ± 35 BP datiert. Alle Daten fallen in das jüngere Holozän.

Den Beobachtungen von Cooke (1984) und Cooke & Verstappen (1984) ist zu entnehmen, dass der Gidikwe Ridge verschiedene Calcrete-Horizonte enthält; auch Breyer (1982) gibt dazu Auskunft. Nach Breyer (1982) wird der Gidikwe Ridge 35 km westlich von Rakops aus Fein- bis Mittelsanden aufgebaut, die in 2 bis 4 m und in 10 bis 30 m Tiefe von weissem Calcrete abgelöst werden. Bei diesem 'weissen Calcrete' handelt es sich um den Kalksandstein Passarges (1904), den dieser unter dem 'Grasland' festgestellt hat. Ob diese Sedimente ein innerwürmzeitliches Alter haben, ist ungewiss. Aus den Befunden geht jedoch hervor, dass der Gidikwe Ridge in seiner Anlage älter als die mittelwürmzeitlichen Brackwasserablagerungen ist und dass der Gidikwe Ridge mehrere Phasen der Bildung aufweist. Er trennt die Brackwasser- und Süsswassergebiete des Ngami-Botletle-Gebiets von denen der Makarikari-Pfannen. Daher zeigen die Profile westlich und östlich des Gidikwe Ridge verschiedene lokale Abfolgen; für das westliche Gebiet (Ngami-Botletle) ergibt sich eine stärkere Differenzierung zwischen Süsswasser- und Brackwasserphasen, vermutlich infolge wiederholter Staus des Botletle-Wassers vor dem (sich während der Brackwasserphasen weiterbildenden) Gidikwe Ridge. Angemerkt sei hier auch, dass Passarge (1904: 392) dieselben Beobachtungen von den Strandwällen westlich des Ngami-Sees beschreibt, wo ebenfalls 'härtere Kalksandsteine' die Böschung bilden.

Shaw (1986) nennt einen ausgedehnten See im Bereich der Ngami-Mababe-Achse, der auf c. 17 000-12 000 BP datiert wird. Da Shaw (1986) methodisch in seinen Arbeiten wie Cooke vorgeht, sollen einige Anmerkungen zu seinen publizierten ^{14}C-Daten gemacht werden. 'Lacustrine bed calcretes' – sie entsprechen den Kalksandsteinen Passarges (1904) –, die von jüngeren Sedimenten bedeckt werden, haben im westlichen Ngami-Becken (Setsau-Bay) (Fig. 5) ein Alter von 23 900 ± 500 BP und in der Goha-Pfanne des Mababe-Beckens 25 850 ± 500 BP. 'High level estuarine sediments' des Savuti-Flusses und des Ngwezumba-Flusses vor Eintritt in die Mababe-Depression wurden auf 11 950 ± 110 BP, 13 070 ± 140 BP und 17 190 ± 210 BP datiert. Schliesslich liegen ^{14}C-Alter für jüngere holozäne 'lacustrine bed calcretes' aus dem Ngami-Becken vor: 6 440 ± 110 BP, 6 445 ± 35 BP und 8 920 ± 140 BP.

Weder Shaw (1986) noch Cooke (1984) nehmen meine Unterscheidung der limnischen Ablagerungen in Brackwassersedimente und Süsswassersedimente vor, wie es Passarge (1904) aufgrund der bis heute umfangreichsten mineralogischen, paläontologischen und mikropaläontologischen Untersuchungen getan hat. Daher kann nicht beurteilt werden, ob die 'lacustrine bed calcretes' Brackwasser- oder/und Süsswasserablagerungen sind, was für deren

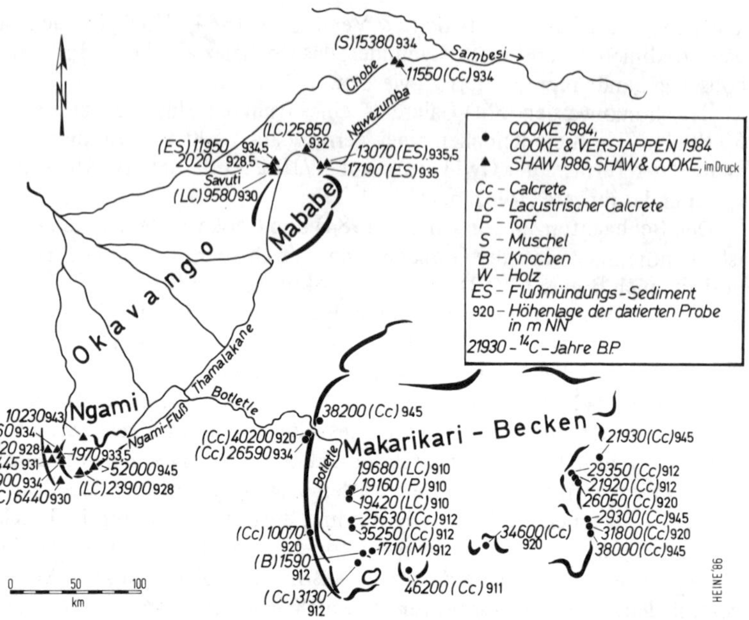

Figur 5. ^{14}C-Alter in Abhängigkeit von Höhenlage und Fundregion, zusammengestellt aufgrund der Arbeiten von Cooke (1984), Cooke & Verstappen (1984), Shaw (1986) und Shaw & Cooke (im Druck)

Deutung hinsichtlich niedriger oder hoher Seespiegel unerlässlich ist. Die ^{14}C-Alter von c. 24 000 und 26 000 BP können somit nicht interpretiert werden. Wenn man sich der Beurteilung von Shaw (1986) anschliesst, wären um 26 000-24 000 BP im Ngami- und Mababe-Becken bereits Brackwasserverhältnisse, d.h. niedrige Seespiegel ausgebildet.

Die 'high level estuarine sediments' am Rande der Mababe-Pfanne, die von Shaw (1986) als Belege für hohe Seespiegel gedeutet werden, die wiederholt zwischen 17 000 und 12 000 BP auftraten, wurden zur Zeit der Brackwassersedimente, die vom Ngami-Fluss und Botletle beschrieben wurden, abgelagert. Die 'high level estuarine sediments' und die jüngsten, bis nahezu 940 m Höhe reichenden Brackwasserablagerungen am Botletle (Fig. 4) gehören dem letzten Hochglacial an; in den Makarikari-Pfannen gibt es keine vergleichbaren Ablagerungen. Damit wird erneut das Problem aufgeworfen, inwieweit – möglicherweise tektonisch bedingt (vgl. Cooke 1980) – das Gebiet des östlichen Okavango-Deltas zwischen Ngami-Becken und Mababe-Depression einen vom Gebiet der Makarikari-Pfannen abweichenden Rhythmus der Seespiegelschwankungen aufwies.

Nicht eindeutig können die 'lacustrine bed calcretes' der Dautsa Flats des

Ngami-Beckens (Fig. 5) interpretiert werden; Shaw (1986) sieht in ihnen Hinweise auf einen niedrigen Seespiegel. Für das Holozän stimmen die Angaben über einen höheren Seespiegelstand sowohl im Ngami- als auch im Mababe-Becken zwischen c. 2 000-1 500 BP mit den o.a. Indikatoren feuchterer Verhältnisse fast überein.

Folgende Hinweise auf Seespiegelschwankungen können den Arbeiten von Cooke (1984), Cooke & Verstappen (1984) und Shaw (1986) entnommen werden, wenn allein die zweifelsfrei paläoökologisch interpretierbaren Befunde bewertet werden:

1. Zwischen 26 000 und 24 000 BP herrschten im Ngami- und Mababe-Becken Brackwasser- und/oder Süsswasserablagerungen vor.
2. Seit 19 000 BP werden im Makarikari-Pfannengebiet keine Süss- und Brackwassersedimente mehr abgelagert.
3. Zwischen c. 17 000 und 12 000 BP werden im Mababe-Randbereich Flussmündungssedimente akkumuliert.
4. Im Holozän werden zwischen 9 000 und 6 000 BP im Ngami-Becken Brack- und/oder Süsswassersedimente gebildet.
5. Zwischen c. 2 000 und 1 500 BP gibt es aus verschiedenen Pfannen Hinweise auf feuchtere Verhältnisse.

Paläoklimaindikatoren aus benachbarten Gebieten

Stalaktiten aus Karsthöhlen der Kwihabe-Berge (Cooke & Verhagen 1977; Shaw & Cooke, im Druck) wurden aufgrund der vorliegenden ^{14}C-Daten vorwiegend zwischen 45 000-37 000 BP, 34 000-29 000 BP und 16 000-13 000 BP und während des Holozäns zwischen c. 6 000 und 750 BP gebildet (Fig. 2).

Archäologische Untersuchungen der Gi-Pfanne nördlich der Aha-Berge geben Aufschluss über Chronostratigraphie der Gi-Pfannensedimente. Zwei Phasen mit lakustrischen Kalkablagerungen werden auf über rund 31 000 BP und über rund 22 000 BP datiert (Helgren & Brooks 1983). Hinweise auf jüngere limnische Sedimente liefert die Gi-Pfanne nicht.

Aus dem Gebiet der südwestlichen Makarikari-Pfannen beschreibt Helgren (1984) Funde von 'Early Stone Age'-Artefakten, die auf der 945 m-Strandlinie liegen; das 'Early Stone Age' endete im südlichen Afrika vor über 200 000 Jahren. Weiterhin werden für das frühe und mittlere Holozän Deltaschüttungen des Botletle im Gebiet zwischen Rakops und dem Xau-See sowie wiederholte seichte Seebildungen und Calcrete-Bildung im 912 m-Niveau genannt. Auch das späte Holozän weist noch – bei etwas stärkerer Austrockung der Pfannen – leichte Schwankungen der Wasserführung des Botletle auf.

Da in der vorliegenden Zusammenstellung die direkten Zeugen jungquartärer Seespiegelschwankungen vergleichend diskutiert werden sollen, wird darauf verzichtet, weitere – vor allem indirekte und nicht hinreichend datierbare – Paläoenvironment-Indikatoren zu nennen.

ZUSAMMENFASSUNG DER BEFUNDE

Die Rekonstruktion jungquartärer Seespiegelschwankungen in der Mittleren Kalahari wurde auf zwei verschiedenen Wegen versucht, die sich methodisch grundsätzlich voneinander unterscheiden:

1. Passarge (1904) und Heine (1978a und b, 1979, 1982) erarbeiten eine Stratigraphie der jungquartären Sedimente. Vor allem auf der Grundlage der detaillierten Angaben von Passarge (1904) kann Heine (1982) eine grobe Chronostratigraphie der limnischen Ablagerungen geben, die pluvialen Seehochständen zugeschrieben werden können. Da die pluvialen jungquartären Seen der Mittleren Kalahari geologisch-paläontologisch-sedimentologisch und stratigraphisch ermittelt wurden, konnten diese bisher nur hypothetisch dem geomorphologischen Formenschatz (wie Strandwälle, Seespiegelniveaus) zugeordnet werden. Aussagen waren über die Ausdehnung der jungquartären Pluvialseen mit gewissen Einschränkungen möglich, nicht jedoch über die Höhe der verschiedenen Seespiegelstände.

2. Cooke (1984), Cooke & Verstappen (1984) und Shaw (1986) ermitteln jungquartäre Pluvialseen in der Mittleren Kalahari mit morphologischen Methoden, d.h. aus der Höhenlage von Calcretes und der paläoklimatischen Interpretation der Calcrete-Bildung. Ergänzend werden lakustrische und Flussmündungssedimente berücksichtigt.

Die sedimentologischen und paläontologischen Untersuchungen haben ergeben, dass in der Mittleren Kalahari im Ngami- und Makarikari-Gebiet einerseits Brackwassersedimente gebildet wurden, und zwar bei ständig sich wiederholenden Überflutungen und Trockenfallen der Pfannen, und andererseits Süsswassersedimente akkumulierten, die längere Phasen relativ starker Süsswasserzufuhr in die Pfannengebiete dokumentieren. Bei Cooke (1984), Cooke & Verstappen (1984) und Shaw (1986) wird nicht zwischen jungquartären Phasen mit Süsswassersee-Bildung und Brackwasserverhältnissen unterschieden (Fig. 2). Heine (1978a und b, 1979, 1982) hat bisher nur die Süsswassersee-Phasen als pluviale Phasen ausgewiesen, während die Brackwasser-Phasen nicht den pluvialen Episoden zugerechnet wurden.

Eine erneute kritische Bewertung der vorliegenden Literatur in Verbindung mit weiteren sedimentologischen Untersuchungen kommt zu wesentlich differenzierten Vorstellungen über die jungquartäre Geschichte der Seespiegelschwankungen in der Mittleren Kalahari und deren paläoklimatischen Informationsgehalt (Fig. 6).

Die ältesten datierten Süsswassersedimente haben ^{14}C-Alter zwischen 30 000 und 19 000 BP. Daraus wurde von mir früher eine pluviale Phase abgeleitet. Aufgrund der heute vorliegenden Erkenntnisse muss diese Anschauung modifiziert werden. Im Makarikari-Pfannen-Gebiet sind Süsswassermuscheln bei Nata und im Profil '910 m' bestimmt und datiert worden; diese Süsswasserfaunen entstammen keinen mächtigen limnischen Schichtpaketen, sondern geringmächtigen Lagen, die bei Nata Uferfazies haben. Es ist nicht auszuschliessen, dass die Süsswassermollusken-Faunen der Makarikari-

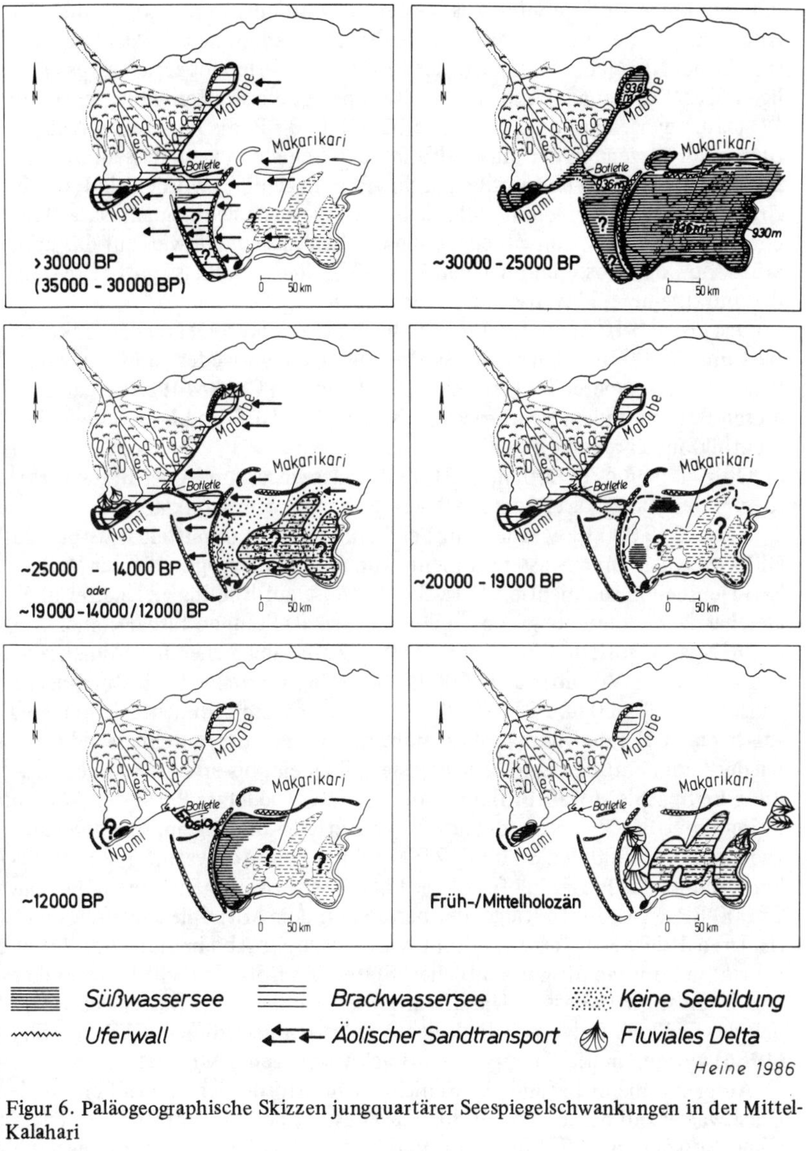

Figur 6. Paläogeographische Skizzen jungquartärer Seespiegelschwankungen in der Mittel-Kalahari

Pfannen keine ausgedehnten Süsswasserseen dokumentieren, sondern als Ablagerungen perennierender Flüsse gedeutet werden müssen. Dagegen spricht bei Nata jedoch die Lage des Profils im Uferlinienbereich eines ehemaligen Sees. Die Fauna des Profils bei Nata belegt Süsswasserbedingungen; die ^{14}C-Daten weisen auf die Zeit um 25 000-28 000 BP; die limnischen Ablagerungen im Liegenden der muschelführenden Schicht zeigen, dass ältere limnische Sedimente in der nördlichen Sua-Pfanne gebildet worden sind. Die bei Grove (1969: 203) erwähnte Middle Stone Age-Jägersiedlung am Nata-Delta ohne wesentliche Sedimentbedeckung ist ebenfalls ein Hinweis auf das prä-MSA-Alter des grösseren pluvialen Makarikari-Sees, und dass post-MSA-zeitlich nur 'kleinere' Pluvialseen existiert haben.

Das Profil '910 m' in der nordwestlichen Makarikari-Depression dokumentiert um 19 000 BP noch Süsswasserbedingungen, entweder eines grösseren Sees oder perennierender Gewässer. Die Daten von Cooke (1984) ergänzen diesen Befund, indem sie um 19 000 BP bei Tsoi Torf- und limnische Sedimentbildung erkennen lassen.

Ob zwischen c. 25 000 und 19 000 BP Süsswasser- und/oder Brackwasserseen in der Makarikari-Pfanne existierten, bleibt vorerst unklar.

Nach c. 19 000 BP wurden nur bei Meno a Kwena unmittelbar östlich des Gidikwe Ridge Brackwassersedimente von c. 2 m Mächtigkeit in der Makarikari-Depression sedimentiert (Fig. 3). Hinweise auf Bildung gleichalter und gleichartiger Sedimente gibt es in den Makarikari-Pfannen bisher nicht.

Im Ngami/Botletle-Fluss-Gebiet tritt uns eine abweichende Sedimentabfolge entgegen. Vor über c. 30 000 BP herrschten Brackwasserbedingungen; zwischen c. 30 000 und 25 000 BP wurden Süsswassersedimente gebildet; es folgte eine Calcrete-Bildungsphase unbestimmter Dauer; anschliessend trat wieder Sedimentbildung im Brackwassermilieu ein, bis erneute Calcrete-Bildung im Spätpleistozän/Frühholozän einsetzte. Die letzte Phase der Akkumulation unter Brackwasserbedingungen im Botletle- und Ngami-Fluss-Gebiet zwischen c. 19 000 und 14 000/12 000 BP hat kein Aquivalent in den Makarikari-Pfanne. Als Erklärung für dieses Phänomen wird angenommen, dass nach 25 000 BP der Gidikwe Ridge, der bereits seit dem Mittelpleistozän bestand (Helgren 1984: 301), erneut erhöht wurde, was zur Abdämmung des Botletle führte; so konnten die gelegentlichen Fluten des Botletle nicht in die Makarikari-Depression abfliessen. Die Bildung wiederholter Brackwasserseen westlich des Gidikwe Ridge war die Folge. Die 'estuarine sediments' i.S. von Shaw (1986) wurden in diese temporären Brackwasserseen geschüttet.

Aufgrund dieser Befunde ist es nicht mehr erforderlich, die Höhenlage der Brackwassersedimente und der 'high level estuarine sediments' mit jeweils einem ausgedehnten See im Makarikari-Ngami-Mababe-Gebiet (Seespiegelhöhe in 945 bzw. 936 m) zu erklären; fragwürdige Wasserhaushaltsberechnungen für das letzte Hochglazial (vgl. u.a. Ebert & Hitchcock 1978, Shaw 1986) werden damit überflüssig.

Nach Bildung der hochglazialen Brackwassersedimente westlich des Gidikwe Ridge konnte – vielleicht mit einer geringen Zunahme des Süsswasserangebot

– der Gidikwe Ridge vom Botletle durchbrochen werden. Der Botletle schuf seine cañon- und kanalartigen Talabschnitte, indem er sich in die Brackwassersedimente (= hangende Kalaharikalke i.S. Passarges (1904)) einschnitt. Im Makarikari-Becken ist diese Phase durch die liegenden, klakreichen und Süss- wie Brackwasserdiatomeen-führenden Kalaharisande belegt, die mit scharfer Grenze den älteren Makarikari-Pfannen-Brackwassersedimenten aufliegen.

Für das Holozän ist eine teilweise Erosion der Kalaharikalke und Kalaharisande in der westlichen Makarikari-Depression belegt, ausserdem die Schüttung verschiedener flacher Schwemmfächer des Botletle in die westliche Depression, sowie Fächerschüttungen des Okwa im Südwesten und Nata im Nordosten. Mehrere kurze limnische Phasen sind für alle Becken (Ngami, Mababe, Makarikari) nachgewiesen; nur um c. 2 000-1 500 BP kann mit einer klimatisch gesteuerten feuchteren Phase gerechnet werden.

Schliesslich sei hervorgehoben, dass die Beobachtungen von Passarge (1904) bereits andeuten, dass die Ngami- und Makarikari-Becken relativ alter Entstehung sind, denn sie sind mit Sedimenten gefüllt, die nicht nur dem Jungquartär angehören. Die Hinweise Passarges (1904) auf das Vorkommen von Botletle-Schichten im Liegenden der grossen Sandwälle (= Strandwälle) bezeugen die alte Anlage der Pfannen und deren Uferwälle. Die 945 m-Uferlinie wird daher mit Recht von Helgren (1984) in ihrer ersten Anlage in das späte Mittelpleistozän (> 200 000 a) gestellt. Die Entstehung der abflusslosen Pfannen der Mittelkalahari kann das Ergebnis pluvialer endorheischer Abtragung in epirogenetisch (Depressionen) und stratigraphisch (Kalahari Beds, vgl. Dingle et al. 1983: 290 ff.) vorgegebener Lage sein, wie es kürzlich Rust (1985) von der Etoscha-Pfanne beschrieben hat.

Abschliessend darf nicht unerwähnt bleiben, dass die vorliegende Zusammenschau die bisherigen Vorstellungen starker hygrischer Klimaschwankungen während des Jungquartärs relativiert. Zweifellos sind die Süsswassersedimente Hinweise auf eine grössere Humidität, die durch veränderte jungquartäre Niederschlags/Evaporations-Verhältnisse bedingt sind; da während der Süsswasserperioden die äolische Morphodynamik unbedeutend war, müssen die ökologischen Verhältnisse humider als heute gewesen sein.

Auch die Brackwasserperioden unterscheiden sich von der gegenwärtigen geoökologischen Situation der Mittel-Kalahari. Einerseits bezeugen sie stärkere Kalkabscheidung in den Brackwasserseen. Passarges (1904) 'Perioden der Kalklösungen' werden durch die Untersuchungen von Watts (1980) an Sedimenten der südwestlichen Makarikari-Pfannen bei Mopipi bestätigt. Inwieweit die Vorgänge der Kalkbildung noch anhalten, ist nicht bekannt.

Eine paläoklimatische Ausdeutung der geschilderten Befunde soll an dieser Stelle vorerst nicht erfolgen. Betont werden muss, dass das Klima im Jungquartär im ganzen semi-arid war; Episoden zeigten mitunter ein etwas humideres oder auch etwas arideres Klima. Ob diese graduellen Schwankungen durch eine Veränderung des thermischen Trends oder durch eine Veränderung des hygrischen Trends oder durch beides während der jeweiligen Phasen verursacht wurden, kann nicht entschieden werden. Feuchtere Verhältnisse scheinen an Phasen abkühlungszeitlicher Intensivierung (30 000-25 000 BP) und erwärmungszeitlicher Abschwächung (14 000-10 000 BP) des globalen thermischen Gradienten im südlichen Afrika gebunden zu sein; das letzte Hochglacial war keine besonders feuchte Phase.

DANKSAGUNG

Ich danke der Deutschen Forschungsgemeinschaft für Reise- und Sachbeihilfen, Herrn M.A.Geyh für zahlreiche ^{14}C-Analysen und Diskussionen, und Herrn R.Huckriede für Mollusken-Bestimmungen. Besonders verpflichtet bin ich Herrn E.M.van Zinderen Bakker Sr. und Frau J.A.Coetzee für mannigfache Hilfe bei meinen Arbeiten und anregende Erörterungen. Schliesslich gilt mein Dank meiner Frau und meinen Kindern, die am Gelingen der Geländearbeiten grossen Anteil hatten.

LITERATUR

Arakel, A.V. 1986. Evolution of calcretes in Palaeodrainages of the Lake Napperby area, central Australia. *Palaeogeogr., Palaeoclimat., Palaeoecol.* 54: 283-303.

Blümel, W.D. 1982. Calcretes in Namibia and SE-Spain. Relations to substratum, soil formation and geomorphic factors. *Catena Suppl.* 1: 67-82.

Bond, G. & R.F.H.Summers 1954. A late Still Bay hunting camp site on the Nata River, B.P. *S.Afr. arch. Bull.* 9: 89-95.

Breyer, J. 1982. Reconnaissance geomorphological terrain classification, lower Boteti region, northern Botswana. *ITC Journal* 1982 (3): 317-323.

Brown, D.S. 1978. Freshwater molluscs. In M.A.Werger (ed.), *Biogeography and ecology of Southern Africa*: 1153-1180. The Hague: Junk.

Cooke, H.J. 1975. The palaeoclimatic significance of caves and adjacent landforms in western Ngamiland, Botswana. *Geogr. J.* 141: 430-444.

Cooke, H.J. 1979. K.Heine: Radiocarbon chronology of Late Quaternary lakes in the Kalahari, Southern Africa. A discussion. *Catena* 6: 107.

Cooke, H.J. 1980. Landform evolution in the context of climatic change and neo-tectonism in the Middle Kalahari of north-central Botswana. *Trans. Inst. Brit. Geogr. NS* 5: 80-99.

Cooke, H.J. 1984. The evidence from northern Botswana of Late Quaternary climatic change. In J.C.Vogel (ed.), *Late Cainozoic palaeoclimates of the southern hemisphere*: 265-278. Rotterdam: Balkema.

Cooke, H.J. & B.Th.Verhagen 1977. The dating of cave development – an example from Botswana. In *Proc. 7th Intern. Spel. Congr. (Sheffield, UK)*: 122-124.

Cooke, H.J. & H.Th.Verstappen 1984. The landforms of the western Makgadikgadi basin in northern Botswana, with a consideration of the chronology of the evolution of Lake Palaeo-Makgadikgadi. *Z. Geomorph. N.F.* 28: 1-19.

Deacon, J., I.N.Lancaster & L.Scott 1984. Evidence for Late Quaternary climatic change in southern Africa. Summary of the proceedings of the SASQUA Workshop held in Johannesburg, September 1983. In J.C.Vogel (ed.), *Late Cainozoic palaeoclimates of the southern hemisphere*: 391-404. Rotterdam: Balkema.

Dingle, R.V., W.G.Siesser & A.R.Newton 1983. *Mesozoic and Tertiary geology of Southern Africa.* Rotterdam: Balkema. 375 pp.

Ebert, J.I. & R.K.Hitchcock 1978. Ancient Lake Makgadikgadi, Botswana: mapping, measurement and palaeoclimatic significance. *Palaeoecol. of Africa* 10: 47-56.

Geyh, M.A. 1983. Physikalische und Chemische Datierungsmethoden in der Quartärforschung. *Clausthaler Tektonische Hefte* 19: 1-163.
Grey, D.R.C. & H.J.Cooke 1977. Some problems in the Quaternary evolution of the landforms of northern Botswana. *Catena* 4: 123-133.
Grove, A.T. 1969. Landforms and climatic change in the Kalahari and Ngamiland. *Geogr. J.* 135: 191-212.
Heine, K. 1977. Beobachtungen und Uberlegungen zur eiszeitlichen Depression von Schneegrenze und Strukturbodengrenze in den Tropen und Subtropen. *Erdkunde* 31: 161-178.
Heine, K. 1978a. Jungquartäre Pluviale und Interpluviale in der Kalahari (südliches Afrika). *Palaeoecol. of Africa* 10: 31-39.
Heine, K. 1978b. Radiocarbon chronology of Late Quaternary lakes in the Kalahari, Southern Africa. *Catena* 5: 145-149.
Heine, K. 1979. Reply to Cooke's discussion of: K.Heine: Radiocarbon chronology of Late Quaternary lakes in the Kalahari, Southern Africa. *Catena* 6: 259-266.
Heine, K. 1982. The main stages of the Late Quaternary evolution of the Kalahari region, Southern Africa. *Palaeoecol. of Africa* 15: 53-76.
Helgren, D.M. 1984. Historical geomorphology and geoarchaeology in the Southwestern Makgadikgadi Basin, Botswana. *Ann. Ass. Amer. Geogr.* 74: 298-307.
Helgren, D.M. & A.S.Brooks 1983. Geoarchaeology at Ģi, a Middle Stone Age and Later Stone Age site in the Northwestern Kalahari. *J. Archaeol. Sci.* 10: 181-197.
Jones, C.R. 1982. The Kalahari of Southern Africa. *Striae* 17: 20-34.
Lancaster, I.N. 1979a. Evidence for a widespread late Pleistocene humid period in the Kalahari. *Nature* 279: 145-146.
Lancaster, I.N. 1979b. Quaternary environments in the arid zone of southern Africa. *Environmental Studies, Univ. of Witwatersrand Occ. Pap.* 22: 1-77.
Lancaster, I.N. 1981. Palaeoenvironmental implications of fixed dune systems in southern Africa. *Palaeogeogr., Palaeoclimat., Palaeoecol.* 33: 327-346.
Mallick, D.I.J., F.Habgood & A.C.Skinner 1981. A geological interpretation of Landsat imagery and air photography of Botswana. *Overseas Geol. and Min. Resources, Inst. Geol. Sci. (London)* 56: 1-35.
Netterberg, F. 1978. Dating and correlation of calcretes and other pedocretes. *Trans. geol. Soc. S.Afr.* 81: 379-391.
Passarge, S. 1904. *Die Kalahari. Versuch einer physisch-geographischen Darstellung der Sandfelder des südafrikanischen Beckens.* Berlin: Reimer. 822 pp. + Kartenband.
Reeves, C.V. 1974. The gravity survey of Ngamiland 1970-71. Part II: Interpretation. Rep. Geol. Surv. Botswana, No. CVR/20/74 (unpublished).
Rust, U. 1985. Die Entstehung der Etoschapfanne im Rahmen der Landschaftsentwicklung des Etoscha Nationaparks (nördliches Südwestafrika/Namibia). *Madoqua* 14: 197-266.
Shaw, P. 1986. The palaeohydrology of the Okavango Delta. Some preliminary results. *Palaeoecol. of Africa* 17: 51-58.
Shaw, P. & H.J.Cooke (im Druck). Geomorphic evidence for the Late Quaternary palaeoclimates of the Middle Kalahari of northern Botswana. *Catena.*
Street, F.A. & A.T.Grove 1976. Environmental and climatic implications of late Quaternary lake-level fluctuations in Africa. *Nature* 261: 385-390.
Street, F.A. & A.T.Grove 1979. Global maps of lake-level fluctuations since 30,000 yr. B.P. *Quaternary Research* 12: 83-118.
Van Zinderen Bakker, E.M. 1982. African palaeoenvironments 18 000 yrs BP. *Palaeoecol. of Africa* 15: 77-99.
Watts, N.L. 1980. Quaternary pedogenic calcretes from the Kalahari (southern Africa): mineralogy, genesis and diagenesis. *Sedimentology* 27: 661-686.

Abbildungen 3 und 4 treffen Sie im Einsteckfach am Ende des Bandes an.

FORMATION AND REACTIVATION OF DUNES IN THE SOUTHWESTERN KALAHARI: PALAEOCLIMATIC IMPLICATIONS

N.LANCASTER
Department of Geology, Arizona State University, Tempe, USA

ABSTRACT

Linear dunes in the southwestern Kalahari are today largely inactive. Rainfall is too high and wind velocities too low for significant sand movement. Boundary conditions for dune reactivation in the region were determined by analysing the relationship between precipitation effectiveness and the duration of sand moving winds. The dunes were probably last active during the period 20 000-17 000 BP in conditions of higher wind velocities and possibly slightly reduced precipitation.

INTRODUCTION

The southwestern Kalahari sand sea has an area of 100 000 km^2 and lies between the southeastern margins of the highlands of Namibia and the Orange River at Upington in South Africa. It consists of a 100-200 km wide belt of dunes, mostly of linear form with NW-SE or NNW-SSE alignments. Typically, the dunes have a spacing of 200-400 m and are 2-15 m high and join in Y junctions open to the northwest. The dunes form the southwestern sector of the massive arc of mostly fixed linear dunes (Fig. 1) developed in the surface Kalahari sands of the interior of southern Africa (Lancaster 1981, Thomas 1984). In many respects, the dunes can be considered almost identical to those in the Simpson-Strzelecki deserts of Australia (Breed & Grow 1979, Buckley 1981).

At the present time, dunes in the southwestern Kalahari are largely inactive except where extensive overgrazing has taken place. The climate of the area is semi-arid to sub-arid, with an annual rainfall of 150-350 mm, increasing to the north and east. The effectiveness of the rainfall is limited, as it occurs mostly during the summer (November-April), when temperatures and evaporation rates are highest, but it is sufficient to promote the growth of vegetation on the dunes. Crestal areas are often bare, especially in the drier southwestern

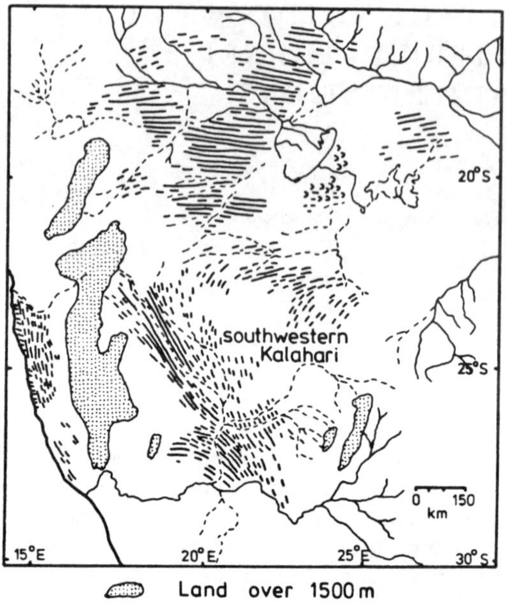

Figure 1. Southern Africa, showing dune systems and location of the southwestern Kalahari region.

parts of the region, whilst dune flanks are partially or completely vegetated, with a cover of scattered small trees, shrubs and grasses. In northern and eastern parts of the region, the vegetation cover on the dunes is largely complete and the dunes are completely stabilised by vegetation, except where overgrazing has occurred.

Winds in the area are from the NE to NW during the winter and spring, with a secondary W or SW mode in the summer They reach their maximum strength and persistence in late winter and spring. Resultant sandflows approximately parallel dune trends in southern and western parts of the area (Breed et al. 1979, Lancaster 1981) but cross dunes at angles of up to 60° in northern and eastern areas of the sand sea. Sand moving wind regimes are obtuse bimodal at Upington and Keetmanshoop, but complex elsewhere in the region. However, as in the Australian dunefields (Ash & Wasson 1983) wind velocities throughout much of the area are too low to generate significant amounts of sand movement (Breed et al. 1979), even if the vegetation cover was sparse enough. Both Lancaster (1981) and Heine (1982) have suggested that the dunes of the area were active during a period of higher wind velocities promoted by a more intense atmospheric circulation over southern Africa. The purpose of this chapter is to discuss the conditions under which the dunes of this region were formed.

DEVELOPMENT OF THE DUNES

Dunes are initiated within the southwestern Kalahari dunefield in three main ways. Firstly, dunes may extend downwind from isolated localities in interdune areas between pre-existing linear dunes by reworking and mobilisation of interdune sands. Secondly, extension of linear dunes downwind from lunette dunes situated on the southeastern margins of pans is an important mechanism for dune initiation, especially in the area of large pans west of the lower Molopo valley (Lancaster 1986a). Thirdly, the lower Molopo valley has probably been an important locality for dune initiation throughout the development of the southwestern Kalahari sand sea and acted in a similar manner as in the Cooper and Diamantina drainages which enter the Simpson-Strzelecki dunefields in Australia to provide an important source of sediments for dune building within the sand sea (Twidale 1972, Wasson 1983). Dune patterns and composition are distinctly different on each side of the lower Molopo and pale coloured linear dunes extend downwind from broad sandy mounds on the southeastern side of the ephemerally flooded valley.

In addition to the growth of new dunes from sediment sources within the sand sea, there is evidence to indicate that many of the linear dunes have been reactivated at intervals in the past. Internal structures reported by Heine (1981) and observed in trenches by the writer show that many dunes have a cap of unconsolidated sands which overlies consolidated, very slightly calcareous sands at a depth of 1·3 m. Some of these sands have calcrete pedotubules developed in them. At Soutpan, near the lower Molopo valley, these pedotubules have a radio-carbon age of 32 500 ± 780 BP (Pta-4124).

AGE OF THE DUNES

A variety of radio-carbon dated evidence provides a framework for the late Pleistocene and Holocene chronology of the southwestern Kalahari sand sea. Radio-carbon dates from pan sediments (Heine 1982, Lancaster 1986b) indicate that periods of high water tables and increased spring discharge occurred in the periods 33 500-30 400 BP and again between 23 000 and 21 900 BP. Climates much more humid than today are inferred for these periods. Dunes were also stabilised by vegetation, as indicated by radio-carbon dates of 32 000-28 000 BP on a fossil soil in an interdune area (Heine 1982) and of 32 500 BP on calcrete pedotubules developed in dunes at Soutpan. In the later parts of this period, at least, the Molopo river experienced perennial or seasonal discharge, which ceased after 19 000 BP (Heine 1982). A further period of humid climates, which was probably not as wet as that prior to 20 000 BP occurred 17 000-12 000 BP. This is evidenced by radio-carbon dates of 16 255-15 790 BP and 12 050 BP on stromatolites and reed beds in pans (Lancaster 1979, 1986b); 17 200-12 800 BP on spring tufas at Gobabis (Deacon et al. 1984) and 16 000-13 000 BP on molluscs in fluvial deposits from the Molopo river (Heine 1982). The intervening period, from 20 000 to 17 000 BP appears to have been characterised by dry conditions and increased aeolian activity (Heine 1982, Lancaster 1986b). It is probable that the last major phase of dune initiation and extension took place during this time, in parallel with similar dunefields in Australia (Wasson 1984).

Heine (1982) suggests that there was a period of Holocene aridity and strong winds during the period 4 500-3 500 BP. Pollen spectra from Kathu Pan, east of the sand sea, similarly indicate dry conditions between 4 500 and 2 700 BP (Beaumont et al. 1984). At Koopan Suid, clay lunette formation, which is indicative of high evaporation rates and strong winds, ceased prior to 1 000 BP (Lancaster 1986b).

CONDITIONS FOR DUNE FORMATION

Various indices have been utilised to determine mobility thresholds for fixed dunes in the Sahel (Talbot 1984) and Australia (Ash & Wasson 1983, Wasson 1984) and from these to suggest the range of conditions under which dunes would have been active. Essentially, the mobility of desert dunes is inversely proportional to their vegetation cover, but directly related to wind velocity. In turn, vegetation cover is a function of the ratio between potential evapotranspiration (PE) and rainfall (P). A useful measure of the sand moving power of the wind is the percentage of the time the wind is blowing above the threshold velocity for sand transport (4.5 m/second). Figure 2 shows the relationship between rainfall effectiveness (PE/P) and the amount of time the wind is able to move sand (W) for nine stations in and around the southwestern Kalahari sand sea and, for comparison, at three stations in the currently active Namib sand sea. There is a gradient in sand mobility as defined by this relationship from the northern and eastern parts of the southwestern Kalahari sand sea to the active Namib sand sea, which accurately parallels the impressions of dune activity gained from observations of the amounts of bare sand and sand movement on dunes in the region. It is difficult to determine a mobility threshold because there are few climatic records from areas of marginally active dunes. However, the relationship:

$$M = 0.25 + (0.10\,W + 1n\,PE/P) = 1$$

seems to discriminate between clearly active and other dunes in southern Africa, in a similar manner to the almost identical relationship derived by Wasson (1984) for Australian sand seas. It is clear from Figure 2 that significan increases in wind velocity or changes in the moisture balance are necessary for dunes in the southwestern Kalahari to be reactivated.

The palaeoclimatic record in the region suggests that the dunes were potentially active in the period 20 000-17 000 BP, coeval with the last Glacial maximum in high latitudes. At this time in southern Africa, mean annual temperatures were some 5° C colder than at present (Heaton et al.

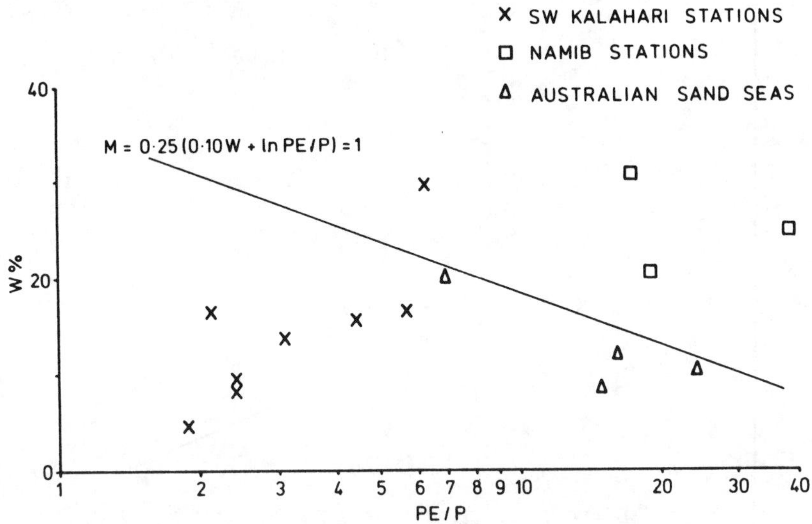

Figure 2. Relationship between the annual percentage of the time the wind is able to move sand (velocity > 4.5 m/sec) and the ratio between mean annual potential evapotranspiration (calculated using Thornthwaite method) and mean annual rainfall for stations in the southwestern Kalahari and Namib. Points for Australian deserts derived from Wasson (1984).

1986). Estimates of past wind speeds are more problematic, but Newell et al. (1981) indicate that wind velocities at the latitude of the southwestern Kalahari would have been 117% of present values during this period. An increase in wind velocity of this magnitude would have been sufficient to substantially increase the sand moving power of winds in the region and resulted in an increase in W from its present mean value of 17% to 37%. Although PE would have been reduced as a result of lower temperatures, the increased windiness in the region would have been sufficient to reactivate dunes over most of the southern parts of the sand sea, as demonstrated by Figure 3. Similar conclusions were reached by Wasson (1984) for Australian sand seas. However, to reactivate dunes in the northern and eastern parts of the southwestern Kalahari sand sea, changes in the moisture balance would have also been necessary. A reduction in mean annual rainfall of 30-40% would have resulted in reactivation of dunes throughout the region during this period. During the mid-late Holocene period of aridity, temperatures in southern Africa may have been 1-2° C warmer than at present (Deacon et al. 1984), but wind velocities were similar to those today. In these conditions, increases in the PE/P ratio due to increased evaporation rates would not have been sufficient to reactivate dunes. Reactivation of dunes could only have taken place with a decrease in rainfall to 30-50% of present values.

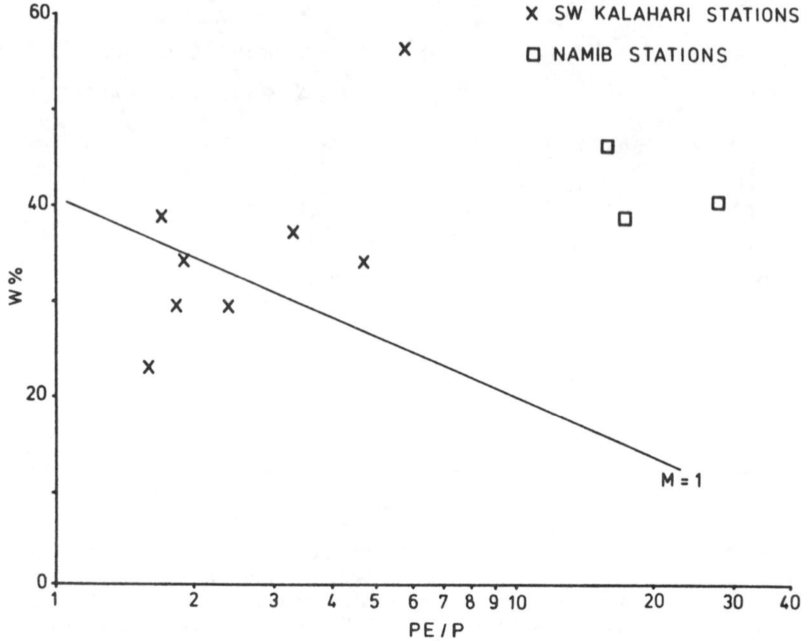

Figure 3. Relationship between annual percentage of the time the wind is able to move sand with mean wind velocities increased to 117% of modern values and ratio between mean annual potential evaporation (with mean annual temperatures 5° C lower than present) and mean annual rainfall (100% of present) as indicative of conditions at the last Glacial maximum.

Lunettes and river valleys are important localities for dune initiation in the southwestern Kalahari. The conditions for lunette formation have been summarised by Bowler & Wasson (1984). They include strong winds, which are required to deflate material from pan surfaces, but high water tables or frequent flooding of pan surfaces, followed by rapid evaporation are required to break up surfaces by salt weathering. Currently, deflation from pans in the southwestern Kalahari is sporadic and lunettes are inactive. Similarly, there is no contemporary dune formation from fluvial sediments in the lower Molopo valley. Dune initiation from these sources in the region implies that a more extreme climate existed in the region, with greater fluctuations in pan and groundwater levels. Flooding of the Molopo river in the period 16 000-13 000 BP and increased groundwater discharge at some pans is not incompatible with a period of increased dune building, as Wasson (1984) has shown for Australian dunefields.

The pattern of alignments of the southwestern Kalahari dunes reflects the interaction of circulations around both the South African and South Atlantic anticyclones (Lancaster 1981) and formation by resultant sandflows from the NW or WNW. Resultant potential sandflow directions approximately parallel dune trends in southern and western parts of the sand sea but cross dunes in the north and east. In winter, when the South African anticyclone is at its most intense, winds are dominantly northerly or northeasterly and all sandflow lines cross the dunes, often at 90°. However, in October, when the relative strengths of the two anticyclonic circulations are comparable and winds are also at their strongest, resultant sandflows throughout the area are from the NW or WNW, or parallel to the dunes. It thus appears that, when the dunes of the southwestern Kalahari were formed, circulation patterns over the year closely resembled the modern October pattern, a situation which often applies in modern drought years.

Support for this model comes also from the alignments of lunette dunes in the region, which are perpendicular to winds from the north or west. It appears that lunette formation took place in periods when dry season winds were more westerly than they are today.

CONCLUSIONS

Dune formation in the southwestern Kalahari occurred primarily in periods of increased wind strength accompanied by climates somewhat drier than the present. A shift in circulation patterns such that there was a greater frequency of NW and W winds, especially in winter months, is also indicated. Increased aeolian activity was a consequence of stronger and more persistent anticyclonic circulations, which in themselves also increased the aridity of the region.

The palaeoclimatic reconstructions put forward in this chapter are highly simplified. For example, they do not take into account the probable increases in evaporation rates that increased windiness would have caused. However, they do indicate the likely minimum degree of climatic change which occurred during periods of dune formation in the region.

ACKNOWLEDGEMENTS

Research in the southwestern Kalahari was supported by grants from the National Programme for Weather, Climate and Atmosphere Research of the CSIR and the Harry Oppenheimer Institute for African Studies, University of Cape Town. I thank Dr J.C.Vogel for provision of radio-carbon dates.

REFERENCES

Ash, J.E. & R.J.Wasson 1983. Vegetation and sand mobility in the Australian desert dunefield. *Zeitschr. für Geomorphol. Supplementband* 45: 7-25.

Beaumont, P., E.M.van Zinderen Bakker & J.C.Vogel 1984. Environmental changes since 32 000 BP at Kathu Pan, northern Cape. In J.C.Vogel (ed.), *Late Cainozoic palaeoclimates of the southern hemisphere*: 329-338. Rotterdam: Balkema.

Bowler, J.M. & R.J.Wasson 1984. Glacial age environments of inland Australia. In J.C.Vogel (ed.), *Late Cainozoic palaeoclimates of the southern hemisphere*: 183-208. Rotterdam: Balkema.

Breed, C.S., S.G.Fryberger, S.Andrews, C.McCauley, F.Lennartz, D.Geber & K.Horstman 1979. Regional studies of sand seas using Landsat (ERTS) imagery. In E.D. McKee (ed.), *A study of global sand seas.* US Geol. Survey, Prof. Paper 1052: 305-398.

Breed, C.S. & T.Grow 1979. Morphology and distribution of dunes in sand seas observed by remote sensing. In E.D.McKee (ed.), *A study of global sand seas.* US Geol. Survey, Prof. Paper 1052: 253-304.

Buckley, R.C. 1981. Parallel dunefield ecosystems: southern Kalahari and central Australia. *J. Arid Environments* 4: 287-298.

Deacon, J., N.Lancaster & L.Scott 1984. Evidence for late Quaternary climatic change in southern Africa. In J.C.Vogel (ed.), *Late Cainozoic palaeoclimates of the southern hemisphere*: 391-406. Rotterdam: Balkema.

Heaton, T.H.E., A.S.Talma & J.C.Vogel 1986. Dissolved gas palaeotemperatures and ^{18}O variations determined from groundwater near Uitenhage, South Africa. *Quat. Res.* 25: 79-88.

Heine, K. 1981. Aride und pluviale Bedingungen während der letzten Kaltzeit in der Südwest-Kalahari (Südliches Afrika). *Zeitschr. für Geomorphol. Supplementband* 38: 1-37.

Heine, K. 1982. The main stages of the late Quaternary evolution of the Kalahari region, southern Africa. *Palaeoecol. of Africa* 15: 53-76.

Lancaster, N. 1979. Evidence for a widespread late Pleistocene humid period in the Kalahari. *Nature* 279: 145-146.

Lancaster, N. 1981. Palaeoenvironmental implications of fixed dune systems in southern Africa. *Palaeogeogr., Palaeoclim., Palaeoecol.* 33: 327-346.

Lancaster, N. 1986a. Pans in the southwestern Kalahari: a preliminary report. *Palaeoecol. of Africa* 7: 59-60.

Lancaster, N. 1986b. *Geomorphic evidence for late Quaternary palaeoclimates in southern Africa.* Final Report to the National Programme for Weather, Climate and Atmosphere Research, CSIR, Pretoria.

Newell, R.E., S.Gould-Stewart & J.C.Chung 1981. Possible interpretation of palaeoclimatic reconstructions for 18 000 BP in the region 60°N to 60°S, 60°W to 100°E. *Palaeoecol. of Africa* 13: 1-19.

Talbot, M.R. 1984. Late Pleistocene dune building and rainfall in the Sahel. *Palaeoecol. of Africa* 16: 203-214.

Thomas, D.S.G. 1984. Ancient ergs of the former arid zones of Zimbabwe, Zambia and Angola. *Transact. Inst. British Geographers, N. Ser.* 9: 75-88.

Twidale, C.R. 1972. Evolution of sand dunes in the Simpson Desert, Central Australia. *Transact. Inst. British Geographers,* 56: 77-110.

Wasson, R.J. 1983. Dune sediment type, sand colour, sediment provenance and hydrology in the Strzelecki-Simpson dunefield, Australia. In M.E.Brookfield & T.S.Ahlbrandt (eds.), *Eolian sediments and processes*: 165-196. Amsterdam: Elsevier.

Wasson, R.J. 1984. Late Quaternary palaeoenvironments in the desert dunefields of Australia. In J.C.Vogel (ed.), *Late Cainozoic palaeoclimates of the southern hemisphere*: 419-432. Rotterdam: Balkema.

LES PHASES D'ARIDITE DU PLEISTOCENE SUPERIEUR ET DE L'HOLOCENE AU SAHARA; ARGUMENTS SEDIMENTOLOGIQUES

P.ROGNON
Géodynamique des milieux continentaux et UA.722 (CNRS) UER des Sciences de la Terre, Université P. et M. Curie, Paris, France

ABSTRACT

Arid periods in the Sahara during Upper Pleistocene and Holocene times; sedimentological data.

For the reconstruction of the palaeoenvironments of the Sahara, many more features are available for the pluvial periods than for the arid ones. The aim of this review is to take stock of the present state of available data on Late Quaternary arid or semi-arid periods. These data are principally derived from sedimentological analysis. The better evidence for major aridity is provided by the dunes and their associated environments. On the southern margins of the Sahara and Sahel, numerous studies are beginning to give a clearer picture of the succession of dune building episodes at about 40 000 BP, from 20 000 to 12 000 BP, for few centuries at about 8 000-7 000 BP and during the recent onset of aridity since 4 500 BP. On the northern margins, the development of sandseas was probably diachronous, but very few data are available, only for the Holocene.

On the other hand, loess deposits have been studied only along the northern margins (South-Tunisia, Negev) whereas they have just been pointed out in few Sahelian regions. Loess testifies to former semi-arid periods, because a steppe plant cover is necessary to trap, the loess and to protect it against erosion. The desert dust components can be found in palaeosols or in alluvial and lacustrine deposits, with the aid of chemical or mineralogical tracers of their Saharan origin. Dust in marine sediments is also a good indicator of arid continental climates, especially in the Eastern Atlantic Ocean and in the Mediterranean Sea: eolomarine dust provides the best data for drawing up a valid chronology of arid episodes, which can be dated with the oxygen isotopic method. Here too the choice of the true tracers is a real problem.

Semi-arid and arid climates can be caused both by a decrease of precipitation and an increase in evaporation rate. Some past changes in aridity can be demonstrated by studies on palaeohydrologic evidence. Thus, calcretes or evaporitic deposits often are good indicators of lacustrine (with large variations of water level) or palustrine (with brackish water) environments which have been subjected to increased aridity. In the same way, the sebkhas and their associated clay dunes represent a wide range of specific hydrologicl environments, but very few studies on the sebkhas around the Sahara have so far been done. On the other hand, the climatic significance of wadi deposits is often a subject for discussion, for instance about the silty or gravelly terraces. Better evidence of aridity on a regional scale is provided when during a certain period all the drainage networks were disintegrated and filled up by massive alluvial deposits, or when the course of large rivers such as the Senegal, the Niger or the Nile was obstructed by dunes.

Many more studies will be needed before it will be possible to reconstruct the fluctuations of the boundaries of the desert during the Late Quaternary. This question is of importance, for the understanding of the mechanisms which control the present-day aridification.

INTRODUCTION

Paradoxalement, le passé du Sahara est surtout connu pour les périodes où il n'a pas été un désert . . . Lorsque les climats ont été plus humides qu'aujourd'hui (définition des 'pluviaux'), ils ont laissé des archives abondantes: dépôts lacustres, paléosols, vestiges biologiques, qui ont été étudiées en priorité. Mais les phases d'aridité, où le Sahara a été un désert, aussi aride ou plus aride qu'à l'heure actuelle, n'apparaissent le plus souvent que sous forme de pointillés sur les courbes ou de hiatus dans les séquences sédimentaires. En effet, les dépôts caractéristiques des paléodéserts sont moins souvent conservés et leur interprétation est rendue difficile par l'insuffisance des connaissances sur les milieux désertiques actuels.

Depuis quelques années pourtant, l'intérêt des chercheurs et même de l'opinion internationale (conférences de l'UNESCO à Nairobi en 1977 et 1984 sur la 'désertification') s'est orienté aussi vers la reconstitution des phases arides en vue de comprendre les mécanismes de l'aridification naturelle et les comparer à ceux de la désertification qui étendent artificiellement les déserts par suite des dégradations anthropiques.

Le but de ce chapitre est de recenser les critères qui permettent de dire que telle période a été caractérisée par l'aridité. Dans l'Actuel les conséquences d'un climat aride sont relativement faciles à définir: prédominance de l'action du vent et de l'évaporation, diminution progressive du nombre des êtres vivants et de l'action de l'eau superficielle etc. Mais on se heurte très vite à la difficulté de préciser des limites entre les climats semi-arides, arides et hyperarides ou entre les déserts et les régions environnantes. En particulier la limite entre les milieux semi-arides et arides est définiée par le passage d'un couvert végétal diffus à une végétation de plus en plus 'contractée' le long des réseaux hydrographiques. Or cette limite n'implique pas un changement dans les unités de végétation: steppe subméditerranéenne à Armoises et Chénopodiacées ou savane sahélienne à Acacia et Panicum. Il a donc paru nécessaire d'inclure les marges semi-arides dans ce chapitre, même si, dans un désert hyperaride comme la Libye orientale, l'existence de mares saisonnières à l'Holocène inférieur est un critère de 'pluvial'. Il n'existe pas en effet, à l'échelle du Sahara, de référence unique pour définir les phases arides ou pluviales. En chaque point, ces termes sont appliqués par comparaison avec le milieu naturel actuel, qui cependant est souvent fortement 'anthropisé'.

DEFLATION ET SEDIMENTATION EOLIENNE CONTINENTALE

La conséquence immédiate de la 'rétraction' du couvert végétal est l'accroissement du rôle du vent qui devient l'agent prépondérant lorsqu'on passe des climats semi-arides à l'hyper-aridité. Le rôle du vent a été longtemps sous-estimé parce qu'il a une capacité de transport moindre que le ruissellement (encore épisodique en régions arides) et qu'il varie en intensité, en turbulence et surtout en direction. Mais le vent s'affranchit des réseaux hydrographiques (de plus en plus désorganisés) et il est responsable de l'organisation de l'espace dans les déserts, comme les glaces le sont en régions froides. Les cartes à petite échelle, puis les images satellitaires ont montré qu'il existait une logique de répartition des zones de déflation ou d'accumulation (les ergs) à l'échelle de l'ensemble du Sahara (Guy & Mainguet 1975). On devrait pouvoir retrouver cette répartition pour des périodes arides plus anciennes. Mais pour l'instant on connaît très peu de traces de ces dunes anciennes, les recherches étant surtout orientées vers la typologie des dunes ou leur position à l'intérieur d'un erg (MacKee 1979) et vers la définition précise des loess péridésertiques (Coudé-Gaussen 1984a).

Les sables éoliens et les dunes

Le transport des sables par saltation et l'édification de dunes est un critère majeur d'aridité. Au Sahel, la migration des sables (favorisée par l'action anthropique) a repris lors de la sécheresse de 1968-84, lorsque la pluviosité est descendue de 300 mm à 150 mm/an. Au Nord du Sahara où les pluies se situent en hiver, la limite de 100 mm est considérée comme caractéristique de la généralisation des phénomènes éoliens. Pourtant en Arabie, vers 25-27°N, les dunes sont considérées comme stabilisées avec seulement 40 mm/an (Anton 1986). Quel que soit le minimum de pluies nécessaire pour la fixation des dunes, l'existence de paléodunes dans des régions où les sables actuels ne sont pas mobiles, est un premier critère de périodes plus arides (Fig. 1).

La succession des paléoergs au Sahel

Une extension de sables éoliens et de paléodunes jusqu'à 4-500 km au Sud de la limite actuelle du Sahara a été observée depuis très longtemps en diverses régions du Sahel.

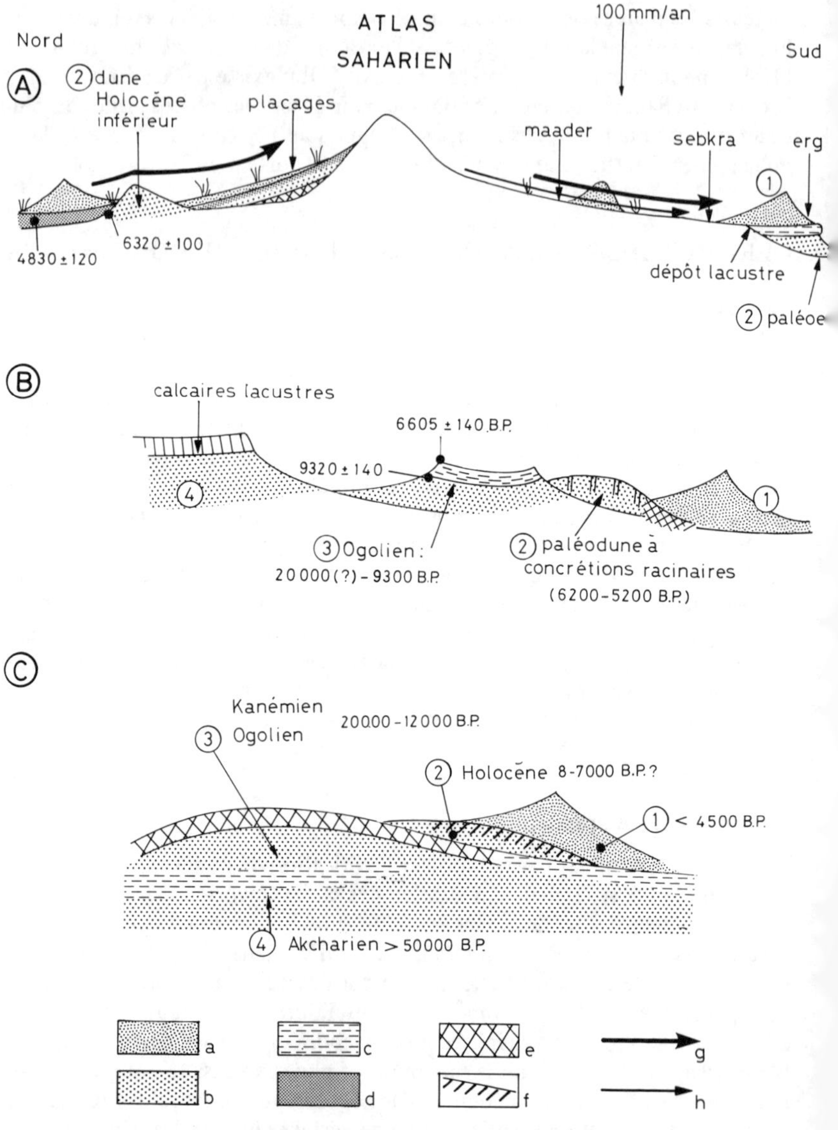

Figure 1. Essai de reconstitution des épisodes dunaires connus: a. sur la bordure septentrionale du Sahara (d'après Ballais et al. 1979, Callot 1984, Fontes et al. 1985); b. au Sahara central-Nord Mali (d'après Riser et al. 1983); c. au Sahel (d'après Servant 1973, Michel 1973, Leprun 1971). Etant donné les incertitudes sur les datations, les épisodes dunaires successifs ont été numérotés de 1 (Actuel) à 4 (l'épisode le plus ancien observé). Légende: a. dune actuelle; b. sable éolien; c. dépôt lacustre ou marécageux; d. limons; e. sol ou paléosol; f. début d'altération; g. migration des sables; h. écoulement hydrique.

a. Une période aride généralisée entre 20 000 et 12 000 BP. Les datations obtenues sur les dépôts des phases humides situés au-dessus et au-dessous de sables éolisés ont permis d'individualiser un premier paléoerg, caractérisé à la fois par un matériel éolien plus grossier que celui des dunes actuelles et par des paléoformes de dunes longitudinales, généralement orientées NE, SW, parfois NNE, SSW ou EW. Cette période a été bien datée au Tchad ('Kanémien'; Servant & Servant-Vildary 1980) et en Arabie (McClue 1976) et corrélée depuis le Sénégal ('Ogolien'; Michel 1973) jusqu'au Soudan ('Low Qoz'; Warren 1970). Mais la datation exacte du début et de la fin d'une période aride est très difficile à préciser car la fixation des dunes ou leur remobilisation sont très sensibles à certains seuils du niveau de la nappe aquifère (Rognon 1980).

Pour l'instant, il n'existe pas d'étude détaillée de la structure exacte de ces paléoergs. Les directions des vents devaient être analogues à celles des vents responsables du modelé des dunes actuelles au Sud-Sahara, mais leur puissance de transport était certainement plus élevée. En effet l'érosion éolienne croît avec le cube de la vitesse du vent et ce paramètre doit être pris en compte dans la reconstitution des paléoaridités. Un vent plus rapide pouvait en effet entraîner des migrations des sables avec des pluies inférieures seulement de 20 à 50% de l'Actuel (Talbot 1984).

b. De brèves crises d'aridité au cours du pluvial holocène (Rognon 1983). Entre 12-10 000 et 5 000 BP environ, ces dunes, fixées, se couvrent de sols semi-arides. Mais cette pédogénèse est interrompue par des reprises, brèves et localisées, de la migration des sables éoliens avec formation de nouvelles dunes, plus petites, formées de matériel plus fin et mieux trié et selon des directions légèrement différentes des précédentes (Leprun 1971). Cette courte phase d'édification de dunes est connue en différents points du Sahel (Sénégal, delta intérieur au Niger etc.) et elle est souvent datée, mais sans grande précision, entre 8 000 et 7 000 BP, en relation avec la baisse des niveaux lacustres (constantée au Sénégal, au Tchad ou dans l'Afar) ou la diminution des pollens caractéristiques de climat humide ou montagnard (Tchad, Ethiopie). Cette crise semble avoir duré de quelques siècles à moins d'un millénaire et n'a pas été sensible partout aux mêmes dates. Ainsi, dans le Nord du Mali, la baisse des paléolacs et la remobilisation des dunes se place dans un bref épisode entre 6 200 et 5 200 BP (Riser et al. 1983) (Pl. I.1).

c. Des paléoergs antérieurs à 20 000 BP. Les périodes arides antérieures sont plus difficiles à mettre en évidence. Elles ont été surtout étudiées sur des sondages. Ainsi dans la cuvette tchadienne, une période aride vers 30 000 BP a été découverte au milieu du pluvial du Ghazalien (environ 40 à 20 000 BP; Servant-Vildary 1978) et une grande période aride serait antérieure à 50 000 BP (Servant 1973). Au Sénégal et en Mauritanie, une cartographie systématique de l'extension des dunes ogoliennes a permis à C. Barbey (1982) de distinguer ces dunes (surtout situées au Nord du 16°) de celles d'un erg nettement plus ancien (Akcharien) s'étendant jusque vers 14°N.

Ces arguments, parmi d'autres, conduisent à réexaminer la succession des épisodes dunaires au Sahel où la couverture sableuse qui s'avance jusqu'à 12°N au Nigéria avait été trop hâtivement attribuée au seul épisode 'ogolien'. La stratigraphie précise de ces ergs successifs sera très difficile à débrouiller car chaque période aride remanie au vieux stock de grains éolisés, dont les caractères sédimentologiques ne varient pas beaucoup d'une période à l'autre (Durand & Lang 1986).

En définitive, cette bordure sahélienne, sur plus de 5 000 km d'Ouest en Est, représente les stades d'avancées de nappes sableuses successives au cours du Quaternaire. On commence à les distinguer pour le dernier cycle climatique, mais on peut espérer en retrouver de plus anciennes. On retrouve ici le même problème qu'à la bordure de l'inlandsis nord-européen: après une première phase exploratoire où l'ensemble des bourrelets morainiques avait été attribué au Weichsélien, la multiplication des sondages et des travaux de cartographie a progressivement fait apparaître plusieurs épisodes glaciaires. De la même façon, les paléodunes sahéliennes sont un des meilleurs terrains d'étude des périodes arides sahariennes.

Autres indices de dépôts dunaires

a. A l'intérieur du Sahara, il existe peu de travaux sur la dynamique interne d'un erg, excepté pour celui de Fachi-Bilma (Mainguet & Callot 1978). Le passage d'un système dunaire à l'autre ou la superposition de deux types de dunes ont très longtemps été interprétés en fonction de combinaisons de directions de vents actuels. Les cartes, établies à partir de la télédétection par E.D.MacKee (1979), ouvrent la voie à de nouvelles recherches en montrant le caractère systématique, à l'échelle régionale, des modifications des directions éoliennes, qui traduiraient un réaménagement de paléo-systèmes dunaires par la dynamique éolienne actuelle. Mais ces méthodes géomorphologiques ne permettront pas de remonter au-delà de la précédente période aride. Parfois la morphologie d'un erg disparu peut être reconstituée par la forme des dépôts hydriques interdunaires qui lui ont succédé en période pluviale. Ces derniers ont, en effet, systématiquement des formes en entonnoirs, parfois perchés à 6-8 mètres d'altitude (par inversion de relief) pour les ergs de la période de 16 000 à 8 000 BP dans le NW du Sahara algérien (Conrad 1969: 269) et des formes allongées et parallèles pour l'erg d'In Sakane dans le Sahara malien pour la période 'ogolienne' (20-12 000 BP) (Riser et al. 1983). Ces observations permettent de conclure à la répétition des mêmes types de dunes (pyramidales au Nord, longitudinales au Sud) lors de ces Arides et de la période aride actuelle.

La stratigraphie des ergs sahariens est encore très mal connue (Pl. I.2). Les travaux récents d'Y.Callot sur le Grand Erg occidental (Callot 1984, Fontes et al. 1985) permettront de reconstituer le paléoerg antérieur à l'Holocène moyen de ce secteur. Au NE du Hoggar, les coupes de l'erg de Tihodaine (Rognon 1967: 508-513), montrent l'existence d'un vieil erg à directions WNW-ESE (directions actuelles: NS), antérieur à un paléolac dont les argiles renferment des diatomées, des Grands Mammifères et des outils du début de l'Acheuléen supérieur. Puis un second erg s'est édifié avant le dépôt de carbonates palustres sans doute contemporain du Paléolithique supérieur. Mais l'étude détaillée de ce gisement reste à faire.

b. Les dépôts dunaires de la bordure sub-méditerranéenne du Sahara sont minces et peu étendus à cause de la permanence des migrations de sables du Nord au Sud du désert dans le sens anticyclonique. J.L.Ballais et al. (1979) ont montré l'intérêt de l'étude des placages éoliens sur des versants au vent du flux de NW dans l'Atlas saharien: trois épisodes de sables, riches en gypse, s'intercalent entre des lits de cailloux de versant qui les ont protégés de l'érosion postérieure. Le plus récent est fossilisé par des dépôts limoneux de l'Holocène moyen (4 500 BP). Une dune voisine dans la même position stratigraphique a pu être datée avec précision puisqu'elle a été ravinée juste après 6 320 ± 100 BP. Cette même période aride à la charnière Pléistocène/Holocène a pu être datée aussi dans le Maroc méridional (Weisrock 1982, Weisrock et al. 1985), où elle est représentée par des paléodunes migrant à partir du littoral atlantique. Enfin, à 100 km du Maroc, dans la plus orientale des îles Canaries, des sables biodétritiques issus des plages ont migré jusqu'à 6-7 km dans l'intérieur de l'île, à trois reprises au moins au cours du dernier cycle climatique (dernière glaciation). Ces sables dunaires indiquent à la fois un climat plus aride que l'Actuel et surtout des vents à forte capacité de transport, de secteur Ouest (alors que le vent dominant actuel est l'alizé de NNE) (Rognon & Coudé-Gaussen 1986).

Les loess désertiques et les limons éoliens

Les poussières soulevées par les tempêtes de sable sont aussi un excellent critère de climat aride. Plus longtemps ignorées que les sables dunaires, elles sont devenues un important sujet de recherches depuis les campagnes océanographiques sur l'Atlantique tropical et grâce à l'observation des lithométéores sur les images de Météosat depuis quelques années (Coudé-Gaussen & Rognon 1983). L'existence de ces aérosols minéraux en suspension est un indice sûr d'une déflation efficace sur les sols, comme l'a bien montré la récente sécheresse du Sahel de 1968 à 1985 (Gac 1985, Prospero & Nees 1986). Les poussières sahariennes entraînées à haute altitude (1 500 à plus de 3 000 mètres) parcourent des distances considérables (10^2 à 10^3 km) et se déposent, de manière diffuse, sur des surfaces, continentales ou océaniques, qui dépassent largement le cadre du Sahara. Ce mode de sédimentation est assez différent de celui des aérosols périglaciaires et explique la relative rareté des dépôts de loess autour du Sahara, comparée à la 'ceinture' homogène de loess autour des inlandsis quaternaires. Mais grâce à l'étude de 'traceurs' de ces aérosols sahariens, il devient possible de retrouver des apports éoliens dans un certain nombre de limons incorporés à des terrasses fluviatiles, des paléosols ou des sédiments lacustres (Fig. 2).

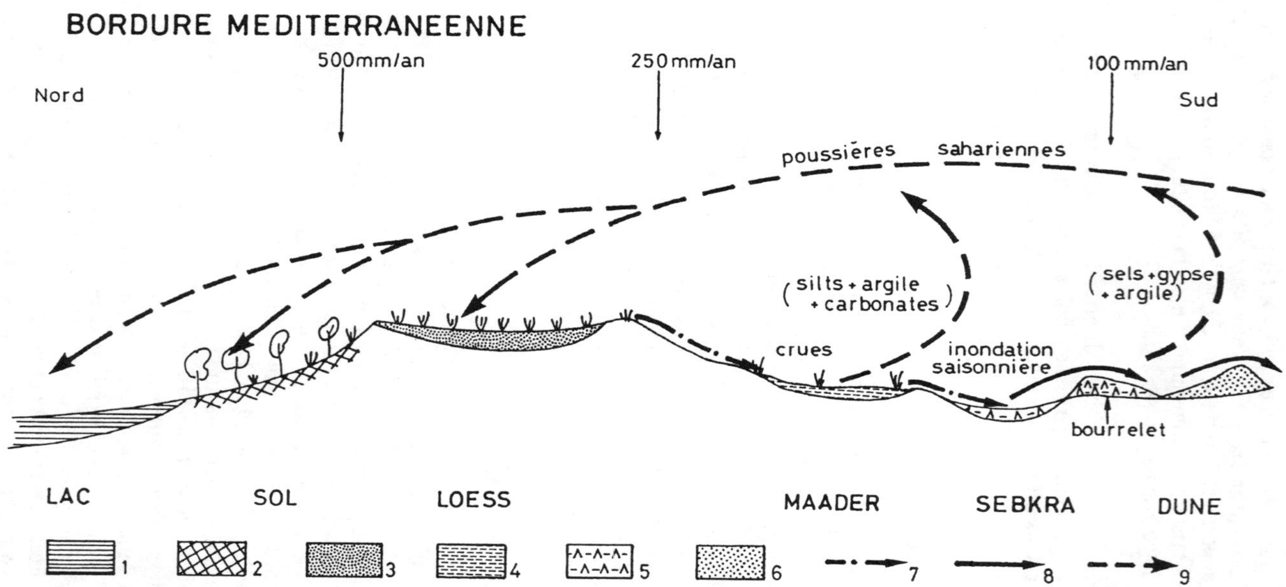

Figure 2. Schéma représentant différents faciès hydriques et éoliens en fonction de l'aridité croissante sur la bordure nord du Sahara: 1. eau; 2. sol; 3. loess; 4. limons argileux; 5. argile et sels; 6. sable éolien; 7. écoulement hydrique saisonnier; 8. migration des sables par saltation; 9. aérosols en suspension. Les limites de pluviométrie sont suggérées à titre indicatif.

Planche I.1. Sédimentation éolienne. Dans la région de l'Erg In Sakane (Nord du Mali), un bref épisode de remobilisation des sables et édification de dunes (entre 6 200 et 5 200 BP) se place entre la phase lacustre de l'Holocène inférieur et une période de fixation des dunes (Holocène moyen). Les vestiges de racines ou de troncs calcifiés jonchent la surface de cette paléodune, érodée par la déflation actuelle.

Planche I.2. Sédimentation éolienne. A l'Est du Cap Blanc (Mauritanie), coupe dans une barkane d'âge Pléistocène dont on aperçoit les litages caractéristiques. Cette dune repose directement sur un paléosol (contenant des Helix et des fragments d'oeufs d'Autruche) qui est masqué par les éboulis.

Planche I.3. Sédimentation éolienne. Coupe dans les loess des Matmata (Sud-Tunisien) sur la piste de Matmata à Tamezret. Ces loess, épais de 10 à 12 mètres, sont constitués surtout par du quartz et ont ennoyé les vallées d'un plateau formé de calcaires mésozoïques (en sombre à l'arrière plan). Plusieurs paléosols sont interstratifiés dans ces loess et ont été datés entre > 30 000 BP et 10-12 000 BP.

Planche II.1. Sédimentation hydrique. Au Sud du Grand Erg occidental, des dépôts d'argile grise et de carbonates tapissent le fond d'une grande dépression en bordure de l'erg. Ils marquent la fin d'un épisode marécageux d'âge Holocène moyen, entre deux périodes dunaires: l'un sous ces dépôts et l'autre représenté par les dunes actuelles à l'arrière plan.

Planche II.2. Sédimentation hydrique. Dans l'Oued Istène (flanc nord de l'Atakor, Hoggar), une vallée creusée dans des plateaux basaltiques a été envahie à l'Holocène moyen (datation: 6 090 ± 300 BP) par une sédimentation argileuse très riche en carbonates de calcium. La présence de Mollusques palustres, adaptés à des salinités plus ou moins élevées et de Roseaux indique l'existence de marécages soumis à une forte évaporation.

Planche II.3. Sédimentation hydrique. A la fin de l'Acheuléen, un brusque assèchement climatique a entraîné le comblement de toutes les vallées de l'Atakor (Hoggar) jusqu'à leur extrémité amont. Ici, à la tête de l'Oued Ilamane, la vallée a été ennoyée par des sables grossiers et des blocs provenant de la destruction de la couverture d'arènes des granites (à droite) et des paléosols basaltiques (à gauche). Cette terrasse graveleuse est la réponse de l'hydrographie au cours d'une phase rapide d'aridification.

Les vraies accumulations de loess sont rares
Les seuls loess étudiés en détail au Sahara sont ceux des Matmata, dans le Sud-Tunisien (Pl. I.3). Leurs caractères granulométriques, minéralogiques et chimiques ont été bien définis (Coudé-Gaussen et al. 1982b, Coudé-Gaussen & Rognon 1987) et leur datation, à la fois au ^{14}C et par thermoluminescence, les situe dans le Pléistocène supérieur entre > 30 000 et 10-12 000 BP (Coudé-Gaussen et al. 1983, 1987b). D'autres séries de loess plus anciens pourront sans doute être datées, un jour, par le paléomagnétisme. Les loess 'récents' sont interstratifiés avec des paléosols qui indiquent un renforcement de la couverture végétale (Coudé-Gaussen et al. 1984), mais des arguments, en particulier isotopiques (^{18}O et ^{13}C), prouvent que cette sédimentation quasi continue de loess sur des dizaines de milliers d'années s'est faite sous un couvert steppique, donc dans un contexte semi-aride (Coudé-Gaussen et al. 1982a). En effet le piègeage des poussières et leur protection ultérieure contre l'érosion nécessite un couvert végétal de buissons assez resserré. La relation entre dépôt de loess et climat semi-aride a été aussi établie sur l'île, actuellement sub-aride de Fuerteventura (Canaries orientales; Rognon & Coudé-Gaussen 1986).

En dehors de la bordure méditerranéenne du Sahara, les données sur les loess en sont encore au stade de l'inventaire. Des loess ont été signalés dans le bassin du Sénégal, le Nord du Mali, le Nord et l'Est du Nigéria, etc. (Maley 1982, Coudé-Gaussen 1986a). Il reste donc encore beaucoup d'archives qui permettraient l'étude des variations des marges semi-arides du Sahara.

La participation des poussières à d'autres sédimentations continentales
Les aérosols sahariens se diluent de plus en plus dans d'autres sédiments en fonction de l'éloignement des zones sources. Dans les zones semi-arides, leur volume global est encore suffisant pour être détecté et interprété en termes de paléoclimatologie. En effet la participation des aérosols à la formation des sols a été invoquée dans le Sud Marocain lorsque des croûtes calcaires épaisses s'étendent uniformément sur toutes les roches, même les quartzites (Oliva 1974), dans un secteur où les poussières actuelles sont riches en carbonates (Coudé-Gaussen et al. 1987a, à paraître). Mais il n'existe pas de preuves évidentes de ces apports, analogues aux travaux menés en Afrique australe sur les calcrètes (Blumel 1981) ou les silcrètes (Summerfield 1982) d'origine éolienne. La contribution des poussières aux sédiments limoneux des terrasses holocènes du Sahara méridional a été envisagée (Rognon 1976), et dans le Sud-Tunisien, il existe de nettes analogies granulométriques et minéralogiques entre les loess (donc les apports de poussière) et les limons de terrasses holocènes du pourtour des Matmata (Coudé-Gaussen 1984b). Enfin les apports éoliens constituent 50% environ des dépôts lacustres dans le lac du Guiers (delta du Sénégal) (Y.Gac, comm. orale), en régions semi-arides.

Pour prouver réellement le dépôt d'aérosols dans ces divers milieux, d'abord qualitativement, puis quantitativement, il est nécessaire de recourir à des traceurs des poussières sahariennes dont l'étude est en cours: grains éolisés de palygorskite (Coudé-Gaussen & Blanc 1985), microfaçonnements des sables fins et des silts analysés au MEB (Coudé-Gaussen 1986b) etc. Des minéraux argileux allogènes ont été utilisés comme traceurs de poussières sahariennes dans les dépôts et paléosols holocènes de l'île de Fuerteventura (Chamley et al. 1987), indiquant à la fois une région source (donc désertique) proche et des directions de vents de secteur Est à Sud (Rognon & Coudé-Gaussen 1986). Certains minéraux allogènes (palygorskite et kaolinite) participent aussi à la sédimentation des lacs montagnards de Corse où leur faible dilution dans une sédimentation autochtone réduite permettra de suivre leurs variations en fonction du temps (Robert et al. 1984).

AEROSOLS SAHARIENS INCORPORES AUX SEDIMENTS OCEANIQUES

Le poids des poussières expulsées du Sahara vers l'Ouest est estimé, actuellement, à 200 millions de t/an dont 150 se sédimentent dans l'Atlantique au cours des 1 000 premiers kilomètres, avec des taux de 8 à 20 cm/millénaire (Schutz et al. 1981). Ces aérosols s'incorporent à la sédimentation marine et, comme les dépôts d'icebergs dans les eaux marines voisines des inlandsis, ils sont d'excellents indicateurs paléoclimatiques: leurs variations peuvent se suivre en continu sur carottes puisque la sédimentation marine ne présente guère de discontinuités. Mais la taille de ces particules (0,5 à 50-80 μm) rend leur distinction bien plus difficile (par rapport aux sédiments marins) que celle des sables et graviers lâchés par les icebergs. Comme pour les sédiments continentaux, il faut d'abord définir les traceurs de ces aérosols sahariens qui varient selon les grandes 'provinces' lithologiques (zones sources) et atmosphériques.

Les études les plus nombreuses sur les phases d'aridité au Sahara proviennent des carottages effectués par différentes missions océanographiques (Meteor, Valdivia, DSDP, etc.) dans l'Atlantique oriental. Ces 'archives' permettent théoriquement de retracer l'histoire du désert jusque dans le Néogène, mais, malgré le nombre considérable des publications, on bute encore sur le choix des critères et leur interprétation. En effet les particules apportées par le vent en suspension sont difficiles à distinguer des apports éventuels des cours d'eau (Sud-Maroc, Sénégal) ou des turbidités, car il n'existe pas ici, comme sur la côte pacifique de l'Amérique du Sud, une fosse profonde, parallèle à la côte, pour piéger les apports continentaux non-éoliens.

Choix des traceurs d'une origine Saharienne

Le pourcentage de quartz

Ce critère est fiable à l'échelle de l'océan Atlantique (Kolla et al. 1979). Il montre l'extension considérable des retombées d'aérosols selon la trajectoire Sénégal/Caraïbes, leur renforcement et leur glissement vers le Sud à 18 000 ans BP et, enfin, des pics très importants de quartz (sur sédiment décarbonaté) à la fin du Würm (20-11 000 BP) et du Riss (> 127 000 BP). Près des côtes, où les apports terrigènes peuvent dépasser 125 μm, L.Diester-Haass (1976) a proposé de ne retenir que les quartz rubéfiés, considérés comme indicateurs de conditions arides. Diverses études ont montré en effet que la rubéfaction pouvait intervenir en milieu désertique, et la décoloration par usure hydrique (oueds; plage) ou par déferruginisation (par exemple durant la phase humide 40-20 000 BP en Mauritanie) (Barbey 1982). Mais la rubéfaction peut aussi être héritée de la pédogenèse tropicale: ainsi dans les dépôts actuels de la baie des Lévriers (21°N), ces quartz ne représentent que 3 à 13% du total des quartz contre 75% à l'embouchure du Sénégal. Pourtant, il paraît très discutable également de considérer, comme M. Sarnthein & B. Koopmann (1980: 248), que les grains pâles sont indicateurs de l'alizé de NE et les grains rubéfiés, de vents de secteur Sud! De nombreuses études, au MEB en particulier, seront encore nécessaires avant de pouvoir utiliser les quartz éolisés comme marqueurs d'aridité.

Les minéraux argileux

Depuis les travaux de G.Coudé-Gaussen sur l'importance des argiles fibreuses dans les aérosols sahariens, on peut considérer la palygorskite comme un traceur d'apports éoliens. Or elle est signalée aussi bien dans les carottes de l'Atlantique oriental surtout vers 21°N (Chamley et al. 1977), que de la Méditerranée occidentale où les pourcentages de palygorskite (ou de sépiolite) sont d'autant plus élevés que l'on se trouve proche des rivages de l'Afrique du Nord (Tomadin et al. 1984, Cossement et al. 1984) ou de la Tripolitaine (Chamley & Millot 1975). Elle est aussi transportée des déserts d'Afrique de l'Est jusqu'en bordure du continent indien (Kolla et al. 1981). La chlorite et la kaolinite sont aussi des marqueurs importants, la première provenant surtout de la bordure méditerranéenne, la seconde de la frange tropicale. Leur répartition dans les poussières actuelles est en effet bien établie sur des transects océaniques (Chester et al. 1972) ou sahariens (Paquet et al. 1984). Le rapport chlorite/kaolinite (Lange 1982) pourrait être un indicateur de la proximité de l'une ou l'autre bordure et donc du déplacement des zones climatiques au cours d'un cycle glaciaire/interglaciaire, si l'on était assuré que les directions des vents n'ont pas varié.

Les indicateurs biologiques

De nombreuses études palynologiques ont tenté de retracer les variations des limites du désert à partir des carottes obtenues par les océanographes. Leurs résultats sont très discordants, faute de connaissances précises concernant la sédimentation actuelle des pollens sur le fond océanique. Une étude sur 100 échantillons répartis de 34 à 6°N fournira bientôt cette échelle de référence (Hooghiemstra 1986). Par contre, les variations des pourcentages de diatomées continentales dans les carottes océaniques (Pokras & Mix 1985) sont bien corrélées avec les périodes arides car la déflation sur les diatomites du Sahara ne peut s'effectuer que lorsque les lacs sont asséchés. Une assez bonne corrélation semble exister aussi pour les pourcentages de phytolites, indicateurs de la savane de Graminées.

La reconstitution des périodes arides

En analysant les pourcentages des divers traceurs le long d'une même carotte, il est possible de suivre le rapprochement ou l'éloignement des régions sources désertiques à condition qu'il n'y ait pas de modifications dans la direction ou l'intensité des paléovents. Pour arriver à des données plus précises sur les variations de l'aridité continentale, il faut faire appel à deux méthodes plus globales.

La granulométrie

Les variations de la taille des grains éoliens le long d'une carotte devraient suivre celles de l'intensité du vent transporteur, à condition de pouvoir isoler ces grains au sein du sédiment marin. Il est donc difficile de retenir des critères qui fixent arbitrairement la limite entre fraction éolienne et 'fluviatile' aux environs de 63 ou 80 μm, ou attribuent au vent les particules > 6 μm avec un mode élevé. Il faudrait systématiquement rechercher des sites comme le forage de Valdivia 13289-½ dans l'Atlantique (18°N et 18°W) sur un relief sous-marin isolé, haut de 300 à 500 mètres, où les turbidités et les apports continentaux par voie hydrique sont exclus. Dans ce cas, on observe que la fraction détritique > 6 μm est bien plus importante à 18 000 BP (66%) qu'à 6 000 BP (< 10%), avec des modes respectifs de 45 et < 25 μm (Sarnthein & Koopmann 1980), ce qui est conforme aux maxima d'aridité (Ogolien) et de pluviosité (Holocène) constatés à cette latitude sur le continent voisin.

Les vitesses de sédimentation

La sécheresse catastrophique du Sahel dans les années 1968-1984 a montré une relation directe entre cette augmentation de l'aridité à ces latitudes et celle de la concentration des aérosols sur l'Atlantique (Prospero & Nees 1986). Il est donc logique de rechercher, à l'aide des carottes obtenues sur les 1 300 km de côtes sahariennes, quelle a été, à chaque époque, la zone d'apports éoliens maximum. Mais faute d'une distinction entre les apports éoliens et les autres, cette méthode ne peut être appliquée que dans des cas très précis. Ainsi, à 800 km de la côte où ne parviennent que des aérosols, Sarnthein & Koopmann (1980) constatent un doublement de la vitesse de sédimentation des silts grossiers à 18 000 BP par rapport à l'Actuel. Mais, pour les carottes proches du talus continental, tout calcul de ce genre est faussé par la présence d' 'eolian sand-turbidities' (Sarnthein & Diester-Haass 1977). Ces turbidités, provoquées par l'avancée des dunes sur le plateau continental lors des régressions glacio-eustatiques, ont en effet fortement modifié les vitesses de sédimentation à ces périodes.

Malgré leur difficulté, ces recherches sont les seules voies qui permettront de caler les périodes arides et leurs conséquences (vitesse des vents, aérosols etc.) par rapport à l'échelle isotopique océanique. C'est ainsi qu'au large du désert péruvo-chilien, le pic des % de quartz précède le signal isotopique de 18 000 BP d'environ 3 000 ans (Molina-Cruz 1977) ou que le dépôt maximum d'aérosols dans le Pacifique oriental est décelé de 5 à 10 000 BP au-delà de ce signal (Janecek & Rea 1985). Il serait très important d'avoir des données identiques pour le Sahara.

PREDOMINANCE DE L'EVAPORATION SUR LES APPORTS HYDRIQUES

Comme le vent, l'évaporation intervient dans tous les milieux arides, entraînant la disparition totale ou saisonnière des eaux superficielles. Ses manifestations sont encore peu étudiées dans les paléosols (concrétionnements, minéraux spécifiques, arrangements micro-morphologiques etc.). Elles commencent à être évaluées à partir des isotopes stables des carbonates (^{18}O et ^{13}C), des sulfates (^{18}O et ^{34}S), des phosphates (^{18}O) ou de la matière organique (^{18}O, ^{13}C et ^{2}H), mais ces études en sont encore au stade expérimental. Les conséquences d'une longue saison sèche ou d'une aridité accentuée sur le couvert végétal ont rarement été mises en évidence à partir de l'analyse des pollens, à la fois à cause de la rareté des sédiments favorables et des difficultés d'interprétations des couverts de steppe (El Moslimany 1982, Brun 1985) ou de Graminées (Gibbs-Russell 1986). Les principaux arguments proviennent donc, actuellement, des données sédimentologiques.

Les dépôts de nappes d'eau peu profondes à régime déficitaire

Pour reconstituer les périodes pluviales, on dispose à la fois de lignes de rivages et de dépôts de fonds lacustres. Quand l'évaporation augmente, le niveau des lacs devient de plus en plus variable et les rivages, très mobiles selon les saisons, ne laissent aucun vestige. Les dépôts de fond s'enrichissent en carbonates, sulfates ou chlorures en fonction des sursaturations successives (Pl. II.1). Enfin, quand le fond est exondé en saison sèche, la déflation intervient et un régime de sebkra s'installe.

Les dépôts chimiques en milieu marécageux

a. Les croûtes calcaires. Les calcrètes sont surtout abondants à la limite entre le Maghreb et le Sahara septentrional. Les très vastes affleurements des 'hamadas' et de la 'torba' d'âge tertiaire ou plio-pléistocène, caractérisés par la calcite, la dolomite et l'attapulgite, correspondent à des faciès lacustres (Conrad 1969), mais les calcrètes quaternaires ont très souvent été assimilés à des horizons B_{ca} de sols, de plus en plus calcarifiés en fonction du temps (Ruellan 1971) et leur interprétation sédimentologique a pris un certain retard (Vogt

1983). Il existe, en fait, une grande diversité de faciès carbonatés qui, mieux connus, seront de bons indicateurs de milieux semi-arides ou même arides. En effet les niveaux B_{ca} à concrétions calcaires des paléosols sont des indices d'une végétation semi-aride d'arbustes ou de steppes, tandis que les croûtes d'origine éolienne correspondent aux mêmes environnements semi-arides que les loess. Au contraire les croûtes 'zonaires', de nature stromatolithique, sont de minces constructions algaires dont les filaments fixent les particules éoliennes: elles fossilisent des surfaces dénudées, sous des conditions plus arides (Vogt 1984).

Beaucoup de croûtes calcaires sont liées, en fait, à la sursaturation des carbonates en milieu marécageux sous l'effet d'une très forte évaporation (Pl. II.2). Ces croûtes renferment, en effet, des coquilles de Mollusques plus ou moins adaptés à la salinité (Rognon 1961) et, parfois, des tiges de Roseaux carbonatées, verticales, fossilisées en position de vie (Conrad 1969: 282, Callot 1984). En lame mince, on observe des restes organiques (bactéries, cyanophycées, champignons) et des structures micromorphologiques (litages, fentes de dessication, ooïdes, pellets etc.) caractéristiques de ce type de milieu (Adolphe 1981, Vogt 1982). Enfin, ces faciès carbonatés se situent toujours à la fin des séquences lacustres, lorsque l'évaporation s'intensifie. Ainsi, les tiges carbonatées autour du Grand Erg occidental sont datées de la fin de la phase lacustre de l'Holocène moyen (Conrad 1969, Callot 1984). De même, au Nord du Mali, les lacs d'eau douce des creux interdunaires de l'Erg In Sakane, apparus vers 9 320 ± 140, s'assèchent vers 6 800-6 500 BP et leur phase d'assèchement est marquée par une carbonatation généralisée, puis limitée aux fentes de dessication du fond marécageux exondé (Riser et al. 1983). Aujourd'hui, l'étude des croûtes calcaires sahariennes s'oriente d'ailleurs vers la mise en évidence de micro- et méga-séquences, montrant que ces croûtes sont un matériel très complexe, polyphasé, capable de fournir des interprétations paléoclimatiques bien plus précises.

Enfin d'autres faciès, plus détritiques, sont liés à une forte évaporation à partir de nappes phréatiques. Leur intérêt paléoclimatique est plus limité. En régions saheliennes, où les dépôts carbonatés sont bien plus rares, leur apparition dans une séquence lacustre est la preuve d'une oscillation ver l'aridité. Ainsi, au Sud du lac Tchad, des dépôts carbonatés datés de 20 600 ± 600 BP, semblables à ceux qui se forment actuellement au Nord du lac en relation avec la nappe phréatique, indiquent la phase d'aridité du Kanémien (Durand & Lang 1986).

b. Les évaporites. Quand l'évaporation devient plus intense, seuls le gypse et les sels parviennent dans les dépressions terminales. Les croûtes de gypse peuvent être soit d'origine éolienne (Trichet 1963), soit liées à la proximité d'une nappe séléniteuse (Conrad 1969: 277-278), soit enfin directement déposées par une nappe d'eau soumise à une évaporation intense. Dans ce dernier cas, la signification climatique est la plus nette et les faciès lités où alternent le gypse, l'halite ou l'anhydrite, ont parfois été interprétés comme la conséquence de variations saisonnières de l'évaporation. Malheureusement, il semble difficile d'exploiter ce type d'information étant donné que les saumures continentales sont très variées et changeantes et que l'équilibre gypse/anhydite est soumis à beaucoup d'interactions (Jauzein 1974). Certains minéraux sont de bons traceurs de ces évaporites, comme la thénardite (Na_2SO_4) ou la mirabilite (Na_2SO_4, 10 H_2O) dans les lits de gypse et d'argile holocènes ou bien la natrojarosite (Na, $Fe_3(SO_4)_2(OH)_6$) dans les ocres des milieux acides, riches en Na et sulfures du Pléistocène supérieur des Pays Bas de l'Ahnet (Conrad 1969: 295, 326). Ces minéraux ont été étudiés jusqu'ici, surtout dans les évaporites actuelles du Tchad (Maglione 1974) ou du Sud-Tunisien (Perthuisot 1976), mais rarement sur des carottages, peu profonds et mal datés. Il n'existe pas, au Sahara, de reconstitution précise des variations de salinité sur 10 ou 20 000 ans, comme celles qui ont été réalisées dans le désert de Thar (Wasson et al. 1984), le NW de l'Iran (Kelts & Shahrabi 1986) etc.

Etant donné l'extrême fragilité de ces évaporites litées à la déflation éolienne, on peut penser que cette sédimentation s'est effectuée sous une nappe d'eau encore permanente, même très mince. Ainsi, dans la région d'Ouargla (NE du Sahara algérien), plus de 7 à 8 mètres de gypse se sont accumulés dans la Sebkra Mellala sous un climat sub-aride (100 à 250 mm/an de pluviosité), alors que sous le climat actuel hyperaride (42 mm/an à Ouargla), ces évaporites sont efficacement érodées par le vent (Boyé et al. 1978). Dans les 2 mètres supérieurs, deux niveaux fossilifères, datés de 9 550 ± 130 et 7 900 ± 110 BP renferment un assemblage très particulier d'organismes à affinités marines ou continentales, adaptés à de très fortes variations de salinité: Cerastoderma, Foraminifères, Ostracodes, Melania, Hydrobia, Melanopsis, ou encore Charophytes. On retrouve des associations très comparables lors de l'assèchement de nombreux paléolacs du Sahara septentrional ou central et A.Levy (1982) a bien posé le problème, très complexe, du mode de diffusion de ces organismes dans les divers 'lacs à Cardium'. En effet, dans le cas de la Sebkra Mellala, ces deux niveaux fossilifères doivent-ils être interprétés comme des épisodes brefs de climat plus humide (semi-aride)? Ou doit-on tenir compte du délai nécessaire à la colonisation d'un milieu déjà favorable par des organismes à migration lente depuis d'autres refuges (Levy 1985)? Ou faut-il envisager une transplantation par des oiseaux migrateurs? On retrouve le même problème dans les petites dépressions fermées (dayas) du Grand Erg occidental (Callot 1984). Une étude interdisciplinaire de ces mêmes organismes (plus les diatomées) le long de coupes détaillées, associée à l'analyse isotopique (^{18}O et ^{13}C) a permis de suivre les variations de salinité d'une paléonappe hydrique entre 9 300 ± 400 et 3 000 BP (Fontes et al. 1985). Mais ces variations ne semblent pas exactement corrélables d'une coupe à l'autre à cause de la complexité du régime hydrologique de ces dayas. D'autres études seront encore nécessaires pour traduire ces variations de salinité en courbe des variations de l'évaporation dans le climat régional.

Signification paléoclimatique des dépôts de sebkras

Les sebkras sont de bons indicateurs d'une évaporation saisonnière très efficace qui entraîne la disparition de la nappe d'eau pendant plusieurs mois. Au cours de la saison sèche, les pseudosables, produits par la floculation des limons et argiles, sont mobilisés par le vent et les plus gros s'accumulent en formant un bourrelet en arc de cercle (lunette, clay dune) du côté sous le vent. La position de ce bourrelet indique la direction dominante des vents de saison sèche à l'époque du fonctionnement de la sebkra: ENE à Bou Ziane à l'Est d'Oran (Boulaine 1954), NW pour toutes les sebkras de la Tunisie orientale (Coque 1979). En Mauritanie côtière, où les alignements de dunes sableuses sont NNE-SSW conformément à l'alizé, les dunes d'argile gypseuse des sebkras de l'Aftout es Sahel indiquent des vents de Sud ou de SE (Tricart 1955): cette opposition est-elle due à deux vents saisonniers différents (alizé et brise de mer) ou à un décalage dans le temps de ces deux actions éoliennes? Dans tous ces cas, il est difficile de fixer à quelle période se rattache la formation de ces bourrelets. En Tunisie, la lunette de Bled el Fahs daterait d'une période plus sèche à l'Holocène (Perthuisot & Jauzein 1975). En Tunisie, R. Coque (1962) remarque que la nature de ces bourrelets est limono-argileuse pour les sebkras du Nord et du Centre, mais gypseuse autour des Chotts de Tunisie méridionale. J.M.Bowler (1973), en Australie, observe également que les bourrelets sont formés de quartz, puis de limons argileux et enfin de gypse quand on va des régions plus humides vers les régions arides. Malheureusement il n'existe qu'une seule étude détaillée sur les faciès d'une dune gypseuse d'Oranie (Trichet 1967), alors qu'en Australie, la succession des épisodes arides a pu être étudiée à partir d'analyses de profils de bourrelets (par exemple: AnZhismeng et al. 1986).

Les difficultés d'interprétation des dépôts alluviaux

Le rôle de l'évaporation est tout aussi important sur l'évolution des cours d'eau puisque déjà dans les régions semi-arides, les écoulements pérennes disparaissent dans la plupart des vallées, excepté quelques grands fleuves allogènes qui atteignent les régions arides.

Les alternances de terrasses graveleuses et limoneuses

L'opposition entre terrasses gravelo-sableuses et sablo-limoneuses est très fréquente dans les régions bordières du Sahara: 'graviers sous berge' et limons holocènes au Sahel, terrasses de galets au début du Soltanien (Pléistocène supérieur) et limons colorés à la fin en Afrique du Nord, etc. Presque toujours, les faciès grossiers et désordonnés ont été attribués aux périodes arides et les limons à des phases d'inondations lentes et prolongées sous climat plus humide. Cette interprétation est possible dans de nombreux cas, mais nécessiterait des études sédimentologiques précises qui sont encore à faire. Le problème est, en effet, bien plus complexe car chaque bassin hydrographique réagit en fonction de sa superficie, de sa lithologie, du contexte tectonique etc.

a. Une terrasse caillouteuse peut être en rapport avec des orages violents et rares, des ruissellements rapides sur des versants dénudés et des crues très brèves, mais puissantes. Elle caractérise alors un climat aride ou semi-aride. Mais, surtout en Afrique du Nord et à la périphérie des reliefs (même au Sahara central) cette charge en galets peut signifier une plus grande activité du gel et de la solifluction en montagne et une compétence accrue des cours d'eau sous l'effet de pluies hivernales. C'est probablement le cas pour la terrasse soltanienne graveleuse. D'ailleurs, selon les secteurs, une période aride ne se traduit pas toujours par un abandon de charge: ainsi, durant la phase plus aride de 14 000 à 6 000 BP, la Saoura s'est enfoncée dans sa terrasse 'saourienne' (Pluvial soltanien) par suite d'une puissance accrue des écoulements brefs et concentrés sur le piedmont de l'Atlas. Lors de la période plus humide de l'Holocène moyen, elle a, à nouveau, construit une terrasse 'guirienne' (Conrad 1969).

b. Une terrasse limoneuse ne correspond pas toujours à une période pluviale. En effet, il existe des secteurs où, en régions arides, les crues déposent essentiellement des limons, dans les parties dégradées de leur cours où la vitesse et le débit diminuent brusquement. C'est le cas des très nombreux maaders actuels, argilo-limoneux, du Sahara septentrional. Cet excès de charge en limons, en période aride, peut être dû à des apports massifs de poussière éolienne (Rognon 1976, 1984) ou à une érosion sélective de loess dans leur bassin versant (Coudé-Gaussen 1984b). Seule une étude minéralogique peut permettre de différencier des limons d'origine pédologique (pluvial) ou désertique (aride).

Le comportement de l'hydrographie à l'échelle régionale

Il est très dangereux de tirer des conclusions à partir des réactions d'un seul bassin versant. Par contre, lorsque l'on constate un ennoyage alluvial généralisé de toutes les vallées d'une région, au même moment, il est clair qu'il s'agit d'un changement climatique. Ainsi le comblement systématique des vallées à la périphérie du massif de l'Atakor (Hoggar) vers la fin de l'Acheuléen s'est produit sur une période relativement brève à la suite d'un brusque assèchement climatique (crise climatique). Les vallées ont été comblées par les apports massifs des débris de la couverture pédologique antérieure, que les crues ont été incapables de transporter: l'ensemble du réseau hydrographique a été totalement désorganisé (Rognon 1967). Cette terrasse 'moyenne' graveleuse, épaisse de 4-5 à 15-20 mètres, représente la réponse de l'hydrographie au cours de la phase d'aridification (Pl. II.3). En effet, sauf conditions tectoniques exceptionnelles, les comblements généralisés des vallées impliquent des apports considérables de matériel sous climat semi-aride plutôt que réellement aride (Rognon 1980).

Un autre exemple est fourni par le comportement des oueds du piedmont et de l'Atlas saharien algérien à l'extrême fin du Pléistocène et au début de l'Holocène. En effet, de nombreuses monographies ont montré le comblement systématique des sorties de gorge (foums) par des cônes alluviaux épais, et/ou par des cordons dunaires vers la fin du Soltanien (Trayssac 1981) ou entre 12 000 et 8 000 BP (Rognon 1987).

Enfin les périodes arides sont enregistrées sur les grands cours d'eau allogènes. L'exemple le plus connu est celui du Niger, barré au niveau de son coude par l'édification de l'erg ogolien entre 20 000 et 12 000 BP (Beaudet et al. 1977). Le haut Niger a édifié une partie de son delta (Macina) en amont du cordon pendant cette période aride, puis un vaste épandage alluvial au Nord de Tombouctou après 8 500 BP, en période pluviale, avant que cède le second barrage dunaire, situé en amont de Gao (Riser 1986). Pendant l'Ogolien également, le Sénégal ne franchissait plus, semble-t-il, la barrière des ergs NNE-SSW en amont desquels il a alors édifié l'épandage alluvial du 'petit remblai' (Michel 1973). Enfin, les écoulements du Nil blanc lui-même ont été très amoindris ou même interrompus vers 13-14°N comme l'indiquent d'abord les dépôts d'évaporites datés de > 40 000 et jusque vers 26 000 BP, puis les barrages dunaires à l'extrême fin du Pléistocène (Williams & Adamson 1980).

CONCLUSION

Peut-on essayer de dresser un bilan des recherches sur les phases arides du Sahara? On constate, tout d'abord, que les méthodes de reconstitution sont, encore, essentiellement sédimentologiques, ce qui s'explique par la prédominance des actions mécaniques, comme en régions froides. Il est possible d'en tirer des données qualitatives sur:

– L'intensité de l'aridité puisque certains dépôts sont caractéristiques du domaine semi-aride (loess, bourrelets de sebkras etc.), aride (croûtes gypseuses) ou hyperaride (déflation, dunes etc.). Il suffirait de fixer des limites climatiques à chacun de ces faciès, comme il en existe pour les températures exigées, en régions périglaciaires, par le permafrost, les fentes de glace, les pingos etc.

– La succession des séquences arides, grâce aux poussières incorporées à la sédimentation océanique (comme les lâchers d'icebergs en mers périglaciaires). Ces épisodes peuvent être relativement bien datés, mais en dehors du contexte des déserts continentaux.

L'abondance des dépôts chimiques, liée à l'intensité de l'évaporation, doit permettre d'envisager des mesures quantitatives sur les variations de l'aridité à une échelle de temps beaucoup plus fine, en utilisant des marqueurs chimiques classiques ou isotopiques. Les carbonates et les évaporites, en particulier, seront des archives à exploiter dans les prochaines années.

Enfin, la rareté des vestiges paléontologiques restera un des points faibles dans l'étude des paléoaridités si on établit une comparaison avec la profusion des données tirées des organismes microscopiques (pollens, diatomées etc.) lors des périodes pluviales. Mais l'étude des microorganismes inclus dans les dépôts hydriques semi-arides ne fait que commencer (ostracodes, foraminifères etc.). On peut espérer que des études sur carottes, en particulier sur les marges du Sahara, permettront, un jour, de reconstituer les variations des limites du désert, comme on a pu restituer celles de la toundra en domaine tempéré froid. Il sera alors possible de comparer les variations naturelles de ces limites (phases d'aridification) avec celles qu'a entraînées, surtout depuis 4 500 BP, la désertification liée à l'intervention croissante de l'homme. Cette question est particulièrement importante au Sahara où la néolithisation commence (au Niger par exemple) avant 9 000 BP, et où l'importance de la dégradation anthropique est très discutée, mais jusqu'ici sans arguments très objectifs.

BIBLIOGRAPHIE

Adolphe, J.P. 1981. Observations et expérimentations géomicrobiologiques et physicochimiques des concrétionnements continentaux actuels et fossiles. Thèse doc. Etat, Paris, no. 81-84, 336 pp.

Anton, D.J. 1986. Morpho-stratigraphy and eolian activity during the Late Quaternary in the Arabian Peninsula. In *Proc. of Symp. 'Changements globaux en Afrique'* (INQUA, ORSTOM): 5-8.

Anzhismeng, J.M.Bowler, N.D.Opdyke, P.G.Macumber & J.B.Firman 1986. Paleomagnetic stratigraphy of Lake Bungunnia: Plio-Pleistocene precursor of aridity in the Murray Basin, SE Australia. *Palaeogeogr., Palaeoclim., Palaeoecol.* 54: 219-239.

Ballais, J.L., A.Marre & P.Rognon 1979. Périodes arides du Quaternaire récent et déplacement des sables éoliens dans les Zibans (Algérie). *Rev. Géog. Phys. et Géol. Dyn.* 21 (2): 97-108.

Barbey, C. 1982. Les ergs du Sud-Ouest de la Mauritanie et du Nord du Sénégal. Thèse doc. Etat, Paris, 2 tomes, 454 pp.

Beaudet, G., R.Coque, P.Michel & P.Rognon 1977. Y-a-t-il eu capture du Niger? *Bull. Assoc. Géog. fr.* 445-446: 215-222.

Blumel, W.D. 1981. Pedologische und geomorphologische Aspekte der Kalkkrustenbildung in Südwestafrika und Südostspanien. *Karlsruher geog.* 10, 228 pp.

Boulaine, J. 1954. La sebkha de Ben Ziane et sa 'lunette'. *Rev. Géom. Dyn. (Paris)* 5 (3): 102-123.

Bowler, J.M. 1973. Clay dunes: their occurrence, formation and environmental significance. *Earth Sci. Rev.* 9: 315-338.

Boyé, M., F.Marmier, C.Nesson & G.Trecolle 1978. Les dépôts de la sebka Mellala (Ouargla, Sahara algérien). *Rev. Géom. Dyn. (Paris)* 28 (2-3): 49-62.

Brun, A. 1985. La couverture steppique en Tunisie au Quaternaire supérieur. *C.R. Acad. Sci. (Paris)* 301, II (14): 1085-1090.

Callot, Y. 1984. Dépôts lacustres et palustres quaternaires de la bordure nord du Grand Erg occidental (Algérie). *C.R. Acad. Sci. (Paris)* 299 (19): 1347-1350.

Chamley, H., G.Coudé-Gaussen, P.Debrabant & P.Rognon 1987. Contribution autochtone et allochtone à la sédimentation quaternaire de l'île de Fuerteventura (Canaries). *Bull. Soc. géol. Fr.* (à paraître).

Chamley, H., L.Diester-Haass & H.Lange 1977. Terrigenous material in East Atlantic sediment cores as an indicator of N.W. African climates. *Meteor. Forsch. Erg.* C 26: 44-59.

Chamley, H. & G.Millot 1975. Observations sur la répartition et la genèse des attapulgites plio-quaternaires de Méditerranée. *C.R. Acad. Sci. (Paris)* D 181: 215-218.

Chester, R., H.Elderfield, J.J.Griffin, L.R.Johnson & R.C.Padgham 1972. Eolian dust along the eastern margins of the Atlantic Ocean. *Marine Geol.* 13 (2): 91-106.

Conrad, G. 1969. L'evolution continentale Post-Hercynienne du Sahara (Saoura, Erg Chech-Tanezrouft, Ahnet-Mouydir) Centre Rech. Zones Arides, *Sér. Géologie* (10): 527 pp.

Coque, R. 1962. *La Tunisie présaharienne. Etude géomorphologique.* Paris: Colin. 476 pp.

Coque, R. 1979. Sur la place du vent dans l'érosion en milieu aride. L'exemple des lunettes de la Tunisie. *Médit.* 1-2: 15-21.

Cossement, P., H.Chamley & L.Pastouret 1984. Considération sur la sédimentation du Quaternaire terminal en mer d'Alboran (Méditerranée occidentale). *Ann. Soc. géol. Nord (Lille)* 104: 17-27.

Coudé-Gaussen, G. 1984a. Le cycle des poussières éoliennes désertiques actuelles et la sédimentation des loess péridésertiques quaternaires. *Bull. Elf. Aquitaine* 8 (1): 167-182.

Coudé-Gaussen, G. 1984b. Mise en place des basses terrasses holocènes dans les Matmata et leurs bordures (Sud-Tunisien). *Bull. Assoc. fr. Et. Quat.* 123: 173-180.

Coudé-Gaussen, G. 1986a. Le problème des loess périsahariens. In *Proc. of Symp. 'Changements globaux en Afrique'* (INQUA, IRSTOM): 79-82.

Coudé-Gaussen, G. 1986b. Observations au MEB de poussières sahariennes récoltées aux Canaries. In *Proc. of Symp. 'Changements globaux en Afrique'* (INQUA, ORSTOM): 83-86.

Coudé-Gaussen, G. & P.Blanc 1985. Présence de grains éolisés de palygorskite dans les poussières actuelles et les sédiments d'origine désertique. *Bull. Soc. géol. Fr.* 4: 571-580.

Coudé-Gaussen, G., C.Hillaire-Marcel & P.Rognon 1982a. Origine et évolution pédologique des fractions carbonatées dans les loess des Matmata (Sud-Tunisien) d'après leurs teneurs en ^{13}C et ^{18}O. *C.R. Acad. Sci. (Paris)* 295 D: 939-942.

Coudé-Gaussen, G., M.N.Le Coustumer & P.Rognon 1984. Paléosols d'âge Pléistocène supérieur dans les loess des Matmata (Sud-Tunisien). *Sc. Géol. (Strasbourg)* 37 (4): 359-386.

Coudé-Gaussen, G., C.Mosser, P.Rognon & J.Tourenq 1982b. Une accumulation de loess du Pléistocène supérieur dans le Sud-Tunisien: la coupe de Téchine. *Bull. Soc. geol. Fr.* 7, 24 (2): 283-292.

Coudé-Gaussen, G., P.Olive & P.Rognon 1983. Datation de dépôts loessiques et variations climatiques à la bordure nord du Sahara algéro-tunisien. *Rev. Géol. Dyn. et Géog. Phys.* 24: 61-73.

Coudé-Gaussen, G. & P.Rognon 1983. Les poussières sahariennes. *La Recherche* 147: 1050-1062.

Coudé-Gaussen, G. & P.Rognon 1987. The Upper Pleistocene loess of southern Tunisia: a statement. *Earth Surf. Proc.* (à paraître).

Coudé-Gaussen, G., P.Rognon & M.Hoyos-Gomez 1987a. Traceurs minéralogiques de poussières d'origine saharienne recueillies sur l'île de Fuerteventura (Canaries). *Estudios geologicos* (à paraître).

Coudé-Gaussen, G., P.Rognon, A.Rapp & T.Nihlen 1987b. Dating of peridesert loess in Matmata (S.Tunisia) by radiocarbon and T.L. methods. *Zeitschr. für Geomorph.* (à paraître).

Diester-Haass, L. 1976. Late Quaternary climatic variations in North West Africa deduced from East Atlantic sediment cores. *Quat. Res.* 6: 299-314.

Durand, A. & J.Lang 1986. Approche critique des méthodes de reconstitution paléoclimatique: le Sahel nigéro-tchadien depuis 40 000 ans. *Bull. Soc. géol. Fr.* 8 (2): 267-278.

El Moslimany, A.P. 1982. The Late Quaternary vegetational history of the Zagros and Taurus Mountains. *BAR. Intern. Series* 133: 343-351.

Fontes, J.C., F.Gasse, Y.Callot, J.C.Plaziat, P.Carbonel, P.A.Dupeuble & I.Kaczmaska 1985. Freshwater to marine like environments from Holocene lakes in northern

Sahara. *Nature* 317: 608-610.
Gac, J.Y. 1985. Le phénomène des brumes sèches au Sénégal en 1984-85. *Veille climatique satellitaire* (Lannion) 7: 31-35.
Gibbs-Russell, G.E. 1986. Significance of different centres of diversity in subfamilies of Poaceae in southern Africa. *Palaeoecol. of Africa* 17: 183-191.
Guy, M. & M.Mainguet 1975. Les courants de transport éolien au Sahara et leurs manifestations au sol. *C.R. Acad. Sci. (Paris)* D: 103-106.
Hooghiemstra, H. 1986. Distribution patterns of pollen in marine sediments from a record for the seasonal wind patterns over NW Africa. In *Proc. of Symp. 'Changements globaux en Afrique'* (INQUA, ORSTOM): 191-194.
Janecek, T.R. & D.K.Rea 1985. Quaternary fluctuations in the northern hemisphere trade winds and westerlies. *Quat. Res.* 24: 150-163.
Jauzein, J. 1974. Les données sur le système Ca SO_4, H_2O et leurs implications géologiques. *Rev. Géog. phys. et Géol. dyn.* 16 (2): 151-159.
Kelts, K. & M.Shahrabi 1986. Holocene sedimentology of hypersaline Lake Urmia, NW Iran. *Palaeogeogr., Palaeoclim., Palaeoecol.* 54: 105-130.
Kolla, V., P.E.Biscaye & A.F.Hanley 1979. Distribution of Quartz in Late Quaternary Atlantic sediments in relation to climate. *Quat. Res.* 11: 261-277.
Kolla, V., J.A.Kostecki, F.Robinson & P.E.Biscaye 1981. Distribution and origins of clay minerals and quartz in surface sediments of the Arabian Sea. *J. Sedim. Petrol.* 51: 563-569.
Lange, H. 1982. Distribution of chlorite and kaolinite in eastern Atlantic sediments off North Africa. *Sedimentology* 29: 427-431.
Leprun, J.C. 1971. Nouvelles observations sur les formations dunaires fixées du Ferlo nord-occidental. *Bull. Ass. Sen. Et. Quat.* 31-32: 69-78.
Levy, A. 1982. Sur la survie de certains Foraminifères dans les eaux continentales et sur ses conséquences. *Bull. Soc. geol. Fr.* 144: 161-172.
Levy, A. 1985. Une nouvelle conception de l'origine énigmatique des Cerastoderma glaucum quaternaires du Sahara. *C.R. Acad. Sci. (Paris)* 301: 437-442.
MacKee, E.D. 1979. A study of global sandstone. *Geol. Surv., Prof. papers* 1052, 429 pp.
Maglione, G. 1974. Géochimie des évaporites et silicates néoformés en milieu continental confiné. Les dépressions interdunaires du Tchad. Thèse doc. Etat, Paris, 331 pp.
Mainguet, M. & Y.Callot 1978. *L'erg de Fachi-Bilma (Tchad-Niger).* Mém. & Doc. CNRS, 18, 184 pp.
Maley, J. 1982. Dust, clouds, rain-types and climatic variations in tropical North Africa. *Quat. Res.* 18: 1-16.
McClure, H.A. 1976. Radiocarbon chronology of Late Quaternary lakes in the Arabian Desert. *Nature* 263 (5580): 755-756.
Michel, P. 1973. *Les bassins des fleuves Sénégal et Gambie. Etude géomorphologique.* Mém. ORSTOM, 3 tomes, 752 pp.
Molina-Cruz, A. 1977. The relation of the Southern trade winds to upwelling processes during the last 75 000 y. *Quat. Res.* 8: 324-338.
Oliva, P. 1974. La distribution des croûtes et encroûtements dans l'Anti-Atlas occidental. In *Etude de certains milieux du Maroc et de leur évolution récente,* tome 2: 47-78.
Paquet, H., G.Coudé-Gaussen & P.Rognon 1984. Etude minéralogique de poussières sahariennes le long d'un transect entre 19 et 35° de latitude Nord. *Rev. Géog. phys. et Géol. dyn.* 25: 257-266.
Perthuisot, J.P. 1976. Une sebkha sulfatée sodique en pays sédimentaire. La sebkha Oum el Krialate (Sud-Tunisien). *Geol. Mediterr. (Marseille)* 3 (4): 265-274.
Perthuisot, J.P. & A.Jauzein 1975. Sebkhas et dunes d'argile: l'enclave endoréique de Pont du Fahs, Tunisie. *Rev. Géog. phys. et Géol. dyn.* 17 (3): 295-306.
Pokras, E.M. & A.C.Mix 1985. Eolian evidence for spatial variability of Late Quaternary climates in Tropical Africa. *Quat. Res.* 24: 137-149.
Prospero, J.M. & R.T.Nees 1986. Impact of the North African drought and El Niño on mineral dust in the Barbados trade winds. *Nature* 320: 735-738.

Riser, J. 1986. Paléohydrographie du bassin d'Araouane. In *Proc. of Symp. 'Changements globaux en Afrique'* (INQUA, ORSTOM): 509-510.

Riser, J., C.Hillaire-Marcel & P.Rognon 1983. Les phases lacustres holocènes. In N.Petit-Maire & J.Riser (eds.), *ahara ou Sahel? Quaternaire récent du Bassin de Taoudeni (Mali)*: 65-86. CNRS.

Robert, C., A.Gauthier & H.Chamley 1984. Origine autochone et allochtone des argiles récentes de haute altitude en Corse. *Géol. médit.* 11 (3): 243-253.

Rognon, P. 1961. Les types d'encroûtements calcaires dans les vallées du Nord de l'Atakor. *Bull. Soc. hist. nat. Af. du N.* 52: 94-103.

Rognon, P. 1967. *Le massif de l'Atakor et ses bordures (Sahara central). Etude géomorphologique.* CNRS, Centre Rech. Zones Arides, 9, 560 pp.

Rognon, P. 1976. Constructions alluviales holocènes et oscillations climatiques du Sahara méridional. *Bull. Assoc. Géog. fr.* 33: 77-84.

Rognon, P. 1980. Pluvial and arid phases in the Sahara: the role of non-climatic factors. *Palaeoecol. of Africa* 12: 45-62.

Rognon, P. 1983. Essai de définition et typologie des crises climatiques. *Bull. Inst. Géol. Bassin Aquitain* no. spécial 'Paléoclimats': 151-163.

Rognon, P. 1984. Signification dynamique et climatique des formations et terrasses fluviatiles en Afrique du Nord et au Proche Orient. *Bull. Assoc. fr. Et. Quat.* 1-2: 161-169.

Rognon, P. 1987. Late Quaternary climatic reconstruction for the Maghreb (North Africa). *Palaeogeogr., Palaeoclim., Palaeoecol.* (à paraître).

Rognon, P. & G.Coudé-Gaussen 1986. Reconstitution paléoclimatique à partir des sédiments du Pléistocène supérieur et de l'Holocène du Nord de Fuerteventura (Canaries). *Zeitschr. für Geomorph.* (à paraître).

Ruellan, A. 1970. *Les sols à profil calcaire différencié des plaines de la Basse Moulouya (Maroc oriental).* Mém. ORSTOM, 54, 302 pp.

Sarnthein, M. & L.Diester-Haass 1977. Eolian sand turbidities. *J. of Sedim. Petrol.* 47 (2): 868-890.

Sarnthein, M. & B.Koopmann 1980. Late Quaternary deepsea record on the North-West African wind circulation. In *Abstracts of Symp. 'Sahara and surrounding seas'* (Mainz): 37-38.

Schutz, L., R.Jaenicke & H.Pietrek 1981. Saharan dust transport over the North Atlantic ocean. In T.L.Péwé (ed.), *Desert dust.* Geol. Soc. Amer., Sp. Paper 186: 87-100.

Servant, M. 1973. Séquences continentales et variations climatiques. Evolution du bassin du Tchad au Cénozoïque supérieur. Thèse doc. Etat, ORSTOM, Paris, 368 pp.

Servant, M. & S.Servant-Vildary 1980. L'environnement quaternaire du bassin du Tchad. In M.A.J.Williams & M.Faure (eds.), *The Sahara and the Nile*: 133-162. Rotterdam: Balkema.

Servant-Vildary, S. 1978. *Etude des diatomées et paléolimnologie du bassin tchadien au Cénozoïque supérieur.* Trav. et Doc. ORSTOM, 2 tomes, 346 pp.

Summerfield, M.A. 1982. Distribution, nature and probable genesis of silcrete in arid and semi-arid southern Africa. *Aridic soils and geomorphic processes. Catena* suppl. 1: 37-65.

Talbot, M.R. 1984. Late Pleistocene rainfall and dune building in the Sahel. *Palaeoecol. of Africa* 16: 203-214.

Tomadin, L., R.Lenaz, V.Landuzzi, A.Mazzucatelli & Vanucci 1984. On wind-blown dusts over the Central Mediterranean. *Oceanologica Acta* 7: 13-24.

Trayssac, J. 1981. *Etude géomorphologique du bassin versant de l'O. Djelfa-Melah (Ouled Naïl, Algérie).* Poitiers: Etudes médit. 3, 241 pp.

Tricart, J. 1955. Nouvelles observations sur les sebkhas de l'Aftout es Sahel mauritanien et du delta du Sénégal. *Rev. geom. dyn.* 6 (4): 177-187.

Trichet, J. 1963. Description d'une forme d'accumulation de gypse par voie éolienne dans le Sud Tunisien. *Bull. Soc. Geol. Fr.* 5: 617-621.

Trichet, J. 1967. Etude des faciès d'une dune gypseuse (Sud d'Oran, Algérie). *Bull. Soc.*

Géol. Fr. 7 (9): 865-875.

Vogt, T. 1982. Sur la présence d'Algues d'eau douce dans une croûte calcaire quaternaire d'Afrique du Nord. *C.R. Acad. Sci. (Paris)* 295: 703-708.

Vogt, T. 1984. Problèmes de genèse des croûtes calcaires quaternaires. *Bull. Elf. Aquitaine* 8: 209-221.

Warren, A. 1970. Dune trends and their implications in the central Sudan. *Zeitschr. für Geomorph.* 10: 154-180.

Wasson, R.J., G.I.Smith & D.P.Agrawal 1984. Late Quaternary sediments, minerals and inferred geochemical history of Didwana Lake, Thar desert. *Palaeogeogr., Palaeoclim., Palaeoecol.* 46: 345-372.

Weisrock, A. 1982. Signification paléoclimatique des dunes littorales d'Essaouira. Cap Sim (Maroc). *Rev. geom. dyn.* 416: 91-107.

Weisrock, A., G.Delibrias, P.Rognon & G.Coudé-Gaussen 1985. Variations climatiques et morphogenèse au Maroc atlantique (30-38°) à la limite Pléistocène-Holocène. *Bull. Soc. géol. Fr.* 8, 1 (4): 565-569.

Williams, M.A.J. & D.A.Adamson 1980. Late Quaternary depositional history of the Blue and White Nile rivers in central Sudan. In M.A.J.Williams & H.Faure (eds.), *The Sahara and the Nile*: 281-304. Rotterdam: Balkema.

HOLOCENE PALAEOHYDROGRAPHY OF THE NIGER

N.PETIT-MAIRE
CNRS Laboratoire de Géologie du Quaternaire, Marseille, France

J.RISER
University of Avignon, France

RESUME

La question d'un déversement récent du Moyen Niger dans l'Azaouad est discutée depuis plus de soixante-dix ans. En 1984, la découverte de poissons Perciformes holocènes au Nord d'Araouane a renforcé cette hypothèse. Des études de terrain, de télédétection et de photointerprétation ont permis de cartographier un vaste delta intérieur, daté sur Mollusques d'eau douce de 8 500 à 3 500 BP, qui s'étendait au Nord et Nord-Est de Tombouctou.

ABSTRACT

For more than seventy years the question has been discussed whether the middle Niger course, in Mali, was temporarily changed to the Azawad area during the Quaternary. In 1984, the discovery of holocene fossil fish North of Arawan supported this hypothesis. Field work and mapping of fluvio-lacustrine deposits from air photographs now present evidence for the existence of a broad interior delta up to c. 100 km to the North and North-East of Timbuktu, between 8 500 and 3 500 BP.

FIELD EVIDENCE AND DISCUSSION OF RESULTS

Upstream from Timbuktu, the Niger flows across a marshy and lacustrine area. Downstream, in contrast, it flows eastwards through a narrow gorge lined with terraces: the Taoussa defile (Fig. 1). This geographic setting, the marked Niger band near Timbuktu and the presence of Quaternary terraces east of Taoussa testify to a capture of the Upper Niger by the Tilemsi system at a period dating from at least the ancient Quaternary (Chudeau 1913, 1919, Furon 1929, Baulig 1950, Radier 1959, Tricart 1965, Blanck 1968, Beaudet et al. 1977). Several authors (Furon 1929, Beaudet et al. 1977, Urvoy 1942, Palausi 1955) consider that the river, prior to its capture, expanded to the North into the vast, practically flat (8 m maximal elevation differences) area,

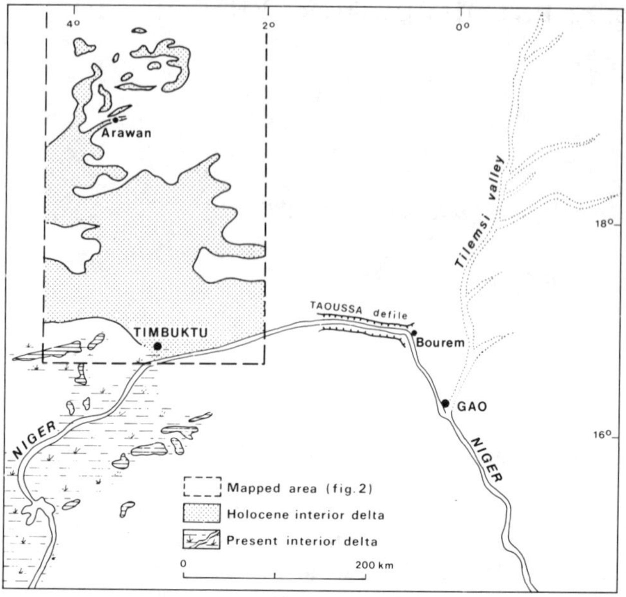

Figure 1.
Locality map

which extends from 16°30′ to 20°N and 2° to 4°W. Their opinions differ as to the reason for the capture. Some of them (Urvoy 1942, Tricart 1965, Blanck 1968, Reichelt 1973, Klein 1979) assume that epeirogenic movements affected the regional drainage. Others (Beaudet et al. 1977) infer climatic changes resulting in the erection or the disappearance of natural barriers: the Niger would have inundated the Arawan basin in the Upper Pleistocene while Ogolian dunes barred it at Taoussa, and the river would have flowed again to the East, in the early Holocene, when strong floods re-opened the passage. We shall see further on that the radio-carbon ages for the Azawad fluvio-lacustrine deposits do not support these conclusions.

Between 1980 and 1986, seven field trips, among which two were especially devoted to the area considered here, were made in northern Mali to gather palaeoenvironmental data. The flat region North of the Niger bend, as well as most of the Taoudenni basin up to 24°N (Petit-Maire & Riser 1981, Petit-Maire et al. 1983, Petit-Maire & Riser 1983), proved indeed to be very rich in fossiliferous silts and clays of fluvial, lacustrine or marshy origins. In the Azawad, these deposits are located in the corridores between the Late Pleistocene (Ogolian) ergs (Rognon 1980, Talbot 1984) while to the North of Arawan they occur within small topographic depressions of varying diameters (3 to 1 500 m), depths (1 to 4 m) and shore slopes (0.1 to 0.4%) (Decobert & Petit-Maire 1985).

A rich invertebrate and vertebrate fauna is associated with these formations, consisting of: Ostracods, molluscs, fish, crocodiles, water turtles and large snakes (Carbonel 1983, Rosso 1983, Gayet 1983, Buffetaut 1983, Broin 1983). A great number of mammalian remains which imply a savanna environment (elephants, hippos, rhinoceros, wart hogs, giraffes, lions, large antelopes and buffaloes) are found along the palaeoshores in large Neolithic middens (Guérin & Faure 1983, Raimbault 1983) which had already been observed by several authors (Gallay 1966, Joleaud 1934, Mauny & Poussibet 1962, Monod 1958, Roman 1934, 1935).

Nine samples of aquatic mollusc shells and fish bones from these formations and two samples of terrestrial Pulmonae (*Limicolaria* sp.) from black palaeosols along shores of palaeolakes were radiocarbon dated from 8 800 to 3 460 BP (Table 1) by Delibrias (Gif lab.) and Hillaire-Marcel (UQ lab.). The four ages from the most northern areas are older than 5 000 BP while the dates younger than 4 000 BP correspond to deposits located at 18°40′, 18°28′ and 17°31′N. These dates perfectly fit the holocene climatic optimum and its regressive phase as observed elsewhere in northern Mali (Petit-Maire & Riser 1983, Petit-Maire 1986).

The existence of such lacustrine areas in the Azawad does of course not imply direct surface connections with the Niger. They may have been fed by rainfall and/or by phreatic underflow and it has been demonstrated that migrating birds may be the vectors of eggs, larva or adult molluscs over long distances (Malone 1965); crocodiles, water turtles, hippopotami may crawl from one pool to the other; even Siluridae fish can survive dessication of irregular channels and lakelets. In contrast, the finding near Hassi-el-Abiod (Fig. 2) of very numerous remains of large (up to 1.60 m) Perciform fish (Petit-Maire & Gayet 1984) brought a new argument in favour of such connections: the two species discovered there (*Lates niloticus* and *L. maliensis* Gayet) require running, well oxygenated, perennial flowing water (Gayet 1983); besides, their living fry of eggs are too fragile to suffer dry transportation at such low, warm latitudes. Therefore, palaeontological remains provided evidence for what has been suggested by sedimentological studies but, to this day, the nature and geometric aspect of the surface water network still remained unknown.

In order to unravel this problem, field investigations were carried out in most of the considered area. Subsequently, all the Holocene lake deposits which are easily characterized by their constant white to greyish colour, were identified by remote sensing (with the help of B.Simon and N.Page). They were then mapped in detail (1/50 000) from air photographs by one of us (J.Riser) who was counselled by J.Marchand. Figures 1 and 2 show the results of that survey.

The southern area, up to c. 100 km North of Timbuktu, belonged to an organized hydrological system. Meandering channels flowed between the Ogolian consolidated ergs lines, typifying a tree-like inland delta. In an intermediate zone, from c. 100 km to 180 km North of Timbuktu, the system shows less dynamism although some channels meander through the interdune

Table 1. Radiocarbon dates from molluscs collected along shores of palaeolakes

Date (BP)	Latitude	Longitude	Material	Notes	Sample nr.	Lab. nr.
3 460 ± 90	18° 28′ 10″N	2° 34′ 30″W	Shells	*Limicolaria* sp.	MN 25	Gif 6459
3 530 ± 90	17° 31′ 40″N	3° 06′ 50″W	Shells	*Melania tuberculata*	MN 2	Gif 6460
3 940 ± 90	18° 40′ 20″N	2° 34′ 10″W	Shells	*Limicolaria* sp.	MN 20	Gif 6463
4 970 ± 60	19° 13′ 00″N	3° 34′ 00″W	Shells	*Caelatura lacoini*	AR 6/1	UQ 370
5 900 ± 200	19° 16′ 00″N	3° 32′ 25″W	Shells	*Biomphalaria pfeifferi*	AR 5	UQ 1078
5 920 ± 100	19° 07′ 50″N	3° 58′ 45″W	Shells	*Melania tuberculata*	MN 32	Gif 6462
6 970 ± 130	19° 07′ 20″N	3° 56′ 35″W	Bones	*Lates maliensis*	AR 7	Gif 5495
8 450 ± 60	19° 06′ 25″N	3° 55′ 35″W	Shells	*Melania tuberculata*	AR 8	UQ 368
8 500 ± 120	17° 25′ 40″N	3° 02′ 40″W	Shells	*Melania tuberculata*	MN 1	Gif 6461
8 700 ± 200	19° 16′ 00″N	3° 32′ 25″W	Shells	*Melania tuberculata*	AR 6/2	UQ 1019
8 800 ± 200	18° 54′ 00″N	3° 28′ 00″W	Shells	*Melania tuberculata*	AR 1	UQ 1021

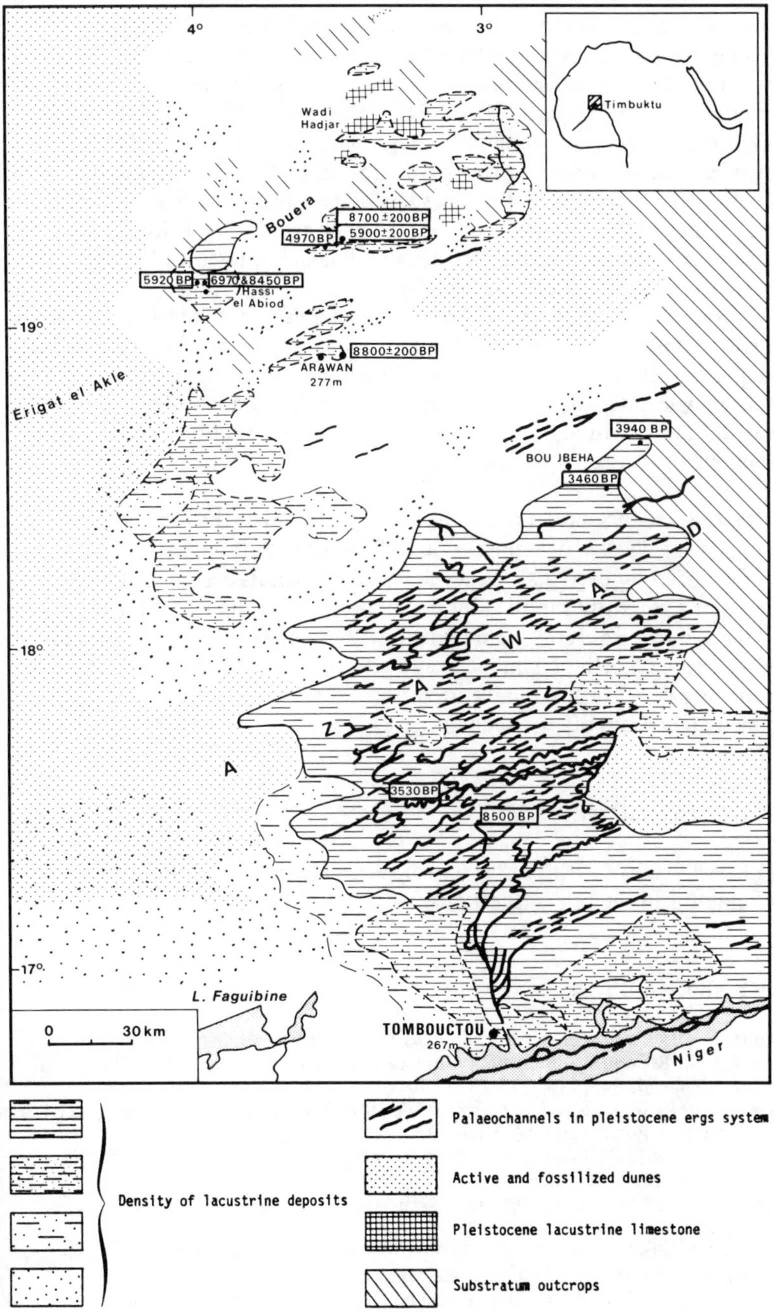

Figure 2. Geomorphology and radiocarbon dates of Holocene interior delta.

corridors; the deposits are still nearly continuous. Further North, up to 330 km from Timbuktu, the lacustrine sediments are scarcer; they occur mainly to the West where channel continuity may at times have been effective as far as Wadi Hadjar, at 19°30′N. In the Erigât and Mreyyé, interdune deposits dot the small topographic depressions, with no indication of organization. We could not check this area in the field, nor its relation with the probable extension of the Lake Faguibine and southern delta. In these areas, phreatic feeding was probably the main factor of surface flooding, perhaps corroborated by important rainfall which is testified by the wide extension of the savanna mammals up North.

It is noteworthy that extensive lacustrine limestones are found along the northern limit of the Holocene lacustrine area. These may be attributed to the pre-Ogolian Late Pleistocene on the grounds of stratigraphic and archaeologic data (they are associated with an Aterian industry). This suggests that the hydrologic situation described here was probably repeated from one or several more ancient models.

From the evidence presented here, it is suggested that the early and mid-Holocene of West Africa must, at least from time to time, have been characterized by extremely heavy seasonal rains, notably exceeding the precipitations presently supplied by the monsoonal system and, occasionally, by the sub-tropical Jet Stream (Rognon 1976, Maley 1983). The interior delta, created North of Timbuktu and covering more than 60 000 km^2, is probably related to a damming up of the Taoussa gorge which could not immediately drain off the sudden floods.

ACKNOWLEDGEMENTS

We thank Professor Rh.Fairbridge for reviewing this text and Professor H.Faure for his helpful suggestions.

REFERENCES

Baulig, H. 1950. *Essais de géomorphologie.* Paris: Les Belles Lettres, 160 pp.

Beaudet, G., R.Coque, P.Michel & P.Rognon 1977. Y-a-t-il eu capture du Niger? *Bull. Ass. Géogr. fr.* 445/446: 215-222 (1 fig.).

Blanck, J.P. 1968. *La boucle du Niger (Mali).* Strasbourg: Centre géog. appliquée, 41 pp., 11 cartes, 1 notice (26 pp.), 1 fasc.

Broin, F.de 1983. Chéloniens. In N.Petit-Maire & J.Riser (eds.), *Sahara ou Sahel?*: 211-233. Marseille: Imprimerie Lamy.

Buffetaut, E. 1983. Crocodiliens. In N.Petit-Maire & J.Riser (eds.), *Sahara ou Sahel?*: 235-238. Marseille: Imprimerie Lamy.

Carbonel, P. 1983. Ostracodes et paléohydrologie. In N.Petit-Maire & J.Riser (eds.), *Sahara ou Sahel?*: 173-180. Marseille: Imprimerie Lamy.

Chudeau, R. 1913. La zone d'inondation du Niger. *Bull. soc. géogr. comm.* 35: 569-587.

Chudeau, R. 1919. La capture du Niger par le Taffassasset. *Ann. géogr.* 151: 52-60.

Decobert, M. & N.Petit-Maire 1985. An early neolithic midden and necropolis in the Malian Sahara. *Nyame Akuma* 26: 26-32 (6 figs.).
Furon, R. 1929. L'ancien delta du Niger (contribution à l'étude de l'hydrographie ancienne du Sahel soudanais et du Sud-saharien). *Rev. géol. dyn. et géogr. phys.* 2: 265-274 (1 carte).
Gallay, A. 1966. Quelques gisements néolithiques du Sahara malien. *J. Soc. africain* 36 (2): 167-208.
Gayet, M. 1983. Poissons. In N.Petit-Maire & J.Riser (eds.), *Sahara ou Sahel?*: 183-210. Marseille: Imprimerie Lamy.
Guérin, C. & M.Faure 1983. Mammifères. In N.Petit-Maire & J.Riser (eds.), *Sahara ou Sahel?*: 239-272. Marseille: Imprimerie Lamy.
Joleaud, L. 1934. Vertébrés subfossiles de l'Azaouad. *C.R. Acad. Sc. Paris* 198 (6): 599-601.
Klein, J.C. 1979. *Détermination des causes des anomalies de la crue du Niger.* Mém. ORSTOM, 77 pp. (14 figs., 3 pl. photos).
Maley, J. 1983. Histoire de la végétation et du climat de l'Afrique nord-tropicale au Quaternaire récent. *Bothalia* 14: 377-389.
Malone, C. 1965. Dispersal of aquatic Gastropods via the intestinal tract of water birds. *Nautilus* 78 (4): 135-140.
Mauny, R. & F.Poussibet 1962. Nouveaux sites à harpons et faune subfossile de l'Azawad (Sahara malien). *Notes afr.* 93: 1-5.
Monod, Th. 1958. *Mâjabât-al-Koubrâ.* Dakar: Mém. IFAN 52, 407 pp.
Palausi, G. 1955. Le fleuve fossile de Tombouctou. *Rev. Géol. dynam.* 5: 217 (2 photos).
Petit-Maire, N. 1986. Palaeoclimates in Malian Sahara (17°-24°N). *Episodes* (in press).
Petit-Maire, N. et al. 1983. The Sahara in northern Mali: man and his environment between 10 000 and 3 500 years BP. *Afr. archaeol. Rev.* 1: 105-125.
Petit-Maire, N. & M.Gayet 1984. Hydrographie du Niger (Mali) à l'Holocène ancien. *C.R. Acad. Sci. Paris* 298: 21-23 (1 fig.).
Petit-Maire, N. & J.Riser 1981. Holocene lake deposits and palaeoenvironments in northeastern Mali. *Palaeogeogr. Palaeoclim., Palaeoecol.* 35: 45-61.
Petit-Maire, N. & J.Riser 1983. *Sahara ou Sahel? Quaternaire récent du Bassin de Taoudenni (Mali).* Marseille: Imprimerie Lamy, 473 pp. (79 figs., 58 tables, 55 pl.).
Radier, H. 1959. Contribution à l'étude géologique du Soudan oriental. *Bull. serv. géol. et prosp. min.* Dakar.
Raimbault, M. 1983. Industrie lithique. In N.Petit-Maire & J.Riser (eds.), *Sahara ou Sahel?*: 317-342. Marseille: Imprimerie Lamy.
Reichelt, R. 1973. Permanence des mouvements épéirogénétiques dans la boucle du Niger, République du Mali. Influence sur la géographie physique. Réflexion sur la dynamique du manteau. *C.R. Acad. Sci. Paris* 82: 417-419 (1 fig.).
Rognon, P. 1976. Essai d'interprétation des variations climatiques au Sahara depuis 40 000 ans. *Rev. Géol. dyn. et Géogr. phys.* 18: 251-281.
Rognon, P. 1980. Une extension des déserts (Sahara et Moyen-Orient) au cours du Tardiglaciaire, 18 000-10 000 BP. *Rev. Géol. dyn. et Géogr. phys.* 22 (4/5): 313-328.
Roman, F. 1934. Sur quelques vertébrés subfossiles du Sahara occidental. *Bull. Soc. géol. Fr.* 5 (2): 13-14.
Roman, F. 1935. Sur une faunule de vertébrés et sur des pièces néolithiques du Sahara occidental. *Bull. Ass. rég. Pal. Préh. Lyon* : 1-13.
Rosso, J.C. 1983. Mollusques. In N.Petit-Maire & J.Riser (eds.), *Sahara ou Sahel?*: 157-172. Marseille: Impremerie Lamy.
Talbot, M.R. 1984. Late Pleistocene rainfall and dune building in the Sahel. *Palaeoecol. of Africa* 16: 203-214 (2 figs.).
Tricart, J. 1965. *Reconnaissance géomorphologique de la vallée moyenne du Niger.* Mém. IFAN 76, 196 pp.
Urvoy, Y. 1942. *Les bassins du Niger.* Dakar: Inst. fond. Afr. noire, 140 pp. (36 figs., VI pl., 4 cartes h.t.).

DIE HOLOZÄNE VEGETATION DER ZENTRALEN SAHARA (N-MALI, N-NIGER, SW-LIBYEN)

E.SCHULZ
Geographisches Institut, Universität Würzburg, F.R.Germany

ZUSAMMENFASSUNG

Pollenanalytische Untersuchungen holozäner Sedimente aus N-Mali, N-Niger und SW-Libyen belegen für die Zeit von 8 000-6 000 BP eine Nordverschiebung der Savannengrenzen auf c.19/22°N und daher eine Einengung der Wüste von Süden her. Holzkohlenbefunde aus der zentralen Ténéré ergeben ebenfalls eine Nord-Verlagerung sudanischer Vegetation um c. 300-400 km.

Die Gebirgsbereiche trugen eine Savannenvegetation, bei der in der südlichen Sahara ein reicherer und ärmerer Typ ausgeprägt waren. Die zentrale Sahara hatte dagegen eine Vegetation aus saharo-sindische Elementen. Die Ausdehnung einer reinen, kontrahierten, Wüstenvegetation war gering.

Ab c. 6 000 BP ist eine graduelle Auflichtung und Verarmung der Vegetation zu vermuten, die c. 4 000 BP den heutigen Grundtyp erhalten hat.

SUMMARY

Pollen analytical investigation of holocene sediments from N. Mali, N. Niger and SW. Libya give evidence of a northward shifting of the desert-savanna boundary to about 19-22°N. Findings of charcoals in the central Ténéré also give proof for a northward extension of sudanian vegetation of about 300-400 km.

Mountain areas were covered with a savanna vegetation which was present in the southern Sahara in a poorer and a richer type. The central Sahara was covered by a saharo-sindic vegetation and the extension of the true, contracted, desert vegetation was certainly restricted.

From about 6 000 BP onward there was a gradually impoverishment and reduction of the plant cover which developed its present character from about 4 000 BP onward.

RESUME

Des analyses polliniques des sédiments holocènes des régions septentrionals du Mali et du Niger ainsi que du sudouest de la Libye montrent que pendant la période de 8 000-6 000 BP la limite de la savane septentrionale a remonté à environ 19-22°N réduisant ainsi l'espace désertique à partir du sud. Des analyses des charbons de bois du Ténéré central montrent également que la végétation soudanienne a remonté d'environ 300-400 km.

Les régions montaigneuses étaient à cette époque couvertes d'une végétation de savane, pour laquelle on peut distinguer dans le Sahara méridional un type plus riche et un type plus pauvre. Le Sahara central par contre avait une végétation saharo-sindique. L'espace couvert d'une végétation désertique pure et contracté était certainement réduit.

A partir d'environ 6 000 BP un éclaircissement et un appauvrissement progressif de la végétation a probablement commencé qui à partir de c. 4 000 BP a abouti à un type fondamental d'aujourdhui.

EINLEITUNG

Einen entscheidenden Landschaftswandel im Holozän der zentralen Sahara belegen Arbeiten von Baumhauer (1986), Gabriel (1985), Pachur & Röper (1984), Petit-Maire & Riser (1983) und Skowronek (1984), sei es durch die Analyse limnischer und fluviatiler Befunde, sei es durch pedologische, prähistorische oder faunistische Untersuchungen.

Tropische Faunenreste werden dabei häufig als Anzeichen für die Ausdehnung von Savannenformationen angenommen, obwohl die floristischen Befunde zumeist keine weitreichenden Schlussfolgerungen tragen. Im Gegensatz stehen dazu die Aussagen von Cour & Duzer (1976) und Dumont (1982), die ein Ausdauern des hyperariden Klimas im Holozän der zentralen Sahara beziehungsweise eine ausgeprägte aride Phase im mittleren Holozän der südlichen Sahara postulieren.

Bisher fehlte aber aus der zentralen Sahara ein kontinuierliches Pollenprofil als Beleg, ebenso wie Makroreste sehr selten waren (vergl. Neumann, dieser Band).

Deutungen von marinen Bohrkernen vor der Küste Westafrikas zur Verschiebung von Landschafts- und Klimazonen (Agwu & Beug 1982, Sarnthein et al. 1982) sind nun einerseits durch die geringere zeitliche Auflösung der holozänen Sedimente, anderseits auch durch die unterschiedliche Einschätzung des jeweiligen Einzugsgebietes für die inneren Teile des Kontinentes vorerst nur schwer zu übertragen.

Ritchie et al. (1983) berichten aus dem Nord-Sudan über eine Ausdehnung von sahelischen Savannen im Frühholozän und nehmen ein Nordwärtswandern der Vegetationszonen um c. 300-400 km an. Damit ist nun erneut die Frage aufgeworfen, ob es wirklich zu einem Uberschreiten der Schwelle zur Savannenbildung in der zentralen Sahara kam – also dem Ubergang von der permanent kontrahierten zur dauerhaft diffusen Vegetation –, oder ob die Anderungen im physischen Milieu nur zu einer Verdichtung der Vegetation führten ohne dass sich ein grundsätzlich anderes ökologisches System einstellen konnte.

Pollenanalytische Untersuchungen von Sedimenten der Depression von Taoudenni (N-Mali) geben nun Gelegenheit, diese Fragen im Vergleich mit Arbeiten aus dem Niger und aus Libyen erneut zu diskutieren. Diese Untersuchungen sind Teil des Programmes des Geographischen Instituts Würzburg in Zusammenarbeit mit der Universität Niamey, dem Laboratoire de Géologie du Quaternaire von Marseille und dem Archäologischen Institut der Universität Rom. Für die Region zwischen Taoudenni und Tombouctou ist in Fortsetzung davon eine Serie vollständiger Bohrungen geplant.

Eine erste Auswertung der Holzkohlenbefunde aus der zentralen Ténéré (Fachi-Dogonbolo) wird zusätzliche Argumente liefern.

HEUTIGES PHYSISCHES MILIEU

Aus der beigegebenen Vegetationskarte (Fig. 1) ist ersichtlich, dass die Sahara nach Norden zu den Halbwüsten und nach Süden zu den Savannen klar begrenzt ist (Schulz 1986b). Ohne auf die verworrenen Darstellungen der sahelischen Vegetation einzugehen (vergl. dazu Monod 1986) soll darauf hingewiesen werden, dass die Sahara im Süden an ein diffuses Baum-Grasland grenzt, das in verschiedenen Formationen gegliedert als Savanne anzusprechen ist.

Alle hier behandelten Lokalitäten liegen im Bereich der reinen Wüstenvegetation. Die Depression von Taoudenni besitzt nur eine sehr spärliche Chenopodiaceae-Vegetation, während die anderen Fundstellen von den *Acacia-Panicum* beziehungsweise *Tamarix-Stipagrostis* Komplexen eingenommen werden.

Klimatisch liegt das Untersuchungsgebiet im Aktionsbereich des Monsuns, wobei der Djebel Acacus (SW-Libyen) stärker von zyklonalen Fronten beeinflusst wird. Die Niederschläge überschreiten nicht die 20 mm und sind äusserst unregelmässig in ihrer Verteilung.

Taoudenni, Seguedine und Fachi-Dogonbolo haben vergleichbare topographische Situationen. Es handelt sich jeweils um eine Depression vor einer verkarsteten Sandsteinstufe, in der holozäne Seeablagerungen unterschiedlicher Ausdehnung zu finden sind. Aus dem Djebel Acacus wurden verschiedene datierte Sedimente der Grabungen in den Wadis von Ti-n-Torha und Uan Muhuggiag zum Vergleich herangezogen.

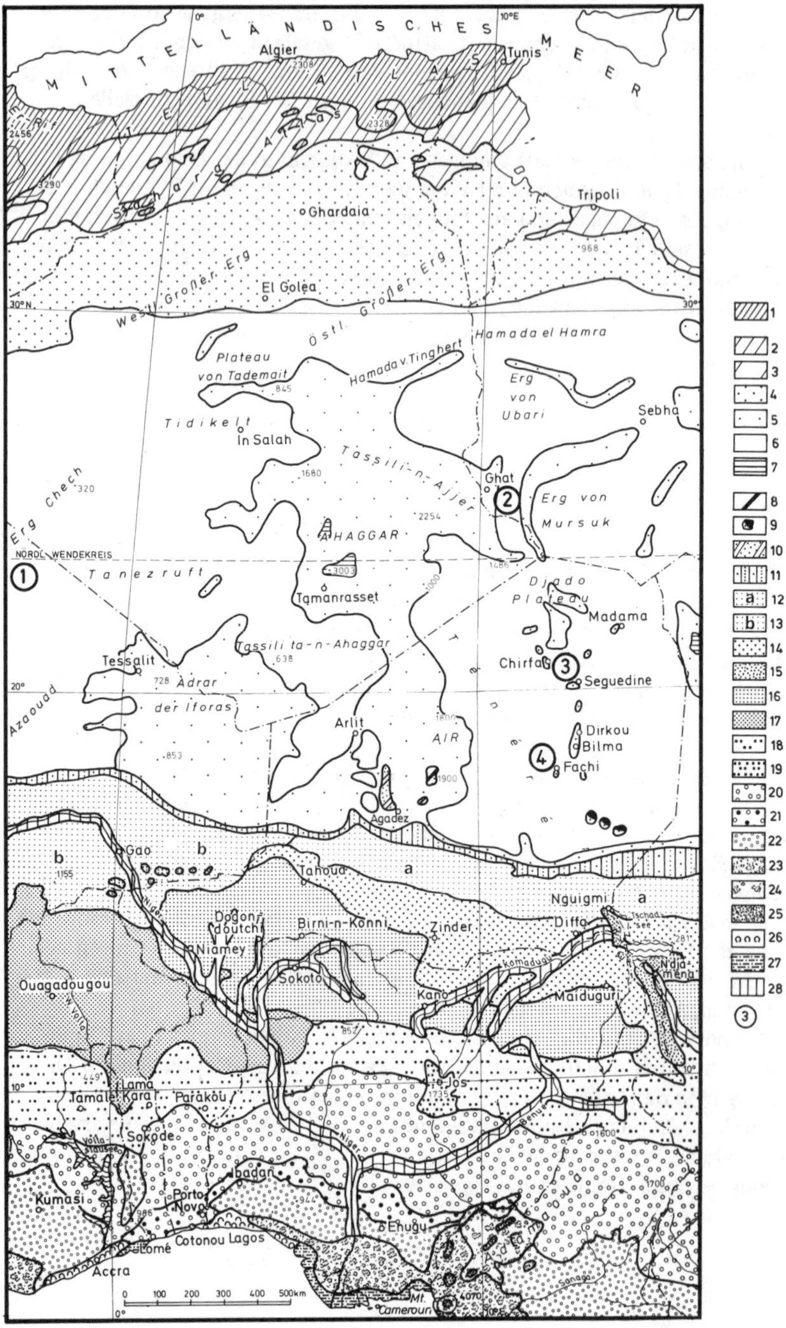
MITTELLÄNDISCHES MEER
Algier
Tunis
TELL ATLAS
Sahara Atlas
Ghardaia
Tripoli
Westl. Großer Erg
Östl. Großer Erg
El Golea
Plateau von Tademait
Hamada v. Tinghert
Hamada el Hamra
Erg von Ubari
Sebha
Tidikelt
In Salah
Tassili-n-Ajjer
Ghat
Erg von Mursuk
Erg Chech
AHAGGAR
NÖRDL. WENDEKREIS
Tanezruft
Tamanrasset
Djado Plateau
Madama
Tassili ta-n-Ahaggar
Tessalit
Chirfa
Seguedine
Adrar der Iforas
Azaouad
Arlit
AIR
Dirkou
Bilma
Fachi
Ténéré
Agadez
Gao
Tahoua
Nguigmi
Diffa
Tschad-See
Dogondoutchi
Birni-n-Konni
Zinder
Niamey
Niger
Sokoto
Kano
Komadugu
Maiduguri
Ndjamena
Ouagadougou
W. Volta
Jos
Lama
Tamale
Kara
Parakou
Benue
Volta-stausee
Sokode
Ibadan
Kumasi
Porto Novo
Enugu
Cotonou
Lagos
Lomé
Accra
Mt. Cameroun
Sanaga
0 100 200 300 400 500 km
30°N
20°
10°
0°
10°E
1 2 3 4 5 6 7 8 9 10 11 12 13 14 15 16 17 18 19 20 21 22 23 24 25 26 27 28

POLLENANALYTISCHE BEFUNDEN

Taoudenni-Agorgott (22° 39'N, 4° 05'E)

Vergleiche Cour & Duzer (1976), Petit-Maire (1985), Petit-Maire & Riser (1983).

Von den zahlreichen Seeablagerungen der Depression von Taoudenni erwiesen sich nur die beiden Profile der Salinen von Agorgott als genügend reich in ihrem Pollengehalt. Die hier dargestellten Profile MT 7 und MT 7b (Fig. 2) sind in den oberen Teilen als Aufschlussgrabungen beprobt worden, wohingegen der untere Abschnitt als Kernbohrung gewonnen wurde, die aber aus technischen Gründen abgebrochen werden musste. Die Korrelation beider Profile, die im Gelände c. 500 m weit entfernt sind, erfolgte nach den Angaben von Fabre (freundliche schriftliche Mitteilung), der auch der Probenautor ist.

Nach den ^{14}C-Daten umfassen die beiden Profile den Zeitraum von c.

Figur 1. Schematische Karte der Vegetationseinheiten des zentralen Teils von Nord- und Westafrika und Lage der behandelten Fundorte (nach Schulz 1986b)

1. Mediterrane Wälder und Gebüsche (Quercus, Ceratonia, Pistacia, Cupressus, Pinus, Cedrus)
2. Mediterrane Steppe (Stipa, Lygeum)
3. Chenopodiaceae-Fluren der Küsten und Schotts
4. Halbwüste (Artemisia, Ephedra, Chenopodiaceae)
5. Kontrahierte Vegetation der Wüste (Acacia-Panicum, Tamarix-Stipagrostis)
6. Ephemere Vegetation. Achab-Floren
7. Diffuse Artemisia-Ephedra-Pentzia-Heiden der Hochgebirge der zentralen Sahara
8. Acacia-Commiphora-Rhus-Savanne der Hochplateaus des südlichen Air
9. Maerua-Savannen der Plateaus des Südostniger
10. Semi-diffuse Acacia-Panicum-Vegetation. Erweiterte Wadi-Vegetation
11. Diffuse Acacia-Panicum-Vegetation. Ubergang Wüste/Savanne
12. Commiphora-Acacia-Savannen
13. Acacia-Leptadenia-Commiphora-Savannen
14. Piliostigma-Bauhinia-Acacia-Savannen
15. Acacia-Dornbusch auf den Uberschwemmungsebenen um den Tchadsee
16. Combretaceae-Savannen
17. Parkia-Butyrospermum-Terminalia-Savannen
18. Isoberlinia-Daniellia-Pterocarpus-Savannen
19. Isoberlinia-Carissa-Ficus-Savannen auf dem Josplateau
20. Afzelia-Lophira-Savannen und Trockenwälder
21. Albizzia-Milletia-Fagara-Savannen
22. Terminalia-Piptadeniastrum-Wälder und -Savannen
23. Tieflandsregenwald (Lophira-Piptadeniastrum-Caesalpiniaceae)
24. Submontane Carapa-Pentadesma-Ficus-Wälder und -Savannen
25. Montane Adenocarpus-Gnidia-Hypericum-Wälder und -Heiden
26. Küstenwälder und -gebüsche (Ficus-Celtis, Cynometra)
27. Mangroven (Avicennia-Rhizophora)
28. Flussbegleitende Vegetation in verschiedenen Formationen

Lage der behandelten Fundorte:

1. Taoudenni; 2. Dj. Acacus; 3. Seguedine; 4. Fachi.

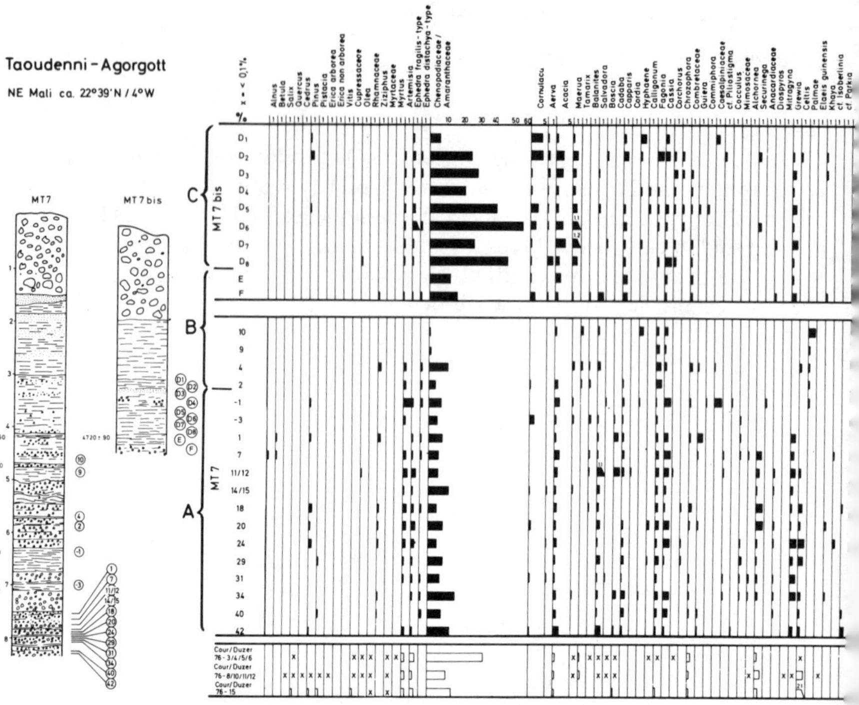

Figur 2. Pollendiagramm von Taoudenni-Agargott

8 000-2 000 BP. Sie bestehen aus wechsellagernden Tonen, Sanden und Salzschichten und haben in ihren unteren lamellierten Lagen teilweise recht hohe Gehalte an organischem Material. Funde von *Pediastrum*-Algen in diesem Bereich weisen auf Süsswasser oder wenig brackisches Wasser hin. Die geochemischen und sedimentologischen Untersuchungen deuten dabei auf mittelgrosse Seen mit wechselnden Wasserständen.

Proben aus dem oberen Bereich dieser Ablagerungen, die einem anderen Aufschluss in den Minen entstammten, wurden schon von Cour & Duzer (1976) untersucht.

Die Sedimente wurden sämtlichst in der Kombination der klassischen Behandlungen (HF, HCl, Acetolyse) aufbereitet und bis zur Konstanz der Prozentwerte in den Pollenspektren ausgezählt. Nach dem Pollengehalt lassen sich die Profile in drei Abschnitte gliedern.

Der untere Teil, A (c. 8 000-6 000 BP), fällt durch den hohen Gehalt an Gramineae und Cyperaceae auf, unter denen der *Cyperus conglomeratus*-Typ überwiegt. Chenopodiaceae/Amaranthaceae haben dagegen nur geringe Werte.

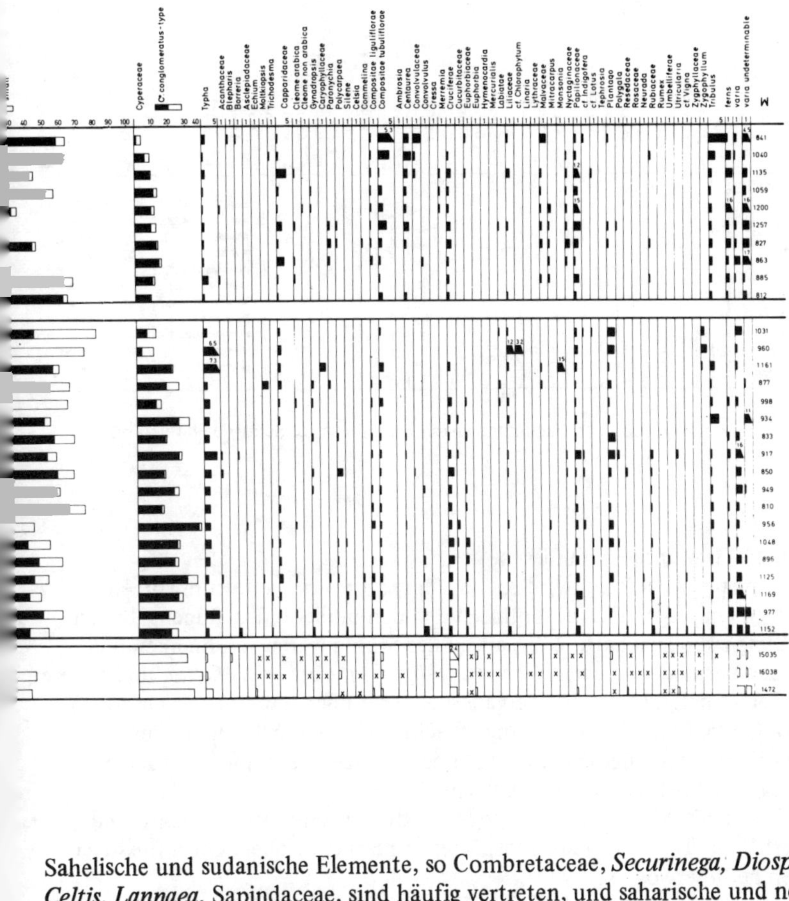

Sahelische und sudanische Elemente, so Combretaceae, *Securinega, Diospyros, Celtis, Lannaea,* Sapindaceae, sind häufig vertreten, und saharische und nordsahelische Elemente wie *Acacia, Maerua, Salvadora, Capparis, Cadaba, Fagonia, Cassia* oder *Grewia* zeigen eine gleichmässige Verteilung in den Spektren. Temperierte und mediterrane beziehungsweise mediterran-präsaharische Florenelemente kommen mit Ausnahme von *Artemisia* und *Ephedra* nur sporadisch vor.

Ein Vergleich mit dem heutigen Polleniederschlag der zentralen Sahara (Cour & Duzer 1976, Schulz 1984) zeigt, dass *Acacia, Grewia, Capparis* in Kombination mit *Cassia* und *Fagonia* deutlich höhere Prozentwerte in den Holozän-Proben zeigen. Man wird also mit einer dauerhaften Präsenz dieser Elemente rechnen müssen. Weiterhin weist das Zusammenspiel hoher Werte von Gramineae und Cyperaceae auf Bedingungen hin, wie sie heute am Südrand der Sahara in den grossen Dünengebieten herrschen (cf. Fig. 5). In diesen Regionen ergibt sich also die Möglichkeit, den Grenzraum Sahara-Sahel im Polleniederschlag zu fassen, während in den Fels- und Gebirgsbereichen

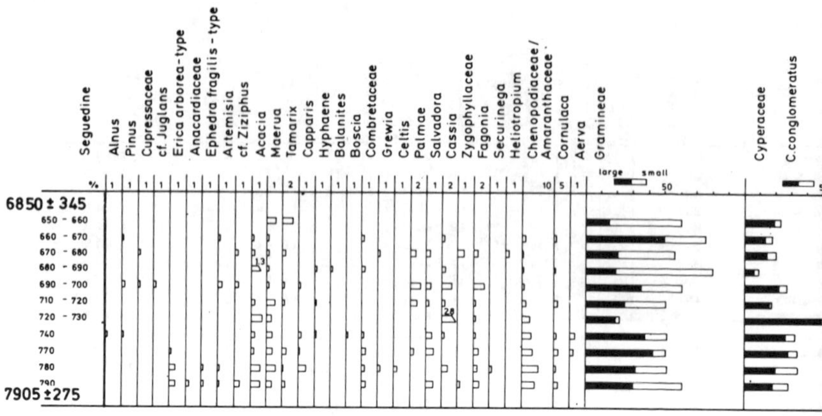

Figur 3. Pollendiagramm von Seguedine (nach Baumhauer & Schulz 1984 verändert)

allein die Gramineae dominieren.

Im mittleren Holozän wird so die Depression von Taoudenni von einer lockeren Grasvegetation bestimmt, die in den umliegenden Dünengebieten viel *Cyperus* in sich hatte. Felspartien der Stufen waren von Bäumen und Büschen bestanden (*Acacia, Grewia, Indigofera, Capparis, Maerua* und andere). Das Gebiet befand sich im Ubergangsbereich zu den Savannen des nördlichen Sahel. Ob sich nun der vollständige Wechsel zu einer lockeren Savanne noch in der Depression oder erst einige Zehner von Kilometern weiter nach Süden vollzogen hat, ist hier nicht zu entscheiden.

Die sudanischen temperierten oder mediterranen Florenelemente sind sicherlich als Teil eines Ferntransportes anzusehen, wobei der Einfluss von Süden her überwog.

Im mitteleren Abschnitt, B (c. 6 000-4 000 BP), lichten sich die Spektren allmählich. Der Anteil der Cyperaceae geht zurück, die Gramineae steigen dagegen zusammen mit *Typha* in ihren Werten an. *Acacia, Maerua, Cassia, Fagonia* und *Grewia* sind weiterhin konstant vertreten, während her Einfluss des Ferntransportes in seiner Bedeutung zurückgeht.

In diesem Zeitraum dürfte sich die Vegetation in die Seerandbereiche mit Schilf- und *Typha*-Beständen und einer offenen Grasvegetation mit einzelnen Baum- und Gebüschgruppen im Fels gegliedert haben. Der Grenzraum zur Savanne ist jedoch weiter im Süden anzunehmen.

Der obere Teil der Sequenz, C (c. 4 000-), zeigt nun ein völlig anderes Bild der Verteilung in den Pollenspektren. Die Chenopodiaceae/Amaranthaceae steigen bis auf 30% an und verhalten sich antagonistisch zu den Gramineae, wie es typisch für den saharischen Bereich ist (Court & Duzer 1976). Auch *Cornulaca* als typischer saharischer Sandbewohner erreicht hohe Werte.

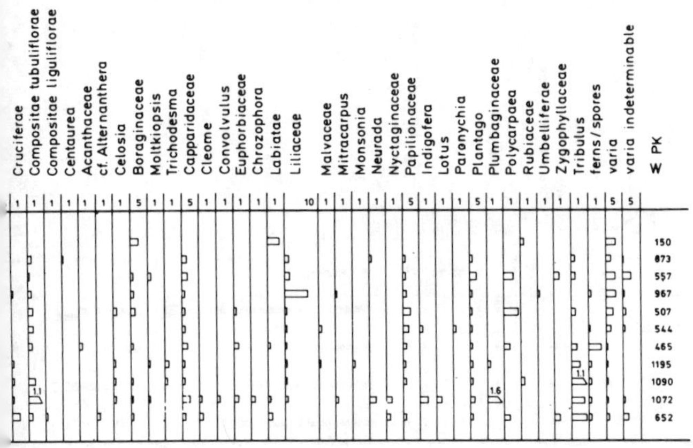

Acacia, Grewia, Cassia, Maerua sowie *Capparis* sind weiterhin Zeichen einer lokalen Vegetation in den Felsbereichen, wobei *Maerua* an Bedeutung gewinnt. Unter den Kräutern fallen jetzt die Anteile der Compositae, Papilionaceae und *ribulus* auf, die für die heutige Vegetation der Sahara in ihrem Polleniederschlag charakteristisch sind.

Die Bedeutung des Ferntransports von Süd nach Nord ist gering geworden. In dieser Periode war somit der Ubergang zu einer reinen Wüstenvegetation endgültig vollzogen, der seinen Anfang in den oberen Teilen des Abschnittes B genommen hat.

Insgesamt lässt sich aus diesen Profilen also ableiten, dass im mittleren Holozän die Zone der sahelischen Savannen bis maximal 22°N gewandert war und danach etappenweise nach Süden gedrängt wurde. Der Ubergang zu einer saharischen Wüstenvegetation dürfte im ausklingenden mittleren Holozän begonnen haben, wobei die Depression von Taoudenni immer noch eine relativ bedeutende Vegetation aufwies. Die Elemente des Fernflugen zeigen, dass die Region sowohl von nördlichen als auch von südlichen Luftmassen beeinflusst wurde, dass aber der monsunale Anteil eindeutig überwog.

Ein Vergleich mit den Spektren von Cour & Duzer (1976), die separat als Diagramm dargestellt wurden (Fig. 2), zeigt die prinzipielle Ubereinstimmung der oberen Teile der Profile. Die Vegetation, die in diesen Lagen repräsentiert ist, war im allgemeinen aus den gleichen Elementen wie die heutige Pflanzendecke zusammengesetzt. Ein Teil der heftigen Reaktionen auf die Deutungen von Cour & Duzer (Rognon 1976) dürfte sich somit erübrigt haben.

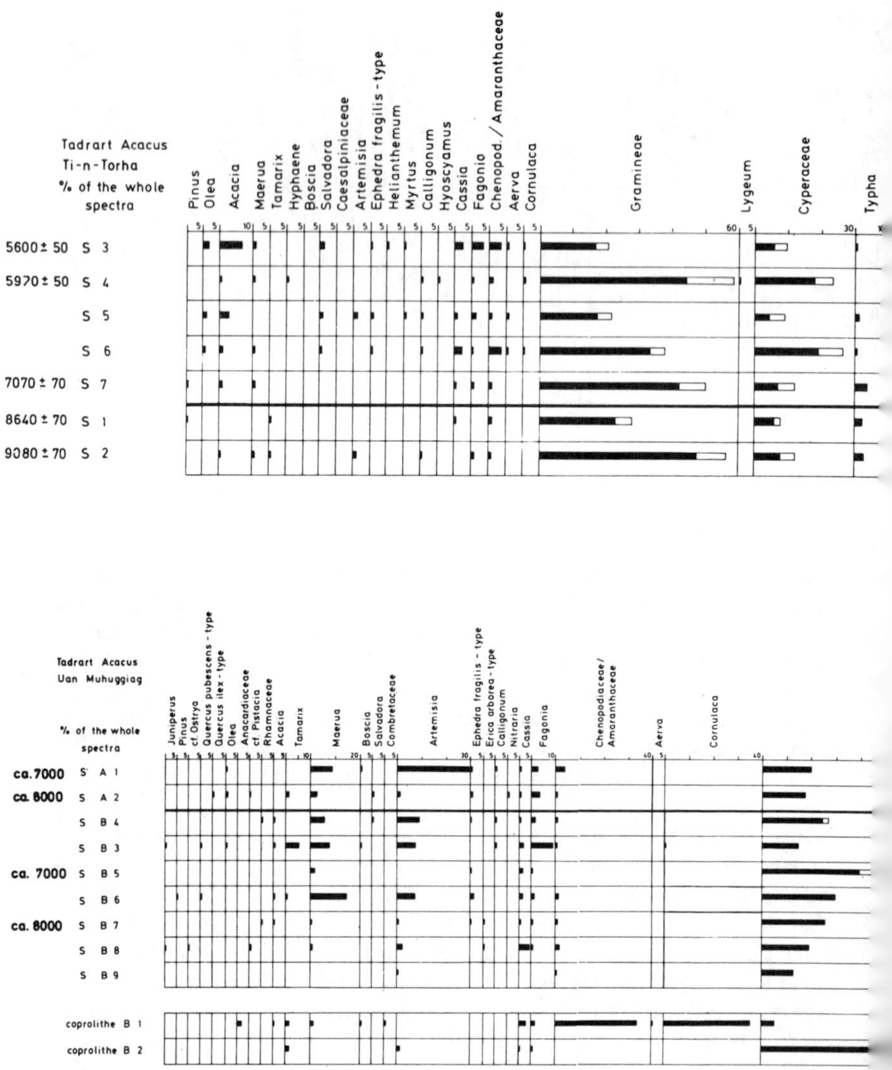

Figur 4. Pollendiagramme von Uan Muhuggiag und Ti-n-Torha, Acacus (nach Schulz 1985)

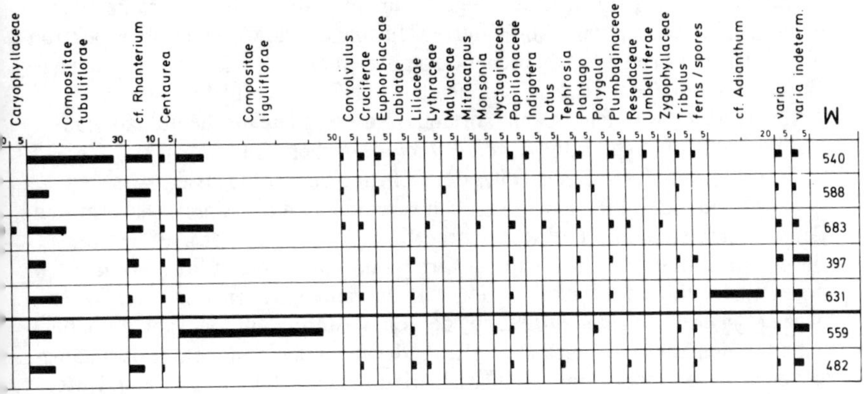
Caryophyllaceae
Compositae tubuliflorae
cf. Rhanterium
Centaurea
Compositae liguliflorae
Convolvulus
Cruciferae
Euphorbiaceae
Labiatae
Liliaceae
Lythraceae
Malvaceae
Mitracarpus
Monsonia
Nyctaginaceae
Papilionaceae
Indigofera
Lotus
Tephrosia
Plantago
Polygala
Plumbaginaceae
Resedaceae
Umbelliferae
Zygophyllaceae
Tribulus
ferns / spores
cf. Adianthum
varia
varia indeterm.
W
540
588
683
397
631
559
482

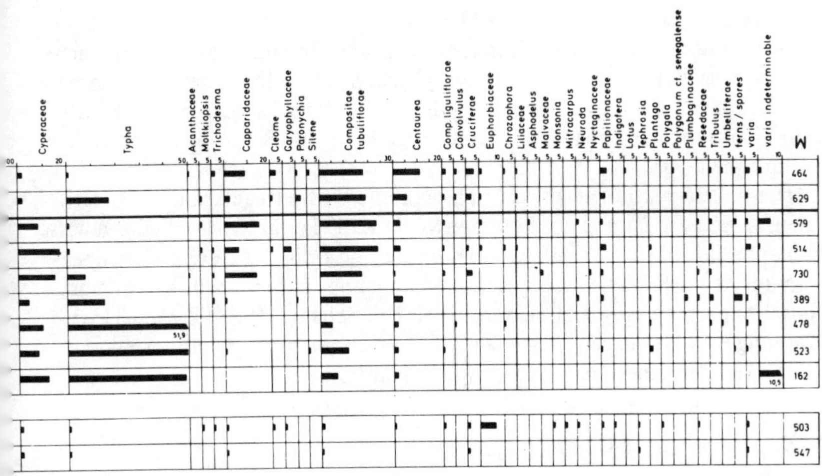
Cyperaceae
Typha
Acanthaceae
Maltkiopsis
Trichodesma
Capparidaceae
Cleome
Caryophyllaceae
Paronychia
Silene
Compositae tubuliflorae
Centaurea
Comp. liguliflorae
Convolvulus
Cruciferae
Euphorbiaceae
Chrozophora
Liliaceae
Asphodelus
Malvaceae
Monsonia
Mitracarpus
Neurada
Nyctaginaceae
Papilionaceae
Indigofera
Lotus
Tephrosia
Plantago
Polygala
Polygonum cf. senegalense
Plumbaginaceae
Resedaceae
Tribulus
Umbelliferae
ferns / spores
varia
varia indeterminable
W
464
629
579
514
730
389
478
523
162
503
547
51,9
10,5

Seguedine (20° 10'N, 12° 57'E)

Vergleiche Baumhauer (1986), Baumhauer & Schulz (1984).

Beobachtungen zur heutigen Vegetation am Südrand der Sahara im Niger geben ein Modell für eine modifizierte Deutung der holozänen Pollenspektren ab. Quasi als Vorposten sahelischer Savannen in einer trockenen Ausprägung finden sich reine *Maerua* Formationen auf den Sandsteinplateaus des SE-Nigers (Fig. 1), während in den Vorländern dieser Plateaus die übliche kontrahierte Wüstenvegetation (*Acacia-Panicum*) vorherrscht.

Die spektren des unteren Teiles der Bohrung Seguedine lassen zwischen c. 8 000 und 7 000 BP einen allmählich verlandenden Süsswassersee vermuten, der von dichten Schilf- und Röhrichtbeständen umgeben war. Auf den umliegenden Stufen muss man eine lockere Baum- und Strauch-Formation aus *Maerua, Acacia* und *Capparis* annehmen. Die hohen Prozentwerte der Cyperaceae (*Cyperus conglomeratus*-Typ) deuten wiederum auf die Nähe des Ubergangs zu den Savannen des Sahel. Die Anderungen des Sedimentationsbedingungen zu einer Sebcha um c. 7 000 BP erlauben jedoch keine weiteren Untersuchungen zur Vegetationsentwicklung.

Djebel Acacus (25° 45'N, 10° 35'E/25° 10'N, 10° 30'E)

Vergleiche Barich (1978), Schulz (1985).

Sedimentproben aus den Grabungen im südlichen und nördlichen Acacus stammen nach den ^{14}C-Daten aus den gleichen Abschnitten des Holozäns wie die Profile von Taoudenni und Seguedine. Allerdings ist eine direkte Korrelation nicht möglich, da sie in lockerer Folge in neolithischen Siedlungsplätzen entnommen wurden.

Die Pollenspektren zeigen eine ausgeprägte Wüstenvegetation des saharischen *Acacia-Panicum* Typs in dieser Region, die im Süden wohl eine gewisse Höhenstufung durch Anteile von *Artemisia* und *Ephedra* aufwies. Die Spektren aus dem nördlichen Acacus deuten dabei nur auf eine sehr offene und floristisch arme Vegetation. Koprolithen von Schaf (B2) und Ziege (B1) der Grabung Uan Muhuggiag zeigen eher das unterschiedliche Weideverhalten dieser Tiere als eine verschiedenartige Vegetation.

Fachi-Dogonbolo (18° 18'N, 11° 23'E)

Vergleiche Servant (1983), Neumann & Schulz (dieser Band).

Zusätzliche Informationen zur holozänen Vegetation kommen nun von den ersten Auswertungen der Holzkohlen, die in den Seesedimenten der Depression von Fachi-Dogonbolo geborgen wurden. Diese Makroreste wurden zu 7 010 ± 180 BP (Hv 14022) datiert.

Die Kombination der sudanischen Elemente von *Terminalia,* Sterculiaceae, *Crataeva* etc. weist auf einen weit nach Norden vorgeschobenen Bereich sudanischer Vegetation, deren genaue Zusammensetzung erst nach weiteren Untersuchungen geklärt werden kann. Jedoch ergibt sich damit ein Gradient der Vegetationstypen des mittleren Holozäns der ein weites Vordringen verschiedener Savannen-elemente nach Norden hin zeigt.

SCHLUSSFOLGERUNGEN

Aus den pollenanalytischen Untersuchungen des Sequenzen von Taoudenni, Seguedine und Djebel Acacus geht hervor, dass die Grenze der sahelischen Savannen in der Zeit von c. 8 000-6 000 BP bis auf c. 19/22°N verschoben war.

Ein Vergleich mit der heutigen Vegetation am Südrand der Sahara, deren Dynamik man als Modell wohl noch ins Holozän transponieren kann, führt zu der folgenden Verteilung der grossen Vegetationseinheiten (Fig. 6).

In den Dünengebieten westlich und östlich der grossen saharischen Gebirge war bei c. 22°N der Grenzraum der diffusen permanenten Vegetation erreicht. Die Pflanzendecke von Horstgräsern und Cyperaceae ging nach Süden hin in eine lockere *Acacia-Maerua-Capparis* Savanne über.

Die Gebirgsbereiche der südlichen Sahara (Adrar der Iforas, Air) trugen eine Savannevegetation aus sahelischen und sudanischen Elementen (*Acacia, Commiphora, Celtis, Tamarindus* etc.), wobei Akazienbestände insgesamt vorgeherrscht haben dürften. Der heutige Ubergang zu den sahelischen Savannen in der Höhenstufung des südlichen Air (Schulz & Adamou 1985) macht dieses Modell wahrscheinlich und muss wohl auch frühere Annahmen relativieren (Morel & Schulz 1981).

Nach Makroresten in holozänen Sedimenten der Grabungen im Tassili-n-Ajjer (Aumassip 1985) und Djebel Acacus (Pasa & Pasa-Durante 1962) zu urteilen, reichte dieser Vegetationstyp in seinen nördlichen Ausläufern bis an den Südrand dieser Gebirge. Die SW-Teile des Tibesti und des Hoggar trugen sicherlich eine ähnliche Vegetation.

Eine floristisch und wohl auch von der Dichte der Bestände her ärmere Savannenformation wird die Sandsteinplateaus von Kaouar, Djado, Mangueni sowie Messak Mellet und Settafet eingenommen haben. Nach dem heutigen Modell des Se-Nigers bestanden diese Savannen aus *Maerua, Acacia* und *Capparis,* wie es sich in den Pollenspektren aus dieser Region wiederspiegelt.

In den Gebirgen der zentralen Sahara nahmen die charakteristischen *Acacia-Panicum* und *Tamarix-Stipagrostis* Formationen weite Gebiete ein und zogen sich entlang der Wadis bis weit in die Vorländer. Eine mediterrangeprägte Höhenstufung der Vegetation in diesen Gebirgen hatte sicherlich zu einer Ausdehnung der *Artemisia-Ephedra* Halbwüsten geführt, mit einem häufigen Einschlag von *Olea-Myrtus* oder *Cupressus-Erica* Gebüschen. Ob aber das heutige Vorkommen von *Olea* im Air Gebirge (Quezel 1960) auf eine ähnliche Höhenstufung im Air schliessen lässt, ist auch wegen der zoochoren Verbreitungsweise von *Olea* sehr ungewiss.

Für die umliegenden Flachlandsbereiche ist ein Mosaik aus Horstgras- und Strauchformationen anzunehmen, wobei entlang der Wadis *Acacia-Panicum*

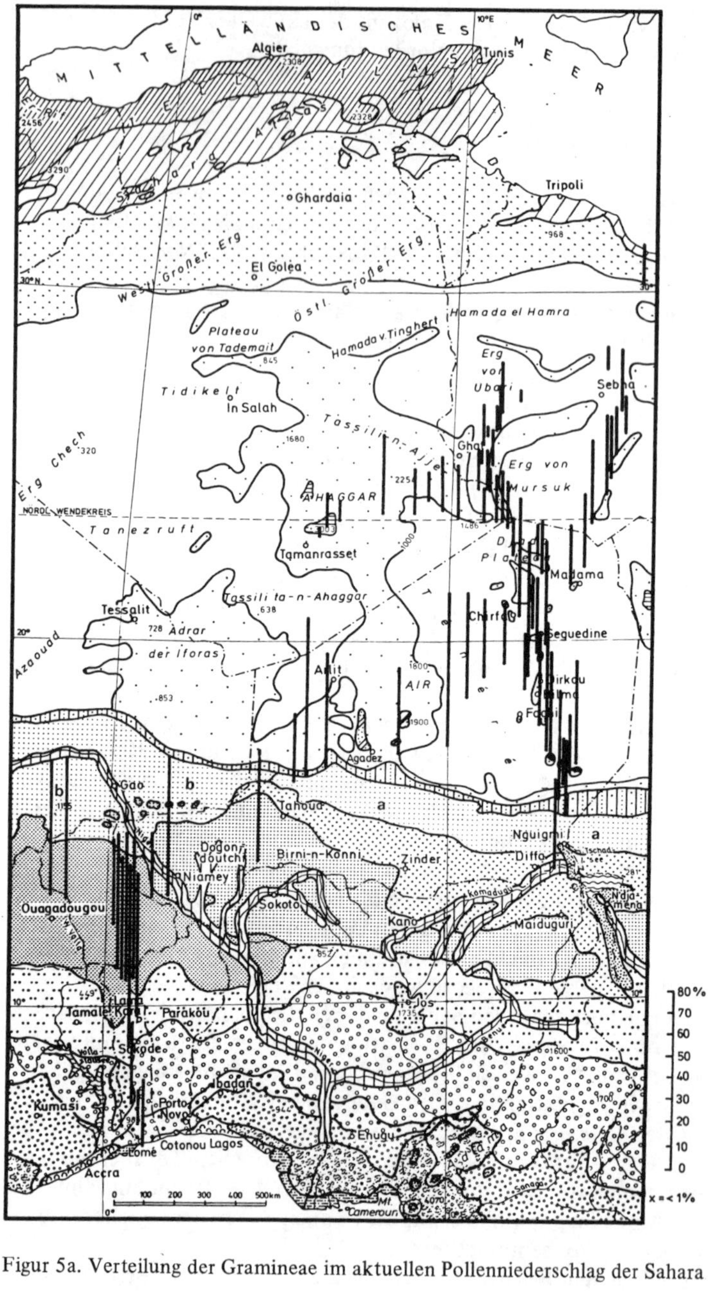

Figur 5a. Verteilung der Gramineae im aktuellen Pollenniederschlag der Sahara

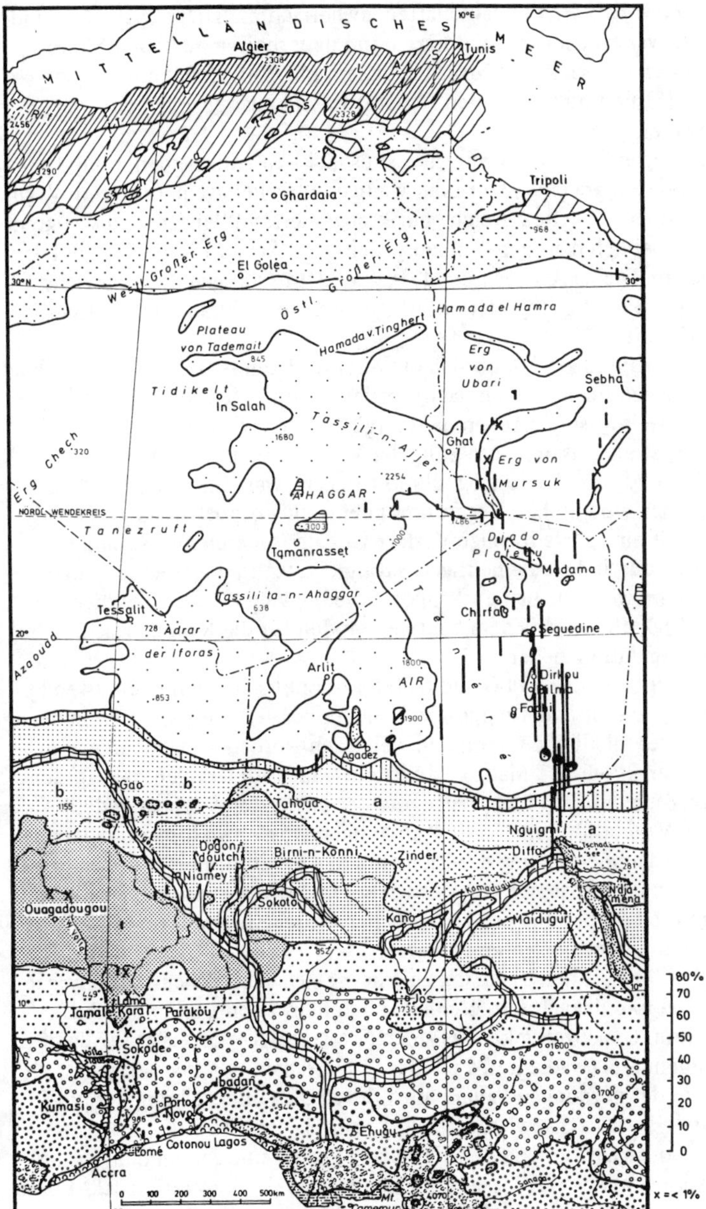

Figur 5b. Verteilung der Cyperaceae im aktuellen Pollenniederschlag der Sahara

und *Tamarix-Stipagrostis* Gesellschaften vorherrschten. Dichte Schilfbestände umgaben die verschiedenen temporären und dauerhaften Seen und Sümpfe, wie es leicht an den vielen Rhizomabdrücken und Wurzelphänomenen zu ersehen ist. Darüberhinaus werden Achab-Floren häufiger und weiter verbreitet gewesen sein.

Beim heutigen Stand der Kenntnis est es nicht zu entscheiden, ob die gesamte zentrale Sahara eine mehr oder weniger ausgedehnte permanente Vegetation trug, oder ob weite Bereiche nur kontrahierten Pflanzenwuchs zeigten, dafür aber erheblich häufigere und ausgedehntere Achab-Floren hatten. Die Ergebnisse von Djebel Acacus und Messak Mellet bestätigen die saharische Grundkomposition der Vegetattion, deuten jedoch nicht auf eine weitläufig diffuse Ausbreitung.

Es ist weiterhin ungewiss, ob sich ein direkter Kontakt oder eine Verschneidung der Akazien-Formationen nach Norden him mit den Strauchgesellschaften der Halbwüsten ergeben hat (vergl. Delibrias et al. 1982).

Ebenfalls ist die Zusammensetzung der Vegetation nach Süden hin noch unklar. Die Funde von Fachi Dogonbolo belegen zwar eine weite Nordverschiebung einer sudanischen Vegetation in einigen Vorposten, damit ist aber noch nicht auf eine gürtelförmige Verlagerung der sudanischen Savanne zu schliessen. Dadurch dass die heutigen Savannen in Zusammensetzung und Ausdehnung im wesentlichen anthropogen bedingt sind, kann man hier nur begrenzte Rückschlüsse aus dem heutigen Modell für die Rekonstruktion der holozänen Vegetation ziehen.

Für die Zeit zwischen 8 000 und 6 000 BP ergibt sich zusammenfassend eine klare Vegetationsänderung, die von Süden her gesteuert war und in der die Grenzen der sahelischen Savannen weit nach Norden verschoben waren. Durch Pollenanalysen und Makroreste ist ein Süd-Nord-Gradient belegt, der eine Verlagerung der Savannengrenze und der äussersten Vorposten der sudanischen Vegetation um c. 300-400 kn nach Norden wahrscheinlich macht.

Aus der Zusammensetzung der Elemente des Ferntransportes in den Spektren von Taoudenni lässt sich das Vorherrschen des Monsuns in der Periode von 8 000 bis 6 000 BP ablesen, wobei sicherlich eine Interaktion mit zyklonalen Fronten nur zurückgedrängt war.

Um c. 4 000 BP ist der Ubergang zu einer reinen saharischen Wüstenvegetation vollzogen und danach erfolgten nur noch Schwankungen im Rückgang der Vegetation. Gleichzeitig kam es zu einer starken Regionalisierung, in der sich neben dem S-N-Gradienten auch der W-E-Gradient stärker herausgestellt hat, wie es in der nördlichen Sahara auch durch die Untersuchung der holozänen Grundwässer nachgewiesen wurde (Sonntag et al. 1980).

Diese Rekonstruktion, von der die Karte (Fig. 6) eine Zusammenfassung des Diskussionsstandes gibt, basiert auf den Resten der Vegetation selbst und der Einschätzung der heutigen Dynamik der Pflanzendecke dieser Region. Dass sich der Charakter der Landschaften der Sahara im Holozän weitgehend verändert hatte, geht auch aus den vielen Befunden der Limnologie, Sedimentologie, Pedologie, Geomorphologie sowie der Prähistorie hervor, wofür die

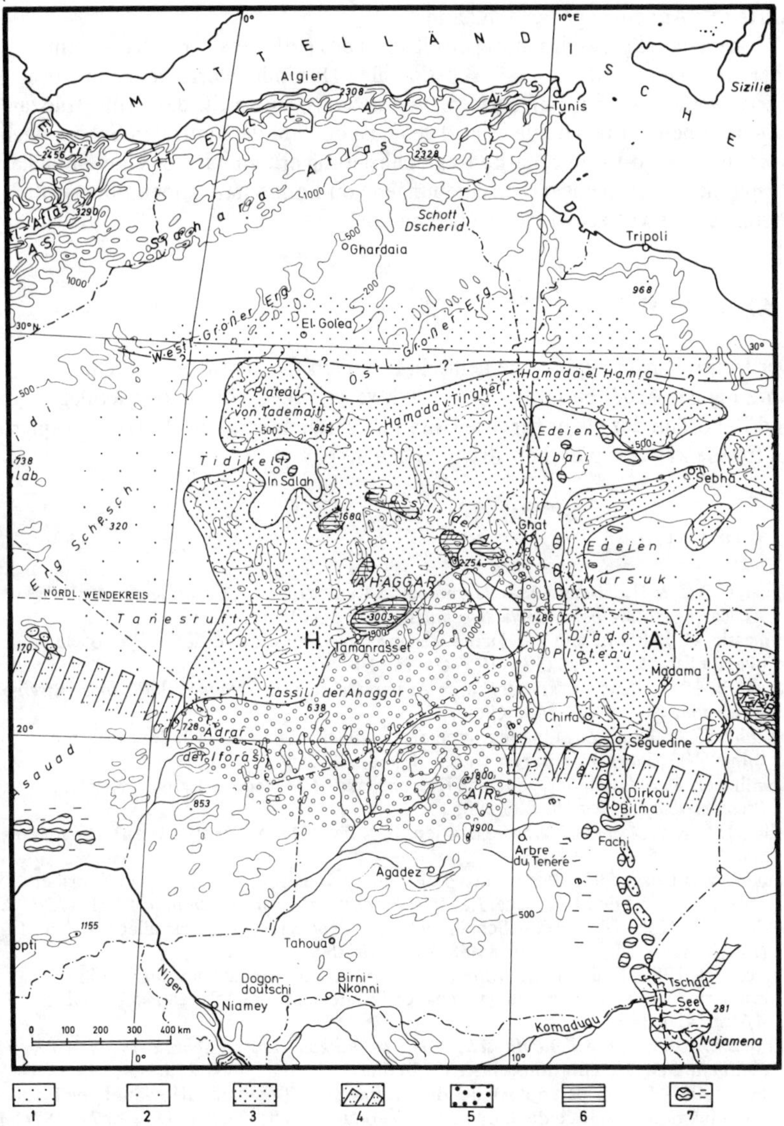

Figur 6. Rekonstruktion der mittelholozänen Vegetation der zentralen Sahara
1. Halbwüsten
2. Horstgrasfluren, ephemere Vegetation
3. Maerua-Acacia Savannen bzw. Acacia-Panicum/Tamarix-Stipagrostis Formationen
4. Grenzbereich der sahelischen Acacia-Maerua-Capparis Savannen
5. Acacia-Commiphora Celtis Savannen
6. Artemisia-Ephedra Halbwüsten der zentralsaharischen Gebirge
7. Seen/Sümpfe

Karte nur Andeutungen geben kann.

Eine generelle Beschreibung des Klimas des Holozäns der Sahara kann aber erst in einer intensiven Diskussion aller Disziplinen erfolgen. Ein Nordwärtswandern der Savannengrenze, wie sie im N-Sudan für das Früh-Holozän angenommen wird (Ritchie et al. 1985), ist auch in der zentralen Sahara nachweisbar. Ebenso ist eine starke Regionalisierung erkennbar, die eine einfache Annahme gürtelförmiger Verschiebungen von Landschaftszonen über weite Distanzen hinweg schwierig erscheinen lässt.

DANKSAGUNG

Der Deutschen Forschungsgemeinschaft bin ich für die finanzielle Unterstützung der Arbeiten im Niger zu Dank verpflichtet, ebenso den Kollegen und Behörden des Niger für ihre Hilfe und Zusammenarbeit. Herrn K.Wepler sei für die kartographischen Arbeiten gedankt.

LITERATUR

Agwu, C.O.C. & H.-J.Beug 1982. Palynological studies of marine sediments off the West African coast. *Meteor Forsch. Ergebnisse,* C 36: 1-30.

Aumassip, G. 1985. Le site Ti-n-Hanakaten et la néolithisation sur les marges occidentales du Sahara oriental. *Cah. ORSTOM* (im Druck).

Barich, B. 1978. Neue Ausgrabungen im Acacus Gebirge. In *Sahara: 10000 Jhre zwischen Weide und Wüste*: 222-245. Köln.

Baumhauer, R. 1986. Zur jungquartären Seenentwicklung im Bereich der Stufe von Bilma (NE-Niger). *Würzburger Geogr. Arb.* 65, 235 pp.

Baumhauer, R. & E.Schulz 1984. The Holocne lake of Seguedine, Kaouar, Niger. *Palaeoecol. of Africa* 16: 283-290.

Celles, J.-C. & R.Maniere 1983. Végétation. In N.Petit-Maire & J.Riser (eds.), *Sahara ou Sahel?*: 29-35. Paris: CNRS.

Cour, P. & D.Duzer 1976. Persistance d'un climat hyperaride du Sahara méridional et central au cours de l'Holocène. *Rev. Géogr. phys. et Géol. dynamique* XVII; 175-198.

Delibrias, G., N.Petit-Maire & E.Schulz 1982. Les dépôts récents de la vallée du Shati. In N.Petit-Maire (ed.), *Le Shati*: 86-88. Paris: CNRS.

Dumont, H. 1982. Relict distribution patterns of aquatic animals: another tool in evaluating late Pleistocene climate changes in the Sahara and Sahel. *Palaeoecol. of Africa* 14: 1-24.

Gabriel, B. 1985. *Die östliche libysche Wüste im Jungquartär, vornehmlich nach neueren Feldbefunden.* Habilitationsschrift TU Berlin.

Maley, J. 1981. Etudes palynologiques dans le bassin du Tchad et paléoclimatologie de l'Afrique nord tropicale de 30000 ans à l'époque actuelle. *Trav. et Doc. de l'ORSTOM* 129, 586 pp.

Monod, Th. 1986. The Sahel zone north of the equator. In Evenari et al. (eds.), *Hot deserts and arid shrublands*: 203-243.

Morel, A. & E.Schulz 1981. Recherches palynologiques et géomorphologiques dans le massif de l'Air (Niger). Note préliminaire. *Rev. de Géogr. alpine* LXIV (4): 583-592.

Pachur, H.-J. & H.J.Röper 1984. The Libyan (western) desert and northern Sudan during the late Pleistocene and Holocene. *Berliner Geowiss. Abh.* (A) 50: 249-284.

Pasa, A. & M.V.Pasa-Durante 1962. Analisi paleoclimatiche nel deposito di Uan Muhuggiag, nel massica dell' Acacus (Fezzan meridionale). *Mem. Mus. Civ. di Storia Naturale* (Verona) X: 251-255.

Petit-Maire, N. 1984. Le Sahara de la steppe au désert. *La recherche* 15: 1374-1382.

Petit-Maire, N. 1985. Palaeoclimates in the Sahara of Mali. A multidisciplinary study. *Episodes* 9 (1): 7-15.

Petit-Maire, N. & J.Riser (eds.) 1983. *Sahara ou Sahel?* Paris: CNRS. 473 pp.

Quezel, P. 1960. A propos de l'olivier de laperrine de l'Adrar Greboun. In H.Hugot (ed.), *Mission Berliet Ténéré-Tchad*: 329-332.

Quezel, P. 1965. *La végétation du Sahara.* Stuttgart: Fischer. 333 pp.

Ritchie, J.C., C.H.Eyles & C.V.Haynes 1985. Sediment and pollen evidence for an early to mid-Holocene humid period in the eastern Sahara. *Nature* 314: 352-359.

Rognon, P. 1976. Essai à interprétation des variations climatique au Sahara depuis 40000 ans. *Rev. Géogr. phys. et de Géol. dynamique* XVIII: 252-283.

Sarnthein, M., J.Thiede, U.Pflaumann, H.Erlenkeuser, D.Fütterer, B.Koopmann, H. Lange & E.Seibold 1982. Atmospheric and oceanic circulation patterns off Northwest Africa during the past 25 million years. In U.von Rad, K.Hinz, M.Sarnthein & E.Seibold (eds.), *Geology of the Northwest African continental margin*: 545-604. Heidelberg: Springer.

Schulz, E. 1984. The recent pollen rain in the eastern central Sahara. A transect between northern Libya and southern Niger. *Palaeoecol. of Africa* 16: 245-253.

Schulz, E. 1985. Holocene vegetation of Tadrart Acacus, Fezzan-Libya. The pollen record of two ceramic sites. In. B.Barich (ed.), *Archaeology and environment in the Libyan Sahara: Excavations at Tadrart Acacus (1978-1983).* (Im Druck).

Schulz, E. 1986a. Present pollen rain between the Mediterranean and the Atlantic ocean. A Libya-Togo transect. In *INQUA/ASEQUA Conf. 'Changements globaux en Afrique durant le Quaternaire' (Dakar)*: 431-433.

Schulz, E. 1986b. Zur Vegetation der Sahara und Ihrer Randgebiete. *Würzburger Geogr. Manuskripte* 16: 52-64.

Schulz, E. & A.Adamou 1985. Das Bagzan Massiv im Air (N-Niger). *Mitt. Bayer. J. Max. Univ. Würzburg* 3 (19): 3-11.

Servant, M. 1983. Séquences continentals et variations climatiques; évolution du bassin du Tchad au Cénozoïque supérieur. *Trav. et Doc. de l'ORSTOM* 159, 573 pp.

Skowronek, A. 1984. *Bodengeographische-palaeopedologische Untersuchungen zum Problem der klimagenetisch-geomorphologischen Landschaftsentwicklung in der zentralen Sahara.* Habilitationsschrift Würzburg, 254 pp.

Sonntag, C., U.Thorweihe, J.Rudolph, E.P.Löhnert, K.O.Junghans, K.O.Münnich, E. Klitzsch, E.M.el Shazly & E.M.Swailem 1980. Isotopic identification of Saharan groundwaters; groundwater formation in the past. *Palaeoecol. of Africa* 12: 159-175.

MIDDLE HOLOCENE SAVANNA VEGETATION IN THE CENTRAL SAHARA – PRELIMINARY REPORT

K.NEUMANN/E.SCHULZ
Botanisches Institut, J.W.Goethe Universität, Frankfurt am Main, F.R.Germany

ABSTRACT

The investigations of charred wood from Holocene lake sediments of Dogonboulo (Erg de Ténéré) yielded the surprising result that the vegetation around 7 000 BP consisted of elements which were completely different from the recent plant cover. The wood, probably charred by a bush fire, furnishes the first direct evidence for a northward shift of Sahelian vegetation of at least 400 km, and for a rich Middle Holocene savanna in an area which, at present, is one of the most arid parts of the Sahara.

INTRODUCTION

The depression of Fachi (North Niger), situated in the heart of the hyperarid Erg de Ténéré, is known for its series of Holocene lake sediments (Servant 1983). At Dogonboulo (ca. 18°18′N, 11°33′E), the lacustrine diatomites on the western side of the escarpment were found to contain a 40 cm thick layer of charred wood embedded in sandy and clayey deposits. Preliminary studies of the charcoal by means of scanning electron microscopy (for methods see Neumann, this vol.) yielded the surprising result that the Middle Holocene vegetation of the lake shore consisted of elements completely different from the recent plant cover. Nowadays, the surrounding sandplains of the Ténéré bear only short-lived ephemeral plant communities. Perennials are restricted to depressions and wadi-beds (Boudouresque & Schulz 1981). At Dogonboulo, only solitary specimens of *Acacia raddiana* and a small cluster of *Salvadora persica,* intermixed with *Panicum turgidum* and *Stipagrostis vulnerans,* are growing at the lowest point of the depression. Three radiocarbon dates of the charcoal layer are consistent with the dates given by Servant (6 850 ± 850, S 4116): 7 010 ± 180 (Hv 14022); 7 125 ± 290 (Hv 14020) and 7 655 ± 225 (Hv 14021).

BRIEF DESCRIPTIONS OF WOOD ANATOMY

Up to now, the following taxa have been identified (information on habitat and distribution from Aubréville 1950, Peyre de Fabrègues & Lebrun 1976, Maydell 1983):

Crateva adansonii (Capparidaceae)
Growth rings not discernible; vessels mostly solitary or in short radial multiples (2-4 elements), with simple perforations; parenchyma vasicentric; fibres thin-walled; rays homogeneous, composed of long, procumbent cells, 2- to 6-seriate.

Sudanian element; in the Sahel zone only near rivers and lakes, on humid soils which may be seasonally inundated.

Ximenia americana (Olacaceae)
Growth rings not discernible; vessels exclusively solitary, with simple perforations; parenchyma apotracheal, diffuse; ground tissue consisting of fibre-tracheids with distinctly bordered pits; rays heterogeneous, composed of procumbent cells with one or more marginal rows of square or upright cells, 2- to 3-seriate, uniseriate rays composed only of upright cells.

Savanna tree with a wide distribution. On stony ground and skeleton soils, but also on clay soils near lakes.

cf. *Sterculia setigera* (Sterculiaceae)
Growth rings not discernible; vessels solitary and in small clusters, with simple perforations; parenchyma apotracheal, banded, also vasicentric; fibres frequently septate; rays heterocellular, composed of procumbent, square and upright cells, 2- to 7-seriate.

The charcoal pieces are from branchwood, thus making identification uncertain. Not consistent with the reference material of adult wood of *Sterculia setigera* are the small, round vessel pits and the non-storeyed structure of the parenchyma.

Tree of a mainly Sudanian distribution, in southern Niger common on skeleton soils.

cf. *Terminalia* sp. (Combretaceae)
Growth rings not discernible; vessels solitary and in short radial multiples, with simple perforations; parenchyma aliform-confluent; rays mostly 2- to 3-seriate, sometimes uniseriate, weakly heterocellular, composed of procumbent and square cells.

Reference material of only a few *Terminalia* species was available for identification. The features of the charcoal resemble the recent wood of *Terminalia macroptera,* but no vessels with vestured pits (a typical feature of Combretaceae) could be observed. *Bauhinia rufescens,* on the other hand, possessing a similar wood structure as described above, without vestured pits, has a much higher proportion of uniseriate rays.

Terminalia species are the dominant trees in the Combretaceae savanna and in the *Parkia-Butyrospermum-Terminalia* savanna (see vegetation map in Schulz, this volume). The most important species are *T.macroptera,* growing on humid soils near rivers and lakes, and *T.avicennoides,* growing on sandy and laterite soils.

Euphorbia sp. (Euphorbiaceae)
Growth rings not discernible; vessels solitary or in short radial multiples, with simple performations and very large pits; parenchyma apotracheal, diffuse; fibres thin-walled, frequently septate; rays uniseriate, homocellular, composed of upright cells, some multiseriate and with included phloem.

No recent reference material of *E.balsamifera,* the only Northwest African woody species of *Euphorbia,* was available. The recent East African *E.candelabrum* shows the same features as the charcoal, but with a much sparser distribution of vessels.

In the northern Sahel, *E.balsamifera* is growing mostly on non-fixed dunes and in disturbed habitats.

CONCLUSIONS

With the exception of *Euphorbia* sp., the main distribution area of all taxa is the southern Sahel and the Sudanian zone. Thus, the vegetation of Dogonboulo around 7 000 BP was of a savanna type which, at present, can be found 400 km further south. Because of the record of more than one species and of the assemblage, a transport by man from the south can be excluded. The trees and shrubs must have been growing at the edge of the lake and were presumably charred by a bush fire and drifted together by a seasonal inundation of the lake.

Thus, the results of the charcoal examination furnish the first direct evidence for a Middle Holocene northward shift of the Sahelian vegetation. Unfortunately, the diatomite contained almost no pollen which could give us further information on other contemporary woody and herbal species. However, 250 km further north, the Middle Holocene pollen flora of Séguédine leads us to assume that the prevailing vegetation type of that area was a transition between desert and northern Sahelian savanna.

ACKNOWLEDGEMENTS

Thanks are due to Professor H.Gottwald for his help in identification of the charred wood, and to Professor K.U.Leistikow for correcting the manuscript.

REFERENCES

Aubréville, A. 1950. *Flore forestière soudano-guinéene.* Paris: Soc. d'éditions géogr., marit, et colon., 523 pp.

Boudouresque, E. & E.Schulz 1981. The flora and vegetation of NE Niger (Djado, Kaouar and Ténéré). *Willdenowia* 11: 363-394.

Maydell, H.-J. von 1983. *Arbres et arbustes du Sahel.* Eschborn: Office Allemand de la Coopération Technique. 531 pp.

Neumann, K. (this vol.). Middle Holocene vegetation of the Gilf Kebir, SW-Egypt.

Peyre de Fabrègues, B. & J.-P.Lebrun 1976. *Catalogue des plantes vasculaires du Niger.* Institut d'élevage et de médecine vétérinaire des pays tropicaux. Maisons Alfort. 433 pp.

Servant, M. 1983. *Séquences continentales et variations climatiques: évolution du Bassin du Tchad au Cénozoique supérieure.* Paris: Trav. et Doc. de l'ORSTOM, no. 159, 573 pp.

HOLOZÄNE LIMNISCHE AKKUMULATIONEN IM BEREICH DER STUFEN VON ZOO BABA UND DIBELLA (NE-NIGER)

R.BAUMHAUER
Geographisches Institut, Universität Würzburg, F.R.Germany

ZUSAMMENFASSUNG

Es werden die vorläufigen Ergebnisse von Untersuchungen an drei prärezenten limnischen Akkumulationen aus den Depressionen der westlichen Vorländer der Stufen von Zoo Baba und Dibella (NE-Niger, Erg von Bilma) vorgestellt.

Die stratigraphisch-morphologische Situation und ein Radiocarbondatum zeigen für die Vorlanddepression der Stufe von Zoo Baba eine ausgeprägte limnische Phase, die nach der Zerschneidung der Mittelterrasse begann und gegen 5 370 ± 100 BP (HV 14024) endete. Zu dieser Zeit hatte der See eine Wassertiefe von 25-30 m bei etwa 10 km^2 Fläche. Limnische Gastropoden und Diatomeenfloren kennzeichnen einen perennierenden Süsswassersee.

In der Stufenvorlanddepression von Dibella konnte eine frühholozäne Seephase nachgewiesen werden, die nach einer stärkeren Bodenbildung (10 230 ± 245 BP, HV 14026) gegen 9 785 ± 75 BP (HV 14025) begann. Die Diatomeen charakterisieren starke Schwankungen der Salinität des Sees, der eine Nord/Süd-Erstreckung vor der Stufe von etwa 5 km und eine Wassertiefe um 20 m gehabt haben dürfte.

SUMMARY

The first results of sedimentological and palaeoecological investigations of three limnic accumulations in the depressions in the western forelands of the cuestas of Zoo Baba and Dibella (NE-Niger, Erg of Bilma) are presented in this chapter.

In the foreland depression of Zoo Baba the stratigraphical and geomorphological situation and a radiocarbon date show a limnic phase which started after the incision of the Middle Terrace and ended at 5 370 ± 100 yrs BP (HV 14024). At this time the lake had a water depth of 25-30 m and an area of about 10 km^2. Limnic gastropods and diatoms characterize a perennial fresh water lake.

In the foreland depression of Dibella an early Holocene lake phase could be identified. The limnic stage started, after a pedogenetic process (10 230 ± 245 yrs. BP, HV 14026) at 9 785 ± 75 yrs. BP (HV 14025). The diatoms characterize fluctuations in the water salinity, which extended from north to south in front of the cuesta over about 5 km and its depth was about 20 m.

EINLEITUNG

Zur Rekonstruktion vorzeitlicher Landschaftstypen und Klimate in der Sahara können Paläoseen als Klimaindikatoren, sei es in der Betrachtung von Wasserspiegelschwankungen oder durch die Zuordnung von Merkmalen und Befunden aus der exakten Beschreibung von Typus, Ausdehnung und Dauer verschiedener Seephasen wertvolle Indizien und Hilfen sein.

Auf einer Forschungsreise des Geographischen Institutes der Univeristät Würzburg in den Grand Erg de Bilma im Frühjahr 1984 konnten, anknüpfend an vorausgegangene Geländeunterschungen in der südwestlichen zentralen Sahara, u.a. zahlreiche limnische Akkumulationen erstmals untersucht werden.

TOPOGRAPHISCHE SITUATION

Die Untersuchungsgebiete liegen im zentralen und südlichen Bereich des Grand Erg de Bilma. Sie umfassen typische Ausschnitte der morphologischen Grundstrukturen des südwestlichen Teiles der Zentralen Sahara.

Weitgespannte Sandtennen, die nach Süden hin in ausgedehnte, durch NE/SW-gerichtete Längsdünenzüge charakterisierte Ergbereiche übergehen, werden durch N/S-streichende, teilweise in Inselbergmassive oder in flache Plateaubereiche aufgelöste Schichtstufen gegliedert. Diese vermitteln zwischen den Plateaulandschaften der Umrahmung des Murzuk-Beckens und den, zum Einflussbereich des Tchadsees überleitenden Dünenregionen im Süden. Die westlichen Vorländer der Stufenbereiche sind durch abflusslose Depressionen und teilweise starke Versandung gekennzeichnet. Weite Bereiche der Sandtennen. eine Vielzahl von Dünendepressionen und Teile der Stufenvorlanddepressionen werden von jungquartären limnischen Akkumulationen bedeckt.

DIE UNTERSUCHTEN PROFILE

In zwei derartigen Depressionen, im westlichen Vorland der Schichtstufen von Zoo Baba und Dibella wurden limnische Akkumulationen näher untersucht. Dabei stellen die hier vorgestellten Geländebefunde und Ergebnisse nur eine erste Interpretation dar. Insgesamt wurden über zehn Profile limnischer Akkumulationen aus dem zentralen und südlichen Bereich des Grand Erg de Bilma au genommen.

Die Schichtstufe von Zoo Baba liegt bei 18°13′ nördlicher Breite und 13°4′ östlicher Länge, etwa 70 km südlich der Oase Bilma. Ihre Stufenstirn weist nach Westen. Unterschiedlich mächtige kretazische bis paleozäne (?) tonige und teilweise karbonatreiche Silt- und Sandsteine bauen die Stufe auf, die trotz stellenweiser starker Einsandung infolge der Ubertiefung ihres Vorlandes eine relative Höhe bis zu 200 m erreicht und dabei auf 578 m ü. NN ansteigt. Nur in diesem Abschnitt, auf etwa 2,5 km bietet die Stufe einen mauerartigen Eindruck, nach Norden und Süden löst sie sich stark auf und besteht dort nur noch aus einzelnen, z.T. stark versandeten Zeugen- und Inselbergen. Die Vorlanddepression erstreckt sich von E nach W über 1,5 km, wobei der Brunnen der kleinen Oase Zoo Baba, im Nordwesten der Senke, mit 370 m ü. NN die tiefste Stelle markiert. Auf der stufenzugewandten Seite wird die Depression durch einen relativ flachen Stufenunterhang begrenzt, in dem ein ausgeprägtes Gerinnenetz ausgeprägt ist, mit durchgehend, bis in das Depressionsinnere zu verfolgenden Mittel- und Niederterrasse. Der Westrand der Vorlandsenke wird sanft ansteigend zum in Passatwindrichtung längsdünenbedeckten Grand Erg de Bilma begrenzt. Der Anstieg erfolgt dabei nicht in Form einer gleichmässigen Rampe, sondern in mehreren Stufen, die durch Verebnungen voneinander getrennt werden.

Das Profil ZB 1 entstammt einem Wadiunterlauf am Stufenwärtigen Unterhang der Schichtstufe, etwa 300 m südöstlich des Brunnen von Zoo Baba (vgl. Fig. 1).

Profil ZB 1

Karte:	République du Niger, Fond Topographique 1:200 000, NE-33-XIV-Zoo Baba (1972).
Luftbild:	NE-33-XIII-XIV-400.
Länge:	13°4′ 40″E.
Breite:	18°13′N.
Höhe:	c. 385 m.

1.	150 cm	hellgrauer Feinsand, schluffig, ungeschichtet;
2.	5 cm	hellgrauer feinsandiger Schluff, ungeschichtet;
3.	6 cm	mittelgrauer schluffiger Feinsand, verfestigt, lagig, Süsswassermollusken;
4.	8 cm	hellgrauer feinlagiger Diatomit, vereinzelt Fe- (Hydr-)oxidausfällungen;
5.	31 cm	hellgrauer, lagiger, kaum verfestigter Diatomit, kluftig, vereinzelt Wurzelröhren;
6.	25 cm	hellgrauer Diatomit, feinlagig, bei 15-25 cm rezente Durchwurzelung, Niederterrassenschuttstreuauflage.

Die stratigraphisch-geomorphologische Situation des Profils deutet die Einlagerung der Sedimente in ein postmittelterrassenzeitlich zerschnittenes

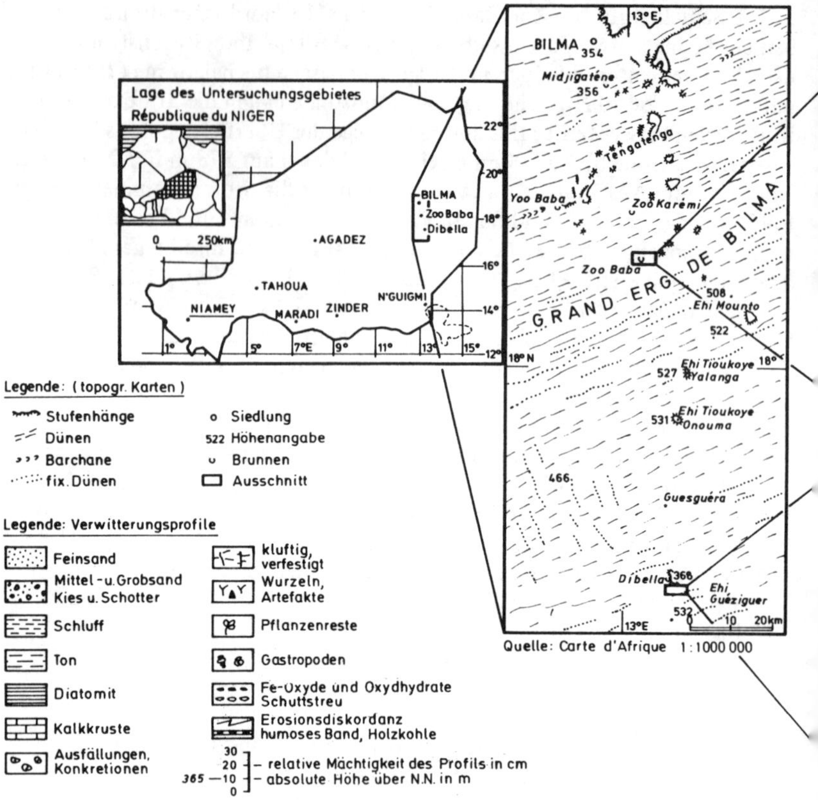

Figur 1. Lage des Untersuchungsgebietes und der untersuchten Profile

Relief an. Da die Mollusken (Horizont 3) noch nicht datiert werden konnten, ist dies auch der einzige Fixpunkt für eine zeitliche Einordnung der Akkumulationen.

Die Basissande (Horizont 1) liegen direkt auf dem tonigen Siltstein der Kafraformation (Faure 1966). Sie weisen keinerlei Schichtung auf und sind fossilfrei. Körnung und morphoskopische Quarzkornanalyse zeigen innerhalb des Horizontes keine Kontinuität. Das lässt auf differenzierte geogene Prozesse während des Transportes und der Sedimentation schliessen. Ob die in der Farbe erkennbare Vergleyung synsedimentär ist, oder später erfolgte, ist nicht mehr zu bestimmen. Aufgrund ihrer relativ uneinheitlichen Fazies ist zu vermuten, dass diese Sande zur gleichen Zeit und unter ähnlichen Bedingungen transportiert und akkumuliert wurden, wie die heterogenen Sedimente im südlichen Kawar, die teilweise zwischen den, die frühholozäne und

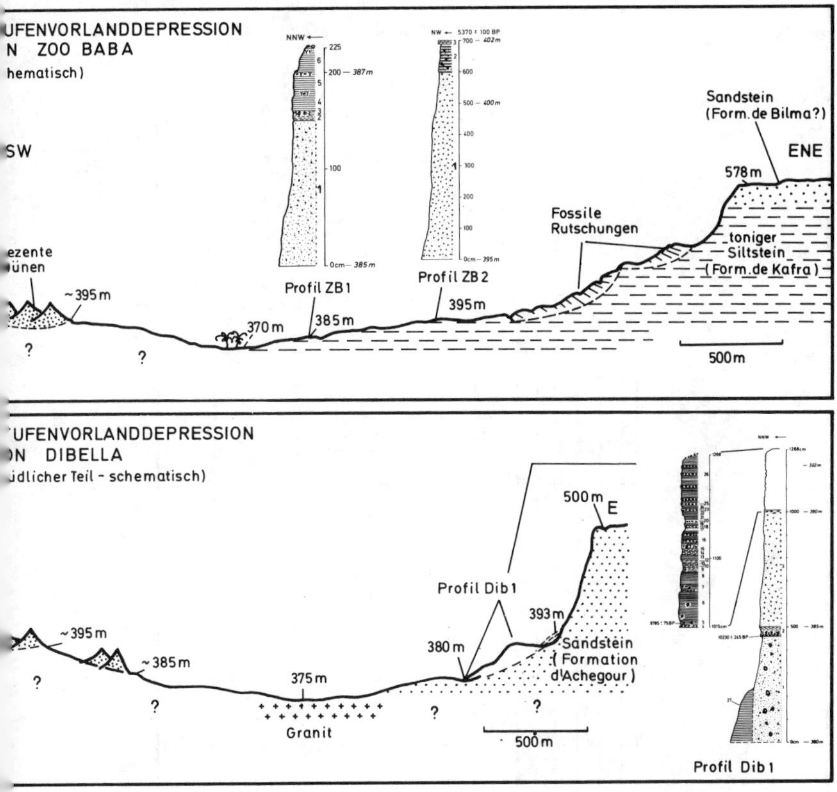

mittelholozäne Seephase anzeigenden, Daitomiten eingelagert sind. Im südlichen Kawar zeigen ausschliesslich diese zwischengelagerten Akkumulationen eine derart uneinheitliche Fazies (Baumhauer 1986).

Im Hangenden der Sande nimmt der Schluff- und Tonanteil zu und es fanden sich limnische Schneckenarten sowie Daitomeen. Sie kennzeichnen den Beginn eines limnischen Stadiums. Die bestimmten Gastropoden, dominant *Melanoides tuberculata* und *Biomphalaria pfeifferi*,[1] deuten auf einen Süsswassersee hin. Andererseits zeigt die untersuchte Diatomeenflora (vgl. Fig. 2), vorherrschend benthische *Campylodiscus clypeus* und untergeordnet epiphytische *Epithemia zebra*, den Beginn limnischer Verhältnisse

1. Weitere malakologischen Bestimmungen werden z.Zt. noch durchgeführt.

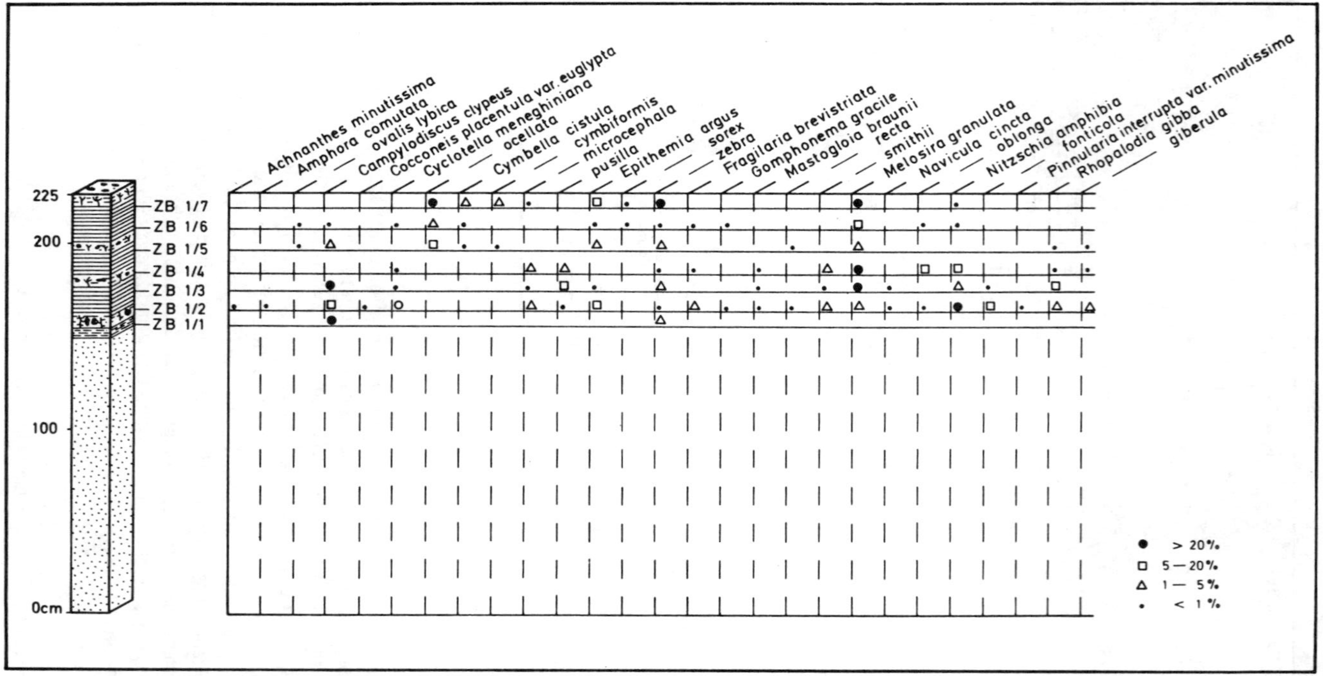

Figur 2. Profil ZB 1, halbquantitative Verteilung der Diatomeen

in einem sumpfigen Milieu mit relativ stark alkalischen Wässern (vgl. u.a. Servant-Vildary 1978). Damit ist die Frage des Milieus zu Beginn der limnischen Phase im Augenblick nur schwer zu beantworten. Zwar ist *Melanoides tuberculata* eine sehr anpassungsfähige Art, die zusammen mit *Biomphalaria pfeifferi* Süsswasserverhältnisse kennzeichnet, aber möglicherweise erfolgten doch über längere Zeiträume Salinitätsschwankungen, an die sich die Gastropoden anpassen mussten (vgl. a. Gabriel & Kröpelin 1983).

Zum Top der Diatomite hin erfolgte ein allmählicher Ubergang der benthischen Floren zu planktonischen (*Melosira granulata, Cyclotella ocellata*) und epiphytischen Arten (*Epithemia zebra, Cymbella cistula*), wobei die relativen Anteile der Gruppen schwanken. Dies deutet die Ausdehnung der limnischen Bedingungen an, mit oligotrophem Milieu und relativ kühlen Wässern. Die Wassertiefe bleibt dabei gering, da epiphytische Diatomeen typische Flachwasserbewohner sind (Servant-Vildary 1978).

Profil ZB 2 entstammt ebenfalls einem Wadi des stufenwärtigen Hanges der Vorlanddepression. Die untersuchte Akkumulation wird in etwa 395 m ü. NN durch das Wadi angeschnitten.

Profil ZB 2

Karte:	République du Niger, Fond Topographique 1:200 000, NE-33-XIV-Zoo Baba (1972).
Luftbild:	NE-33-XIII-XIV-400.
Länge:	13° 4′ 50″E.
Breite:	18° 12′ 55″N.
Höhe:	c. 395 m.

1. 600 cm grauer schluffiger Feinsand, ungeschichtet;
2. 90 cm hellgrauer Diatomit, lagig;
3. 10 cm Kalkkruste.

Wadiabwärts überlagern die Basissande (Horizont 1) die Mittelterrasse, d.h. wie Profil ZB 1 ist auch dieses Profil in ein postmittelterrassenzeitliches zerschnittenes Relief eingelagert worden. Diese Einschätzung wird von dem vorliegenden Radiocarbondatum bestätigt. Die Kalkkruste (Horizont 3) besitzt ein ^{14}C-Modellalter von 5 370 ± 100 BP (HV 14024).

Die Basissande gleichen in ihrer heterogenen Fazies den Basissanden von Profil ZB 1. Eine Vergleichbarkeit wird jedoch durch fehlende Fossilien erschwert. Die bisher durchgeführten Diatomeenuntersuchungen bestätigen die topographische Situation des Profils: vorherrschend epiphytische Floren (*Epithemia zebra, E.argus, Cymbella cistula*) weisen auf eine geringe Wasserbedeckung hin. Identifizierte Phytolithe begünstigen die Entwicklung der Epiphyten und zeigen Ufernähe an. Gegen Ende der limnischen Phase deutet die Kalkkruste auf starke Verdunstung hin.

Insgesamt scheint m.E. eine Parallelisierung der beiden Profile möglich.

Sie kennzeichnen eine ausgeprägte limnische Phase, die nach der Zerschneidung der Mittelterrasse einsetzt und gegen 5 370 ± 100 BP endet. Exaktere Angaben zum Beginn der Seephase sind erst nach der Datierung der limnischen Schnecken zu erwarten. Profil ZB 1 wurde anschliessend, vermutlich während der niederterrassenzeitlichen Zerschneidung gekappt – darauf deutet die Niederterrassenschuttstreu auf dem Top des Profils hin. Verbreitung und Höhenlage der untersuchten Akkumulationen erlauben eine Abschätzung von Ausdehnung und Tiefe der Seephase. Interpoliert man die, mittels Aneroid-Barometer gemessenen Höhen der Sedimente, so erhält man über dem heutigen tiefsten Teil der Vorlandsenke eine Wassertiefe von etwa 25-30 m bei etwa 10 km^2 Fläche. Die bestimmten Diatomeenfloren charakterisieren einen perennierenden Süsswassersee mit relativ konstantem Sauerstoffgehalt. Brackwasserverhältnisse scheinen auf den Beginn und das Ende der Seebildung beschränkt gewesen zu sein.

Etwa 120 km südlich Bilma, im äussersten südlichen Randbereich des Grand Erg de Bilma, befindet sich bei 17°32′N und 13°8′E die Schichtstufe von Dibella. Direkt dem präkambrischen Sockel aufliegende unterkretazische Sandsteine (Faure 1966) bauen die Stufe auf, die eine absolute Höhe von 500 m ü. NN erreicht. Die teilweise stark eingesandete Vorlanddepression westlich der Stufe erstreckt sich von NNW nach SSE über deren gesamte Länge (5 km). Sie weist an ihrer tiefsten Stelle, im nördlichen Bereich der Senke eine Höhe von 368 m ü. NN auf. Der nach Westen exponierte Hang der Dibellastufe ist in seinem nördlichen Teil von z.T. mächtigen Dünensandrampen bedeckt, in die limnische Akkumulationen eingeschaltet sein können. Nach Süden nimmt die Versandung ab und bis in eine Höhe von 392 m ü. NN reichende diatomitische Sedimente sind direkt der Stufe angelagert. Aus diesem Bereich wurde ein Profil untersucht (vgl. Fig. 1). Dabei sei erwähnt, dass das Profil noch nicht vollständig analytisch bearbeitet ist; es sollen hier erste Ergebnisse vorgestellt werden.

Profil DIB 1

Karte: République du Niger, Fond Topographique 1:200 000, NE-33-VIII-Dibella (1972).
Luftbild: NE-33-VII-VIII-552.
Länge: 13°8′ 30″E.
Breite: 17°32′ 20″N.
Höhe: c. 380 m.

1. 460 cm gleyfarbener Feinsand;
2. 40 cm schluffiger Feinsand, Holzkohle, Gräserreste;
3. 515 cm schluffiger mittlerer Feinsand, Fe- (hydr-)oxidausfällungen;
4. 2 cm schluffiger mittlerer Feinsand, Fe- (hydr-)oxidausfällungen, stark verfestigt;
5. 5 cm schwärzlich/grauer Diatomit, Blattreste;

6. 30 cm weisslicher Diatomit, feinlagig, Pflanzenreste;
7. 13 cm weisslicher Diatomit, feinlagig, stark verwittert;
8. 22 cm weisslicher Diatomit, feinlagig;
9. 3 cm feinsandiger Diatomit, ockerfarben;
10. 1 cm feinsandiger Schluff;
11. 1 cm schluffiger Feinsand;
12. 1 cm schluffiger Diatomit;
13. 14 cm weisslicher Diatomit, feinlagig;
14. 2 cm schluffiger Diatomit;
15. 3 cm gleyfarbener schluffiger Feinsand;
16. 31 cm Diatomit, feinlagig, wechselweise weissliche und rostrote Lagen, vereinzelt Fe-oxidbänder, Rhizome;
17. 4 cm gleyfarbener Feinsand, feinkiesig;
18. 1 cm gleyfarbener Feinsand, Fe-oxidbändchen;
19. 2 cm Diatomit, feinlagig, umbrafarben;
20. 3 cm gleyfarbener schluffiger Feinsand, feinkiesig, Fe- (hydr-)oxidausfällungen;
21. 5 cm Diatomit, umbrafarben, stark verwittert;
22. 4 cm Diatomit umbrafarben;
23. 1 cm Fe-oxidband;
24. 4 cm Diatomit, umbrafarben;
25. 1 cm Fe-oxidband;
26. 100 cm umbrafarbener Diatomit, vereinzelt Bänder von Fe-oxiden;
27. 400 cm angelagerter Diatomit (vgl. Fig. 1 und 3).

Das Top des Profils ist mit grobem Schutt und neolithischen Artefakten bedeckt, die Oberfläche fällt mit etwa 10° nach SSW ein.

Mittels Korngrössenanalyse und morphoskopischer Sandanalyse lässt sich eine äolische Zulieferung der Basissande (Horizont 1) mit einiger Sicherheit nachweisen. Gegen Ende der Dünensandakkumulation setzte offensichtlich nach einer ausgeprägten Bodenbildungsphase ein erstes limnisches Stadium ein (Horizont 2). Einen zeitlichen Fixpunkt hierfür gibt ein Radiocarbondatum von 10 230 ± 245 BP (HV 14025). Das aus einem Huminsäureauszug ermittelte ^{14}C-Modellalter dürfte dabei nicht den Zeitpunkt der Einschwemmung des organischen Materials oder den Beginn der Bodenbildung anzeigen, sondern das Alter der Huminsäure vor der Fossilierung. Danach müssten pedogenetische Prozesse, die zur Bildung des organischen Materials führten, bereits früher eingesetzt haben.

Die Bedingungen unter denen die diesen Horizont überlagernden schluffigen mittleren Feinsande abgelagert wurden, sind schwieriger zu rekonstruieren. Heterogene Körnung und ein grosser Anteil rund glänzender Quarzkörner (vgl. Pachur 1966) machen eine, zumindest zeitweise, Bewegung in fliessendem Wasser wahrscheinlich.

Etwa gegen 9 785 ± 75 BP (HV 14026) beginnt eine durch Diatomite gekennzeichnete Seephase. Die bereits untersuchten Diatomeen (vgl. Fig. 3)

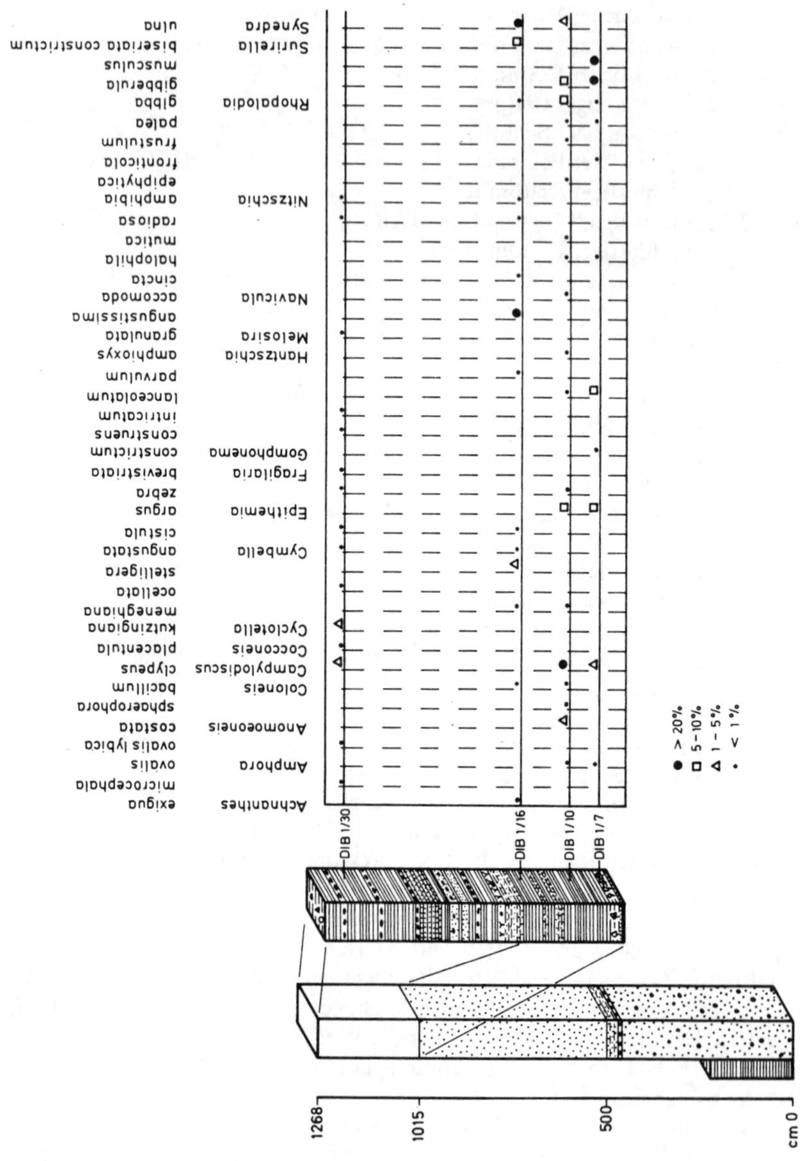

Achnanthes exigua
microcephala
Amphora ovalis
ovalis lybica
Anomoeoneis costata
sphaerophora
Caloneis bacillum
Campylodiscus clypeus
Cocconeis placentula
Cyclotella kutzingiana
meneghiana
ocellata
stelligera
Cymbella angustata
cistula
Epithemia argus
zebra
Fragilaria brevistriata
Gomphonema constrictum
construens
intricatum
lanceolatum
parvulum
Hantzschia amphioxys
Melosira granulata
Navicula angustissima
accomoda
cincta
halophila
mutica
radiosa
Nitzschia amphibia
epiphytica
fronticola
frustulum
palea
Rhopalodia gibba
gibberula
musculus
Surirella biseriata constrictum
Synedra ulna
DIB 1/30
DIB 1/16
DIB 1/10
DIB 1/7
> 20%
5 – 10%
1 – 5%
< 1%
1268
1015
500
cm 0

erwiesen sich als recht gute Milieuanzeiger. Danach dominieren an der Basis (Horizont 5) der Diatomite *Rhopalodia gibberula* und *R.musculus*. Beide zeigen Natriumcarbonatwässer an (vgl. Iltis 1973). In Horizont 6 erfolgt eine Zunahme der Alkalität der Wässer. Die Vorherrschaft von *Campylodiscus clypeus* weist auf starke Sulfat- oder Chloridzunahme hin (vgl. Baudrimont 1973). Die identifizierten Floren sind typische Anzeiger eines Brackwassermilieus in relativ eng begrenzten Tümpeln. Dies widerspricht der topographischen Situation des Profils. Danach wäre ein See mit einer Wassertiefe von etwa 20 m anzunehmen. Möglicherweise existierten zu Beginn der frühholozänen limnischen Phase innerhalb der Depression mehrere kleine Seen in verschiedener Höhe, die durch Sandbarrieren getrennt wurden. Im Hangenden des Profils (Horizont 8) zeigen die Diatomeen eine deutliche Anderung dieser Bedingungen an. In diesem Profilabschnitt dominieren mit *Cyclotella stelligera, Synedra ulna* und *Melosira granulata* var. *angustissima* Süsswasser anzeigende Diatomeenfloren, typisch für den littoralen Bereich grösserer Seen (vgl. Richardson 1966).

LITERATUR

Baudrimont, R. 1973. Les paléosalinités des formations lacustres quaternaires du Sahara algérien d'après l'étude des diatomées. *C.R. Acad. Sci.* 296: 3681-3684.

Baumhauer, R. 1986. Zur jungquartären Seenentwicklung im Bereich der Stufe von Blima (NE-Niger). *Würzburger Geogr. Arb.* 65: 244 S.

Faure, H. 1966. Reconnaissance géologique des formations sédimentaires postpaléozoïques du Niger oriental. *Mém. BRGM* 47: 630 S.

Gabriel, B. & S.Kröpelin 1983. Jungquartäre limnische Akkumulationen im NW-Sudan. *Z. Geomorph.* N.F. Suppl.

Iltis, A. 1973. *Le phytoplancton des eaux natronées du Kanem (Tchad). Influence de la teneur en sels dissous sur le peuplement algal.* Thèse d'Etat, Université de Paris VI. 313 S.

Pachur, H.-J. 1966. Untersuchungen zur morphoskopischen Sandanalyse. *Berl. Geogr. Arb.* 4: 35 S.

Richardson, J.L. 1966. Changes in level of Lake Naivasha, Kenya, during postglacial times. *Nature* 09 (5020): 290-291.

Servant-Vildary, S. 1978. Etude des diatomées et paléolimnologie du bassin tchadien au Cénozoïque supérieur. *Trav. Doc. ORSTOM* 84: 346 S.

MIDDLE HOLOCENE VEGETATION OF THE GILF KEBIR / SW EGYPT – A RECONSTRUCTION

K.NEUMANN
Botanisches Institut, J.W.Goethe Universität, Frankfurt am Main, F.R.Germany

ABSTRACT

Excavations of prehistoric sites from two wadis in the Gilf Kebir (Libyan Desert) yielded a number of charred wood remains which were examined by means of SEM-photography. The species combination (*Tamarix* sp., *Ziziphus lotus, Acacia* sp., *Maerua crassifolia* and *Balanites aegyptiaca*) suggest an arid climate with annual precipitations not exceeding 100 mm between 8 000 and 4 000 BP. The isolated presence of *Acacia albida,* dated at 6 150 ± 180 BP, indicates a slightly moister period of unknown duration. In spite of the arid climate prevailing during the Middle and Late Holocene, the Gilf Kebir offered a favourable environment for prehistoric nomads because of the seasonal availability of surface water and the comparatively rich wadi vegetation.

RESUME

La fouille de sites préhistoriques du Gilf Kébir (Désert de Libye) a produit des restes de bois carbonisé qui sont examinés à l'aide de microscope électronique à balayage. La combination des espèces (*Tamarix* sp., *Ziziphus lotus, Acacia* sp., *Maerua crassifolia* et *Balanites aegyptiaca*) suggère un climat aride et des précipitations annuelles n'excédant pas 100 mm entre 8 000 et 4 000 BP. La présence isolée d'*Acacia albida,* datée à 6 150 ± 180 BP, indique une période un peu plus humide de durée indéterminée. En dépit de l'aridité climatique qui prévalait pendant l'Holocène moyen et tardif, le Gilf Kébir constituait un environnement favorable pour les nomades préhistoriques, en raison de la disponibilité saisonnière d'eaux de surface et de la végétation relativement riche des wadis.

INTRODUCTION

The Gilf Kebir, a plateau of 'Nubian' sandstone, is situated in the heart of the Libyan desert (between 22°40′ and 25°N, 25° and 26°30′E, Fig. 1), one of

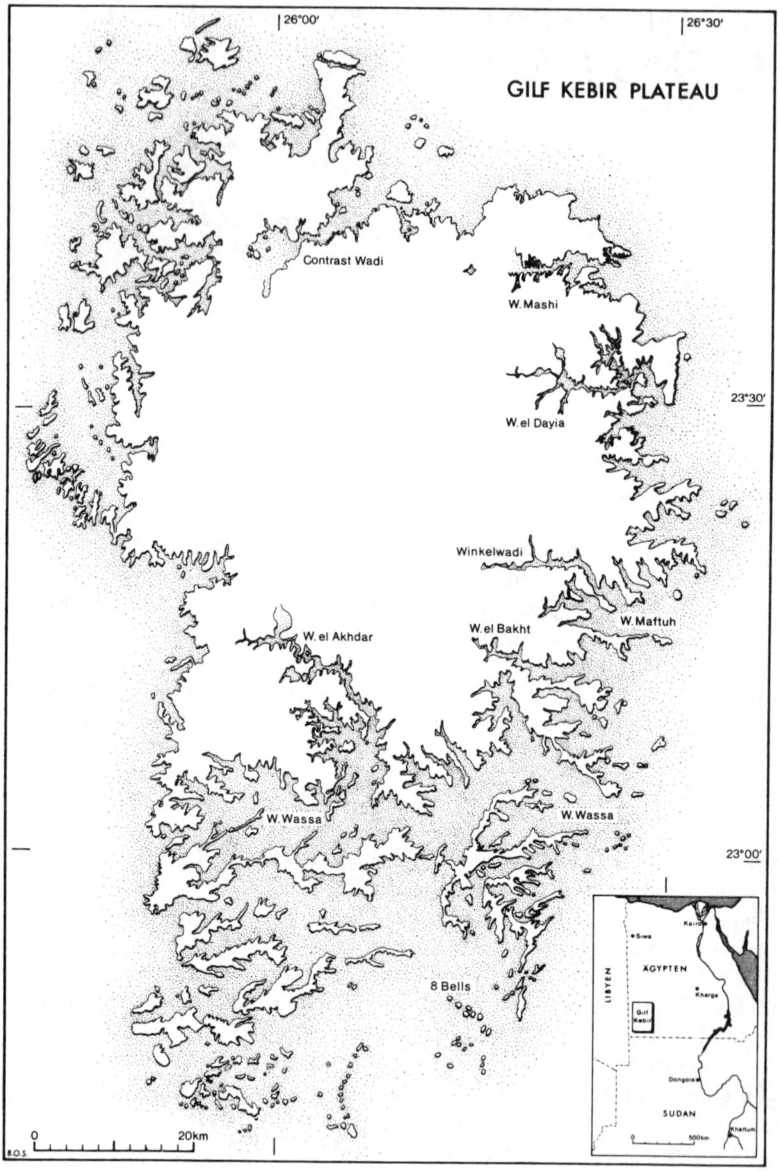

Figure 1. The Gilf Kebir plateau.

the most arid regions of the world. The edge of the plateau, especially in the southeast, is dissected by several wadis, two of which, Wadi el Akhdar and Wadi Bakht, were the object of archaeological investigations by the research group BOS ('Besiedlungsgeschichte der Ost-sahara') (Kuper 1981). Numerous excavations yielded a number of charred wood remains which could be identified, sometimes to species level, by their anatomical features, Charcoal identification, a method commonly used by European archaeobotanists, has so far rarely been applied to African material. It gives us the opportunity to reconstruct the woody vegetation of a given area, to reflect upon the palaeoclimate and, moreover, to form an idea of the plant cover as the economic basis of prehistoric people.

MATERIAL AND METHODS

The charcoal pieces with a size from 1 mm to 10 mm were fractured manually into the diagnostic relevant transverse, tangential and radial planes. After fixing on aluminium stubs with conductive carbon cement and coating with gold, the charcoal pieces were examined under the scanning electron microscope (SEM) and photographed on Ilford Pan F films.

Only samples from radiocarbon dated sites were selected. Samples which yielded sufficient organic material for radiocarbon dating and, at the same time, were well enough preserved for identification, are marked with an asterisk in the list (see Tables 1 and 2). 'Younger than'/'older than' points to the relative position of the sample above or under the dated stratum.

Some of the samples were so badly preserved that they fell to pieces when touched. They were not counted, but the presence of the species was noted in the tables by an x. Three of the samples (23, 316 and 321) were so rich that they could only be treated in part. For this, ten random sub-samples, altogether 10 per cent of the identifiable material, were chosen from each sample for identification. From the studies of Van der Veen & Fieller (1982), it is clear that random sampling can be expected to produce representative results for the whole sample.

The full description of the material with photographs of all planes, together with illustrations of the reconstructed plant communities, will be published elsewhere (Neumann, in press).

RECENT CLIMATE AND VEGETATION

No continuous climatic records for the Gilf Kebir area exist. The average annual precipitation in Kharga, 500 km to the northeast, is as low as 1 mm. Kufra, 350 km to the northwest, has a higher value of 7 mm/year. Alaily et al. (in press) estimate that the Gilf Kebir witnesses one short rain-storm with 5 to 10 mm in a day or less every two years.

When, in 1932, Almásy discovered the Wadi Abd el Malik in the north-

Table 1. Wadi el Akhdar – results of charcoal identification

Site	Radiocarbon date(s) of the site or the stratum of the profile	Laboratory number	Sample number	*Tamarix* sp.	*Ziziphus lotus*	*Acacia* sp.	*Maerua crassifolia*	indet.
80/7-1	7 700 ± 60	KN 2878	3, 381, 382, 384,	x	–	–	–	–
Stratum D	7 670 ± 75	KN 2934	385, 386					
			165	7	–	–	–	–
			371	2	–	–	–	1
			387	10	–	–	–	–
			390	3	–	–	–	–
			376	–	–	1	–	–
80/7-1	5 780 ± 80	KN 2935	7	1	9	38	–	5
Stratum A	5 670 ± 75	KN 2882						
81/4	5 670 ± 65	KN 3016	226, 275	x	–	–	–	–
Stratum 2	5 650 ± 130	KN 3381	273 (older than dated stratum)	5	–	–	–	–
			276	7	–	–	–	–
			274	70	76	6	1	7
80/7-2	5 610 ± 60	KN 2936	42	x	–	–	–	–
81/2	5 440 ± 60	KN 3106	142	x	–	–	–	–
			143*	x	–	–	–	–
80/12-1	5 420 ± 65	KN 2881a	26, 37	x	–	–	–	–
	5 360 ± 210	KN 2881b						
	5 150 ± 125	KN 2933						
80/7-5	5 150 ± 70	KN 2880b	23 (p)	572	–	–	–	–
80/14	4 200 ± 1 000	KN 3085	201	9	–	–	–	–
	4 150 ± 55	KN 3173	202	x	–	–	–	–
	3 950 ± 55	KN 2926	203	9	–	–	–	–
			204	19	–	–	–	–

(p) = Only 10% of the sample was examined.
* = Radiocarbon date and charcoal identification were obtained from the same sample.

west of Gilf Kebir, he thought it to be the legendary oasis Zerzura. In this wadi he found a vegetation zone of 24 km in length, with acacias and some *heglig* trees (*Balanites aegyptiaca*) (Almásy 1939). Later, the northwestern wadis were visited by Bagnold (Bagnold et al. 1939) and Léonard (1969, 1971), none of which gives us detailed information on the plant species growing there. Léonard only mentions that the woody vegetation consisted of acacias and *Maerua crassifolia,* but most of the trees were found to be dead.

With the exception of four specimens of *Acacia tortilis* subsp. *raddiana* in Contrast Wadi (Back 1981), no trees are at present growing in the southern and eastern wadis of the plateau. Here a very scattered vegetation of not more than fifteen species of ephemeral grasses and shrubs appears after the extremely rare precipitations (Alaily et al., in press, Back 1981, Bagnold et al. 1939, Boulos 1980a, 1981).

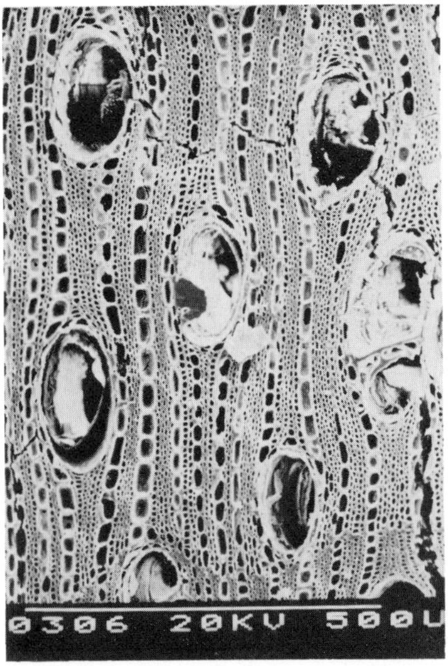

Photo 1. *Ziziphus lotus,* sample no. 314. Transverse fracture.

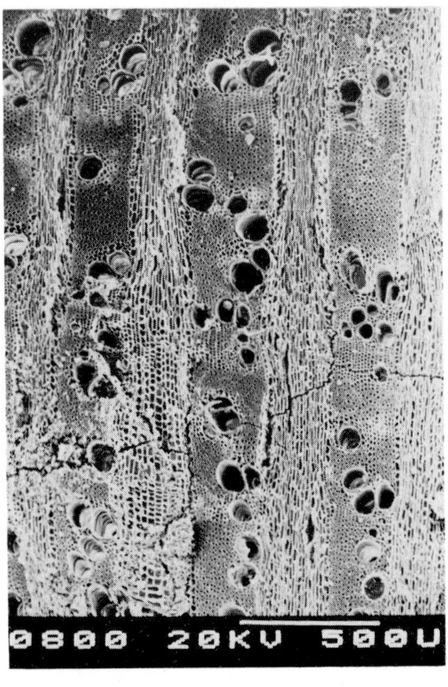

Photo 2. *Tamarix* p., sample no. 274. Transverse fracture

Table 2. Wadi Bakht – results of charcoal identification

Site	Radiocarbon date(s) of the site or the stratum of the profile	Laboratory number	Sample number	*Tamarix* sp.	*Ziziphus lotus*	*Acacia* sp.	*Acacia albida*	*Maerua crassifolia*	*Balanites aegyptiaca*	indet.
82/21-2	6 600 ± 300	KN 3410	380*	-	1	1	-	-	-	-
82/22	6 150 ± 180	KN 3328	94*	-	-	-	74	-	-	12
			316 (p)	-	-	-	762	-	-	72
82/13	6 080 ± 420	KN 3179	64	3	-	-	-	-	-	-
			312	x	-	-	-	-	-	-
			314 younger	-	2	-	-	-	-	1
			315 than dated stratum	-	3	-	-	-	-	-
82/15	5 070 ± 60	KN 3149	71*	x	-	-	-	-	-	-
82/18	4 880 ± 390	KN 3182	85*	x	-	-	-	-	-	-
82/16	4 820 ± 60	KN 3098	78	14	-	3	-	-	-	2
			81	6	1	-	-	-	-	-
82/19	4 770 ± 130	KN 3184	321 (p)	-	431	-	-	-	-	-
82/24	5 030 ± 150	KN 3591	474	-	-	4	-	-	-	18
			491	1	-	1	-	5	-	20
			493	-	-	-	-	-	1	-
			495*	-	-	19	-	-	-	10

(p) = Only 10% of the sample was examined.
* = Radiocarbon date and charcoal identification were obtained from the same sample.

BRIEF DESCRIPTION OF WOOD ANATOMY

1. *Acacia albida* (Mimosaceae)

Growth rings not discernible; vessels solitary or in small clusters, with simple perforations; parenchyma confluent-banded; rays homogeneous, composed of long, procumbent cells, uni- and biseriate, storeyed.

Easily distinguishable from the wood of the other North African *Acacia* species by the banded parenchyma and the uni- and biseriate, storeyed rays.

Widely distributed in the African savannas, preferably on alluvial soils (Wickens 1969). Some relict specimens in the Tibesti (Scholz 1967), Hoggar (Quézel 1954) and Tassili n'Ajjer (Leredde 1957).

2. *Acacia* sp. (Mimosaceae)

Growth rings not discernible; vessels solitary or in small clusters, with simple perforations, pits vestured; parenchyma aliform-confluent, with short strands of crystal cells; rays homogeneous, composed of long, procumbent cells, 1- to 8-seriate.

The anatomical features of the *Acacia* species are very variable so that identification to species level is in most of the cases impossible. The charcoal probably comes from *Acacia tortilis* subsp. *raddiana,* one of the most abundant trees of the Sahara.

3. *Ziziphus lotus* (Rhamnaceae) (Photo 1)
Growth ring boundary marked by a change in fibre wall thickness and a thin band of terminal parenchyma; diffuse-porous; vessels solitary or in short radial multiples, perforations simple; parenchyma vasicentric, sparse; rays heterocellular, composed of short procumbent and square cells, uni- and biseriate.

Similar anatomical features occur in *Z.spina-christi* and *Z.mauritiana,* but ray frequency is 18-24/mm in *Z.lotus,* whereas only 11-13/mm in the two other species; tangential width of the ray cells is < 25 μm in *Z.lotus,* < 42 μm in *Z.spina-christi* and *Z.mauritiana.*

Floristic category: Irano-Turanian region. Recent distribution of the subsp. *saharae*: Central Sahara; of the subsp. *lotus*: Morocco to Cyrenaica and in Palestine.

4. *Maerua crassifolia* (Capparidaceae)
Growth rings not discernible; vessels solitary and in short radial multiples, small (< 65 μm); parenchyma vasicentric, sparse; rays heterocellular, composed of square and short procumbent cells, 1- to 5-seriate.

A similar wood structure is found in other Capparidaceae, especially *Capparis decidua,* but vessel diameter is much larger in the latter (< 220 μm), and radial arrangement of the vessels is not as distinct.

Widely distributed in the Sahara.

5. *Tamarix* sp. (Tamaricaceae) (Photo 2)
Growth rings not discernible; vessels solitary or in clusters, with simple perforations; parenchyma vasicentric, storeyed; rays very wide (up to 24 cells), heterocellular, composed of procumbent, square and upright cells.

The features of the investigated specimens resemble the recent wood of *T.articulata.* According to Messeri (1938), the wood of *T.articulata* should be distinguishable from *T.nilotica, T.passerinoides* and *T.africana.* Yet, as the Egyptian flora includes nine species of *Tamarix* (Baum 1978), we cannot be sure if other species of *Tamarix* do not show the same characteristics.

Most of the *Tamarix* species belong to the Saharo-Sindian and the Irano-Turanian region. In the Eastern Sahara, the recent *T.articulata* acts as a sand binder and is forming hills up to several meters at places where some groundwater occurs.

6. *Balanites aegyptiaca* (Balanitaceae)
Only one sample contained a very small piece of wood of this species. The characteristic elements are, nonetheless, discernible: apotracheal, uniseriate bands of parenchyma; parenchyma and fibres storeyed; rays very wide; vessels thick-walled.

Widespread in the drier parts of Africa. Floristic category: Sudano-Zambezian and Saharo-Sindian regions (Wickens 1975).

DISCUSSION

All the species recorded in the samples are part of the modern Saharan flora. Therefore, the following reconstruction is based on comparison with recent plant communities of this area (Santa 1958/59). It is highly probable that the dominant species of these communities will be found in the archaeological samples, especially when they provide good firewood.

As for the palaeoclimatic conclusions inferred from the presence of a given species, we should be very careful. The distribution of trees does not depend only on climatic factors, but, even more, on soil properties and relief (Smith 1949). In interpreting the results, I shall, therefore, consider sedimentological data and, moreover, try to formulate minimum conditions for the occurrence of the assumed plant communities.

The continuous presence of *Tamarix* wood in the samples suggests an arid climate with precipitation not exceeding 100 mm around 7 700 BP and between 5 700 and 4 000 BP. The drier phases of this period witnessed a comparatively sparse tree cover, whereas, during the moister phases, the vegetation may have consisted of dense stands of tamarisks (*Tamarix articulata*?) in the lower parts of the wadis. These 'gallery forests' are known today from some wadis in the Hoggar and Tibesti (Quézel 1965: 179f.) and northern Egypt (Kassas 1952, Kassas & Iman 1954) under an average annual rainfall of 50-100 mm. The fact that other arborescent species rarely occur in this community, could serve as an explanation for the small number of species in the samples. However, this may also be due to human selection because tamarisks make excellent firewood (Boulos 1980b). Although there is no evidence of a permanent groundwater table in Wadi el Akhdar and Wadi Bakht (Kröpelin 1985: 82 f., 162 f.), the sandy layers of the playa sediments were capable of water storage and supplying the trees with sufficient moisture. Temporary pools, indicated by thick pelitic playa layers (Pachur & Röper 1984, Kröpelin 1985), probably carried at their margins dense stands of sedges and other hydrophillous herbs mixed with *Tamarix*. Today, we find this association around shallow guelti in the wadis of the Central Saharan mountains (Quézel 1965: 203 f.).

The second dominant species in the samples, *Ziziphus lotus*, is a low, spiny bush, and because of this habit not very attractive for collecting firewood. An explanation for its use could be the scarcity of other woody species under rather arid conditions. A corresponding recent plant community could be found in the 'dayas' of the northwestern Sahara under a mean annual rainfall of 30-40 mm. Just as the playas of Wadi el Akhdar and Wadi Bakht, the daya sediments consist mainly of fine sand and clay although they are much shallower.

Notwithstanding the different thickness of the Gilf Kebir playas and the daya sediments, both habitats are unfavourable for plant growth, because of the low amount of available water. Therefore, the bushes probably grew at considerable distances from each other. A combination of *Ziziphus lotus* with tamarisks, as it is found repeatedly in the samples, was observed by Schulz (pers. comm.) in the huge wadis filled with fine sands between Tit and In Ekker (90-130 km KW of Tamanrasset/Hoggar).

Like many other Mediterranean and Irano-Turanian plants, the recent *Ziziphus lotus* has a disjunct distribution. The species is missing in the 'Sinaitic-Libyan gap', i.e. in Egypt and on the Sinai peninsula (Zohary 1951). The Gilf Kebir finds are evidence for at least a partial connection between the two sub-areas up to Middle Holocene times.

In contrast to other contemporary samples from the Eastern Sahara, the Gilf Kebir samples rarely contain *Acacia* wood. Since *Acacia tortilis* subsp. *raddiana,* the main constituent of the actual 'desert savanna', is restricted to habitats with coarse sands, I assume that this community, including *Maerua crassifolia* and *Balanites aegyptiaca,* was growing in the alluvial plains off the plateau, while tamarisks and *Ziziphus lotus* were predominant on the fine-grained playa sediments in the upper parts of Wadi el Akhdar and Wadi Bakht.

It is important to notice that there is a crucial difference between a true savanna and the 'desert savanna': the latter always occurs as contracted vegetation in the lower parts of the relief and consists of scattered trees intermixed with some tufts of grasses (mainly *Panicum turgidum*). To avoid misunderstandings, some authors prefer the term Acacia-Panicum-association (Leredde 1957, Schulz 1980).

The isolated presence of *Acacia albida,* dated at 6 150 ± 180 BP, indicates a slightly moister period, confirming the beginning of a semi-arid phase which, after Kröpelin (1985) lasted from 6 000 to 5 000 BP in the Gilf Kebir. The precipitation in this phase probably did not exceed 100-150 mm, which is the minimum condition under which the recent *Acacia albida* lives in the Central Sahara. On the duration of the 'semi-arid' period, results slightly different from the sedimentological data of Kröpelin were obtained by our botanical investigations. According to these, the climate was arid between 5 700 and 4 000 BP, with rainfall between 30 and 100 mm, at best. This is even less than the 200 mm postulated by Pachur & Röper (1984) for the Holocene of the Gilf Kebir. In any case, a savanna vegetation under a mean annual rainfall of 200-600 mm, as assumed by McHugh (1971), can be excluded. The Middle Holocene northward shift of Sahelian vegetation, hypothetically stated by Wickens (1975) and confirmed by the investigations of Ritchie et al. (1985) and Neumann & Schulz (this vol.), did not reach the Gilf Kebir area. As in the surrounding Northeastern and East Central Sahara (Haynes et al. 1979, El Hadidi 1980, Schulz 1980, Mehringer n.d. & 1982), the plant cover of the Gilf Kebir consisted more or less of the same elements as today, though with a higher density.

In spite of the arid climate the Gilf Kebir offered a favourable environment for prehistoric nomads because of the seasonal availability of surface water. A slight increase in precipitation created extensive stands of ephemeral grasses and herbs and thus furnished a basis for nomadic cattle keeping and for hunting and gathering activities. The episodic visits of nomads to the Gilf Kebir continued up to the hyperarid 20th century (Almásy 1939), even though the surrounding lowlands then were a hostile and almost barren environment.

ACKNOWLEDGEMENTS

I wish to thank the German Research Foundation (DFG) for funding the project BOS, especially the expeditions to the Gilf Kebir. Thanks are due to Professor K.U.Leistikow for correcting the manuscript, furthermore to the members of the project and to Dr E.Schulz, for critical remarks and fruitful discussions.

REFERENCES

Alaily, F., R.Bornkamm, H.-P.Blume, H.Kehl & H.Zielinsky (in press). Ecological investigations in the Gilf Kebir (SW-Egypt). *Phytocoenologia.*

Almásy, L.E.de 1939. *Unbekannte Sahara. Mit Flugzeug und Auto in der Libyschen Wüste.* Leipzig: Brockhaus, 214 pp.

Back, H.-E. 1981. Beitrag zur Faunistik und Zoogeographie der Lepidopteren in der östlichen Sahara. *Nota Lepid.* 4 (4): 111-124.

Bagnold, R.A., O.H.Myers, R.F.Peel & H.A.Winkler 1939. An expedition to the Gilf Kebir and Uweinat. *Geogr. J.* 93: 281-313.

Baum, B.R. 1978. *The genus Tamarix.* Jerusalem: The Israel Academy of Science and Humanities. 209 pp.

Boulos, L. 1980a. Botanical results of the expedition. In Farouk el Baz et al. (eds.), Journey to the Gilf Kebir and Uweinat, South West Egypt, 1978. *Geogr. J.* 146: 68-71.

Boulos, L. 1980b. *Firewood crops – shrubs and tree species for energy production.* Washington, D.C.: Natural Academy of Science. 234 pp.

Boulos, L. 1981. The flora of Gebel Uweinat and neighbouring regions, Southwest Egypt. *Ann. Geol. Surv. Egypt* 11: 341-347.

Hadidi, N. el 1980. Vegetation of the Nubian Desert (Nabta Region). In F.Wendorf & R.Schild (eds.), *Prehistory of the Eastern Sahara*: 345-351. New York: Academic Press.

Haynes, C.V., P.J.Mehringer & El Sayed Abbas Zaghloul 1979. Pluvial lakes of North-Western Sudan. *Geogr. J.* 145 (3): 437-445.

Kassas, M. 1952. Habitats and plant communities in the Egyptian desert. 1. Introduction. *J. Ecol.* 40: 342-351.

Kröpelin, S. 1985. Untersuchungen zum Sedimentationsmilieu von Playas im Gilf Kebir (SW-Ägypten). M.A. thesis, Technical University, Berlin.

Kröpelin, S. (this vol.). Palaeoclimatic evidence from Early to Mid-Holocene playas in the Gilf Kebir (Southwest Egypt).

Kuper, R. 1981. Untersuchungen zur Besiedlungsgeschichte der östlichen Sahara. Vor-

bericht über die Expedition 1980. *Beitr. Allg. Vergl. Arch.* 3: 215-275.
Léonard, J. 1969. Expédition scientifique Belge dans le désert de Libye. *Africa-Tervuren* 15 (4): 101-136.
Léonard, J. 1971. Aperçu de la flore et de la végétation du Jebel Uweinat (Désert de Libye). Résumé. *Mitt. Bot. Staatssamml. München* 10: 476-477.
Leredde, C. 1957. *Etude écologique et phytogéographique du Tassili n'Ajjer.* Alger: Institut de recherches Sahariennes de l'Université d'Alger. Mission scientifique au Tassili des Ajjer, 2 (1949). 455 pp.
Maire, R. 1940. Etude sur la flore et la végétation du Sahara Central, 3ième partie. *Mém. Soc. Nat. d'Afr.* 3: 273-433.
McHugh, W.P. 1971. Late prehistoric cultural adaptation in the Southeastern Libyan Desert. Ph.D. dissertation, 373 pp., Dept. Anthrop., University of Wisconsin.
Mehringer, P.J. n.d. Investigation of Holocene lake sediments, Bir Merga, Sudan, and the Western Desert, Egypt. *Nat. Geogr. Soc. Res. Rep. 1975 Projects*: 487-499.
Mehringer, P.J. 1982. Early Holocene climate and vegetation in the Eastern Sahara: the evidence from Selima oasis, Sudan. 95th annual meeting of the Geological Society of America, New Orleans.
Messeri, A. 1938. Studio anatomico-ecologico del legno secondario di alcune painte del Fezzan. *Nuov. Giorn. Bot. Ital.* 45 (3): 267-356.
Neumann, K. (in press). Holzkohlen aus prähistorischen Fundstellen des Gilf Kebir (SW-Ägypten). In R.Kuper (ed.), *Forschungen zur Prähistorie Afrikas* (vol. 1). Bonn.
Neumann, K. & E.Schulz (this vol.). Middle Holocene savanna vegetation in the Central Sahara – preliminary report.
Pachur, H.-J. & H.P.Röper 1984. The Libyan (Western) Desert and Northern Sudan during the late Pleistocene and Holocene. *Berliner Geowiss. Abh.* 50: 249-284.
Quézel, P. 1954. *Contribution à l'étude de la flore et de la végétation du Hoggar.* Alger: Institut de recherches Sahariennes de l'Université d'Alger. Monographies régionales 2, 164 pp.
Quézel, P. 1965. *La végétation du Sahara (du Tchad au Mauritanie).* Stuttgart: Fischer. 333 pp.
Ritchie, J.C., C.H.Eyles & C.V.Haynes 1985. Sediment and pollen evidence for an early to mid-Holocene humid period in the Eastern Sahara. *Nature* 314: 352-355.
Santa, S. 1958/59. Essai de reconstitution de paysages végétaux quaternaires d 'Afrique du Nord. *Libyca* 6/7: 37-77.
Scholz, H. 1967. Baumbestand, Vegetationsgliederung und Klima des Tibesti-Gebirges. *Berliner Geogr. Abh.* 5: 11-17.
Schulz, E. 1980. *Zur Vegetation der östlichen zentralen Sahara und zu ihrer Entwicklung im Holozän.* Würzburg: Würzburger Geogr. Arb. 51.
Smith, J. 1949. Distribution of tree species in the Sudan in relation to rainfall and soil texture. *Ministry of Agriculture Bull.* 4, Khartoum. 68 pp.
Van der Veen, M. & N.Fieller 1982. Sampling Seeds. *J. Arch. Sci.* 9: 287-298.
Wickens, G.E. 1969. A study of Acacia albida Del. (Mimosoideae). *Kew Bull.* 23 (2): 181-202.
Wickens, G.E. 1975. Changes in the climate and vegetation of the Sudan since 20 000 BP. *Boissiera* 24a: 43-65.
Zohary, M. 1951. The arboreal flora of Israel and Transjordan and its ecological and phytogeographical significance. Oxford: Imperial Forestry Institute of the University of Oxford, Oxford: Paper No. 26, 59 pp.

PALAEOCLIMATIC EVIDENCE FROM EARLY TO MID-HOLOCENE PLAYAS IN THE GILF KEBIR (SOUTHWEST EGYPT)

S.KRÖPELIN
Institut für Physische Geographie, Freie Universität Berlin, F.R.Germany

SUMMARY

Investigations of the sediments of early to Mid-Holocene playas in valleys of the Gilf Kebir Plateau (c. 23°N/26°E) resulted in new data on the palaeoclimate in the centre of the largest hyperarid region of the earth. From detailed stratigraphic evidence, scanning electron microscopy, more than forty radiocarbon-dates and interpretations of prehistoric floral and faunal remains, the following climatic development was reconstructed: Up to about 9500 years BP an extremely arid climate, similar to the present, prevailed. The period between c. 9500 and 6000 BP had an arid climate with rare heavy rainfalls (approximately four events per 100 years), feeble soil formation and sparse plant growth in the surroundings of the temporary rainpools, and high wind activity. Between 6000 and 5000 BP conditions tended towards moderate aridity with an estimated maximum annual rainfall of 100 mm. This millennium seems to represent the main phase of Neolithic settlement. The playa deposits contain no evidence of the subsequent increase in aridity about 4800 BP. On the whole, there is growing evidence that in this part of the eastern Sahara – in contrary to more southerly regions of the Libyan Desert – the climate was arid during the entire Holocene, with only a slight, sporadic tendency towards semi-arid conditions.

RESUME

Des études au sujet du milieu de sédimentation des playas dans les vallées du plateau de Gilf Kébir (environ 23°N/26°E) ont apporté des données paléoécologiques concernant le paléoclimat du centre de la région hyperaride la plus étendue de la terre pendant l'Holocène inférieur et moyen. D'après les résultats microstratigraphiques en rapport avec une quarantaine de datations à radiocarbone jusqu'ici disponible, des études en microscopie électronique ainsi que des interprétations des restes de la flore et de la faune préhistoriques, il en résulte le déroulement climatique suivant: Jusqu'à environ 9 500 BP régnait un climat extrêmement aride, comparable à celui de nos jours. La période entre environ 9 500 et 6 000 BP semble être caractérisée par un climat aride, accompagné des averses rares (environ quatre événements par 100 ans), avec une formation de sol faible et une végétation modeste dans les environs des playas et une activité éolienne importante. De 6 000 à 5 000 BP les conditions climatiques ont eu une tendance vers le sémiaride avec des précipitations annuelles éstimées à 100 mm maximum. C'est à cette époque que la colonisation Néolithique principale à apparemment eu lieu. L'aridification qui suivit après environ 4 800 BP n'est plus documentée dans les accumulations de playa. En résumé, il semble que cette région du Sahara oriental était caractérisée par une aridité persistante pendant tout l'Holocène, contrairement à la partie plus au sud du Désert Libyque.

ZUSAMMENFASSUNG

Untersuchungen zum Sedimentationsmilieu früh- bis mittelholozäner Playas in Tälern des Gilf-Kebir-Plateaus (ca. 23°N/26°E) erbrachten paläoklimatisch ausdeutbare Daten aus dem Zentrum des ausgedehntesten hyperariden Raumes der Erde. Aus den mikrostratigraphischen Befunden, rasterelektronen-mikroskopischen Ergebnissen, über 40 bisher vorliegenden ^{14}C-Altern sowie Interpretationen der prähistorischen Floren- und Faunenreste wird folgender Klimaablauf abgeleitet: Vor etwa 9 500 BP herrschte ein den rezenten Verhältnissen vergleichbares extrem arides Klima. Der Zeitraum von etwa 9 500 BP bis 6 000 BP scheint durch ein von seltenen Starkniederschlagsereignissen (ca. vier Ereignisse/100 Jahre) geprägtes arides Klima mit ansatzhafter Bodenbildung und spärlichem Pflanzenwachstum im Umfeld der Playas sowie hoher äolischer Aktivität gekennzeichnet gewesen zu sein. Von etwa 6 000 bis 5 000 BP tendierten die Verhältnisse zu einer abgeschwächten Aridität mit geschätzten Jahresniederschlägen von maximal 100 mm. In diesem Jahrtausend sind anscheinend auch die neolithischen Hauptbesiedlungsphasen anzusetzen. Die in der Folgezeit nach etwa 4 800 BP einsetzende zunehmende Trockenheit ist nicht mehr in den Playaakkumulationen dokumentiert. Insgesamt verdichtet sich somit der Eindruck eines Raumes in diesem Teil der östlichen Sahara, der – im Gegensatz zu den südlicher gelegene Teilen der Libyschen Wüste – während des gesamten Holozäns von einem ariden und nur phasenhaft zum Semiariden hin tendierenden Klima geprägt war.

INTRODUCTION

The results presented here were obtained during several field seasons in SW Egypt and NW Sudan from 1980 onwards. The investigations formed part of the interdisciplinary research project 'Prehistoric Settlement of the Eastern

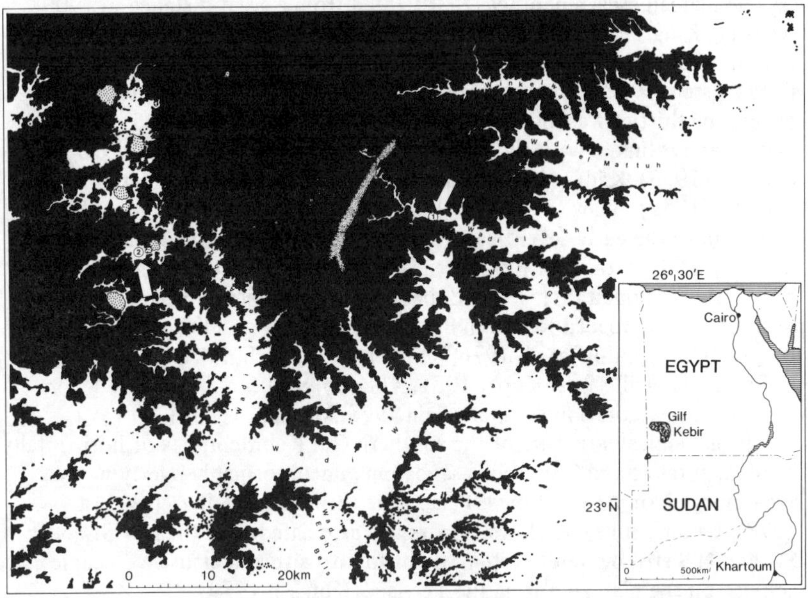

Figure 1. Southeastern part of the Gilf Kebir Plateau with localities. 1. Playa Wadi el Bakht; 2. Playa Wadi Ard el Akhdar; dotted areas: basaltic intrusions; shaded area: wind eroded depression along major fault. Source of data: Landsat and Apollo VII imagery.

Sahara' (BOS) with the title 'Cultural development, landform evolution and life styles in relation to climatic change in the eastern Sahara'. In this connection archaeological evidence has shown that during the early and middle Holocene the playas of the Gilf Kebir were favourite Neolithic settlement sites owing to their temporary water supply in this nowadays totally uninhabited part of Egypt's Western Desert. The objective of the present study was to investigate the sedimentary environments of the playa deposits and thus to clarify the boundary conditions for settlement at the time of deposition. Almost inevitably, this also yielded data contributing to the reconstruction of the palaeoclimate.

THE GILF KEBIR

The playas dealt with here are located in two valleys of the Gilf Kebir Plateau, which is situated at the Tropic of Cancer at about 26°E in the SW-Egyptian part of the Libyan Desert. The Gilf Kebir is a sandstone plateau c. 5800 km^2 in size, dissected by numerous valleys and with steep cuesta-like edges at approximately 1000 m a.s.l. (Fig. 1).

The plateau escarpment was first sighted from a great distance in 1909 (Harding-King 1912), but it was observed at close range and given its name only in 1926 (Gilf el Kebir = 'great scarp' in Arabic; Ball 1927, Kemal El Dine Hussein 1928). The actual pioneer exploration of the remote plateau during the thirties was motivated by geodetic, archaeological and military factors and is linked with the names of Almásy (1936), Bagnold (1939), Clayton (1933), Kádár (1937), Myers (1939), Peel (1939), Penderel (1934), Rhotert (1952), Shaw (1936) and Winkler (1939).

Not until the early seventies, after a thirty year break, field research was taken up again in the Gilf Kebir area (Expédition Belge, Geological Survey of Egypt, Combined Prehistoric Expedition, Technical and Free University of Berlin, BOS, Special Research Project Arid Areas; cf. Misonne 1969, Issawi 1978, Wendorf et al. 1976, Klitzsch 1978, Pachur & Gabriel 1980, El-Baz & Maxwell 1982, Kuper 1981, Klitzsch, Said & Schrank 1984; for a history of exploration see Meckelein 1975, McHugh 1981a).

The geological structure of the southern Gilf Kebir consists of horizontally bedded, Jurassic and Cretaceous sandstone, unconformably overlying the Pre-Cambrian or various Lower Palaeozoic formations (Klitzsch & List 1978). Basaltic hills and flows, apparently associated with the NW-SE, NE-SW and N-S striking fault system, point to intrusive and effusive volcanic activity on the plateau during the Tertiary (Klitzsch 1979).

Most likely the existing valley morphology with its box canyons and abrupt headcuts was controlled by the highly resistant quartzitic caprock of the southern plateau (Peel 1939). Haynes (1981) presumed that the initial fluvial dissection of the plateau cut into overlying, less resistant shales which were later completely stripped by aeolian erosion. Other processes such as subsurface drainage with undercutting of cliffs may have contributed to the formation of steep valley sides (Maxwell 1981), just as morphodynamic activity in mainly arid and semiarid conditions during Quaternary times remodelled the entire valley morphology (McCauley, Breed & Grolier 1981).

The region of the Gilf Kebir is situated in the centre of what is today the vastest hyperarid area of the world (based on the Budyko-Ratio; Henning & Flohn 1977). Half-way between the Mediterranean winter rainfall zone in the north and the summer rainfall zone of the wet-dry tropics an equatorial trade wind desert climate of strong continentality with an estimated mean precipitation of less than 5 mm/year prevails in this part of the eastern Sahara, the increased hyperaridity being due to the cross circulation of the easterly jet (Flohn 1964). Because of the almost complete absence of rainfall and the lack of any waterholes the Gilf Kebir is virtually inanimate except for a few isolated localities with sparse vegetation (Back 1981).

WADI EL BAKHT

Wadi el Bakht ('lucky valley' in Arabic; Bagnold et al. 1939) is one of the longest valleys cutting into the Gilf Kebir plateau. About 12.5 km below the valley head the psammitic-psephitic facies of characteristic wadi deposits is gradually succeeded by a pelitic facies. This is the location of the c. 66000 m^2 pluvial playa (Fig. 1, loc. 1). The importance of this locality was stressed by McHugh (1980: 65) when he wrote: 'This locale has outstanding potential for contributing to the study of the late prehistoric palaeoenvironments and cultural adaption in the area.'

The position of the playa and the thick valley fill up-stream are due to the previous complete blockage of the valley by a still partly preserved fossil dune, covered in part by a recent falling dune (Photo 1). The long profile is almost horizontal over 3 km above the dune barrier, which indicates that the local base level remained stable over a long period. We measured a gradient of only 4.15 m/950.5 m, i.e. 0.0044; Maxwell (1980: 79) gave a gradient of 0.0028 over about 1 775 m.

Mud crack polygons of different orders with diameters between 1 and 10 m cover the playa surface and document the final desiccation. The NE corner of the playa was lowered 3 m by increased deflation probably caused by turbulences at the foot of the falling dune. Here several yardangs about 1 m high are preserved at different levels and provide the best evidence of the presence of old roots, most likely mainly of reed (*Phragmites communis*). The central part of the fossil playa is covered by a 400 m^2 sized, 3 cm thick layer of curled clay flakes, the remains of recent playa deposits.

Similar deposits with a maximum thickness of 10 cm were found 11 km northward, in the 'Winkelwadi', the northern branch of Wadi el Maftuh, as well as in parts of Wadi Wassa (cf. Fig. 1). These pelitic deposits, known as 'Daq' in the Persian deserts, are evidence of secular rainstorms, which occur even during the present hyperarid climate. Field evidence suggests that water columns more than 2 m high can be attained in such short-lived rainpools (Kröpelin 1985). Cemented-in 1930s petrol cans in contemporary 'clay-flake-playas' in the western Wadi Wassa are proof of their recent age.

The prehistoric implements on the slopes of the fossil dune barrier are remarkably abundant (McHugh 1981b, Schön 1984, Wendorf & Schild 1980: 217-222). This dense layer of hundreds of thousands of stone tools and flakes seems to have preserved the fossil dune beyond the consolidated shore zone. However, it is likely that projection or slope processes contributed to the concentration of artifacts. Similar observations were also made at Wadi Howar in NW Sudan where Neolithic implements have stabilized early Holocene parabolic dunes which are surrounded by mobile barchan chains (Gabriel et al. 1985).

The minor channel incision of the upstream portion of the wadi floor also continues into the playa deposits. At its eastern edge it turns into a step-like gorge which descends 9 m down to the original valley floor over a horizontal distance of only 20 m (Photo 2). At this site the dune barrier was breached, thus terminating the former local base level.

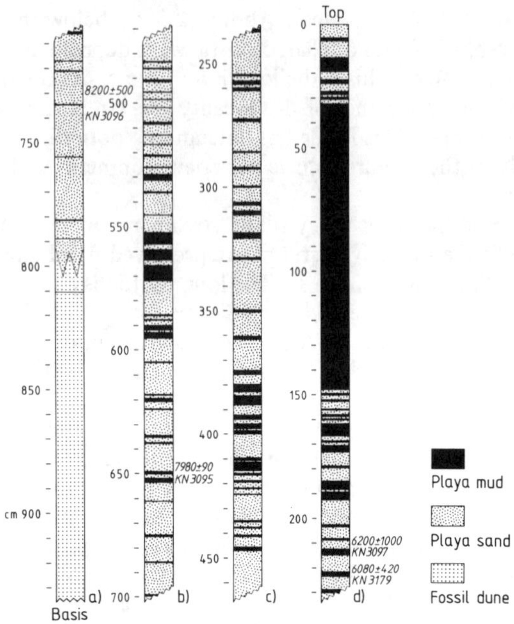

Figure 2. Wadi el Bakht, section 82/13.

Stratigraphy

The smoothened outcrop at the former playa-lake shore gives an excellent insight into the sequence of sedimentary processes (Photo 2). As already established by Peel (1939: 306), the deposits consist of thin alternating layers of mud and partially cemented red sand, overlain by a 6 ft (= 183 cm) thick mud cover (Peel mentions 6 inches = 15.24 cm; as this is obviously a slip of the pen, the correct value of 6 ft has been given here). Maxwell (1980) called for detailed stratigraphic investigations in order to clarify sedimentary conditions. These studies were carried out during our stay in January/February 1982. Figure 2 shows the micro-stratigraphic column of section 82/13. Both the scale and the few symbols demonstrate that it is a sequence of homogeneous clastic sediments with no indications of erosional disconformities.

The underlying bed is the former western slope of the fossil blocking dune and consists of cross-bedded medium sand with sharp grain-size differences between individual laminae and is undoubtedly aeolian in origin. The internal structure strikes 170° and dips 12° W on average. The fossil dune sands are orange (Munsell 5 YR 6/8 dry) and show numerous calcified root casts in their upper parts. Between 810 and 790 cm below the surface the underlying dune sands interfinger with the overlying psammites, which will be referred to here as 'playa sands' in the sense of sands that have been deposited in a playa area. This is the first occurrence here of the reddish yellow playa sands (Munsell 5 YR 6/6-8 dry). Due to a non-calcareous pelitic matrix, the medium to coarse grained sands are well consolidated.

At 788 cm the first clayey silty deposit occurs, a layer of 5 mm thick. Up to the top these deposits recur 89 times, with thicknesses up to a maximum of more than 1 m (Fig. 2). Pachur & Röper (1984: 254) quote a grain size distribution from the opposite wall of 54% silt, 45% clay and 1% sand, classifying the sediments as very clayey silts. 77% of these strata are less than 2 cm thick, 22% are 2-7 cm thick, and only the near-surface layer is 116 cm thick. Disregarding this top layer, the mean thickness is 1.43 cm with a standard deviation of 1.26 cm. On the whole, the pelites form 30% (242 cm) of the vertical thickness of the profile above the fossil dune sands.

An important argument in favour of the playa sands having been deposited in a wet environment is that load casts (Kuenen 1953) occur at almost all the basal sides of the sand layers. These pressure marks are evidence of plasticity of the still saturated pelitic layers during or shortly after sedimentation. Curled clay flakes, on the other hand, which would be evidence of complete surficial desiccation, were not found at all, although the clay content of the fossil pelites is somewhat lower than in the recent playa deposits mentioned above.

What is remarkable, however, is the total lack of signs of 'anthropoturbation'. If either humans or animals had walked over the wet, unconsolidated mud at or after playa deposition the thinly interbedded layers would have been disrupted to a greater extent. Possibly the reeds hindered entry to the swampy environment.

Of particular interest are several finds of pottery up to depths of approximately 5 m in the playa deposits. One sherd was found in the profile 82/13 at about 4 m below the surface. It is of the same type as that of a vessel found in the main profile 80/7-1 in bed D at Wadi Ard el Akhdar which was dated at about 8 000 BP (R.Kuper, pers. comm. 1982; cf. Fig. 3).

Scanning electron microscopy and energy-dispersive analysis of playa sediments

Analogous to what has been described by Walker (1979) the consolidation of parts of the fossil dune is attributed to mechanically infiltrated clay minerals (cf. Crone 1975) and extremely finely distributed iron oxides. The playa sands and pelites were analysed, taking into account results by El-Baz & Prestel (1981) who investigated the coatings of loose dune sand from Wadi el Bakht, and by Potter & Rossman (1979) who established the significance of clay minerals in the Mn-Fe-oxide deposition. Scanning electron microscopy and semi-quantitative energy-dispersive X-ray microanalysis (EDAX) of the

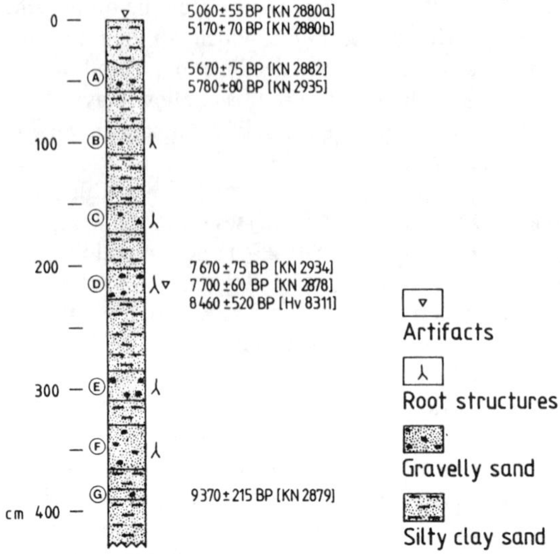

Figure 3. Wadi Ard el Akhdar, section 80/7-1 (modified after Kuper 1981).

playa sands and pelites yielded the following results:

a. Both the coatings and matrix of fossil dune and playa sands seem to consist of the same material.

b. Both coatings and matrix consist of detritic alumosilicates, mainly kaolinite, and hematite. The latter is evidence of oxidization during and after formation.

c. The water necessary for redistributing both clay minerals and iron oxides originated from the playa and not (as Walker (1979) proposes for similar features in Libya) directly from rainfall, as consolidation of fossil dune sands and playa sands only occurs in the intake area of the playa; fossil dune sands outside the shore zone of the former playa lake are still loose today. This may be a further variant in the formation of red beds, i.e. in the vicinity of playas.

d. The clay minerals and matrix of aeolianite and playa sands originated in the fluvially deposited pelites of the playa which in turn came from the cementing material of the surrounding 'Nubian' sandstone. Hence the material forming the matrix does not consist of aeolian dust, contrary to the findings of Walker (1979) in Libya.

e. Evidence of authigenic mineral formation (? Halloysite) within the porous playa sands is proof of distinctly higher humidity, as under the recent and subrecent hyperarid conditions such clay mineral formation would be impossible.

f. The micromorphological comparison between playa sands and aeolianite indicates that the playa sands were deposited in a humid environment thus confirming the conclusions drawn from the load casts mentioned above and related microstratigraphic features. The playa sands are therefore not to be interpreted as wind-blown sand deposited after total desiccation of the playa-lake.

g. The pelites have the typical structure of clastic silicate mud deposits with a high kaolinite content.

h. Furthermore the total absence of any micro- and nannofossils emphasizes that the environment within the playa was hostile to life.

WADI ARD EL AKHDAR

In the upper reaches of Wadi Ard el Akhdar ('wadi with the green floor'; Bagnold et al. 1939) the valley widens out in several places, as the satellite image interpregation map shows (Fig. 1). The playa is located in a depression which is 2 200 m long, 750-1 500 m wide, with a total area of about 3 km^2, and lies about 35 km upvalley from the junction with Wadi Wassa. In the eastern part the playa surface of about 160 000 m^2 is covered with a network of small channels. They run more or less radially from the edges into a 50-80 m wide main channel which cuts up to 8 m into the playa deposits. The narrow mouth of the depression is situated at the eastern end of the main channel, below a basaltic intrusion recently dated at 38 ± 2 Ma (Franz, Puchelt & Pasteels 1986; cf. Fig. 1, dotted signature E of loc. 2). This constriction is about 600 m long and only 50 m wide, and the valley floor is reduced to about 10 m width by a recent, N-S striking dune.

The complete blockage of this constriction in earlier times was due to a dune barrier formed in the lee of the basaltic hill, as the present situation suggests (McCauley, Breed & Grolier 1981, Pachur & Braun 1982). This blockage which seems to have been initiated by an apparently sudden fall of metre-size boulders from the southern slope (Kröpelin 1985) caused the depression to fill with thick, loose sediments over a considerable span of time.

Stratigraphy

The most significant section 80/7-1 is located 640 m upvalley from the fossil dune barrier on the north side of the main channel. Here, pottery sherds were found about 2 m below the surface. Associated charcoal was dated at 8 460 ± 520 BP (Hv 8311) so that the fragments are among the oldest pottery in the entire Sahara (Pachur & Gabriel 1980). Figure 3 shows seven alternations of pale and dark layers. The darker strata, marked A to G, mostly contain root structures and bioturbational features, and have distinctly more coarse sand and fine, angular gravels than the comparatively homogeneous paler layers (Kuper 1981: 232).

Detailed chemical analyses and grain-size distributions of this profile have been published by Pachur & Röper (1984: 253). They confirm the predominantly psammitic nature of all the layers which show a content of sand between 41.4 and 81.5% ($\overline{X}$ = 63.2%), silt between 3.4 and 31.6% ($\overline{X}$ = 19.1%) and clay between 6.7 and 25.3% ($\overline{X}$ = 13.5%) with poor sorting and, in part, bimodal distribution. Higher gravel contents of 20.8% and 8.4% were found only in the two samples from the top, the others have less than 3.3%. The organic carbon content varies between 0.03% and 0.09%. As the radiocarbon dates in Figure 3 show, section 80/7-1 reflects the course of sedimentation in part of the depression from about 9 400 to 5 000 BP; an erosion-induced unconformity is only perceptible above the A layer.

The material was probably washed by sheetfloods or rill erosion over a short distance only from the margins into the deeper, central part of the depression, a process which has been verified in other sections by corresponding sediments (basalt and quartzite particles; W.Schön 1984 in litt.). Both the poor sorting and the high degree of consolidation are typical of such fanglomerates. Better sorting occurs only in a position further downvalley, where there are few or no psephite components in the over-9-m-thick, fine sandy-silty deposits which represent the core of the playa and occupy c. 12 000 m^2. This means that at the time of deposition site 80/7-1 was situated at the periphery of the former playa lake. This interpretation, based on independent evidence, corresponds in the main to the conclusions drawn by Pachur & Röper (1984: 252) from this profile: that it is a sedimentary environment characterized by short and heavily loaded sediment transport, with no features of genuine limnic sedimentation.

In fact, the upper Wadi Ard el Akhdar contains almost no purely pelitic stillwater deposits, i.e. playa sediments in the narrow sense (with grain sizes $< 63\ \mu m$), comparable to the pelitic layers in Wadi el Bakht. This is the main facies difference to the accumulations there. Several sections and test pits confirmed the predominantly lateral-colluvial character of the Holocene deposits, which reveal a comparatively high energy level.

PREHISTORIC FAUNA AND FLORA

The faunal remains obtained from Wadi el Bakht and Wadi Ard el Akhdar suggest a meagre palaeofauna and, indirectly, as well a poor palaeoflora (Table 1). They were found mainly on or in surface layers at some distance from the playas and belong exclusively to non-aquatic organisms. Since it is by no means certain that the bone only tentatively identified as ?*Elephas* did in fact come from Wadi el Bakht or Gilf Kebir (see note to Table 1), the only significant palaeoecological evidence of wild fauna consists of two finds of giraffe bones (Pachur & Röper 1984; this chapter). Since giraffes subsist on the leaves, buds and twigs of trees, mainly acacias, these scanty relics are the only faunal indications of tree vegetation. The other wild mammals found here are desert animals accustomed to extreme aridity rather than typical

Photo 1. Wadi el Bakht. View WNW upvalley with residual fossil dune responsible for damming the valley and causing the accumulation of stillwater sediments between c. 9 000 and 5 000 BP. Valley width: 650 m, difference in elevation between plateau edge and playa surface: 110 m. Two Unimogs show scale of photo. (Photo by S.Kröpelin, 29/1/1982)

Photo 2. Wadi el Bakht, section 82/13. Natural outcrop in the gorge breaching the fossil dune barrier (see centre of Photo 1). Date of breach probably about 4 800 BP. The playa sediments dip W upvalley and are over 8 m thick at this site near the former playa lake shore. (Photo by S.Kröpelin, 3/2/1982)

Table 1. Faunal remains from Wadi el Bakht and Wadi Ard el Akhdar, Gilf Kebir

Name	Specific name	Source
Wadi el Bakht		
? Elephant	? *Elephas* sp.	McHugh 1975*
Big bovide	*Bos* sp.	McHugh 1975*
Addax	*Addax* sp. *nasomaculatus*	McHugh 1975*
Damaliscus	*Damaliscus* sp.	McHugh 1975*
Gazelle	*Gazella* sp.	McHugh 1975*
? Goat	? *Capra* sp.	McHugh 1975*
Jackal, ? dog	*Canis* cf. *anthus*	McHugh 1975*
Ostrich	*Struthio camelus*	McHugh 1975*
Wild donkey	*Equus asinus*	McHugh 1975*
Ostrich	*Struthio camelus*	Gautier 1980
Big dog	*Canis lupus* f. *familiaris*	Gautier 1980
Dorcas gazelle	*Gazella dorcas*	Gautier 1980
Sheep or	*Ovis ammon* f. *arres* or	Gautier 1980
goat	*Capra aegragus* f. *hircus*	Gautier 1980
Domestic cattle	*Bos primigenius* f. *taurus*	Gautier 1981
Big bovides	*Bos* sp.	Gautier 1981
Giraffe	*Giraffa camelopardalis*	Pachur & Röper 1984 (det. Uerpmann, Tübingen)
Buffalo	*Bubalus* sp.	
Antelope	?	
Gazelle	?	
Giraffe	*Giraffa camelopardalis*	This chapter (det. Uerpmann, Tübingen; Kuper 1984 in litt.)
Cattle	*Bos* sp.	
Mice	?	
Wadi Ard el Akhdar		
?	? *Bos* sp.	Gautier 1981
Giraffe	*Giraffa camelopardalis*	This chapter (det. Uerpmann, Tübingen; Kuper 1984 in litt.)
Dorcas gazelle	*Gazella dorcas*	
Gazelle	*Gazella* sp.	
Kamma antelope	*Tragelaphus oryx*	
Rock rabbit	*Procavia capensis*	
Ostrich	*Struthio camelus*	
Bovide	*Bos* sp.	
Big buffalo	*Bubalus* sp.	
Capride	?	
See above	See above	McHugh 1975*

*These faunal remains were collected by O.H.Myers in 1938 (Myers 1939) and only preliminarily identified by D.M.A.Bate & J.Johnson, London. Their provenance from Wadi el Bakht or Wadi Ard el Akhdar or even from the Gilf Kebir is not ascertained though it is likely that some of them were associated with artifact assemblages recorded by Myers in the upstream portions of these valleys (Gautier 1981). Therefore parts of the material listed under Wadi el Bakht could possibly have come from Wadi Ard el Akhdar.

savanna species. Addax, Dorcas gazelle and Damaliscus live on grass and herbs only and can exist for months, years or even entirely without water other than the fluid intake from plants and dew (Haltenorth & Diller 1977). Furthermore, Almásy (1936: 45) mentions finding the fresh carcass of an *Addax nasomaculatus* on the western side of the Gilf Kebir as late as 1933.

Owing to the absence of pollen, diatoms and identifiable opal phytoliths, until recently the only direct palaeofloral evidence consisted of root structures (Fig. 3), a few root and stem fillings with diameters up to 10 cm, and the close-set stone casts of roots and stalks, which were identified as reeds (*Phragmites communis*) by their position, cross-sections, and diameters of about 1 cm. In addition, grass steppe fires were deduced from the widespread minuscule traces of charcoal found well away from hearth sites.

Recently, however, the results of extensive palaeoxylotomic studies of charcoal remains with the aid of scanning electron microscopy have been presented (Neumann, this volume, and 1986), the great majority of which have been dated by radiocarbon. They provide independent confirmation of conclusions on palaeoflora and early and mid-Holocene rainfall drawn from sediment facies, electric conductivity logging, the absence of micro- and nannofossils, geomorphological mapping and other evidence (Kröpelin 1985). Relicts of almost all the identified genera and species, mainly acacia (*Acacia tortilis* subsp. *raddiana* and *Acacia ehrenbergiana*), *Maerua crassifolia* and *Balanites aegyptica*, still occur in Gilf Kebir; *Tamarix articulata* is found in the neighbouring Jebel Uweinat. Only *Acacia albida* and *Ziziphus lotus* no longer occur here today, but are found in the mountains of the central Sahara (Neumann 1986).

DISCUSSION AND PALAEOCLIMATIC CONCLUSIONS

Playas which are not groundwater-fed represent valuable indicators of climate because their occurrence is due only to surficial run-off, contrary to lakes in the Sahara which are always dependent on specific conditions in which local, non-climatic factors may play a major part (Rognon 1980). The evidence presented above strongly suggests that the early to mid-Holocene playas in the Gilf Kebir were such 'dry playas' (Motts 1965), i.e. temporary pools containing water for days, weeks or months only.

Of course, local factors are also important to a certain degree in the formation of both fossil and recent playas in the Gilf Kebir. The necessary conditions for the origin of basins and thus for the accumulation of playa sediments are the result of a combination of W-E trending valleys with constant winds blowing roughly N-S (trade winds), and a source of sand in the north (Great Sand Sea). Additional factors were residual terraces ('Winkelwadi') or valley constrictions (Wadi Ard el Akhdar). On the other hand, such special local influences would not have had any great effect had they not been accompanied by climatic parameters considerably different from the present day.

The sections of Wadi el Bakht and Wadi Ard el Akhdar (82/13 and 80/7-1) show the following climatic development: Blockage of the valleys, which created the conditions necessary for playa accumulation must have occurred, or at least begun, before 9 370 ± 215 or 8 200 ± 500 BP (cf. Table 2); the latter seems to be the case in Wadi Ard el Akhdar. An age of about 9 400 BP is more likely to be correct since the date from Wadi el Bakht is from material overlying the earliest deposits in the playa centre. The climate seems to have been even more arid than today, since actual dunes are not able to cause total blockage of the valleys of Wadi el Bakht and Wadi Ard el Akhdar. However, different non-climatic boundary conditions may have prevailed, causing for example an increased supply of sand.

For the subsequent period up to about 6 200 BP the section 82/13 (Fig. 2) is more instructive because the playa in Wadi el Bakht had a larger catchment area and therefore, due to the amplifying effect, provides more information on the course of sedimentation than the Wadi Ard el Akhdar playa where lateral influences dominated. The number of pelitic strata documented in section 82/13 between 8 200 ± 500 BP and 6 080 ± 420 or 6 200 ± 1000 BP (cf. Fig. 2) could be explained by the fact that rainfall events were only twice as frequent (approximately 70 events/2 000 years) as in the present hyperarid climate (approximately 2 events/100 years = 40 events/2 000 years). Likewise the thicknesses of fossil deposits (less than 2 cm on average) roughly correspond to recent rates of sedimentation in the 'Winkelwadi' or in Wadi Wassa. Nevertheless, neither the spatial extent and volume of the deposits, nor the composition of the sediments are conceivable under the actual climatic regime. This suggests that there must have been both heavier or lengthier rainfall and, above all, different conditions for material supply. Particularly the texture of the pelites can only be explained by more intense weathering or soil formation in the source area. There are signs of pedogenesis at numerous outcrops especially in the surroundings of the playa in Wadi Ard el Akhdar where there are root structures, bioturbation features, still measurable organic carbon, higher proportions of clay, at least some of which probably formed in situ, and red beds in various stratigraphic positions.

Hence, the strata A to G in section 80/7-1 (Fig. 3) cannot be interpreted as representing pluvial phases of climate, since they probably resulted from single fanglomeratic depositions after exceptional rainfall. However, they contain pedogenetic features which provide palaeoecological evidence for the entire phase between two depositional events. At any rate, this period, roughly lasting from 9 000 to 6 000 BP, was characterised by uneven, discontinuous sedimentation, which points to an arid climate with single heavy showers of rain.

Only after 6 200 ± 1 000 or 6 080 ± 420 BP does the series of pelitic layers in section 82/13 become denser, and there is a more or less continuous deposition of the thick pelitic top, lasting up to about 5 000 BP (cf. Fig. 2). Some evidence of earlier human occupation exists; however, radiocarbon dates

Table 2. ^{14}C-dates from Wadi el Bakht and Wadi Ard el Akhdar, Gilf Kebir

No.	Age BP	Site/Stratigraphic level	Material	Source
Hv 11644	8 715 ± 920	10 m W of section 82/13, 310 cm below top	Charcoal	Pachur & Röper 1984
KN 3096	8 200 ± 500	Section 82/13, 720 cm below top	Charcoal	Kuper 1984
KN 3095	7 980 ± 90	Section 82/13, 640 cm below top	Charcoal	Kuper 1984
Hv 11648	7 585 ± 80	Fossil dune	Calc. rt.	Pachur & Röper 1984
SMU 273	6 980 ± 80	Northeastern part of playa surface	Ostr.	Haynes in litt. 1981
KN 3410	6 700 ± 300p	Site 82/21-2, artifact layer	Charcoal	Kuper 1984
KN 3097	6 200 ± 1 000	Section 82/13, 220 cm below top	Charcoal	Kuper 1984
KN 3328	6 150 ± 200	Site 82/22, ditch	Charcoal	Kuper 1984
KN 3179	6 080 ± 420	Section 82/13, 240 cm below top	Charcoal	Kuper 1984
KN 3079	5 180 ± 60	Site 82/15, artifact layer	Charcoal	Kuper 1984
KN 3149	5 070 ± 60	Site 82/15, hearth	Charcoal	Kuper 1984
KN 3182	4 880 ± 390	Site 82/18, hearth	Charcoal	Kuper 1984
KN 3096	4 820 ± 60	Site 82/16, hearth	Charcoal	Kuper 1984
KN 3184	4 770 ± 130	Site 82/19, hearth	Charcoal	Kuper 1984
Wadi Ard el Akhdar				
KN 2879	9 370 ± 215	Section 80/7-1, layer G	Charcoal	Kuper 1981
Hv 8312	8 695 ± 560	Root horizon	Calc. rt.	Pachur & Gabriel 1980
Hv 8311	8 460 ± 520	Section 80/7-1, layer D	Charcoal	Pachur & Gabriel 1980
KN 2878	7 700 ± 60	Section 80/7-1, layer D	Charcoal	Kuper 1981
KN 2934	7 670 ± 75	Section 80/7-1, layer D	Charcoal	Kuper 1981
KN 3191	6 510 ± 220	Site 80/32, hearth	Charcoal	Kuper 1984
KN 3176	5 940 ± 230	Site 81/8, surface artifact layer	Charcoal	Kuper 1984
KN 3358	5 800 ± 450	Section 81/4, layer III	Charcoal	Kuper 1984
KN 2935	5 780 ± 80	Section 80/7-1, layer A	Charcoal	Kuper 1981
KN 2882	5 670 ± 75	Section 80/7-1, layer A	Charcoal	Kuper 1981
KN 3016	5 670 ± 65	Section 81/4, layer II	Charcoal	Kuper 1984
KN 3381	5 650 ± 130	Section 81/4, layer II	Charcoal	Kuper 1984
KN 2936	5 610 ± 60	Site 80/7-2, artifact layer	Charcoal	Kuper 1984
KN 3187	5 500 ± 60	Site 81/4, hearth	Charcoal	Kuper 1984
KN 3106	5 440 ± 60	Section 81/2	Charcoal	Kuper 1984
KN 3103	5 430 ± 65	Site 81/8, hearth	Charcoal	Kuper 1984
KN 2881a	5 420 ± 65	Site 80/12-1, hearth	Charcoal	Kuper 1984
KN 2881b	5 360 ± 210	Site 80/12-1, hearth	Charcoal	Kuper 1984
KN 3104	5 250 ± 140	Site 80/15-2, hearth	Charcoal	Kuper 1984
KN 2880b	5 170 ± 70	Site 80/7-5, surface artifact layer	Charcoal	Kuper 1984
KN 2933	5 150 ± 125	Site 80/12-1, hearth	Charcoal	Kuper 1984
KN 2880a	5 060 ± 55	Site 80/7-5, surface artifact layer	Charcoal	Kuper 1984
Hv 8313	4 365 ± 95	?	Calc. rt.	Pachur & Gabriel 1980
KN 3085	4 200 ± 1 100	Site 80/14, hearth	Charcoal	Kuper 1984
KN 3173	4 150 ± 55	Site 80/14, southern annex	Ostr.	Kuper 1984
KN 2926	3 950 ± 55	Site 80/14, right concentration	Ostr.	Kuper 1984
KN 2925	3 860 ± 60	Site 80/14, left concentration	Ostr.	Kuper 1984
Hv 8242	3 830 ± 365	Hearth	Charcoal	Pachur & Gabriel 1980

calc. rt. = calcified root casts; ostr. = ostrich egg-shells; p = preliminary date show that the main settlement phases occurred between 6 000 and 5 000 BP, both in Wadi el Bakht and Wadi Ard el Akhdar (Cziesla 1986, Schön 1984; cf. Table 2). This could also be evidence of comparatively favourable climatic conditions.

Not long after c. 5 000 BP the deposition of playa sediments stopped in both valleys, as the two main sections 82/13 and 80/7-1 (Figs. 2 and 3) and numerous surface dates show, even if a certain amount of deflation is taken into account. This suggests that the dunes blocking Wadi el Bakht and Wadi Ard el Akhdar were breached during this period. There are two possible explanations: Either the breaches were due to an unprecedented, exceptionally high water pressure or level, pointing to a climatic optimum with comparatively high precipitation or a unique millennial rainfall event. Or the sediments had reached the same height as the dune barriers so that overflow and breaching were inevitable. The fact that the occurrence of thick pelitic deposits (section 82/13) coincides with the cessation of playa sedimentation 'shortly afterwards' and the roughly contemporary breaching of the blockages in both valleys makes the first explanation more likely.

We have now reached the open question of whether accumulation or erosion was the prevailing process during the Saharan pluvials (cf. Butzer 1963, Chavaillon 1964, Hagedorn 1980, and, with reference to Wadi el Bakht, Pachur & Röper 1984). However, it would be going too far here to equate playa formation in the Gilf Kebir with arid phases and fluvial erosion with humid conditions. A possible approach to solving this apparent contradiction may be to consider the distribution of rainfall rather than its annual mean, i.e. the rainfall regime, as Gabriel (1977) suggested in the case of the Tibesti mountains. Thus a west-wind-induced type of occasional winter rainfall with steady rains could have alternated with a regime of secular monsoonal-convective summer rainstorms triggered by exceptional northward surges of the surface inter-tropical convergence zone (cf. Pedgley 1974). The former is more likely in the case of the more or less continuous pelitic deposits at the top of section 82/13, and the latter in the case of the preceding and subsequent 'wet' phases. At any rate, after about 5 000 BP no more playa sediments were deposited and hence, there is no evidence of the renewed climatic deterioration which is underlined by the fact that the following fluvial erosion obviously had no significant effect on their degradation, at least in Wadi el Bakht. However, this interpretation of events implies that the main gully in the upper Wadi Ard el Akhdar already existed during the late settlement phase around 4 000 BP.

In attempting to estimate the approximate rainfall level we do not intend to repeat the above mentioned conclusions which were drawn from the sedimentary structure of the playa deposits, the absence of micro- and nannofossils, the faunal remains of non-aquatic organisms and palaeofloral evidence. Nevertheless, botanical results established by Walther (1979: 124) will be quoted which bear on the question of palaeoevegetation – and thus indirectly of the palaeofauna – in the valleys from the desert near Helwân with rainfall about 25 mm/year. Given that 40% of the rainfall flows down into the deepest parts of the relief which cover about 2% of the whole area there, the plants at these favourable sites would receive the same amount of water as those in the plain with a rainfall of 500 mm/year. Yet we do not have to go that far to explain the occurrence of reeds and resistant species of acacia in the Gilf Kebir valleys. The fact that the vegetation in the valleys was apparently not abundant, suggests that the plant cover on the surrounding plateau was very sparse.

It is true that Neolithic implements in this nowadays completely uninhabited area are surprisingly frequent. However, they are few compared with the abundance of archaeological remains found in the more elevated Tibesti mountains (Gabriel 1977) or in the Wadi Howar (Gabriel & Kröpelin 1986, Kuper 1981). Thus it seems unlikely that the area was permanently occupied even during the most favourable periods.

On the basis of the evidence described above and compared to the conditions under the present Sahel climate (cf. Ibrahim 1980, Mensching 1984, Rzóska 1961, Wickens 1976, and others), the amount of rainfall in the Gilf Kebir during the 'Neolithic' pluvials more likely lay within the lower range of 50-100 mm/year estimated by Butzer (1959: 77) and Hester & Hobler (1969: 164) on general considerations for the area. Here, the higher values should be restricted to the mid-Holocene period approximately between 6 000 and 4 500 BP, lasting from the onset of the top pelitic layer at Wadi el Bakht to the time after the barrier breaches.

McHugh (1974: 13) based his estimate of 200-800 mm/year rainfall mainly on the normal ecological requirements of the giraffe, ostrich, oryx and domestic cattle depicted by rock art (cf. Rhotert 1952, Winkler 1939). However, it is not necessary to postulate such a high maximum to account for Neolithic pastoralists migrating seasonally or sporadically at times of good grazing with their herds from the climatically more favourable Jebel Uweinat with its waterholes to the southern Gilf Kebir, 150 km away. At the beginning of this century cattle were still being driven 200 km from Kufra oasis to graze in the northwestern Gilf Kebir (now known as Abu Ras Plateau) after exceptional summer rainfall (Almásy 1939: 131 f.). Thus, the conclusions drawn here lie even somewhat below the palaeoclimatic inferences drawn by Pachur & Gabriel (1980) and Pachur & Röper (1984) from the 'montane variant of pelitic sediments' in the Libyan desert, according to which a semi-arid climate with high wind activity and rainfall around 200 mm/year prevailed at the Gilf Kebir between 8 000 and 4 000 BP.

On the whole, it becomes more evident that climatic conditions during the entire Holocene in this central part of the eastern Sahara were arid, with only a sporadic tendency towards semi-aridity (cf. Cour & Duzer 1976, Meckelein 1982, Wendorf & Hassan 1980). It seems possible that such wetter phases may be over-interpreted in some cases.

ACKNOWLEDGEMENTS

The BOS project is supported by the German Research Council ('Deutsche Forschungsgemeinschaft'). I am particularly grateful to Dr Rudolph Kuper, Cologne, and Dr Baldur Gabriel, Berlin, as well as to the other members of the BOS team for their friendly and constructive cooperation during the months of field seasons. Valuable information was provided by Professor Dr H.-J.Pachur, Berlin, on the occasion of a joint inspection of the Wadi Ard el Akhdar. Scanning electron microscopy was carried out at the Central SEM Unit of the Technical University Berlin (ZELMI). Furthermore I wish to thank Mrs Katharina Neumann, Frankfurt, for permission to include her latest palaeoxylotomic results, Dr H.-P.Uerpmann, Tübingen, for osteological identifications, and Mrs Anne Beck for translating the manuscript.

REFERENCES

Almásy, L.E. de 1936. Récentes explorations dans le désert Libyque (1932-1936). *Publ. Spéc. Soc. Roy. Géogr. Egypte,* Cairo, 100 pp.

Almásy, L.E. de 1939. *Unbekannte Sahara – Mit Flugzeug und Auto in der Libyschen Wüste.* Leipzig: Brockhaus, 214 pp.

Back, H.E. 1981. Beitrag zur Faunistik und Zoogeographie der Lepidopteren in der östlichen Sahara. *Nota lepid.* 4: 111-125.

Bagnold, R.A., O.H.Myers, R.F.Peel & H.A.Winkler 1939. An expedition to the Gilf el Kebir and Uweinat, 1938. *Geogr. J.* 93: 281-313.

Ball, J. 1927. Problems of the Libyan desert. *Geogr. J.* 70: 21-38, 108-128, 209-224.

Butzer, K.W. 1959. Environment and human ecology in Egypt during predynastic and early dynastic times. *Bull. Soc. Géogr. Egypte* 32: 43-88.

Butzer, K.W. 1963. Climatic-geomorphologic interpretation of Pleistocene sediments in the Eurafrica subtropics. In F.C.Howell & F.Bourlière (eds.), African ecology and human evolution. *Viking Fund Publ. Antropol.* 36: 1-27.

Chavaillon, J. 1964. Etude stratigraphique des formations quaternaires du Sahara nord-occidental (Colomb-Béchar à Reggane). *Publ. Centre Rech. Zones Arides, CNRS, Sér. Géol.* 5, 393 pp.

Clayton, P.A. 1933. The western side of the Gilf Kebir. *Geogr. J.* 81: 254-259.

Cour, P. & D.Duzer 1976. Persistance d'un climat hyperaride au Sahara central et méridional au cours de l'Holocène. *Rév. Géogr. phys. Géol. dynam.* 18(2): 175-198.

Crone, A.J. 1975. Laboratory and field studies of mechanically infiltrated matrix clay in arid fluvial sediments. Ph.D. thesis, Colorado Univ., 162 pp.

Cziesla, E. 1986. Der Fundplatz Wadi el Akhdar 80/14. *Forschungen zur Prähistorie Afrikas* 1 (in press).

El-Baz, F. et al. 1980. Journey to the Gilf Kebir and Uweinat, southwest Egypt, 1978. *Geogr. J.* 146: 51-93.

El-Baz, F. & T.A.Maxwell (eds.), 1982. *Desert landforms of southwest Egypt: A basis for comparison with Mars.* NASA Contr. Rep. 3611, 372 pp.

El-Baz, F. & D.Prestel 1981. Coatings on sand grains from southwestern Egypt. *Ann. Geol. Surv. Egypt* 11: 175-188.

Flohn, H. 1964. Investigations on the Tropical Easterly Jet. *Bonner Meteor. Abh.* 4, 83 pp.

Franz, G., H.Puchelt & P.Pasteels 1986. Petrology, geochemistry and age relations of Triassic and Tertiary volcanic rocks from SW-Egypt and NW-Sudan. *J. African Earth Sc.* 4 (in press).

Gabriel, B. 1977. Zum ökologischen Wandel im Neolithikum der östlichen Zentralsahara. *Berliner Geogr. Abh.* 27, 111 pp.
Gabriel, B. 1986. Die östliche Libysche Wüste im Jungquartär. *Berliner Geogr. Stud.* 19 (in press).
Gabriel, B. & S.Kröpelin 1984. Holocene lake deposits in northwest-Sudan. *Palaeoecol. of Africa* 16: 295-299.
Gabriel, B. & S.Kröpelin 1986. Habitats dunaires néolithiques au Soudan septentrional et implications paléoclimatiques. In *Proc. of INQUA-ASEQUA Symp. Intern. 'Changements globaux en Afrique durant le Quaternaire: passé – présent – futur', Dakar, Senegal,* Editions de l'ORSTOM, Coll. Travaux et Documents 197: 157-160.
Gabriel, B., S.Kröpelin, J.Richter & E.Cziesla 1985. Parabeldünen am Wadi Howar – Besiedlung und Klima in neolithischer Zeit im Nordsudan. *Geowissenschaften in unserer Zeit* 3: 105-112.
Gautier, A. 1980. Contribution to the archaeozoology of Egypt. In F.Wendorf & R.Schild (eds.), *Prehistory of the Eastern Sahara*: 317-344. New York: Academic Press.
Gautier, A. 1981. Neolithic faunal remains in the Gilf Kebir and the Abu Hussein dune-field. *Ann. Geol. Surv. Egypt* 11: 335-339.
Hagedorn, H. 1980. Fluvial processes in the Sahara. *Palaeoecol. of Africa* 12: 115-123.
Haltenorth, T. & H.Diller 1977. *Säugetiere Afrikas und Madagaskars.* Munich: BLV, 403 pp.
Harding-King, W.J. 1912. Travels in the Libyan desert. *Geogr. J.* 39: 133-137.
Haynes, C.V. 1981. The Darb El-Arba'in desert: a product of Quaternary climatic change. *Ann. Geol. Surv. Egypt* 11: 91-117.
Henning, D. & H.Flohn 1977. Climate Aridity Index Map. In *Proc. of UN Conf. 'Desertification', UNEP, Nairobi, Kenya,* UN Doc. A/CONF. 74/31.
Hester, J.J. & P.Hobler 1969. *Prehistoric settlement patterns in the Libyan desert.* Univ. Utah Pap. Anthropol. 92, 171 pp.
Ibrahim, F.N. 1980. Desertifikation in Nord-Darfur. *Hamburger Geogr. Schr.* 35, 175 pp.
Issawi, B. 1978. New findings in the Gilf-Uweinat area. *Ann. Geol. Surv. Egypt* 8: 275-293.
Kádár, L. 1937. La morfologia dell'altipiano del Gilf Kebir. *Boll. R. Soc. Geogr. Italiana* 7: 485-503.
Kemal El Dine Hussein 1928. L'exploration du désert Libyque. *La Géographie* 50: 171-183, 320-336.
Klitzsch, E. 1978. Geologische Bearbeitung Südwest-Agyptens. *Geol. Rdsch.* 67: 509-520.
Klitzsch, E. 1979. Zur Geologie des Gilf Kebir Gebietes in der Ostsahara. *Clausthaler Geol. Abh.* 30: 113-132.
Klitzsch, E. & F.K.List (eds.) 1978. Geological Interpretation Map Egypt/Sudan 1:500 000, Sheets 2523 Gilf Kebir, 2521 Gebel Uweinat. TFH Berlin.
Klitzsch, E., R.Said & E.Schrank (eds.) 1984. Research in Egypt and Sudan. Sonderforschungsbereich 69: Results of the Special Research Project Arid Areas, period 1981-1984. *Berliner geowiss. Abh. (A)* 50, 457 pp.
Kröpelin, S. 1985. Untersuchungen zum Sedimentationsmilieu von Playas im Gilf Kebir (SW-Agypten). M.A. thesis, Techn. Univ. Berlin, 194 pp.
Kuenen, P.H. 1953. Significant features of graded bedding. *Amer. Assoc. Petrol. Geol. Bull.* 37: 1044-1066.
Kuper, R. 1981. Untersuchungen zur Besiedlungsgeschichte der östlichen Sahara. *Allg. Vergl. Archäol. Beitr.* 3: 215-275.
Kuper, R. 1984. The eastern Sahara from north to south: data and dates from the BOS project. Symp. 'Late prehistory of the Nile basin and the eastern Sahara', Poznán (in press).
Maxwell, T.A. 1980. Geomorphology of the Gilf Kebir. *Geogr. J.* 146: 76-83.
Maxwell, T.A. 1981. Erosional patterns of the Gilf Kebir Plateau and implications for

the origin of Martian canyonlands. *Ann. Geol. Surv. Egypt* 11: 281-300.
McCauley, J.F., C.S.Breed & M.J.Grolier 1981. The interplay of fluvial, mass-wasting and eolican processes in the eastern Gilf Kebir region. *Ann. Geol. Surv. Egypt* 11: 207-239.
McHugh, W.P. 1971. Late prehistoric cultural adaption in the southeastern Libyan desert. Ph.D. dissertation, Univ. of Wisconsin, Dept. Anthropology, 434 pp.
McHugh, W.P. 1974. Late prehistoric cultural adaption in southwest Egypt and the problem of the Nilotic origins of Saharan cattle pastoralism. *J. American Res. Center Egypt* 11: 9-22.
McHugh, W.P. 1975. Some archaeological results of the Bagnold-Mond expedition to the Gilf Kebir and Gebel Uweinat, southern Libyan desert. *J. Near Eastern Stud.* 34: 31-62.
McHugh, W.P. 1980. Archaeological sites of the Gilf Kebir. *Geogr. J.* 146: 64-68.
McHugh, W.P. 1981a. A chronicle of research and exploration in the Gilf Kebir – Uweinat mountain region. *Ann. Geol. Surv. Egypt* 11: 21-26.
McHugh, W.P. 1981b. Archaeological investigations in the Gilf Kebir and the Abu Hussein dunefield. *Ann. Geol. Surv. Egypt* 11: 301-334.
Meckelein, W. 1975. Progress in the exploration and in the scientific research of the Libyan desert since Gerhard Rohlfs. *Ann. Geol. Surv. Egypt* 5: 47-60.
Meckelein, W. 1982. The question of permanent aridity in the Central Sahara since the Pliocene. *Bull. Assoc. Géogr. Franç.* 484: 39-42.
Mensching, H. (ed.) 1984. *Beiträge zur Morphodynamik im Relief des Jebel-Marra-Massivs und in seinem Vorland (Darfur/Republik Sudan).* Hamburg: 253 pp.
Misonne, X. 1969. Expédition scientifique Belge dans le désert de Libye. La faune. *Africa-Tervuren* 15: 117-119.
Motts, W.S. 1965. Hydrologic types of playas and closed valleys and some relations of hydrology to playa geology. In J.T.Neal (ed.), *Geology, mineralogy and hydrology of US playas.* AFCRL Environm. Res. Pap. 96: 73-104.
Myers, O.H. 1939. The Sir Robert Mond expedition of the Egypt Exploration Society, *Geogr. J.* 93: 287-291.
Neumann, K. 1986. Holzkohlen aus prähistorischen Fundstätten des Gilf Kebir (SW-Agypten). *Forschungen zur Prähistorie Afrikas* 1 (in press).
Pachur, H.-J. & G.Braun 1980. The palaeoclimate of the Central Sahara, Libya, and the Libyan desert. *Palaeoecol. of Africa* 2: 351-363.
Pachur, H.-J. & G.Braun 1982. Aspekte paläoklimatischer Befunde in der östlichen Zentralsahara. *Geomethodica* 7: 23-54.
Pachur, H.-J. & B.Gabriel 1980. Quartärklimatische Befunde aus der Libyschen Wüste (Agypten). Unpubl. ms., 24 pp.
Pachur, H.-J. & H.P.Röper 1984. The Libyan (western) desert and northern Sudan during the late Pleistocene and Holocene. *Berliner geowiss. Abh. (A)* 50: 249-284.
Pedgley, D.C. 1974. An exceptional desert rainstorm at Kufra, Libya. *Weather* 29: 64-71.
Peel, R.F. 1939. The Gilf Kebir. *Geogr. J.* 93: 295-307.
Penderel, H.W.G.J. 1934. The Gilf Kebir. *Geogr. J.* 83: 449-456.
Potter, R.M. & J.R.Rossman 1979. The manganese and iron oxide mineralogy of desert varnish. *Chem. Geol.* 25: 79-94.
Rhotert, H. 1952. *Libysche Felsbilder.* Darmstadt: Wittich, 146 pp.
Rognon, P. 1980. Pluvial and arid phases in the Sahara: The role of non-climatic factors. *Palaeoecol. of Africa* 12: 45-62.
Rzóska, J. 1961. Observations on tropical rainpools and general remarks on temporary waters. *Hydrobiologia* 17: 265-286.
Schön, W. 1984. New results from two playa sites in the Gilf Kebir. Symp. 'Late prehistory of the Nile basin and the eastern Sahara', Poznań (in press).
Shaw, W.B.K. 1936. An expedition in the southern Libyan desert. *Geogr. J.* 87: 193-221.
Vahrson, W.-G. 1986. Aspekte bodenphysikalischer Untersuchungen in der Libyschen Wüste – ein Beitrag zur Frage spätpleistozäner und Holozäner Grundwasserbildung. *Berliner Geogr. Abh.* 43 (in press).

Walker, T.R. 1979. Red color in dune sand. In E.D.McKee (ed.), *A study of global sand seas.* US Geol. Survey Prof. Pap. 1052: 61-81.
Walther, H. 1979. *Vegetation und Klimazonen.* Stuttgart: Ulmer, 342 pp.
Wendorf, F. & F.A.Hassan 1980. Holocene ecology and prehistory in the Egyptian Sahara. In M.A.Williams & H.Faure (eds.), *The Sahara and the Nile.* Rotterdam: Balkema, 407-419.
Wendorf, F. & R.Schild 1980. *Prehistory of the Eastern Sahara.* New York: Academic Press, 414 pp.
Wendorf, F., R.Schild, R.Said, C.V.Haynes, A.Gautier & M.Kobusiewicz 1976. The prehistory of the Egyptian Sahara. *Science* 193: 103-114.
Wickens, G.E. 1976. The flora of Jebel Marra (Sudan Republic) and its geographical affinities. *Roy. Bot. Gard. Kew Bull. Add. Ser.* 5, 368 pp.
Winkler, H.A. 1939. *Rock drawings of Southern Upper Egypt,* II. London: Oxford Univ. Press, 40 pp., 61 pl.

ZUR BODENSTRATIGRAPHIE IM NÖRDLICHEN AFRIKA

A.SKOWRONEK
Geographisches Institut der Universität Würzburg, F.R.Germany

SUMMARY

Based on a pedostratigraphical comparison the author tries to draw palaeoecological conclusions for the Cainozoic of Africa north of the equator. For this purpose pedological literature of each of the present climatic zones was stratigraphically evaluated, own investigations concerning the central Sahara were added. Deep-sea sediments off NW Africa and late Quaternary oscillations of Lake Tchad were included as well. Figure 1 shows the chronological comparison.

Particularly with various intensities of soil formation, visible in different taxonomic soil units, it can be demonstrated that, already during the Tertiary, there were different bioclimatic conditions in the lower latitudes of Africa. A significant alternation of drier and humid climates must be considered in the central Sahara since the Oligocene. The relief sequence in the permanent humid tropics is in principle similar to the sequence in the humid extratropics. Formations of glacis and soils in dry areas (tropics and subtropics), however, can still not be correlated with each other.

As in pure degradation areas important indicator soils are closely correlated with morphological elements of plains. Here we have to take pedostratigraphical gaps into account. It is possible that morphological plains were destroyed during the complete development of the relief when there were times with different intensities of planation processes.

EINFUHRUNG

Böden und deren Erosionssedimente eignen sich zur Kennzeichnung ihrer bioklimatischen Bildungsbedingungen, wenn man pedogene Prozesse und Merkmale ursächlich mit bestimmten Klimakonstellationen verknüpfen kann. Blume (1985) hat die mögliche Klimaabhängigkeit der Intensität bodenbildender Prozesse in terrestrischen Böden der Niederen Breiten beschrieben und in saharischen Vorkommen überprüft.

Weiterhin erscheint es möglich, auf bodenstratigraphischer Grundlage einige paläoökologische Schlussfolgerungen für das afrikanische Känozoikum abzuleiten. Zu diesem Zweck hat Skowronek (1984) einschlägige Literatur und eigene Untersuchungsergebnisse chronologisch eingestuft und ein pedostratigraphisches Schema für Afrika nördlich des Aquators erstellt.

Figur 1 enthält die chronostratigraphischen Positionen von Holozänböden, pleistozänen und tertiären Paläosolen und Sedimenten, geordnet nach ihrer Lage in den heutigen Klimazonen bzw. in besonderen Faziesbereichen.

TROCKENE SUBTROPEN UND TIEFSEE-SEDIMENTE VOR NW-AFRIKA

Während sich in den feuchteren mediterranen Subtropen der letztinterglaziale und der spätglazial-holozäne Lössboden ungestört und über einen längeren Zeitraum entwickeln konnte (Brunnacker 1980), zeigen die Boden-Sediment-Serien am Cap Rhir (SW-Marokko) einen lebhaften und differenzierten Wechsel von Bodenbildung und Geomorphodynamik seit Beginn des Würm an (Rohdenburg & Sabelberg 1980).

Altere pleistozäne Paläosole Marokkos werden als 'Rote Mediterrane Böden' bezeichnet, 'Eisenhaltige tropische Böden' sollen noch nach dem Pliozän entstanden sein (Schoen 1966). Auch die Terra rossa im algerischen Atlas wird einem neogenen und altquartären 'climat tropical' zugeschrieben (J.H.Durand 1959).

Die Bodengeschichte des terrestrischen Quartär wird in ihren Grundzügen von den Tiefsee-Sedimenten vor NW-Afrika bestätigt (Diester-Haass 1979, Diester-Haass & Chamley 1978). Darüber hinaus existieren noch ältere Feuchtzeit-Indikatoren in den Bohrkernen, welche zeitlich mit plio-pleistozänen und miozänen Fossilböden der zentralen Sahara parallelisiert werden können.

ZENTRALE SAHARA

Die exhumierte und in der Oberkreide bzw. im Alttertiär weitergebildete Pediplain des Hoggar-Gebirges trägt unter den Atakor-Vulkaniten Verwitterungsreste die Rognon (1967) einer 'altération faible de type tropical sec' zuordnet. Diese ist in umgelagertem Continental Intercalaire (?) und auf dem präkambrischen Sockel ausgebildet.

Insgesamt acht Lagen fossiler 'Sols ferrallitiques' scheidet Girod (1971) in den älteren Basalten aus, welche auf Grund von K/Ar-Datierungen (19.9 u. 16.7 MA) in das Miozän gestellt werden. Im Gegensatz zu Girod verfügt Dutil (1971) über zahlreiche Bodenanalysen und kann daher die o.g. Bildungen sicher als 'sols ferrugineux tropicaux' ansprechen. Die jüngeren Böden des Pliozän indizieren eine deutliche Abnahme der Humidität. Interessant ist das Vorkommen von roten Bodensedimenten unter 1.5 MA alten Basalten (Rognon et al. 1981), weil danach nur noch braune Böden auftreten.

Das Pleistozän im hohen Atakor ist gekennzeichnet durch drei Feuchtzeiten ('sols bruns') und zwei morphodynamische Aktivitätsphasen, in denen es zu (glazialer u.) periglazialer Abtragung, zur Einschneidung und zur Glacisbildung kam. Der jüngste 'sol brun' korrespondiert mit der Mittelterrasse (MT), die mit ihren Schluff-, Ton- und $CaCO_3$-Gehalten vegetations- und bodenbedeckte Hänge anzeigt. Gegenüber der Niederterrasse (NT) und der zeitlich nicht genau fixierbaren Oberterrasse (OT) enthält diese 'biostatic Mittelterrasse' (Rognon 1980a) feinkörnige Bodensedimente, auf denen jeweils eine neue Pedogenese ansetzen konnte (Rohdenburg 1977, Vogg 1981). Dies belegen auch Untersuchungen aus dem Air-Gebirge, wo in der 'terrasse moyenne' ein fossiler 'paléosol rubéfié' auftritt (Morel & Schulz 1981).

Der bodenstratigraphische Nachweis einer altholozänen Flächenbildung gelang in der südlichen Umrahmung des Mourzouk-Beckens, wo die spätglazial-holozäne Bodenoberfläche zu Pedimenten (Glacis d'érosion) abgetragen und zur Mittelterrasse (Glacis d'accumulation) aufgeschüttet wurde (Skowronek 1985b). In den einzelnen MT-Ablagerungen liefen noch die Prozesse der Vertisolierung ab, bevor jüngere – im Neolithikum pedogen überprägte – Bodensedimente und schliesslich die Niederterrasse die altholozänen Pedimente überdeckten.

Bei einem grossräumigen Vergleich hat Skowronek (1985a) in Analogie zur Pedostratigraphie des Atakor und nach lokalen Datierungen die roten Oberflächenböden des Air- und Hoggar-Gebirges zeitlich in das Plio-Pleistozän, die braunen in das Quartär gestellt. Das würde bedeuten, dass weite Areale der zentralen Sahara noch ein neogen-altquartäres Bodenerbe besitzen, d.h. die pleistozäne und holozäne Abtragung hat hier noch nicht alles Lockermaterial entfernt. Sicher gilt das für die meisten (polygenetischen) Böden auf der Dachfläche des Mourzouk-Beckens und für das Plateau du Tademait.

Hinsichtlich der Bodenbildungsintensität, besonders auf den kristallinen Basement-Gesteinen, müsste man dagegen noch die Möglichkeit prüfen, ob nicht topographische, also nicht-klimatische Faktoren ausschlaggebend waren für eine stärkere Pedogenese, so z.B. am Unterhang (Flussfläche) von Inselbergen (Rognon 1980b). Das hiesse, dass unterschiedliche Bodenfazies nicht immer auch verschiedenes Bodenalter besitzen. Problematisch bleibt auch die Alterseinstufung der Saprolith-C-Horizonte, die nicht zeitgleich mit dem darauf entwickelten Solum (B-Horizont) gebildet sein müssen, und die in der Nähe der Tassili bei Djanet noch aus der Zeit vor dem Paläozoikum stammen können (Barbier 1967).

Während für klimageomorphologische Fragestellungen eine umfassende Bodenstratigraphie sehr hilfreich wäre, konzentriert sich die Forschung bei der Rekonstruktion saharischer Paläoklimate derzeit mehr auf den ^{14}C-Zeitraum und auf nicht-pedogene Klimazeugen (Giessner 1981). Hauptanliegen der vorliegenden Ubersicht (Fig. 1) ist es daher, ein bodengeschicht-

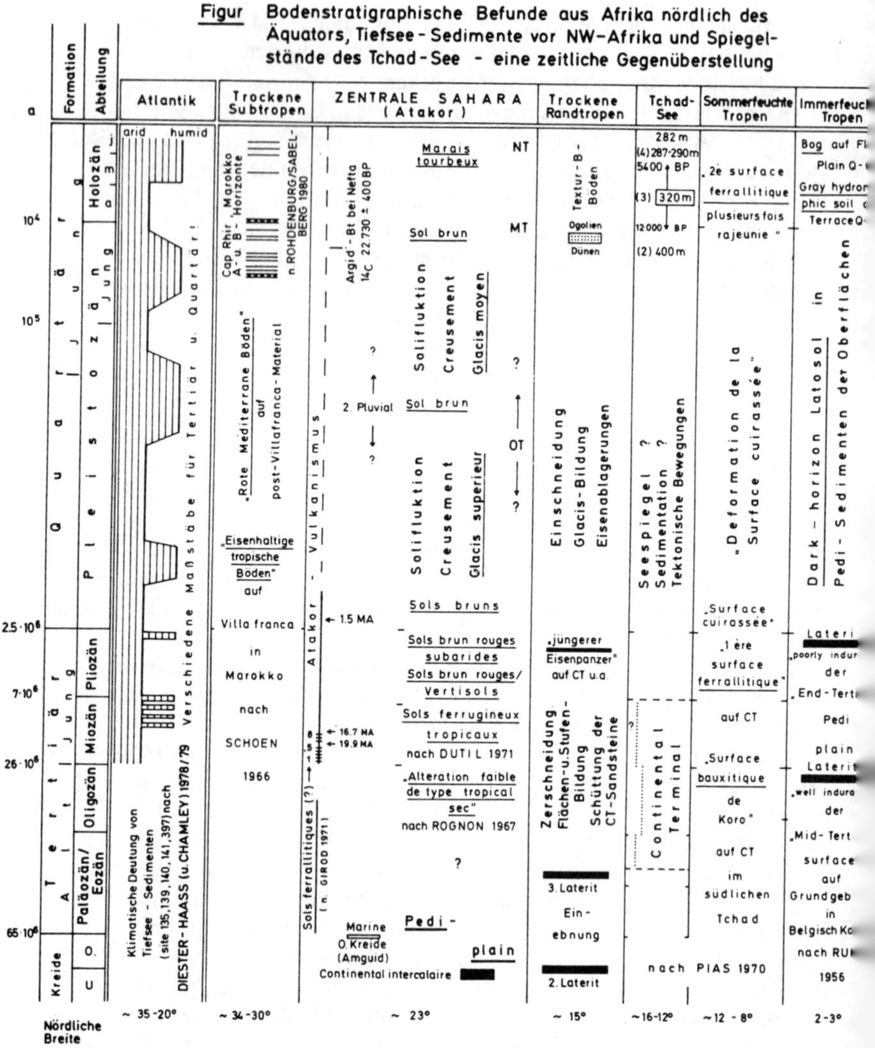

Figur 1. Bodenstratigraphische Befunde aus Afrika nördlich des Aquators, Tiefsee – Sedimente vor NW-Afrika und Spiegelstände des Tchad-See – eine zeitliche Gegenüberstellung.

liches Bild der zentralen Sahara zu zeichnen, womit bereits Büdel (1963) begonnen hatte, und darüber hinaus diesen heute ariden Grossraum pedostratigraphisch mit den nördlich und südlich anschliessenden Klimazonen zu vergleichen und ggf. zu verknüpfen.

TCHAD-SEE UND TROCKENE RANDTROPEN

Die Entwicklung des Tchad-Sees spielt insofern eine wichtige Rolle, als seine Spiegelstände die Erosionsbasis für die endorheische Entwässerung der südlichen Zentralsahara bildeten und zugleich höhere Niederschläge bezeugen. Letzteres gilt aber nicht unbedingt auch für die nördlich anschliessende Seenkette, wo die Hochstände grundwassergespeister Seen mit ariden Phasen zusammenfallen können und dann Niederschläge von 'Vorgestern' repräsentieren (Baumhauer 1986). Von drei jungquartären Tchad-Seen (A.Durand et al. 1983) variiert das 320 m-Niveau zeitlich sehr stark. Auffällig ist weiterhin, dass der 400 m-Stand chronostratigraphisch gut mit einem tondurschschlämmten Sandboden (Scharpenseel & Zakosek 1979) am Nordrand der Sahara bei Nefta in Tunesien korreliert.

Die Holozänböden der Trockenen Randtropen sind unter Dornsavanne auf Ogolien-Dünen rubefiziert und lessiviert (Felix-Henningsen 1983), am Südrand der Trockensavanne auch auf Granit (Boquier 1973).

Pedostratigraphische Leithorizonte aus dem übrigen Quartär sind nicht bekannt, erst im Tertiär spielen Eisenpanzer und Laterite eine Rolle. So könnte der dritte Laterit aus dem westafrikanischen Sahel (Michel 1980) dem Eozän-Laterit entsprechen, den Busche (1983) um Mourzouk annimmt, der aber als Laterit sensu strictu in der übrigen Zentral-sahara fehlt. Auch ist dabei die paläogene Gradnetzlage Nordafrikas zu berücksichtigen, auf Grund derer Bardossy (1981) für das west- und nordsaharische Eozän und Mitteloligozän ein trockenes Paläoklima rekonstruiert.

SOMMERFEUCHTE TROPEN

Die schon bei c. 14°N in Resten auf CT vorkommende '1ère surface ferrallitique' des Pliozän ist die letzte stärkere Bodenbildung in den heutigen wechselfeuchten Tropen unter Feuchtsavanne. Intensive bauxitische Verwitterung ging dem Miozän voraus, die holozäne ist schwach ferrallitisch und meist auf lockeren Pleistozän-Sedimenten entwickelt (Pias 1970a und b).

IMMERFEUCHTE TROPEN

Eine den feuchten Ektropen prinzipiell ähnliche Abfolge von Reliefelementen scheint in den immerfeuchten Tropen von ehemals Belgisch Kongo (heute

Zaire) vorzuliegen. An Böden scheidet Ruhe (1956) zwei – unterschiedlich indurierte – Laterite auf einer mitteltertiären bzw. auf einer endtertiären Pediplain aus, während die vier Pedimente des Quartär (Q 5-2) Latosole tragen. Auf der jungpleistozänen Terrasse Q-1 besitzt der Boden noch hydromorphe Merkmale, im holozänen Hochflutbett Q-0 ist er ein Torfmoor.

ERGEBNIS

Geht man davon aus, dass die pedostratigraphische Ubersicht die wichtigsten Leitböden enthält, können m.E. folgende paläoökologische Schlussfolgerungen abgeleitet werden:

1. Schon im Tertiär gab es in den Niederen Breiten Afrikas Unterschiede hinsichtlich Intensität und Dauer der Pedogenese. Das wird besonders deutlich an dem verschiedenen Alter bauxitisch-lateritischer Verwitterung.
2. Im saharischen Neogen existieren – teilweise fossilierte – feuchtzeitliche Böden, die zumindest im Miozän keine südlichen Aquivalente besitzen, die sich aber generell in den Tiefsee-Sedimenten wiederfinden.
3. In der zentralen Sahara (Atakor) zeigt sich seit dem Oligozän ein deutlicher Wechsel von trockenen und feuchten Klimazuständen, welche auch aus den verschiedenen Bodengruppen ableitbar sind.
4. Das Känozoikum entwickelt in den feuchten Tropen Reliefabfolgen wie sie als Rumpfflächen, Pedimente und Terrassen bereits in den humiden Winterregengebieten wiederkehren.
5. Randtropen und zentrale Sahara bringen im Quartär auf den wenigen Glacis nur schwache Eisenablagerungen bzw. braune Böden hervor, welche aber untereinander noch nicht befriedigend parallelisiert werden können.

In den längerfristigen Abtragungsgebieten wird man allerdings mit dem Verlust morphologischer Elemente und der zugehörigen Böden rechnen müssen, da während einer so weit zurückreichenden Reliefentwicklung sicher eine Auslese von Flächen (Pedimente, Terrassen usw.) stattgefunden hat, was letzlich auf die unterschiedliche Intensität und Dauer der einzelnen Abtragungsphasen zurückzuführen ist (Rohdenburg 1983).

LITERATUR

Barbier, R. 1967. Nouvelles réflexions sur le problème des 'pains de sucre' à propos d'observations dans le Tassili N'Ajjer (Algérie). *Trav. Lab. Géol. (Grenoble)* 43: 15-21.

Bardossy, G. 1981. Paleoenvironments of laterites and lateritic bauxites – effect of global tectonism on bauxite formation. In *Proc. Intern. Sem. Lateritisation Processes* (Trivandrum, India, 11-14 December, 1979): 287-294. Rotterdam: Balkema.

Baumhauer, R. 1986. Zur jungquartären Seenentwicklung im Bereich der Stufe von Bilma (NE-Niger). *Würzb. Geogr. Arb.* 65: 235 S.

Blume, H.-P. 1985. Klimabezogene Deutung rezenter und reliktischer Eigenschaften von Wüstenböden. *Geomethodica* = Veröff. 10. BGC 10: 91-121.

Boquier, G. 1973. Genèse et évolution de deux toposéquences des sols tropicaux du

Tchad. Interprétation biogéodynamique. *Mém. ORSTOM* 62: 350 S.
Brunnacker, K. 1980. Young Pleistocene loess as an indicator for the climate in the Mediterranean area. *Palaeoecol. of Africa* 12: 99-113.
Büdel, J. 1963. Die pliozänen und quartären Pluvialzeiten der Sahara. *Eiszeitalter und Gegenwart* 14: 161-187.
Busche, D. 1983. Silcrete in der zentralen Sahara (Murzuk-Becken, Djado-Plateau und Kaouar; Süd-Libyen und Nord-Niger). *Z. Geomorph.* N.F., Suppl.-Bd. 48: 35-49.
Diester-Haass, L. 1979. DSDP Site 397: Climatological, sedimentological and oceanographic changes in the Neogene autochthonous sequence. *Initial Reports of the Deep Sea Drilling Project* XLVII: 647-670.
Diester-Haass, L. & H.Chamley 1978. Neogene paleoenvironment off NW Africa based on sediments from DSDP Leg 14. *J. Sed. Petrol.* 48: 879-896.
Durand, A., J.Lang, A.Morel & J.-P.Roset 1983. Evolution géomorphologique, stratigraphique et paléoclimatique au Pléistocène supérieur et à l'Holocène de l'Air oriental (Sahara méridional, Niger). *Rev. Géol. Dyn. Géogr. Phys.* 24: 47-59.
Durand, J.H. 1959. *Les sols rouges et les croûtes en Algérie.* Serv. des Etudes Scien., Birmandreis.
Dutil, P. 1971. Contribution à l'étude des sols et des paléosols du Sahara. Thèse Strasbourg, 346 S.
Felix-Henningsen, P. 1983. Zur Genese und Vergesellschaftung von Böden auf den Altdünen der nördlichen Goz-Region im Sudan. *Mitt. Dtsch. Bodenkundl. Ges.* 38: 485-490.
Giessner, K. 1981. Die jungquartäre Klimageschichte der Sahara – Stand und Vergleich der bisherigen Forschungsergebnisse. *Würzb. Geogr. Arb.* 53: 111-127.
Girod, M. 1971. *Le massif volcanique de l'Atakor (Hoggar, Sahara algérien). Etude pétrographique, structurale et volcanologique.* Paris: Centre National de la Recherche Scientifique. 158 S.
Michel, P. 1980. Vergleichende Reliefentwicklung in der südlichen Sahara, im Sahel und in Südwest-Afrika. *Tübinger Geogr. Stud.* 80: 95-111.
Morel, A. & E.Schulz 1981. Recherches palynologiques et géomorphologiques dans le massif de l'Air (Niger): note préliminaire. *Rev. Geogr. Alpine* 69: 583-592.
Pias, J. 1970a. Les formations sédimentaires tertiaires et quaternaires de la cuvette tchadienne et les sols qui en dérivent. *Mém. ORSTOM* 3.
Pias, J. 1970b. Notice explicative No. 41. Carte pédologique du Tchad à 1/1.000.000. Vol. I, II. Paris: Centre ORSTOM de Fort-Lamy.
Rognon, P. 1967. *Le massif de l'Atakor et ses bordures (Sahara Central).* Paris: Centre National de la Recherche Scientifique. 559 S.
Rognon, P. 1980a. Comparison between the Late Quaternary terraces around Atakor and Tibesti. In M.J.Salem & M.T.Busrewill (eds.), *The geology of Libya,* Volume III: 815-821. London.
Rognon, P. 1980b. Pluvial and arid phases in the Sahara: the role of nonclimatic factors. *Palaeoecol. of Africa* 12: 45-62.
Rognon, P., Y.Gourinard & Y.Bandet 1981. Un épisode de climat aride dans l'Atakor (Hoggar) vers 1,5 Ma (datations K/Ar) et sa place dans le contexte paléoclimatique du Plio-Pléistocène africain. *Bull. Soc. géol. France* 23: 313-318.
Rohdenburg, H. 1977. Beispiele für holozäne Flächenbildung in Nord- und Westafrika. *Catena* 4: 65-109.
Rohdenburg, H. 1983. Beiträge zur allgemeinen Geomorphologie der Tropen und Subtropen. Geomorphodynamik und Vegetation, klimazyklische Sedimentation, Panplain/Pediplain-Pediment-Terrassen-Treppen. *Catena* 10: 393-438.
Rohdenburg, H. & U.Sabelberg 1980. Northwestern Sahara margin: terrestrial stratigraphy of the Upper Quaternary and some paleoclimatic implications. *Palaeoecol. of Africa* 12: 267-275.
Ruhe, R.V. 1956. *Landscape evolution in the High Ituri, Belgian Congo.* Publ. INEAC 66.

Scharpenseel, H.W. & H.Zakosek 1979. Phasen der Bodenbildung in Tunesien. *Z. Geomorph.* N.F., Suppl.-Bd. 33: 118-126.
Schoen, U. 1966. *Beitrag zur Kenntnis der Tonminerale im marokkanischen Boden.* Habil.-Schrift, Göttingen.
Skowronek, A. 1984. *Bodengeographisch-paläopedologische Untersuchungen zum Problem der klimagenetisch-geomorphologischen Landschaftsentwicklung in der zentralen Sahara.* Habil.-Schrift, Würzburg.
Skowronek, A. 1985a. Zur känozoischen Klimaentwicklung der zentralen Sahara nach bodenstratigraphischen Befunden. *Geomethodica* = Veröff. 10. BGC 10: 123-151.
Skowronek, A. 1985b. Zur Morpho- und Pedostratigraphie der zentralen Sahara. *Z. Geomorph.* N.F., Suppl.-Bd. 56: 69-87.
Vogg, R. 1981. *Bodenressourcen arider Gebiete. Untersuchungen zur potentiellen Fruchtbarkeit von Wüstenböden in der mittleren Sahara.* Stuttgarter Geogr. Stud. 97: 224 S.

WINDSCHLIFFE UND WINDKANTER IN DER WESTLICHEN ZENTRAL-SAHARA

HELGA BESLER
Geographisches Institut, Universität zu Köln, F.R.Germany

SUMMARY: WIND FLUTING AND VENTIFACTS IN THE WESTERN CENTRAL SAHARA

Distinct traces of aeolian abrasion in rocks and the statistical distribution of ventifacts on land surfaces of the serir type were investigated along a N-S profile across the western central Sahara. Wind fluting everywhere may be explained by modern meteorological conditions.

After analyzing the literature on ventifacts, the model of formation perpendicular to the dominant wind direction is favoured for ventifacts from small pebbles. Triquetous wind-faceted stones (Brazil-nut shape) in this case are formed by rotation which is attributed to fluvial activity. The distribution of ventifacts in general is in accordance with modern wind conditions. The varying percentage of Brazil-nut shapes, however, cannot be correlated to the modern precipitation pattern and therefore seems to indicate surfaces of different ages.

EINFUHRUNG

Von Anfang Februar bis Anfang April 1984 fand unter der Leitung von W.Meckelein die Stuttgarter Geowissenschaftliche Sahara-Expedition statt. An diesem interdisziplinären Projekt der Universitäten Stuttgart und Hohenheim nahmen Wissenschaftler und Studenten der Fachrichtungen Geomorphologie, Geologie, Boden- und Hydrogeographie, Klimatologie und Biologie teil. Die Route führte durch Tunesien und Algerien und weiter durch Mali nach Niger. Im Frühjahr 1985 unternahm die Verfasserin, die vor allem die äolische Formung bearbeitete, eine Anschlussreise in die Republik Niger, um im Gebiet von Ténéré und Aïr vergleichende Untersuchungen durchzuführen. Eines der behandelten Themen war die Fragestellung: Wie stark – im Vergleich – werden die grossen Flàchen der Sahara vom Wind geformt?

Zu diesem Komplex waren schon 1978 auf der Tanezrouft-Expedition des Geographischen Instituts der Universität Stuttgart, ebenfalls unter der

Leitung von W.Meckelein, Untersuchungen in der Tanezrouft, in der Region Tidikelt und weiter nördlich durchgeführt worden, deren Ergebnisse zum Vergleich herangezogen werden (Besler 1982, 1984; Meckelein 1982, 1984). Behandelt wird also ein Nord-Süd-Profil durch die westliche Zentral-Sahara.

WINDSCHLIFFE

Bestandsaufnahme

Entlang der Route und im Bereich der Arbeitsschwerpunkte wurden alle deutlichen Korrosionsspuren in Festgesteinen nach Art und Schliffrichtung untersucht und kartiert. In den meisten Fällen handelt es sich um deutlich in einer Richtung vorherrschende Fazetten oder Rillen, die eine Richtungsbestimmung des effektiven Windes erlauben. Fig. 1 zeigt die kartierten Windschliffe mit ihrer Richtung, wobei ein Pfeil häufig mehrere Befunde repräsentiert. Es handelt sich ausschliesslich um Schliffe an Ausbissen auf freien Flächen oder an exponierten grösseren Relief-einheiten wie Stufenkanten oder auch grossen verfestigten Altdünen. Nicht berücksichtigt wurden Windschliffe in starkem Kleinrelief (z.B. am Ostrand des Aïr), deren Richtungen häufig lokale Wirbel widerspiegeln. Natürlich ist die Karte nicht vollständig, da nicht alle Gebiete mit starkem Windschliff, wie z.B. nördlich des Aïr-Gebirges (siehe Mainguet & Canon 1976), an der Route lagen.

Die meisten Windschliffe fanden sich auf Kalkgestein im weiteren Sinne (auch kristalline Kalke, Dolomite, Kalksandstein, Seekreide). Zu den Ausnahmen gehören gipshaltige Krusten des Schott Merouane. Sowohl Gips als auch Kalke werden besonders schnell von Sandwind korrodiert (vgl. Miotke 1979, Besler 1980). Auch die Äolianite am Niger gehören zu den leicht schleifbaren Materialien. Windschliffe fanden sich aber auch auf anstehendem Kristallin des Präkambriums in der südöstlichen Tanezrouft und besonders in den quarzitischen Metamorphiten am Rande des Djado-Plateaus, wo auch grosse Yardangs geschliffen wurden. Letztere Vorkommen liegen also in widerständigerem Material. Mit Ausnahme des Djado-Plateaus sind die Windschliffe jedoch überall relativ kümmerlich und nicht mit dem äolischen Relief am Tibesti (s. Hagedorn 1968, 1971; Mainguet et al. 1980) oder auch mit Korrosionsformen in der Lybischen Wüste Ägyptens zu vergleichen.

Auf der Windschliffkarte (Fig. 1) fällt eine deutliche Zweiteilung in Schliffe aus westlichen Richtungen (SW bis NW) im Norden und Schliffe aus östlichen Richtungen (NE bis SE) im Süden auf. Die Grenze verläuft durch das zentrale Hoggar-Gebirge, biegt aber an seinem Westrand stark nach dem Norden aus und zieht über das Plateau von Tademait nach dem Nordwesten. Besonders auffallend ist die Scheide zwischen den benachbarten Landschaften von Tidikelt mit östlichen und Tinrhert mit westlichen Schliffrichtungen. Südlich des Hoggar überwiegt die Passatrichtung mit Ausnahme der südöstlichen Tanezrouft. Die Südostrichtung hier könnte dadurch entstehen, dass der NE-Passat das Hoggar-Massiv auf der Südseite umströmt, während im wieder ansteigenden Gelände des Adrar des Iforas die normale Passatrichtung wieder wirksam wird. In der inneren Tanezrouft, im Passat-Lee des Hoggar, war schon 1978 die Armut an Windschliffen aufgefallen (Besler 1982, 1984).

Vergleich mit meteorologischen Daten und Literatur

Obwohl es in der Sahara durchaus Messstationen gibt, sind Winddaten immer noch ein Problem. Sie werden in den Klima-Atlanten stiefmütterlich behandelt und sind selbst in den Tabellen des *World Survey of Climatology* nur durch mittlere Geschwindigkeiten und vorherrschende Richtungen vertreten. Beide Angaben sind für die morphologische Wirksamkeit von Winden völlig irrelevant (vgl. Besler 1987). Die brauchbarsten Angaben findet man daher immer noch in den Werken von Dubief (1952, 1953).

Schon 1952 veröffentlichte Dubief seine Karten mit den im Januar und Juli vorherrschenden Windrichtungen in der Sahara. Vergleicht man diese mit den Korrosionsrichtungen der Fig. 1, dann fällt sofort auf, dass nur die Winterkarte Ubereinstimmung zeigt. Am Niger z.B. machen sich die sommerlichen monsunalen SW-Winde nicht in der Landschaft bemerkbar. Auch in der nördlichen Sahara bleiben die Etesien offenbar ohne Korrosionswirkung. Dagegen entsprechen die Schleifspuren sehr gut den Winterwinden. Die Grenze zwischen den ektropischen Westwinden und den Passaten verläuft auch bei Dubief nördlich des Tidikelt. Sie biegt allerdings nicht so stark nach S zum Hoggar aus wie der Korrosionsbefund. Viel differenzierter ist Dubiefs Karte 3 mit jährlichen Sandwindresultanten. Hier zeichnet sich auf den ersten Blick eine perfekte Ubereinstimmung mit den Windschliffrichtungen ab, da die Grenze zwischen östlichen und westlichen Resultanten ebenfalls nach S ausbiegt und durch den Hoggar verläuft. Kleinere Diskrepanzen ergeben sich bei El Golea und Timimoun. Dagegen entspricht die Sandwindresultante in der südlichen Tanezrouft (Bidon V) genau den in der SE-Tanezrouft gefundenen Schliffspuren aus SE.

In der geographischen Literatur existiert eine Sandbewegungskarte von I.G.Wilson (1971, Fig. 5), die neben den meteorologischen Resultanten nach Dubief Sandflusslinien für fast die gesamte Sahara enthält. Die Sandflusslinien wurden nach Dünenrichtungen (auch Yardangs) konstruiert. Mit dieser Karte stimmen die Windschliffe in Festgesteinen sehr gut überein. Auch bei Wilson reicht die Wirksamkeit der nördlichen SW-Winde bis an den Hoggar und spart die Region Tidikelt mit effektiven NE-Winden aus. In der südlichen Tanezrouft lässt Wilson seine Flusslinien allerdings nicht der meteorologischen Resultanten aus SE nach NW folgen, vielleicht weil hier keine Dünen oder ähnliche grössere Vollformen existieren, aus denen die Bewegungsrichtung entnommen werden könnte. Die Ubereinstimmung von Dubiefs Karte der Sandwindresultanten und Wilsons Karte der Sandbewegungen ist recht gut. Die Meteorologie spiegelt sich also in der Dünenmorphologie

Figur 1. Windschliffrichtungen in Festgesteinen auf dem N-S-Profil durch die westliche Zentral-Sahara.
[Directions of wind-fluting in rocks along the N-S-profile across the western central Sahara.]

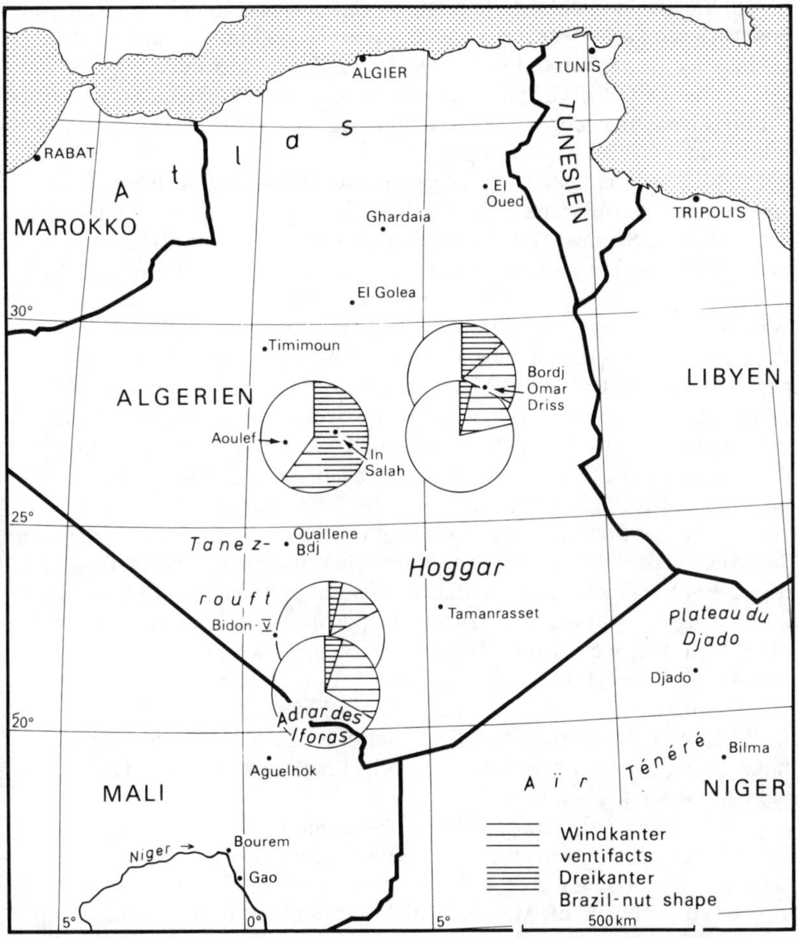

Figure 2. Windkanter-Statistik auf grossen Serir-Flächen: Tinrhert (direkt am S-Rand des Grossen Ostlichen Erg), Igharghar (bei Ta-n-Elaq), Tidikelt (zwischen Aoulef und In Salah), SE-Tanezrouft (Nähe Oued Tamanrasset, dreimal auf Ebene, einmal auf Bodenwelle).
[Statistical analysis of ventifacts on surfaces of the serir type: Tinrhert (southern border of the Great Eastern Erg), Igharghar (near Ta-n-Elaq), Tidikelt (between Aoulef and In Salah), SE-Tanezrouft (near Oued Tamanrasset, three times on plain = one circle, once on rise).]

wider. Wie die Windschliffkarte zeigt, scheint auch die äolische Formung an Festgesteinen die Meteorologie widerzuspiegeln. Dadurch werden einige Lücken in dünenarmen Gebieten geschlossen. Ein Widerspruch zeigt sich nur bei Timimoun und El Golea, die allerdings sowohl bei Dubief als auch bei Wilson (hier stark gebogene Sandflusslinien) nahe der Scheide zwischen westlichen und östlichen Winden liegen.

Die kleinmassstäblichen Karten geben natürlich die tatsächlichen Richtungen nur ungenau wieder. Ein Vergleich der Windschliffe mit Zahlenangaben über die Sandwindresultanten bei Dubief (1952, Tab. VII: Richtungsgrade und Häufigkeit) erlaubt detailliertere Vergleiche. Für Nord-Niger kann als Ergänzung eine Angabe von Mainguet & Callot (1978) herangezogen werden, wonach die Station Bilma-Dirkou für Geschwindigkeiten über 25 km/h (mit Sicherheit Sandwinde) NNE-E-Winde anzeigt (Mittel 56°). Abgesehen von El Golea und Timimoun, wo Schliffe und Resultanten nahezu entgegengesetzt liegen, weichen die Schliffrichtungen bis 30° nach rechts (Ft. Flatters = Bordj Omar Driss) und bis 20° nach links (Aguelhok) ab. Dabei überwiegen die Rechtsabweichungen. Es kann sich also nicht um die normale Reibungsablenkung der höheren Stationswinde am Boden handeln. Ob die Stationswerte überhaupt für die zum Teil entfernt gelegenen Untersuchungsareale relevant sind, muss eine Korrelations-überprüfung zeigen. Ein Vergleich zwischen Stationsentfernungen und Schliffrichtungsabweichungen ergibt nach Spearman eine positive Rang-Korrelation (r_s = + 0,55), die aber nicht signifikant ist. Lässt man jedoch die sehr ungenaue Angabe von Bilma-Dirkou unberücksichtigt, so zeigt sich eine stark positive Korrelation (r_s = + 0,79) auf dem Signifikanz-Niveau 0,05. Je näher die Schliffstelle an der Station liegt, desto besser ist die Richtungsübereinstimmung. Dieses Ergebnis zeigt auch, dass Richtungsangaben ohne Gradzahl nicht aussagekräftig genug sind.

Es fiel auf, dass die geringste Abweichung mit 4° bei In Salah (25 km Entfernung) mit den weitaus meisten Sandwindstunden in Richtung der Resultanten gemessen wurde. Bei Aoulef war dagegen die Abweichung grösser (10°), obwohl die Messstelle dicht bei der Station lag. Danach kann vermutet werden, dass auch die Sandwindhäufigkeit eine Rolle für die Abweichung spielt. In der Tat besteht nach Spearman eine deutliche negative Rang-Korrelation (r_s = - 0,7) auf den Signifikanz-Niveau 0,05. Eine Erklärung könnte in der Tatsache vermutet werden, dass mit zunehmender Windstärke die Windrichtung konstanter wird.

Jedenfalls kann aus diesen geringen und offensichtlich systematischen Abweichungen der Schliff- von der heutigen Windrichtung nicht auf Paläoformen geschlossen werden.

WINDKANTER

Windkanter in der Literatur

Veröffentlichungen zum Thema Windkanter sind zahlreich. Schon früh wurden zu diesem Phänomen auch Laborversuche gemacht (z.B. Bryan 1931, Schoewe 1932). Die Ergebnisse waren jedoch nie besonders einheitlich, so dass sich auch heute noch viele Wissenschaftler mit Versuchen beschäftigen (z.B. Miotke 1979, Whitney 1983). In den meisten Fällen wurde dabei kantiges Ausgangsmaterial untersucht. Die Angaben über Windkanter aus Geröllen oder gerundeten Steinen sind weitaus seltener; aber nur diese werden hier diskutiert. So uneinheitlich wie die Ergebnisse sind die verwendeten Namen, wobei leider verschiedene Formen mit derselben Bezeichnung belegt werden. Bryan, der sich 1931 zusammenfassend mit der Terminologie auseinandersetzte, unterschied drei Haupttypen: Pyramidalkanter ('pyramidal wind-faceted') mit drei Flächen, Firstkanter ('ridge-shaped') mit einem Grat zwischen zwei Flächen und Dreikanter ('triquetous') von der Form einer Paranuss mit drei leicht gebogenen Flächen. Bei Schoewe (1932) scheinen die Einkanter den Firstkantern Bryans zu entsprechen, und die Paranussform wird übereinstimmend beschrieben. Die von ihm erwähnten Kuenenschen Driekanter sind jedoch laut Abbildung Pyramidalkanter. Whitney (1983) unterscheidet non noch zwischen Dreikanter ('triquetous') und Paranussform ('Brazil-nut'), wobei letztere massiger als Dreikanter sind und vier oder mehr Flächen besitzen. In dieser Untersuchung werden die Begriffe von Bryan verwendet. d.h. die Dreikanter zeigen mehr oder weniger ausgebildete Paranussformen. Die Einkanter entsprechen Bryans Firstkantern.

Die nächste Unklarheit herrscht über die Entstehung der Kanten oder Grate zwischen den Flächen. Die von Bryan gesichtete Literatur liess die Flächen – und damit Kanten – durch Sandschliff senkrecht zur Windrichtung entstehen. Bryan selbst fand durch Versuche, dass bei grösseren Steinen durch Windstromteilung auch Flächen parallel zur Windrichtung entstehen können. Schoewe legte in seinen Versuchen die Längsachsen stets quer zum Wind und erhielt entsprechende Kanten. Auch er erwähnt parallele Flächen nur bei grossen Blöcken. Higgins (1956) meint die These von der Parallelität von Wind- und Kantenrichtung falsifizieren zu können, weil er noch im Konglomerat festsitzende kleine Kiesel mit Windkanten quer zur Hauptwindrichtung fand. Czajka (1972) berichtet aus der argentinischen Puna, dass sich kleinere Steine bei Störung immer wieder quer zur Windrichtung lagern. Whitney & Dietrich (1973) beobachteten in Windtunnelversuchen Grate quer, parallel und subparallel zum Wind. Bei Miotke (1979) entstanden die Kanten laut Abbildung quer zur Windrichtung. Und bei Whitney (1983) wurden Grate laut Abbildung sogar durch zwei schräg auftreffende Winde gebildet. Da auf den untersuchten Flächen die Windkanter aus kleinen Kieseln entstanden waren, die ein dichtes Steinpflaster bildeten und teilweise im Boden eingebettet lagen, wird hier die Bildung der Kanten quer zur Windrichtung favorisiert.

Dies erklärt aber nur die Entstehung von Einkantern und nicht von Dreikantern. Nach Whitney (1983) entstehen Dreikanter in situ ohne Umlagerung und sind in unimodalen Windsystemen selten. Higgins (1956) fand an Kieseln noch im Konglomeratverband neben vielen Einkantern auch wenige Dreikanter. Die bei Schoewe abgebildeten Windkanter aus den Versuchen von Kuenen mit wechselnden Windrichtungen entsprechen nicht den in der Natur häufiger vorkommenden Formen. Auch Pyramidalkanter wurden sehr selten gefunden. Aus Schoewes Versuchen mit konstanter Windrichtung resultieren dagegen aus den verschiedensten Ausgangsformen Einkanter. Wurde dann die Basis nach oben gedreht, so entstanden Paranussformen mit drei Kanten. Schoewe vermutet, dass in der Natur die Steine durch äolische Unterminierung, starke Böen, Frostwechsel, Tiere, Bodenkriechen, Lösungsvorgänge im Boden u.a. umgedreht werden könnten. Auch Bryan erwähnt aus der gesichteten literatur die Entstehung von Paranussformen durch Drehung.

Nun sind streng unimodale Windsysteme praktisch nicht existent. Es gibt aber stark dominierende Winde in der Sahara. Eine Umströmung der gesamten Kiesel nach dem Whitney-Modell wäre durch die Einbettung im Boden nicht möglich. Nirgends wurde an den Kieseln äolische Unterminierung beobachtet. Die Kanten im Kieselpflaster wiesen die unterschiedlichsten Richtungen auf. Alles deutet auf Entstehung der Dreikanter nach dem Schoewe-Modell hin. Für die Umlagerung kämen auch Vorgänge in Frage, die Schoewe nicht berücksichtigt, die aber im Gelände überall ihre Spuren hinterlassen haben. Selbst in grossen Teilen der Tanezrouft überzieht ein Netz von kleinen Gerinnen mit Materialsortierung die Fläche. Auch heute kommt es hier sicher zu Materialumlagerung (Meckelein 1984). Die so unberührt wirkenden Steinpflaster können sich danach sehr schnell wieder durch Deflation bilden, wie Symmons & Hemming (1968) in der südlichen Sahara bei Feldexperimenten feststellten.

Auch über die Bildungsgeschwindigkeit von Windkantern findet man unterschiedliche Angaben. Laborversuche lieferten hohe Korrosionsraten (Miotke 1979). Nach Kuenen (1960) können in Holland Windkanter in siebzig Jahren entstehen. Diese Aussagen wurden jedoch über grössere Steine und kantiges Ausgangsmaterial gemacht. Kuenen, der auch mit kleineren Korngrössen arbeitete, stellte fest, dass die Abrasion mit zunehmender Korngrösse und zunehmender Rundung abnimmt. Aus der allgemeinen Sedimentologie-Literatur ist bekannt, dass Kornbearbeitungen gleich welcher Art an grossen Körnern stets wirksamer sind. Ein hoher Windkanteranteil auf kleinsteinigen Serirflächen könnte also durchaus Indiz für höheres Alter sein.

Die Methode und der Aussagewert der Ergebnisse

Schon auf der Tanezrouft-Expedition 1978 waren auf Regoder Serirflächen Windkanter-Untersuchungen durchgeführt worden (Besler 1982, 1984).

Dabei handelt es sich um die Auszählung von Windkantern in einer zufallsbedingten Testfläche von 0,5 oder 1 m^2. Zu diesem Zwecke wird ein Rahmen oder Winkel auf die Kieselfläche geworfen, und auf der so bestimmten Fläche werden alle Silikatgesteine über 2 cm Durchmesser als Windkanter oder Nichtkanter ausgezählt. Andere Steine werden nicht berücksichtigt, da die Formungsgeschwindigkeit sehr unterschiedlich sein kann (z.B. sehr gross bei Kalken). Ebenfalls aus Vergleichsgründen muss die Ausgangsfläche von ehemals zugerundetem Material bedeckt sein; es muss sich also um Serire handeln. Scharfkantige Verwitterungsbruchstücke können nach leichter Kantenüberarbeitung Windkanter vortäuschen. Da viele hunderte Steine im Gelände ausgezählt werden müssen, kann nicht mit Laborgenauigkeit gearbeitet werden, mit der man auch ehemals kantiges Material begutachten könnte.

Auf der Sahara-Expedition 1984 wurde diese Methode in der südöstlichen Tanezrouft auch auf ihre statistische Signifikanz geprüft. Zu diesem Zwecke wurden etwa 10-20 m voneinander entfernt drei verschiedene Testflächen untersucht (je 1 m^2). Die Ergebnisse zeigt Tab. 1. Es wurden Nichtkanter, Einkanter und Dreikanter unterschieden. Die Variabilität der Dreikanter ist mit bis zu 67% Abweichung vom arithmetrischen Mittel (= 100%) sehr hoch. Dies erklärt sich jedoch aus der geringen Absolutzahl der Dreikanter. Die Abweichung der Nichtkanter vom Mittelwert beträgt dagegen nur bis zu 5%. Werden also Einkanter und Dreikanter als Windkanter zusammengefasst und gemeinsam bewertet, so können die Ergebnisse einer einmaligen Zählung als durchaus repräsentativ für eine grössere Fläche gelten. Voraussetzung ist, dass sich die Topographie nicht ändert. Auch dies wurde in der SE-Tanezrouft überprüft. In der Nähe eines flachen Rückens aus präkambrischem Kristallin, also in topographisch höherer und damit exponierterer Lage, ergab eine vierte Zählung einen signifikant höheren Anteil an Windkantern.

Damit scheint diese Methode geeignet zu sein, auf den ausgedehnten monotonen Flächen der Sahara – soweit es sich um Serirflächen handelt – unterschiedliche äolische Bearbeitung diagnostizieren zu können.

Der Befund

Fig. 2 zeigt den Befund der Windkanteruntersuchungen in vier Regionen im räumlichen Vergleich (siehe auch Tab. 1). In der Karte sind die drei Testzählungen in der SE-Tanezrouft als Mittelwert in einem Kreis repräsentiert. Sofort fallen die grossen räumlichen Unterschiede in der äolischen Formung der Serirkiesel auf. An der Spitze steht das Plateau von Tidikelt, wo die meisten Windkanter als Dreikanter notiert wurden (aber nicht ausgezählt). Mit grossem Abstand folgt das Plateau von Tinrhert, wo etwa die Hälfte aller Windkanter als Dreikanter vorlag. Es folgt die kristalline Bodenwelle der SE-Tanezrouft mit vergleichbarer Gesamtzahl von Windkantern, aber verschwindendem Dreikanteranteil. Deutlich geringer ist der Windkanteranteil in der weiten Talung des Oued Igharghar (SSW von Bordj Omar Driss). Das Schluss-

Tabelle 1. Windkanter-Statistik (Statistical analysis of ventifacts)

Lokalität	Testfläche m^2	Ausgezählte Steine	Nichtkanter %	Einkanter %	Dreikanter %
Plateau von Tinrhert	1	366	67	19	14
Oued Igharghar	1	325	78,5	18,5	3
Tidikelt	0,5	107	41	59	
Se-Tanezrouft, Ebene	1	486	81	14	5
	1	399	87	11	3
	1	354	81	16	3
Mittel	1	413	83	14	3
SE-Tanezrouft, Bodenwelle	1	321	67	28	5

licht bildet schliesslich die monotone SE-Tanezrouft. Der Vollständigkeit halber muss bemerkt werden, dass die zentrale Tanezrouft so wenige Windkanter aufwies, dass eine Auszählung nicht als sinnvoll erschien. Nach den Tagebuchaufzeichnungen von 1978 nehmen die Windkanter in der Tanezrouft vom Nord- und Südrand gegen die Mitte ab. Das Zentrum steht damit am eigentlichen Ende der Rangfolge. In anderen Gebieten, z.B. in Mali und Niger, wurden entweder keine Serirflächen gefunden oder nur Feinserire in Form von Kiesdecken, deren Komponenten zu klein waren (z.B. nördliche Ténéré).

Diskussion

Für die Auswertung der Befunde in der Sahara ist folgendes festzuhalten:

1. Windkanter aus Kieseln können allgemein Indiz für verschiedene Sachverhalte sein: entweder starke Windwirkung oder (und) alte Flächen.

2. Liegen viele Windkanter als Dreikanter vor, so ist dies ein Indiz für wiederholte Umlagerungsprozesse und daher wahrscheinlich auch ältere Flächen.

Ob es sich um starke aktuelle Windwirkung oder um alte Flächen handelt, ist nur durch einen Vergleich mit meteorologischen Daten zu klären. Dubiefs Karte 3 (1952) der jährlichen Sandwindresultanten ist leider insofern für einen Vergleich nicht vollständig, als im Oued Igharghar keine Station angegeben und in der südlichen Tanezrouft bei Bidon V nur eine nicht durch Zahlenwerte gestützte Resultante eingetragen ist. Die anderen Regionen sind in seiner Tab. VII vertreten, die mittlere Tanezrouft notfalls durch Ouallene. Danach ergeben sich abnehmende Sandwindhäufigkeiten in der Reihenfolge Tidikelt → (S-Tanezrouft) → Tinrhert → Mitte Tanezrouft. Wird die Ungenauigkeit bei Bidon V berücksichtigt, so stimmt diese Folge mit abnehmender Windkanterzahl überein. Die Talung des Oued Igharghar ist sicher windgeschützter als das Plateau von Tinrhert. Die zentrale Tanezrouft im Lee des Hoggar ist ausgesprochen windarm (vgl. Besler 1982).

Der besonders hohe Dreikanteranteil im Tidikelt lässt an alte Flächen denken. Verglichen hiermit müssten die anderen untersuchten Flächen jünger sein. Nun könnte man aber auch ableiten, dass auf diesen Flächen wegen zu häufiger Umlagerung die Windkanterbildung überhaupt gestört worden sei. Gegenargumente liefert die Tanezrouft: An den Zählstellen im SE waren Windkanter deutlich häufiger als weiter westlich in Abflussrichtung des Oued Tamanrasset. Mit einiger Wahrscheinlichkeit nehmen Abflussereignisse und fluviale Umlagerung mit Entfernung vom Hoggar ab. Man müsste nach obiger Hypothese also mehr Windkanter im Westen finden. Da das Gegenteil der Fall ist, kann eine für Windkanterbildung zu häufige Umlagerung ausgeschlossen werden.

Gute Vergleichsobjekte sind die Fläche von Tinrhert und die Bodenwelle auf Kristallin im SE der Tanezrouft, die bei gleicher Gesamtzahl von Windkantern sehr unterschiedlichen Dreikanteranteil aufweisen. Spielt tatsächlich die aktuelle fluviale Umlagerung für die Bildung von Dreikantern aus Einkantern eine entscheidende Rolle, so müssen am S-Rand des Grossen Ostlichen Ergs mehr Abflussereignisse stattfinden als in der SE-Tanezrouft. Eine Uberprüfung auf statistische Zusammenhänge zwischen Dreikanteranteil und Niederschlägen ergab sowohl die Niederschlagsmengen als auch die Intensitäten betreffend keine signifikanten Rang-Korrelationen nach Spearman (Niederschlagsdaten aus Tabellen und Karten bei Dubief 1963). Danach sind maximale Niederschläge und Intensitäten in Tinrhert und SE-Tanezrouft vergleichbar. In allen Fällen ist die Diskrepanz im Tidikelt mit den geringen Niederschlägen besonders gross. Hohe Dreikanteranteile können also nicht durch häufigere Umlagerungen im selben Zeitraum erklärt werden. Damit könnte der Dreikanteranteil tatsächlich ein Indiz für das Alter der Flächen sein. Dies spiegelt sich in den Verhältnissen der SE-Tanezrouft gut wider: Der unterschiedliche Windkanteranteil wird durch die mehr oder weniger windexponierte Lage aktuell erklärt. Der trotzdem vergleichbare Dreikanteranteil dagegen lässt sich durch das gemeinsame Flächenalter erklären, das hier offensichtlich nicht besonders hoch ist.

FAZIT

Nach den Untersuchungen stimmen Windschliffe an Festgesteinen mit den aktuellen Windverhältnissen überein. Windkanter aus kleinen Kieseln werden quer zum dominierenden Wind gebildet, Dreikanter durch zusätzliche fluviale Umlagerung. Die verwendete Methode der statistischen Windkanteranalyse erlaubt Aussagen über die unterschiedliche äolische Bearbeitung grosser Serirflächen. Während der Windkanteranteil ebenfalls mit den aktuellen Bedingungen zu korrelieren scheint, kann der Dreikanteranteil durchaus ein Indiz für unterschiedliches Flächenalter sein.

LITERATUR

Besler, H. 1980. Die Dünen-Namib: Entstehung und Dynamik eines Ergs. *Stuttgarter Geogr. Stud.* 96: 1-241.

Besler, H. 1982. A contribution to the aeolian history of the Tanezrouft. *Bull. de l'Ass. de Géogr. Français* 59 (483-484): 55-59.

Besler, H. 1984. Verschiedene Typen von Reg, Dünen und kleinen Ergs in der algerischen Sahara. *Erde* 115: 47-79.

Besler, H. 1987. Dünenstudien am Nordrand des Grossen Ostlichen Erg in Tunesien. *Stuttgarter Geogr. Stud.* 100 (im Druck).

Bryan, K. 1931. Wind-worn stones or ventifacts – a discussion and bibliography. *Repr. & Circ. Ser. of the Nat. Res. Council* 98: 29-50.

Czajka, W. 1972. Windschliffe als Landschaftsmerkmal. *Z. f. Geom.* 16: 27-54.

Dubief, J. 1952. Le vent et le déplacement du sable au Sahara. *Trav. de l'Inst. de Rech. Sahariennes* 8: 123-164.

Dubief, J. 1953. Les vents de sable au Sahara français. *Coll. Int. CNRS Paris* 35: 45-70.

Dubief, J. 1963. *Le climat du Sahara* (T. II, F. I). Algcr. 275 pp.

Hagendorn, H. 1968. Uber äolische Abtragung und Formung in der SE-Sahara. Ein Beitrag zur Gliederung der Oberflächenformen in der Wüste. *Erdk.* 22: 257-269.

Hagedorn, H. 1971. Untersuchungen über Relieftypen arider Räume an Beispielen aus dem Tibesti-Gebirge und seiner Umgebung. *Z. f. Geom. Suppl.* 11: 251 pp.

Higgins, C.G. 1956. Formation of small ventifacts. *J. Geol.* 64: 506-517.

Kuenen, P.H. 1960. Experimental abrasion: 4. Eolian action. *J. Geol.* 68: 427-449.

Mainguet, M. & Y.Callot 1978. L'erg de Fachi-Bilma (Tchad-Niger). Contribution à la connaissance de la dynamique des ergs et des dunes des zones arides chaudes. *Mém. et Docum. NS* 18: 184 pp.

Mainguet, M. & L.Canon 1976. Vents et paléovents du Sahara. Tentative d'approche paléoclimatique. *Rev. de Géogr. Phys. et de Géol. Dyn.* 18: 241-250.

Mainguet, M., L.Canon & M.C.Chemin 1980. Le Sahara: géomorphologie et paléo-géomorphologie éoliennes. In M.A.J.Williams & H.Faure (eds.), *The Sahara and the Nile*: 17-35. Rotterdam: Balkema.

Meckelein, W. 1982. The question of permanent aridity in the central Sahara since the Pliocene. *Bull. de l'Assoc. de Géogr. Français* 59 (484): 39-42.

Meckelein, W. 1984. Zur physischen Geographie der Tanezrouft (Algerische Sahara). *Erde* 115: 19-46.

Miotke, F.D. 1979. Die Formung und Formungsgeschwindigkeit von Windkantern in Victoria-Land, Antarktis. *Polarforschg.* 49: 30-43.

Schoewe, W.H. 1932. Experiments on the formation of windfaceted pebbles. *Amer. J. Sci.* 24 (139): 111-134.

Symmons, P.M. & C.F.Hemming 1968. A note on wind-stable stone-mantles in the southern Sahara. *Geogr. J.* 134: 60-64.

Whitney, M.I. 1983. Eolian features shaped by aerodynamic and vorticity processes. *Developmts. in Sed.* 38: 223-245.

Whitney, M.I. & R.V.Dietrich 1973. Ventifact sculpture by windblown dust. *Bull. Geol. Soc. of America* 84: 2561-2581.

Wilson, I.G. 1971. Desert sandflow basins and a model for thc development of ergs. *Geogr. J.* 137: 180-199.

EAST AFRICAN PALAEOENVIRONMENTS

PALYNOLOGICAL INTIMATIONS ON THE EAST AFRICAN MOUNTAINS

J.A.COETZEE
Botany Department, University of the Orange Free State, Bloemfontein, South Africa

ABSTRACT

A number of important pollen diagrams from high altitudinal sites in East Africa cover the time of the Last Glacial Maximum (LGM) and provide similar data on the palaeoenvironment of these regions for that period. However, the interpretation of the results and the explanations of the causes of these environmental changes have been the source of some debate during the last decades. The present review, based on comparison of present-day and LGM climatic data, indicates that although rainfall is rightly considered to be an important parameter in biographic discussion, it is temperature that is the dominant primary cause of major palaeoenvironmental change. The considerable drops in sea surface temperature (SST) during LGM times in both the Benguela and Agulhas Currents strongly influenced the advection of humidity to Central and East Africa causing severe aridity on plateaus and high mountains. Cloud formation and the distribution of precipitation on the high mountains are temperature dependent. In LGM times with lower evaporation of the ocean surface, the mountains were generally drier, the cloud belt developed at lower altitudes and the harsh diurnal climate extended downward, lowering the tree line and reducing the extent of the forest. This phenomenon is clearly demonstrated by the total fossil pollen assemblages from each of the six high altitudinal sites in which similar elements were recorded. During LGM times the sites were in most cases surrounded by alpine desert indicating aridity and colder conditions at these altitudes.

INTRODUCTION

As the greater part of Africa is exposed to more or less arid climates the preservation of pollen bearing sediments on the continent is generally very poor. This pessimistic perspective for pollen analytical studies is corroborated and aggravated by the fact that severe aridity occurred during the last glacial maximum (LGM) even in regions which at present enjoy a very favourable climate. It is no wonder that pollen analysts working in Africa concentrated their efforts on high mountainous areas where humidity is generally much more conducive to pollen preservation. The consequence of this situation is that by far most pollen research in Africa has been done on the high mountains of East Africa.

The explanation of the first results of these pollen studies was hampered by preoccupation with either the pluvial theory or by untimely correlations with climatic chronologies well known from the Northern Hemisphere. At a later stage a general consensus was reached by palynologists about the interpretation of the important results which have been achieved. However, some arguments concerning the causes of the environmental changes and the interpretation of some pollen assemblages have survived tenaciously for the last twenty years in literature as 'pièces de résistance' in our understanding of the climatic history of East Africa. Some of these arguments in which Bloemfontein is often mentioned, will be discussed here in the hope that some clarification can be reached.

THE CLIMATIC FACTOR

In climatological considerations temperature is a basic factor. This is even more evident if we look at global climatic patterns which depend on the energy budget of the Earth, the difference between the incoming and outgoing radiation, which affects the atmosphere and the Earth's surface. The consequence of this situation is that life on Earth is constrained by changes in this energy budget.

In the explanation of pollen diagrams from the East African mountains and elsewhere the so-called 'Bloemfontein Group' (sic), in this case only consisting of Van Zinderen Bakker and Coetzee, has repeatedly argued that the temperature factor is the primary cause of climatic and major ecological changes. This principle has hardly ever been appreciated and has often been misinterpreted.

Let us first look at the climatic importance of this parameter and in this connection especially at the oceans of the world. Palaeotemperature determinations of ocean sediments have for many years consistently proved that during geological time considerable changes in the temperature of the oceans occurred. The geomorphology of the ocean basins and their geographic position have among other factors caused local and regional differences in ocean temperature at a given time. However, there is agreement about the fact that the oceans of the globe generally had a lower temperature during glacial maxima. The finding that, except for the Indian Ocean, the equatorial ocean surface temperatures (SST) were 4-7°C lower than at present (Flohn & Nicholson 1980) is still not well appreciated in ecological circles. This temperature change drastically reduced the evaporation over the tropical oceans. In 1979 Flohn suggested an 18 per cent reduction of the oceanic evaporation during the LGM with a regional reduction near 70 per cent in the equatorial belt between 10°N and 10°S latitude (Flohn & Nicholson 1980). While negating these important data some authors have argued that the oceans did not change much during glacial times. These facts are, however, of extreme importance for the explanation of the palaeoclimatology of East and Central Africa.

Already more than twenty years ago considerations about temperature changes during glacial times drew the attention of the so-called 'Bloemfontein Group' to the polar regions and especially to the Antarctic. During his participation in SCAR (the Scientific Committee on Antarctic Research) Van Zinderen Bakker was nominated in 1966 as convener of the 'sub-Committee on Quaternary Studies'. This sub-Committee evolved in later years into the 'Group of Specialists on Late Cenozoic Studies' in which Van Zinderen Bakker held the same position. In this capacity he organised three international symposia on the history of the Antarctic ice sheet (Van Zinderen Bakker 1969, 1973, 1978). At these conferences, the influences the evolution of the Antarctic ice sheet had on the world's oceans and on global climate were discussed by glaciologists, climatologists, geologists, oceanographers and biologists. The profound changes which took place in the far south as a consequence of these processes had a tremendous impact on the ecology of the African continent. During Late Cenozoic times the Southern Ocean changed from a temperate ocean into a polar sea. Cold bottom water and frigid surface currents strongly influenced the southern Atlantic, while the northern Atlantic was covered with sea ice in winter during LGM times. In this connection two facts are of great importance for our discussion on the palaeoclimate of East and Central Africa. First, the enhanced upwelling and the greater strength of the Benguela Current during LGM-times caused a considerable drop of the SST along the Namibia coast and much further northwards. A mass of cold water, which was a continuation of the Benguela current namely reached the equator at ~ 30°W longitude (Prell et al. 1976: 261). The surface water in this current has been estimated to have been 8-9°C colder in August than at present, while the drop in temperature in July was only minimal. A second important change took place in the region of the Agulhas Current. Although the SST of the Indian Ocean did not change substantially the water of this current was 4-5°C colder right through the year (Vincent 1972). These drops in temperature of the ocean surface water had a strong influence on the advection of humidity to Central and East Africa as can be concluded from the general wind system of this region. At present an important source of humidity is the so-called Atlantic monsoon which is a continuation of the SE trades which travel over the eastern south Atlantic Ocean and as a consequence of the Coriolus effect turn eastward north of the equator and enter the Zaire basin as southwesterlies. This flow of humid maritime air originates over the region of the Benguela Current and the tropical south Atlantic Ocean. This Atlantic monsoon is during all seasons the main source of the heavy precipitation in the Zaire basin and even on the mountains of the western rift valley, such as Ruwenzori. During equinoctial times easterlies bring maritime air from the Indian Ocean to East Africa and the convergence between these two air flows from the west and the east can cause heavy precipitation even as far east as the coast of the Indian Ocean (Trewartha 1966).

The East African climate is mainly dominated by the seasonally alternating NE and SE monsoons which are the result of the migrating pressure and wind system following the course of the sun (ibid.). The NE monsoon reaches East Africa during the northern winter and only its eastern branch, which has a short track over the Arabian sea, can bring limited precipitation to East Africa. In the alternate season the SE monsoon coming from the SW Indian Ocean advects more humidity to East Africa, although it loses considerable moisture while travelling over Madagascar and entering the African continent.

The total picture of the distribution of rainfall in East Africa is complicated and difficult to understand. However, coming back to the East African mountains it is evident from the present wind system that mountains in Rwanda and eastern Uganda, like Ruwenzori, receive much more rainfall during all seasons than those on the East African plateau. This difference in climate causes considerable differences in vegetation. It can also be inferred from the monsoonal system that the dry NE monsoon is responsible for the dry aspect of the northern face of mountains like Kilimanjaro and Mt Kenya. It is further reasonable to conclude that the SE side of Mt Kenya receives high orographic precipitation from the SE monsoon, while the W slope of the mountain is favoured by the Atlantic monsoon.

If we now turn our attention to the situation during LGM-times we see that the lowering of the SST must have had a considerable influence on the climate of Central and East Africa. The decrease in temperature of the extended cold Benguela water will not only have spread aridity in the Zaire basin and eliminated most of the tropical rain forest, but it will also have affected the climate and vegetation of the plateau and mountains of East Africa. Similarly the reduction in evaporation in the Agulhas region will have had an aridifying effect on East Africa.

Comparison of present-day and LGM climatic data therefore shows that although rainfall is rightly considered to be an important parameter in biographic discussions, theoretically it is the result of a chain reaction in which temperature is a dominating primary factor.

The influence world wide changes in temperature had on the general plant ecology of the high East African mountains will now be briefly considered. The distribution of rainfall on these mountains depends on many factors, the most important of which are the topography, the moisture content of the air masses and the direction and speed of the wind. It is a well-known fact that the summits of the mountains like Mt Kenya and Kilimanjaro are clearly visible in the early morning, while later in the day the tops of these mountains are obscured from sight by cloud formation at high altitude. This condensation in the ascending air during the day together with the aspect of the mountain slopes in relation to the rain bringing winds determine the general plant ecology of the mountains (Troll & Wien 1950, Coetzee 1967). The consequence of this is that, for instance, on the western and southeastern slopes of Mt Kenya the highest rainfall of 1500 mm and even over 2300 mm per annum occurs near the altitude of the cloud cover where humid bamboo forest occupies the mountain (Coetzee 1967: 19). Above the clouds the mountain is, especially in the early morning, exposed to a dry and harsh climate with intense insolation while at night regular frost occurs as a result of a clear sky and strong outgoing radiation.

The altitude of the cloud formation and the distribution of the precipitation depend primarily on the temperature gradient in the atmosphere surrounding the mountain and on the moisture content of the air. Changes in both these temperature depending factors affect the local climate and vegetation on the mountain slopes, especially at the tree line where the ecological boundary between two entirely different habitats occurs (Coetzee 1967). Taking the much lower evaporation of the ocean surface during the LGM into account it is not surprising that the mountains were then generally much drier, the cloud belt developed at lower altitudes and the harsh diurnal climate extended downward. The lowering of the tree line and the profound reduction of the forest on the East African mountains are the most significant indications of the decrease in temperature on the continent during the LGM. This cooling has been demonstrated by pollen analytical studies at the following sites: Mt Kenya, Cherangani, Mt Elgon, Muchoya Swamp in Uganda, Ruwenzori and Kuwasenkoko Swamp in Rwanda (Fig. 1).

POLLEN ANALYTICAL ASPECTS

Surveying this evidence it is important to remember, as already mentioned for the Ruwenzori, that the various sites are located on mountains with a different general humidity. The mountains in SW Uganda and Rwanda are the wettest, Cherangani and Mt Elgon have a much drier climate, while Mt Kenya is more humid than the latter two. The vegetation of these high mountains consequently shows remarkable differences so that fossil pollen spectra from the various sites cannot be subjected to straightforward comparisons.

Another important point which must be appreciated in our review is that many plant taxa occur over a wide altitudinal range. It is self-evident that it is consequently impossible to use the former occurrence of a single plant taxon as a climatic indicator even if the export ability of its pollen is taken into account. For the critical assessment of former climatic conditions only complete pollen assemblages representing a particular community should be considered. The occurrence of such a plant community on a certain part of the mountain slope, as in any habitat, is the consequence of a host of factors. The taxa which are found there are able to germinate, grow and flower under prevailing ecological conditions. Every plant species in this community reacts to these conditions according to its ecological amplitude and its ability to

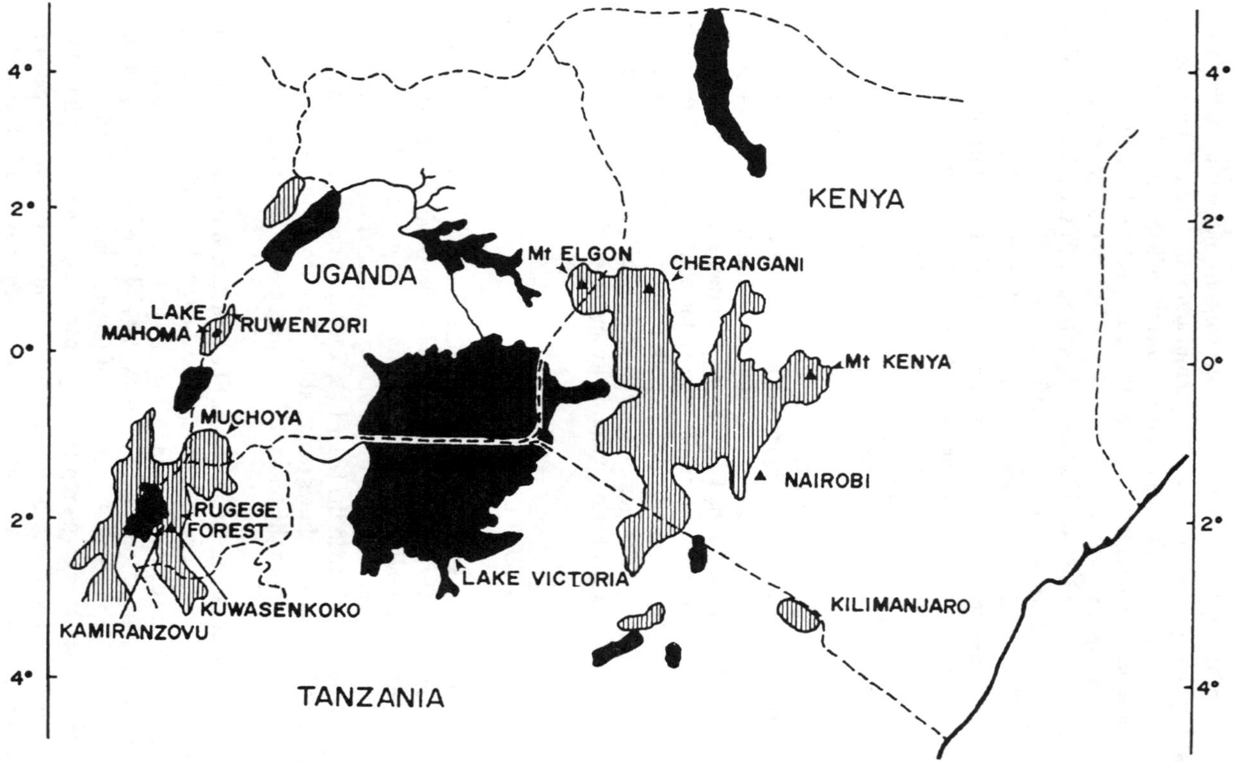

Figure 1. Localities of the pollen sites studied. Areas above 1800 m altitude cross-hatched.

compete and survive. The consequence of this well-known phytosociological situation is that a plant community is established. According to this principle a number of plant species characterise this physiognomic vegetation belt at certain altitudes. When ecological conditions change several taxa may disappear and others may replace them. In pollen diagrams from the mountains it is evident that the dominant species of a pollen assemblage which characterise certain vegetation belts are often replaced by others. This generally indicates environmental changes. Unfortunately this phenomenon has been simplistically described as the movement of vegetation belts. Non-plant ecologists seem to take such changes as an indication that complete vegetation belts or communities migrate up and down the mountain slope. This process is more complicated, but the characteristic species nevertheless give good indications of the environmental changes which occurred.

In evaluating the pollen sequences it has already been pointed out that the position of the upper forest limit forms an important threshold in the ecological conditions on the East African mountain slopes. In this connection it should be stressed that the conditions which determine the position of the tree line in the Northern Hemisphere mountains are entirely different from those of the equatorial East African mountains. In the former mountains seasonal changes and the position of the 10°C July isotherm are the most important limiting factors while in the latter the harsh diurnal climate determines which saplings will survive above the tree limit (Coetzee 1967).

Another aspect concerning the question of the importance of the high percentages of grass pollen recorded in the pollen assemblages of the LGM deposits at these high altitudinal sites has posed some problems in the interpretations of the pollen diagrams. Arguments about the provenance of the fossil Poaceae pollen have recurred several times in literature in the last decade. The possibility is not excluded that the fossil grass pollen deposited during LGM times at the high altitudinal sites also includes pollen of this family from the surrounding plateaus. However, it has so far not been possible on morphological grounds to separate this long distance component from the grass pollen originating in the Afro-alpine and Ericaceous Belts above the forest limit. This problem also exists for most African pollen sites as most plant communities except for example virgin rainforest, montane rainforests and others are of the open type and grasses grow everywhere. At present this family plays an important role in the vegetation belts above the tree line on the drier eastern mountains like Elgon and Mt Kenya while on the wetter Ruwenzori grasses are not so numerous at similar altitudes.

At Sacred Lake on Mt Kenya consideration of the high percentages of Poaceae pollen during the LGM together with well represented pollen of Ericaceous Belt shrubs and very low forest pollen counts were taken to indicate high altitude open vegetation (Coetzee 1967). The same conclusion could be inferred when the grass pollen is excluded from the pollen sum (Van Zinderen Bakker & Coetzee 1972, Hamilton 1982). As a matter of interest very high proportions of grass pollen have been recorded for this period from the other high altitude sites under discussion (see further on). It is now considered irrelevant whether part of the grass pollen of LGM age was transported up the mountain from the plateau.

With these points in mind it is proposed to review the pollen evidence available for temperature changes in the East African mountains during LGM times.

The most complete pollen record from the East African mountains is that from Sacred Lake on Mt Kenya. This lake is situated on the NE slope of the mountain at an altitude of 2 400 m and is at present surrounded by moist montane forest (Coetzee 1967). In the fossil pollen spectra between the calculated ages of 21 800 and 17 000 BP five taxa of the Ericaceous Belt together reach a maximum. Poaceae are very abundant (up to 48%) and pollen of the montane forest is very rare except for *Podocarpus,* the pollen of which has a high export ability. These pollen spectra point to the presence of the higher part of the Ericaceous Belt round the lake. The forest must have been situated at a fair distance below the lake. The climate must have been dry as can be inferred from the relatively high quantities of *Artemisia* pollen. As the present upper forest limit lies at an altitude of c. 3 100 m, the minimum lowering of the boundary must have been 700 m (Hamilton 1982), but this depression can conservatively be estimated to have been between 1 000 and 1 100 m. A lowering of the forest boundary of this measure will represent a lowering in temperature of 5-8°C (Coetzee 1967). This relatively small temperature drop engendered a considerable change in vegetation from moist montane forest to the dry and cold upper part of the Ericaceous Belt. The figures estimated on pollen evidence are in general agreement with those calculated from the descent of the snow line. Livingstone (1980) computed this figure for Mt Kenya at ~ 7.5-9.5°C.

At Lake Rutundu situated above Sacred Lake at 3 140 m on Mt Kenya the pronounced climatic changes during glacial times transferred the surrounding Ericaceous Belt vegetation into a much drier type (Coetzee 1967). Unfortunately we do not possess an age determination close to the LGM. The deposits from below the level with a ^{14}C date of 10 800 BP consist of grey clay which testifies to a dry climate with an open vegetation in the catchment (Hamilton 1982). The spectra contain 45-50% grass pollen and a scattered occurrence of microphyllous shrubs of the Ericaceous Belt except for *Artemisia* and Ericaceae which both reach values > 10%. The forest is only represented by very low percentages of pollen transported from below. It can be concluded that the climate round the lake was much drier than at present before 10 800 BP. Although this grey clay extends 3 m beyond the level of this date it cannot be stated with certainly whether LGM deposits are represented in this core.

The next important site that provided palaeoclimatic information on LGM times is the Kaisungor Swamp which covers the bottom of a long valley at an altitude of 2 900 m on the SW side of the Cherangani Hills. These hills are situated 350 m NW of Mt Kenya on the western side of the eastern Rift

Valley. The Cherangani Hills have a drier climatic setting than Mt Kenya. The vegetation at present surrounding the site is very disturbed, but it can be concluded that the original natural vegetation was dry montane forest. The pollen assemblages with a radiocarbon age of approximately 17 000 BP give the following picture (Coetzee 1967). Poaceae pollen percentages are constantly over 70%, the montane forest is represented by very low values which indicate long-distance transport, the Ericaceous Belt elements *Stoebe, Cliffortia, Anthospermum* and Ericaceae are only occasionally present while other Compositae and *Alchemilla* are of more importance and *Artemisia* slightly less so.

The spectra indicate that during the LGM the site was situated above the Ericaceous Belt which could be consistent with its higher altitude than Sacred Lake. The pollen assemblages also indicate that forest on these Hills had decreased considerably in area. The climate at Kaisungor was arid as the surroundings were covered by an Afroalpine desert as at present found in the high Afroalpine Belt of the drier eastern mountains of East Africa like Mt Kenya and especially Kilimanjaro (Hedberg 1964). The palaeotemperature cannot be estimated as the Cherangani Hills are not high enough to harbour alpine desert at present. The theoretical lowering of the vegetation belts will at least have been in the order of 1 100 m so that the temperature drop must have been very similar to that at Sacred Lake on Mt Kenya.

On Mt Elgon, situated 120 km to the west of the Cherangani Hills in a comparable climatic setting, a pollen profile which contains LGM sediments has been studied at Laboot Swamp by Hamilton (1982). The site is situated at 2 880 m (20 m lower than the Kaisungor Swamp) and is surrounded by remnants of dry montane forest. Mt Elgon is drier than Mt Kenya and natural vegetation of the lower and upper reaches of the mountain slopes has been nearly completely destroyed by dense human settlements and cattle grazing.

The fossil pollen described by Hamilton (1982) consists of two sections and it may well be that part of his pollen zone C is missing. It is, however, very likely that the top part of the lower section above the dated horizon of 23 073 ± 120 BP originated during LGM times. The pollen in these clayey sediments is well preserved and consists of up to 90% Poaceae, while montane forest types, mostly of long distance origin, are nearly absent. Hamilton (1982) concludes that extensive grassland surrounded the swamp and that the reduced montane forest was of a dry type. The climate was too arid to permit much growth of microphyllous thicket (Hamilton 1982: 135). The occurrence of *Alchemilla* could indicate a lowering in temperature as the genus at present grows at about the altitude of c. 3 200 m (ibid.: 134). As on the Cherangani Hills pollen assemblages of this type strongly indicate open high altitude Afroalpine grassland. Such grassland which occurs on the East African mountains at the altitude of ~ 4 000 m points to an arid and cold climate. These present conclusions are based on the totality of the pollen spectra and not on the former rare occurrence of one single pollen type. The lowering in temperature may well have been of the same order as on the Cherangani Hills and on Mt Kenya, but a specific figure cannot be suggested for Laboot Swamp. To draw the conclusion that the climate just became drier is ignoring the basic cause of this aridity and also our knowledge of the ecology of the mountains. A clear shift in climatic belts from warmer, more humid to cold and dry conditions is indicated.

In SW Uganda on the eastern slope of the Western Rift some important pollen diagrams have been described by Morrison (1968) and Hamilton (1972, 1982). Only the diagram from Muchoya Swamp may give information on LGM palaeoclimatic conditions. The swamp originated as a lake which was overgrown by mire vegetation and is at present surrounded by bamboo forest. The mire is situated at about the same altitude as Sacred Lake on Mt Kenya, viz. 2 256 m. The sandy lake mud of pollen zone C of Hamilton (1982) with an age of ~ 13 000 years near the top of the zone probably represents the LGM. The pollen spectra with some 80% of Poaceae, some *Anthospermum, Stoebe* and *Myrica* represent a colder and drier climate (ibid.). Morrison (1968) concluded that the mean annual temperature may have been 5 to 8°C lower than today. Pollen of the trees *Hagenia* (especially in the lower part of the zone), *Olea* and *Podocarpus* is present in low percentages. The total picture again indicates the presence of cold dry grassland with shrubs of the Ericaceous Belt which can at present only occur at a much higher altitude. The only difference with the Laboot and Kaisungor Swamps is that the climate at Muchoya was not as arid which is in agreement with the present-day situation.

In western Uganda just south of the equator pollen studies were made at three localities on Ruwenzori by Morrison and Livingstone (Hamilton 1982). This massive mountain has a very wet climate, as already mentioned, and natural grassland does at present not occur above the montane forest. The pollen profile of Mahoma Lake, at 2 960 m altitude, which is at present surrounded by bamboo forest, is of interest as pollen zone D with an age of ~ 15 000 to 12 700 BP, contains evidence of a glacial palaeoclimate. In this zone grass pollen reaches values of 35-45%, tree pollen is relatively scarce, while some elements of the Ericaceous Belt occur (ibid.). The remarkable fact is that *Dendrosenecio*-type pollen has been found abundantly which points to an altitudinal difference of at least some 700 m (ibid.). Hamilton also argued that the silt content of the deposits indicates a more open vegetation and therefore a drier climate. He explains the abundance of *Myrica* pollen mainly as the result of climatic dryness.

The pollen spectra in general show the same tendency as those discussed from the other mountains, viz. that a colder and drier climate existed during LGM times. Even with the overall higher humidity on Ruwenzori these points are clearly evident.

In SW Rwanda some swamps have been the focus of palynological analyses by Deuse (1966) and Hamilton (1982). Hamilton (ibid.) gives an extensive description of the pollen diagram of Kamiranzovu Swamp, a very large mire in Rugege Forest situated at an altitude of 1 950 m. From the rate of peat accumulation it is concluded that during the period from 37 630 to about 12 600-11 000 BP the temperature was reduced. The huge production of pollen by the swamp vegetation makes this site unsuitable for a detailed examination of forest history (Hamilton 1982: 179). Consideration on vegetational succession and changes in climate led Hamilton to suggest that during the period 20 000-15 800 BP the climate was wetter and far colder (ibid.: 185).

The pollen content from another mire situated some 20 km to the east of Kamiranzovu, the Kuwasenkoko Swamp at 2 340 m altitude has also been described by Hamilton (1982). Only two samples with an age of more than 13 835 years could be examined. The very high percentage of grass pollen together with a representation of some Ericaceous Belt pollen and some pollen of *Podocarpus* and *Myrica* point to a dry climate although it was not as dry as on Mt Elgon during a comparable grass maximum. This is testified by the presence of Urticaceae, fern beans and a number of other pollen types (Hamilton 1982).

In a discussion on the possibility of surface wash of some pollen types (e.g. *Dendrosenecio*-type) in some samples from horizons slightly higher up in the profile, Hamilton (ibid.) concludes that the temperature must have been considerably lower than at present. If this interpretation is correct it can be concluded that grassland under a dry and really cold climate occupied the site at a time prior to ~ 14 000 BP. This represents a considerable change compared with the present-day Rugege forest.

CONCLUDING REMARKS

Surveying the pollen evidence for the LGM climate on the East African mountains (Table 1) it is evident that the data are entirely concurrent. At the six significant sites where pollen analysis has been performed the percentages of similar elements in the pollen assemblages show irrefutably that the vegetation of a higher altitudinal belt had occupied the surroundings. The climate of these higher altitudes was not only much drier during this period but also colder and this habitat could in most cases be described as alpine desert. These temperature results can only in special cases, as on Mt Kenya, be translated into degrees centigrade. On the basis of other evidence from multidisciplinary studies there also exists full general correlation between the global climatic trend, the SST of the ocean, the consequent dryness of the monsoons and the low lake levels in East Africa.

It has never been the intention of the 'Bloemfontein Group' to maintain that fossil records from the mountains can always supply detailed quantitative measures for former temperatures. It has instead been stressed that temperature was the primary cause of the major environmental changes which took place. The gloomy outlook some palynologists had about this problem does not seem to have been justified. Part of the 'misunderstanding' about the in-

Table 1. Data on LGM vegetation and climate in the East African mountains based on pollen analytical studies of high altitudinal sites

East African highlands	Site	Altitude (m)	Surrounding vegetation Present	Surrounding vegetation LGM	Pollen Poaceae %	LGM-climate
Mt Kenya	Sacred Lake	2400	Moist montane forest	Higher part of Ericaceous Belt	48	Dry, 5-8° colder (Coetzee 1967)
Mt Kenya	Lake Rutundu	3 140	Ericaceous Belt	?	45-50	? LGM, dry
Cherangani Hills	Kaisungor Swamp	2 900	Dry montane forest	Afroalpine grassland	70	Dry, cold
Mt Elgon	Laboot Swamp	2 880	Dry montane forest	Afroalpine grassland	90	Dry, cold
SW Uganda	Muchoya Swamp	2 256	Bamboo forest	Open grassland of Ericaceous Belt	max. 78	Dry, 5-8° colder (Morrison 1968)
Ruwenzori	Mahoma Lake	2 960	Bamboo forest	Afroalpine grassland	35-45	Dry, 4.4 C° colder (Hamilton 1982)
SW Rwanda	Kuwasenkoko Swamp	2 340	Montane forest	Afroalpine grassland	60	Dry, cold
SW Rwanda	Kamiranzovu	1 590	Montane forest	?	?	Dry, cold

fluence of the temperature factor may have been caused by the notion that the 'Bloemfontein Group' postulated that the change in temperature as such has a decisive influence on the autecology of the plant taxa and therefore on the variations in ecology which took place. It must, however, be emphasised here that the 'Bloemfontein Group' never maintained this principle except for the situation at the tree line where the temperature reaches threshold values. As has been described by Coetzee (1967) the frost conditions at night, the overheating and physiological drought in the early morning which occur above the forest limit play the decisive role in determining which seedlings will survive the harsh diurnal climatic conditions which prevail there.

It is peculiar that even in recent literature the basic role of temperature is rarely accepted or is half-heartedly supported. Humidity as such has instead mostly been considered to be the basic underlying cause of major changes in vegetation. It is astonishing that in the explanation of such changes that occurred on the mountains ~ 15 000 years ago, the important climatic oscillations have been ignored and that changes in edaphic factors have been preferred.

REFERENCES

Coetzee, J.A. 1967. Pollen analytical studies in East and Southern Africa. *Palaeoecol. of Africa* 3. Cape Town: Balkema. 146 pp.

Deuse, P. 1966. *Contribution à l'étude des tourbières du Rwanda et du Burundi.* Inst. Nat. Recherche Scient. Butare, Rwanda. Public 4, 115 pp.

Flohn, H. & S.Nicholson 1980. Climatic fluctuations in the arid belt of the 'Old World' since the last glacial maximum: Possible causes and future implications. *Palaeoecol. of Africa* 12: 3-21.

Hamilton, A.C. 1972. The interpretation of pollen diagrams from highland Uganda. *Palaeoecol. of Africa* 7: 45-149.

Hamilton, A.C. 1982. *Environmental history of East Africa: a study of the Quaternary.* London: Academic Press. 328 pp.

Hedberg, O. 1964. *Features of Afroalpine plant ecology.* Uppsala: Almqvist & Wiksell. 144 pp.

Livingstone, D.A. 1980. Environmental changes in the Nile headwaters. In M.A.J.Williams & H.Faure (eds.), *The Sahara and the Nile*: 339-359. Rotterdam: Balkema.

Morrison, M.E.S. 1968. Vegetation and climate in the uplands of southwestern Uganda during the Later Pleistocene Period, 1. Muchoya Swamp, Kigezi District. *J. Ecol.* 56: 363-384.

Prell, W.L., J.V.Gardner, A.W.H.Bé & J.D.Hays 1976. Equatorial Atlantic and Carribean foraminiferal assemblages, temperatures and circulation: interglacial and glacial comparisons. In R.M.Cline & J.D.Hays (eds.), *Investigation of Late Quaternary paleoceanography and paleoclimatology*: 247-266. Geol. Soc. Am. Mem. 145.

Trewartha, G.T. 1966. *The earth's problem climates.* London: Methuen. 334 pp.

Troll, C. & K.Wien 1950. Der Lewisgletscher am Mount Kenya. *Geog. Ann. Stockholm* 31: 257-274.

Van Zinderen Bakker, E.M. Sr. (ed.) 1969. *Palaeoecology of Africa* 5. Cape Town: Balkema. 240 pp.

Van Zinderen Bakker, E.M. Sr. (ed.) 1973. *Palaeoecology of Africa* 8. Cape Town: Balkema. 198 pp.

Van Zinderen Bakker, E.M. Sr. (ed.) 1978. *Antarctic glacial history and world palaeoenvironments.* Rotterdam: Balkema. 172 pp.
Van Zinderen Bakker, E.M. & J.A.Coetzee 1972. A re-appraisal of Late Quaternary climatic evidence from Tropical Africa. *Palaeoecol. of Africa* 7: 151-181.
Vincent, E. 1972. Climatic change at the Pleistocene-Holocene boundary in the southwestern Indian Ocean. *Palaeoecol. of Africa* 6: 45-54.

PALAEOENVIRONMENTAL PROBLEMS IN THE EAST AFRICAN MOUNTAINS

W.C.MAHANEY
Department of Geography, Atkinson College, York University, North York, Ontario, Canada

ABSTRACT

Palaeoenvironmental reconstruction on Mount Kenya is possible from detailed analyses of landform distribution, various palaeosol characteristics, and palynological data. Landform distributions provide evidence for high magnitude environmental changes (especially glacial to interglaciation conditions), and quite often they provide the setting for other detailed palaeosol and pollen investigations. Palaeosols, from the upper Afroalpine (~ 4 000 m) to the Bambool Forest (2 900 m), often exist as compound soils, 'staked' in order of development, following episodic emplacement of deposits by various geologic processes including volcanic activity, glaciation, airfall influx, slope wash, colluviation and other mass wasting processes. Many palaeosols contain important physical, chemical, microfloral, and pollen data that often provide the only evidence for palaeoenvironmental conditions during the time of their formation. Investigations over the last decade on Mount Kenya have answered some palaeoenvironmental questions that, in turn, have led to the definition of new problems.

INTRODUCTION

The Quaternary palaeoenvironmental record on Mount Kenya is long and complex. The glacial sequence alone is considered to cover most of the Quaternary period, extending into the Matuyama Chron (Mahaney 1983, 1984a). Buried land surfaces with palaeosols formed in them abound on Mount Kenya, and some of these document times of geologic stability in the pre-Pleniglacial period (Mahaney 1981, 1982, 1983, 1984b and c, 1986). Microflora distributions in these palaeosols provide a method for calculating the amount of biogeochemical contamination which might affect the validity of radiocarbon dates (Boyer & Mahaney 1985, Mahaney & Boyer 1986). Detailed mineralogical analyses of the sand fractions in older palaeosols assist in differentiating palaeopedological alteration from detrital effects

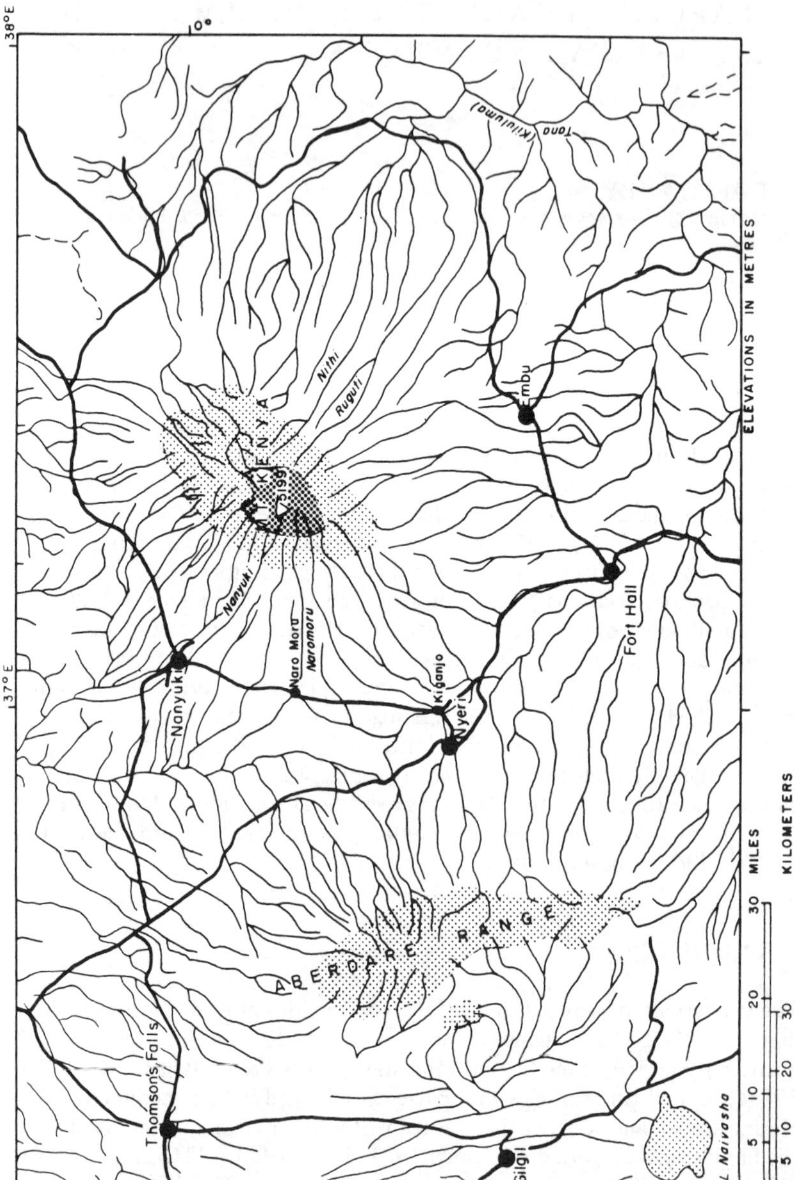
38°E
0°
37°E
Tana (Kiluluma)
Nithi
Ruguti
MT. KENYA
5199
Nanyuki
Naro Moru
Naromoru
Nanyuki
Kiganjo
Nyeri
Embu
Fort Hall
ABERDARE RANGE
Thomson's Falls
Gilgil
L. Naivasha
ELEVATIONS IN METRES
MILES
KILOMETERS
0 5 10 20 30
0 5 10 20 30

providing important palaeoenvironmental information (Vortisch & Mahaney 1986). Pollen sequences correlated with landform chronologies (Mahaney 1985) provide new information useful in understanding the evolution of the Afro-alpine ecosystem from the end of the late glacial to the present. This chapter reviews advances made over the last ten years during the operation of the Mount Kenya Quaternary History Project.

FIELD AREA

Mount Kenya (5 199 m), a dome-shaped volcanic pile (Fig. 1), consists of rocks of basaltic, rhomb prophyritic, phonolitic, and trachytic composition, surrounding a volcanic plug of nepheline syenite and phonolite that is of late Pliocene to early Pleistocene age (Baker 1967, Mahaney 1979). It forms one of the principal high mountains in East Africa and most drainages above 3 000 m are covered with sediments deposited by Quaternary glaciers. Most of the postglacial deposits are of fluvial or mass wasting origin. There are ten active glaciers existing today above 4 200 m, which have contributed sufficient stream load and discharge to emplace large valley train deposits, such as in the Teleki Valley.

Climate

The climate of Mount Kenya is dominated by the Tropical Easterly Wind Belt (Pedgley 1966, Thompson 1966). Rainfall comes with the monsoon: dry conditions prevail from December to March when northeasterly winds flow from the northern desert areas; wet conditions occur from July to October when south-easterly winds bring moisture-laden air masses from the Indian Ocean. Daily weather patterns bring nocturnal katabatic (downhill) and daytime anabatic (uphill) winds, which tend to produce early morning dew and frost and late afternoon thunderstorms. Precipitation at 3 000 m is estimated at 1 525 mm (60 in.), whereas mean annual temperature is 7.2° C, mean maximum is 16.2 and mean minimum is 1.7° C.

Vegetation

Below the heather belt at timberline (between 3 200 and 3 300 m) the Hagenia Woodland merges with the upper Bamboo Forest at 3 000 m. Above the ericaceous belt at timberline, the lower Afroalpine belt dominated by *Senecio brassica* extends upward to about 3 700 m where it merges with the upper Afroalpine belt characterized by *Senecio keniodendron.* The upper Afroalpine belt merges with vegetation of the nival belt in the forelands of several cirques at about 4 500 m (Coe 1967, Hedberg 1964) (Fig. 1).

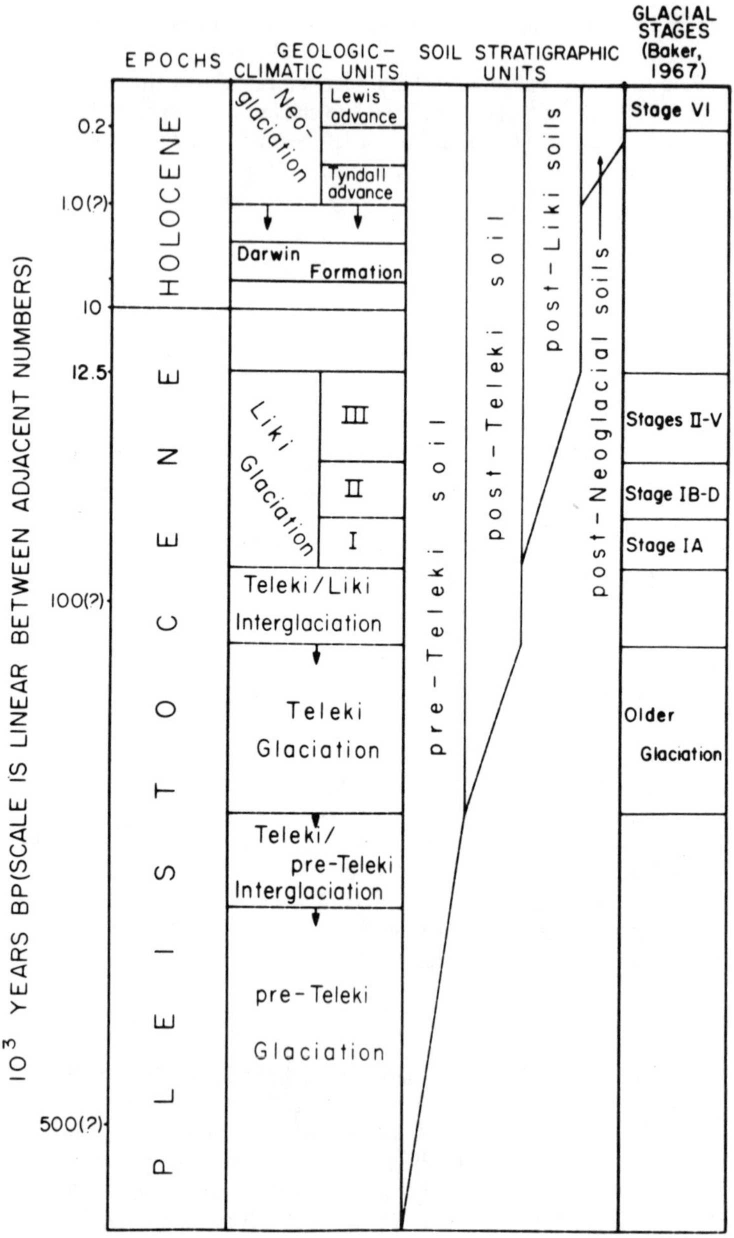

Figure 2. Glacial and postglacial chronology of Mount Kenya. Arrows denote probable direction of future boundary shifts.

RESULTS AND DISCUSSION

Landform record

The most extensive record of palaeoenvironmental change is contained in the landform record. Glacial deposits, recording episodic glaciation, are found as low as 2 900 m in most drainages on Mount Kenya (Mahaney 1972, 1979, 1986). Relative and absolute age determinations of these deposits allow the recognition and subdivision of three principal Pleistocene units in the succession, represented from oldest to youngest, as pre-Teleki, Teleki and Liki glaciations (Fig. 2). No Early or Middle Holocene glaciations occurred, but Neoglaciation starting approximately 1 000 years BP resulted in the readvance of several cirque glaciers, which may have completely disappeared during the preceding postglacial period. The oldest and youngest units in any valley are only rarely exposed for distances greater than 1 to 1.5 km; the oldest units always occur near the lower limit of glaciation, and the youngest in the valley heads. In a few instances multiple glaciation has produced tills in a 'stacked' stratigraphic sequence, complete with intercalated buried palaeosols (Mahaney 1981, 1983, 1984a and b, Boyer & Mahaney 1985). In other sequences, tills are associated with lacustrine and aeolian sediments suggesting that, while environmental influences change over time spans of 10^3 to 10^5 years or less, the intensity of change is sometimes extreme.

An important aspect of the glacial record on Mount Kenya is that it is quite similar to sequences in the Rocky Mountains (Mahaney 1982, Mahaney et al. 1984a and b) and in the Alps. Aside from obvious differences such as equilibrium line altitudes, floristic characteristics in the overlying vegetation cover, and weathering expression in older palaeosols, the valley-to-valley sequences on Mount Kenya are strikingly similar to the Western Alps and the Rocky Mountains. In all these areas complex sequences of end moraines and outwash are found near the lower limit of glaciation, and in the higher valleys only ground moraine predominates, suggesting that withdrawal of late Pleistocene ice was a rapid event, and without stillstand, and readvances in most areas. On Mount Kenya, ice recession from its fullest extent at 3 100 m ($\sim$ 15 000 ^{14}C years BP during the early to late Liki glaciation), to $\sim$ 4 000 m (12 500 ^{14}C years BP) in the high valleys, occurred within 2 500 years (Fig. 2). Following late glacial stillstand in most valleys, ice recession began in full, and ice probably disappeared or remained as small cirque glaciers during most of the Holocene.

In the Neoglacial, starting $\sim$ 1 000 ^{14}C years BP, ice reappeared or resurged in the form of small cirque glaciers in the Central Peaks Area. Equilibrium line altitudes probably reached 4 400 m and some glaciers began to re-enter wide U-shaped valleys originally formed by Pleistocene ice. While this early episode of Neoglaciation is documented only in Tyndall and Lewis cirques (Mahaney 1985), it was followed by the Lewis advance (late Neoglacial) in all cirques when ice once again readvanced with similar amplitudes, ending by 1 900 AD (Gregory 1900). This evidence suggests that Neoglaciation on Mount Kenya is not over, and while equilibrium line altitudes are now high (~ 4 700 m), they will migrate down valley some time in the future.

Buried landscapes and palaeosols

Buried landscapes bearing palaeosols formed in them, abound on Mount Kenya (Vortisch & Mahaney 1986). Buried palaeosols, and the landscapes they formed in, have been identified in nearly every drainage (Mahaney 1979, 1981, 1982, 1983, 1984a, b and c, 1985, 1986). My experience in other alpine areas (e.g. the Western Alps and the Rocky Mountains; see e.g. Mahaney et al. (1984b) and Mahaney & Havorson (1986b)) is that buried palaeosols are rare, and only occasionally preserved. On Mount Kenya frequent mass wasting and fluvial episodes, and a succession of glaciations diminishing in extent and intensity towards the upper Quaternary (Mahaney & Halvorson 1986a), left many surfaces buried by slope deposits and tills in which new palaeosols formed over varying lengths of time.

Buried palaeosols yield important physical, chemical, stratigraphic, microflora, and mineralogical data useful in assessing palaeoenvironmental changes. While clay mineral, variations are not as useful on Mount Kenya, as they are in the Rocky Mountains, for eliciting information on former environments (Mahaney 1984, Mahaney et al. 1984a and b, Mahaney & Halvorson 1986), there are weathering sequences of volcanic glass halloysite (Fig. 3), and vitric tuffs → gibbsite that appear to be time and climate dependent (Vortisch & Mahaney 1986).

Preliminary data available for some older palaeosols suggest that tills on Mount Kenya contain a high percentage of ferromagnetic minerals (mainly magnetite and amphibole (Vortisch & Mahaney 1986) (Fig. 4), as well as glacially-crushed quartz (Fig. 5). Aeolian materials in older palaeosols consist of abundant quartz glass, and ash (Figs. 5 and 6), the former resulting from long-distance transport, and the latter restricted to older (~ 100 000 years BP) sediments. The presence of so many diverse sediments on Mount Kenya proves that palaeoenvironmental changes have long and complex histories.

Palaeosol microflora

Microflora, consisting of bacteria and fungi in palaeosols on Mount Kenya, provide important information on palaeoenvironmental influences and geobiochemical contamination as it affects radiocarbon dates in buried horizons. The data available at present (Boyer & Mahaney 1985, Mahaney & Boyer 1986) suggest that three patterns of depth distribution are common on Mount Kenya. In most cases microflora numbers undergo sequential decline with depth in palaeosols away from the zone of maximum biological activity. That other distributions are possible is apparent from bimodal

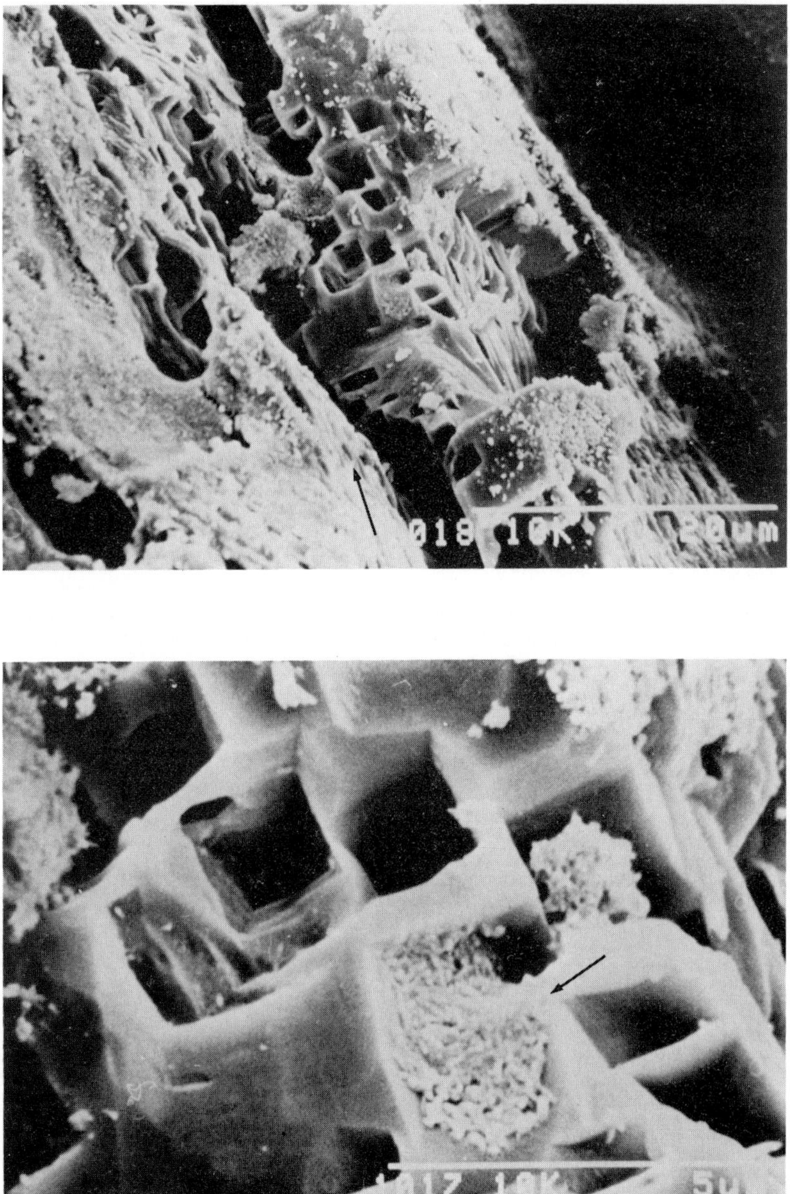

Figure 3. Halloysite/metahalloysite (arrows) formed on etched feldspar surfaces in the pre-Teleki palaeosol (site TV 23 in Teleki Valley): A for wide view, B for enlargement. See Boyer & Mahaney (1985) for location.

Figure 4. Magnetite (M) and glacially-crushed quartz (Qu) in a relatively unweathered state in the pre-Teleki palaeosol (till unit, site TV 23) in Teleki Valley.

Figure 5. SEM micrograph of the fine and very fine sand fractions: in post-Teleki loess (site TV 23, Teleki Valley). This sample consists mainly of volcanic ash (a) of different weathering stages and feldspar (f) (centre).

Figure 6. Aeolian quartz (Qu) and plagioclase (P) in the loess unit of the pre-Teleki soil, site TV 23, Teleki Valley.

and disjunct patterns describing distributions that either decrease with depth, followed by very low numbers, followed again by increases, or distributions that decline with depth to zero and then increase again (Boyer & Mahaney 1985). A threshold value of $13 \times 10^3 g^{-1}$ (for bacteria) and $5 \times 10^2 g^{-1}$ (for fungi) appears to be required in order to produce enough biogeochemical contamination (e.g. from microflora) to affect radiocarbon dates; higher counts produce younger radiocarbon dates, as the difference between true age and ^{14}C age widens (Mahaney & Boyer 1986).

While microflora distributions in palaeosols are not a direct measure of palaeoenvironmental change, they are extremely useful in assessing geobiochemical contamination of ^{14}C dated palaeosols which might affect age controls for buried surfaces, and hence, palaeoenvironmental interpretations.

Postglacial pollen and landform chronologies

The detailed pollen record and inferred environmental changes described by Coetzee (1967) are similar in broad outline to the landform-soil record described here (Fig. 7). Pollen zone Ui is characterized by the first amelioration following the Pleniglacial Period of expanded ice cover on the mountain. During this period, timberline approached 2 400 m and receding glaciers in the major valleys halted at ~ 4 000 m to build large end moraine systems. Warmer periods ensued after 12 500 ^{14}C years BP, culminating in a dense timberline (zone V), and disappearance and/or retreat of the main valley glaciers. From the landform evidence in the Afro-alpine belt, the period between 12 500 and 6 700 years appears to be either variable wet and dry (?) or dry (?), which conflicts somewhat with the pollen record. At least, it appears there was little slope instability and streams were apparently at or near equilibrium, neither aggrading nor degrading their channels.

Climate during the mid Holocene appears to have been wet, bringing first a humid montane forest with warmer temperatures, and later, a lower tree line at 2 400 m with cooler temperatures. In the Afroalpine zone this period is characterized by widespread slope instability, alluvial fan deposition, and minor cycles of stream aggradation and degradation. The widespread slope instability that led to emplacement of the Darwin Formation (Mahaney 1985) coincided with wetter climatic conditions in the Afroalpine and maximum humid forest development and higher temperatures at 2 400 m, supporting the pollen evidence. Pollen zone Y persisted to 3 400 BP (Coetzee, pers. comm. 1986), the upper portion of which is marked by isolated colluvial events (Fig. 2). Neoglaciation which began ~ 1 000 years BP is the culmination of cooler temperatures and drier conditions that began earlier with pollen zone Y.

Differences between the landform-soil evidence and the pollen chronology may be due to variations in climatic changes between Sacred Lake and the Afroalpine zone, or to different timing in the two areas for events of similar magnitude.

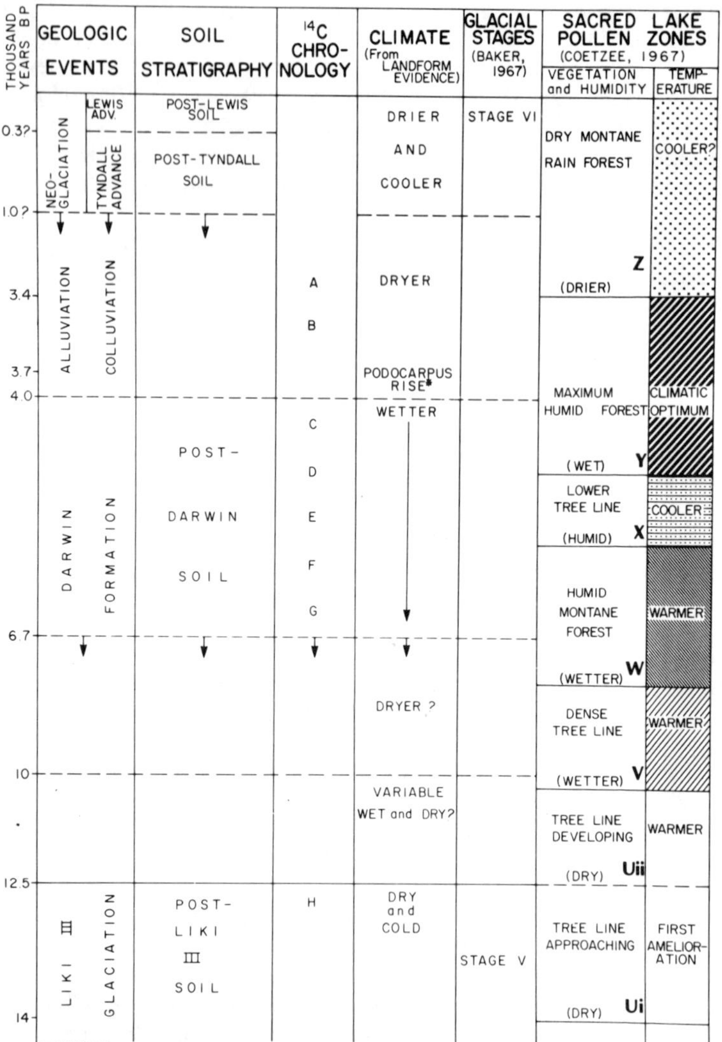

FOOTNOTES

A. Ab horizon (site TV4), 1940±120 yrs. B.P. (Gak-8273)
B. Ab horizon (site LN9), 3430±100 yrs. B.P. (BGS-756)
C. Ab horizon (site TV33), 4310±180 yrs. B.P. (Gak-8214)
D. Plant fragments in alluvial fan (site ONT 3), 4375±100 yrs. B.P. (BGS-753)
E. Ab horizon (site TV33), 4480±160 yrs. B.P. (Gak-8215)
F. Organic-rich sediment at 140 cm depth in alluvial fan (site LN18), 5280±110 yrs. B.P. (BGS-757)
G. Cb horizon with plant fragments in alluvial fan (site TV28), 6670±170 yrs. B.P. (Gak-8276)
H. Bog-bottom material (site LN4), 12590±300 yrs. B.P. (Gak-8275)
* Beginning 3720 yrs. B.P. (Perrott, 1982)

Figure 7. Correlation diagram for late glacial and postglacial deposits on Mount Kenya. Unit boundaries are based on maximum and minimum dates for glacial and non-glacial sediments. Dashes denote boundaries for which there is no radiometric control. Arrows indicate directions in which boundaries are likely to shift as additional dates become available.

Problems and forecasts

As more data are produced from laboratory analyses new problems involving palaeoenvironmental interpretations and reconstructions will probably emerge. More refined stratigraphic work on Mount Kenya may reveal that pre-Teleki drift (Fig. 2) could be subdivided into two or more tills (Mahaney 1986). Indeed, it is probable that the pre-Teleki drifts exposed on the surface, in various drainages, may not derive from a single event. The great thickness of some pre-Teleki moraines (e.g. ~ 100 m in Gorges Valley) suggests they may represent two or more glaciations (Barendregt & Mahaney 1986).

Palaeosols within the pre-Teleki tills will likely yield considerable data when, and if, they are eventually located, described and analyzed. More importantly, pollen in these palaeosols, as in other younger buried palaeosols of Teleki, Liki and postglacial age, will probably yield important data on interglacial, interstadial and postglacial palaeoclimate. Stable isotopes (especially deuterium and ^{18}O) in many Mount Kenya palaeosols may eventually yield important palaeoclimatic information pertaining to precipitation and evapotranspiration.

Mount Kenya is a special case because it has great antiquity (late Pliocene/early Quaternary age) and contains a long well-preserved Quaternary record. Most other mountain areas in East Africa are younger (e.g. Kilimanjaro), but even there at least one pre-Pleniglacial event is preserved in the geologic succession. In other areas, such as in the Ruwenzori, traces of multiple glaciation have been documented by Osmaston (1965), although apparently the detailed succession of deposits, so well preserved on Mount Kenya, was removed in the Ugandan mountains by erosion during the Pleniglacial.

CONCLUSION

Palaeoenvironmental reconstructions on Mount Kenya are beginning now to provide new insights into the magnitude and intensity of change over the whole Quaternary Period. The landform record alone shows that a succession of volcanic, glacial, lacustrine, aeolian, mass wasting, and fluvial processes have shaped the surface of Mount Kenya; reconstruction of these geologic events relies on analysis of sediments using methods employed in a number of diverse fields.

ACKNOWLEDGEMENTS

This research (Mount Kenya Quaternary History Project) is supported in part by grants from the National Geographic Society (Nos. 1576.76, 2306.81, 2672.83), the Natural Sciences and Engineering Research Council of Canada (1983-88), and minor research grants from York University. I am particularly indebted to Linda M.Mahaney for assistance in the field during 1976 and 1983/84, to students in the York University Mountain Geomorphology Field Schools (1976, 1983), and to Larry Gowland (1976 and 1983). Various Quaternarists in laboratories in several countries have assisted in analyzing samples. In particular, Dr Walter Vortisch (University of Marburg, FRG), Mr Tom Edwards and Dr Peter Fritz (University of Waterloo), Dr René W. Barendregt (University of Lethbridge, Alberta), Dr Emile Roche (Museum for Middle Africa, Tervuren, Belgium), Dr Ed Derbyshire (Leicester University, UK), Dr Dave Huntley and Dr Glen Berger (Simon Fraser University, BC), Dr Don Halvorson (University of North Dakota), Dr Nat Rutter (University of Alberta), Dr M.G.Boyer (York University, Toronto), Dr Murray Tamers (Beta Analytic, Florida), and Professors Eduard van Zinderen Bakker and J.A.Coetzee have contributed significantly to the research discussed in this chapter. Professor J.A.Coetzee critically reviewed a draft of this text.

REFERENCES

Baker, B.H. 1967. *Geology of the Mount Kenya area.* Geological Survey of Kenya, Rep. 79, 78 pp.

Barendregt, R.W. & W.C.Mahaney 1986. Reconnaissance paleomagnetic determinations in Liki and pre-Liki sediments on Mount Kenya, East Africa. In W.C.Mahaney (ed.), *Quaternary research on East African mountains.* Rotterdam: Balkema (in press).

Boyer, M.G. & W.C.Mahaney 1985. Some microbiological aspects of surface and buried soils on Mount Kenya, East Africa. *Plant and Soil* 86: 309-319.

Coe, J.J. 1967. *The ecology of the alpine zone of Mount Kenya.* The Hague: Junk, 136 pp.

Coetzee, J.A. 1967. Pollen analytical studies in East and Southern Africa. *Palaeoecol. of Africa* 3, 146 pp.

Gregory, J.W. 1900. The geology of Mount Kenya. *Quarterly J. Geol. Soc.* 56: 205-222.

Hedberg, O. 1964. *Features of Afroalpine plant ecology.* Uppsala: Almquist & Wiksells, 144 pp.

Mahaney, W.C. 1972. Late Quaternary history of the Mount Kenya Afroalpine area, East Africa. *Palaeoecol. of Africa* 6: 139-142.

Mahaney, W.C. 1979. Reconnaissance Quaternary stratigraphy of Mount Kenya. *Palaeoecol. of Africa* 11: 163-170.

Mahaney, W.C. 1980. Late Quaternary rock glaciers, Mount Kenya, East Africa. *J. Glaciol.* 25: 492-497.

Mahaney, W.C. 1981. Paleoclimate reconstructed from paleosols: evidence from the Rocky Mountains and East Africa. In W.C.Mahaney (ed.), *Quaternary Paleoclimate*: 227-248. Norwich, UK: Geoabstracts.

Mahaney, W.C. 1982. Chronology of glacial deposits on Mount Kenya, East Africa: description of type sections. *Palaeoecol. of Africa* 14: 25-43.

Mahaney, W.C. 1983. Chronology of Quaternary glacial deposits on Mount Kenya, East Africa. *Geol. Assoc. Canada Abstracts* 256.

Mahaney, W.C. 1984a. Soils of the Mount Kenya area, their formation, ecological and agricultural significance (with soil map, scale 1:125000): Discussion. *Mountain Research and Development* 4 (3): 284-285.

Mahaney, W.C. 1984b. Late glacial and postglacial chronology of Mount Kenya, East Africa. *Palaeoecol. of Africa* 16: 327-341.

Mahaney, W.C. 1984c. Loess in Pleistocene soils on Mount Kenya, East Africa. *Loess Letter Supplement* 6. Univ. of Waterloo, Dept. Earth Science, 20 pp.

Mahaney, W.C. 1985. Late glacial and Holocene paleoclimate of Mount Kenya, East Africa. *Zeitschr. Gletscherkunde und Glazialgeol.* 21: 203-211.

Mahaney, W.C. 1986. Environmental impact on the western flank of Mount Kenya, East Africa. *Mountain Research and Development* (in press).

Mahaney, W.C., M.G.Boyer, D.L.Halvorson & K.Sanmugadas 1984a. Glacial chronology of the Rocky Mountains: some problems from western Wyoming. In W.C.Mahaney (ed.), *Correlation of Quaternary chronologies*: 225-242. Norwich, UK: GeoBooks.

Mahaney, W.C., J.Piegat, D.L.Halvorson & K.Sanmugadas 1984b. Evaluation of dating methods used to assign ages in the Wind River and Teton ranges, western Wyoming. In W.C.Mahaney (ed.), *Quaternary dating methods*: 355-374. Amsterdam: Elsevier.

Mahaney, W.C., M.G.Boyer & N.W.Rutter 1985. Evaluation of amino acid composition as a geochronometer in buried soils on Mount Kenya, East Africa. In *Seventh Quaternary Symp. Abstracts and Field Guide.* Leithbridge.

Mahaney, W.C. & M.G.Boyer 1986. Microflora distributions in paleosols: a method for calculating the validity of radiocarbon dated surfaces. *Soil Sci.* (in press).

Mahaney, W.C. & D.L.Halvorson 1986a. Quaternary stratigraphy in Gorges Valley, Mount Kenya, East Africa. In W.C.Mahaney (ed.), *Quaternary research on East African mountains.* (in press).

Mahaney, W.C. & D.L.Halvorson 1986b. Rates of mineral weathering in Rocky Mountain glacial sequences: evidence from western Wyoming. In S.M.Colman (ed.), *Rates of chemical weathering of rocks and minerals.* New York: Academic Press (in press).

Osmaston, H. 1965. *The past and present climate and vegetation of Ruwenzori and its neighbourhood.* D.Phil. thesis, Oxford Univ., 300 pp.

Oyama, M. & H.Takehara 1970. *Standard Soil Color Charts.* Japan Research Council for Agriculture, Forestry & Fisheries.

Pedgley, D.E. 1966. Rainfall of Mount Kenya. *Weather* 21: 176-188.

Thompson, B.W. 1966. The mean annual rainfall of Mount Kenya. *Weather* 21: 48-49.

Vortisch, W. & W.C.Mahaney 1986. Lithology and weathering in older paleosols on Mount Kenya, East Africa. *J. Geol.* (Submitted).

ENVIRONNEMENTS BOTANIQUE ET CLIMATIQUE DES HOMINIDES DE L'EST TURKANA, KENYA, ENTRE 2,0 ET 1,4 MILLIONS D'ANNEES: APPORT DE LA PALYNOLOGIE

A.VINCENS
Laboratoire de Géologie du Quaternaire, CNRS, Marseille, France

SUMMARY

In this chapter, all palynological data today obtained from Plio-Pleistocene deposits at East Turkana, Kenya, are synthetised. Twenty-seven pollen spectra with numerous and diversified pollen types have been extracted from sediments of the Koobi Fora Formation dated between 2.0 and 1.4 MY. Each pollen spectrum, well located in the general stratigraphy is situated in the sedimentary and palaeogeographical background. Pollen diagrams show variations of major pollen components. Its interpretation allows reconstruction of local and regional past vegetational and climatic environments. Between 2.0 and 1.4 MY, pollen data show that the vegetation surrounding the lake was of a sub-arid type and could be reconstructed as an *Acacia-Commiphora* steppe, more wooded along the streams, with the same composition as the modern one. An extension of grassland happened around 1.9 MY, just before the KBS Tuff was deposited, in relation to an arid episode, which is also shown in the regional vegetation by an important regression of the montane forests.

INTRODUCTION

Depuis 1968, le bassin du lac Turkana (ex. lac Rudolf) a été l'objet de très importantes recherches paléoanthropologiques, archéologiques, paléontologiques et géologiques (Coppens et al. 1976, Leakey & Leakey 1978, Brown et al. 1985). La découverte en 1976 d'un niveau renfermant des pollens fossiles (Bonnefille 1976) dans les dépôts fluvio-lacustres plio-pléistocènes affleurant en rive nord-est du lac (3 à 4°N, 36°E) a incité à poursuivre les recherches palynologiques dans cette région. Celles-ci s'intègrent dans le programme international 'Koobi Fora Research Project' centré sur la reconstitution du paléoenvironnement des Hominidés fossiles, et ont abouti à la découverte de 27 spectres polliniques fossiles dans les dépôts supérieurs de la Formation de Koobi Fora (Vincens 1982).

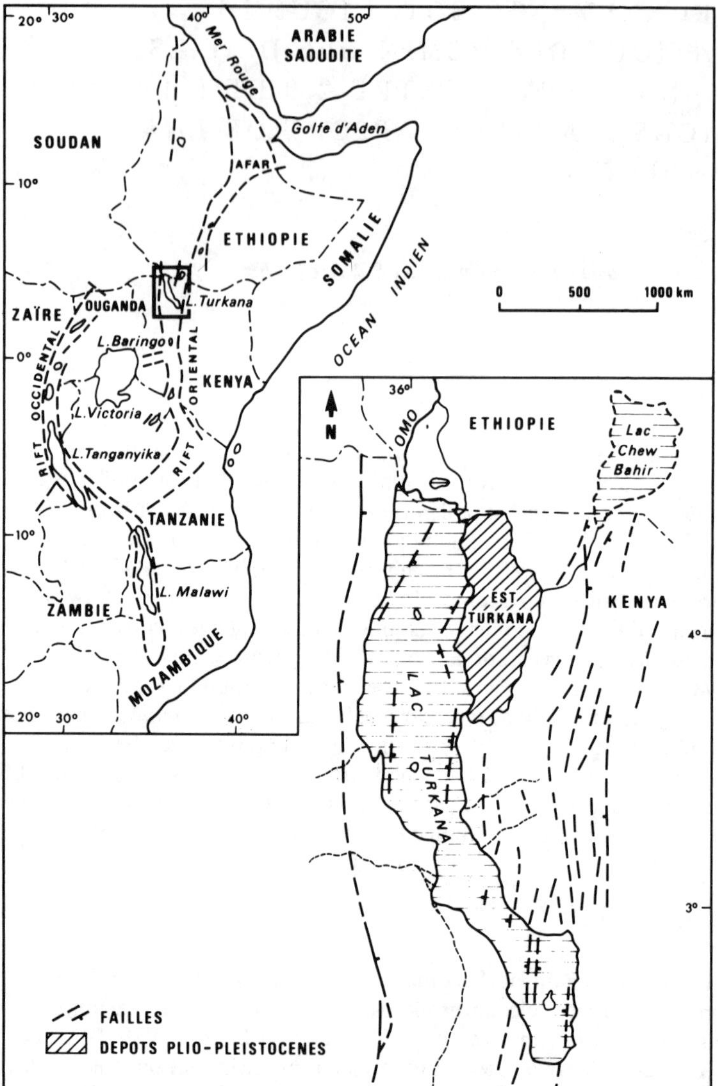

Figure 1. Localisation des dépôts plio-pléistocènes d'Est Turkana dans le Rift Est Africain

Associés aux données fournies par d'autres disciplines, comme la géologie, la sédimentologie et la paléontologie, les résultats palynologiques obtenus apportent un complément d'informations précieuses, en particulier en ce qui concerne les reconstitutions des paysages végétaux dans lesquels sont apparus et ont évolué les Hominidés de l'Est Turkana, mais également la climatologie.

LE CONTEXTE ACTUEL

Le lac Turkana est situé en zone intertropicale (entre 2°24′ et 4°35′ de latitude N, 35°51′ et 36°43′ de longitude E), à une altitude de 375 m, dans une région subdésertique de pluviosité moyenne annuelle inférieure à 400 mm, à deux saisons sèches bien marquées. La végétation locale est de type steppique somalia-masai: steppe subdésertique, steppe boisée à *Acacia, Commiphora* et nombreuses succulentes. Celle-ci passe graduellement en altitude dans le bassin à des savanes et forêts claires soudaniennes puis, au-dessus de 1 500 m, à des groupements afromontagnards: fourrés et forêts sempervirentes, communautés afroalpines (Keay 1959, White 1983).

LE CONTEXTE GEOLOGIQUE

Les dépôts sédimentaires plio-pléistocènes de l'Est Turkana affleurent sur plus de 1 000 km^2 et ont une épaisseur d'environ 450 m (Fig. 1). Ils sont d'origine lacustre, fluviatile ou deltaïque (Behrensmeyer 1970, Bowen 1974, Findlater 1976).

Du fait de la discontinuité des affleurements et des changements brusques de faciès dans l'espace, un certain nombre d'*Area* de collectes paléontologiques ont dû être définies. Mais les corrélations stratigraphiques entre les dépôts de chacune d'elles ont été difficiles à faire. Les premières tentatives de Behrensmeyer (1970), Bowen & Vondra (1973) et Findlater (1976, 1978), en particulier pour les dépôts antérieurs au tuf KBS, ont fait l'objet de révisions ultérieures. Sur la base des travaux biostratigraphiques concernant les Suidés (White & Harris 1977), les Mollusques (Williamson 1982) et plus récemment de travaux géochimiques et isotopiques (Brown & Cerling 1982, Brown & Feibel 1985), l'existence d'une vingtaine de tufs ou complexes tuffacés a été démontrée. Une partie d'entre eux sont actuellement parfaitement datés (McDougall 1981, McDougall et al. 1980, 1985). De plus, les corrélations établies entre la séquence géologique de l'Est Turkana et celle de l'Omo, montrent que les dépôts des sédiments se situent entre 4,1 et 1,4 MA, mais qu'il existe en particulier une lacune sédimentaire importante entre 2,4 et 2,0 MA (Cerling & Brown 1982, Brown & Feibel 1985).

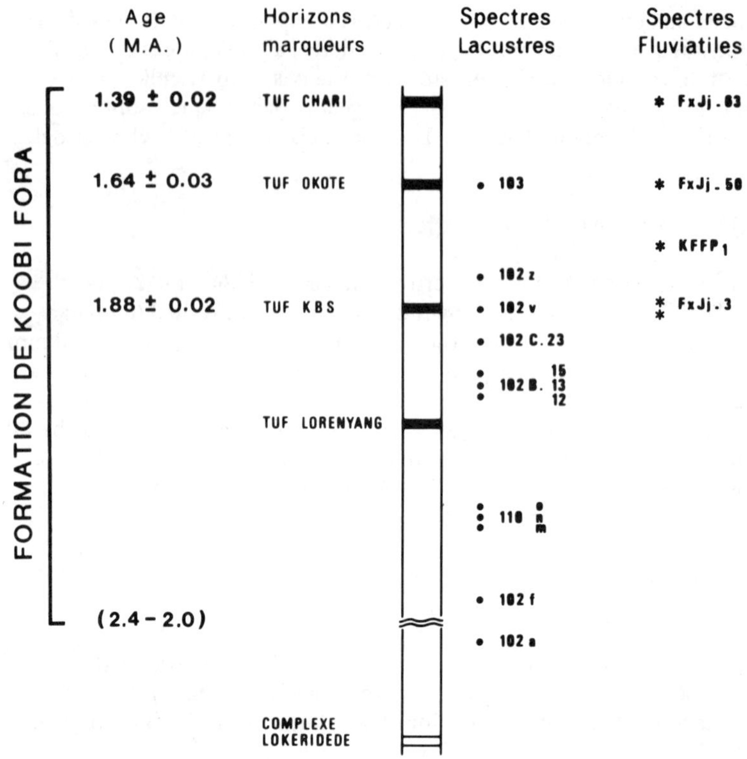

Figure 2. Succession chronologique des spectres polliniques des dépôts plio-pléistocènes d'Est Turkana, Formation de Koobi Fora

LES SPECTRES POLLINIQUES

Un échantillonnage de la séquence géologique de l'Est Turkana (322 échantillons) a permis de mettre en évidence des pollens fossiles dans 27 niveaux. Ceux-ci correspondent à des dépôts de nature sédimentologique variée: sables grossiers, grès, dalles à gastéropodes, silts, argiles.

Parmi les 27 spectres, 20 ont été extraits des affleurements lacustres des *Area* 110, 103, 102 et 116; 7 autres correspondent à des dépôts de plaine d'inondation associés aux sites archéologiques FxJj-3 (*Area* 105), FxJj-50 (*Area* 131) ou à des restes d'Hominidés, spectres KFFP 1 (*Area* 12) (Fig. 2).

Du point de vue stratigraphique, tous les spectres, à l'exception de ceux de l'*Area* 116, sont parfaitement localisés dans le temps. Replacés avec précision par rapport à des niveaux marqueurs tels que tufs, niveaux à gastéropodes, surface d'érosion, leur succession chronologique a été établie sur la base des travaux géologiques de Brown & Cerling (1982). Ils appartiennent

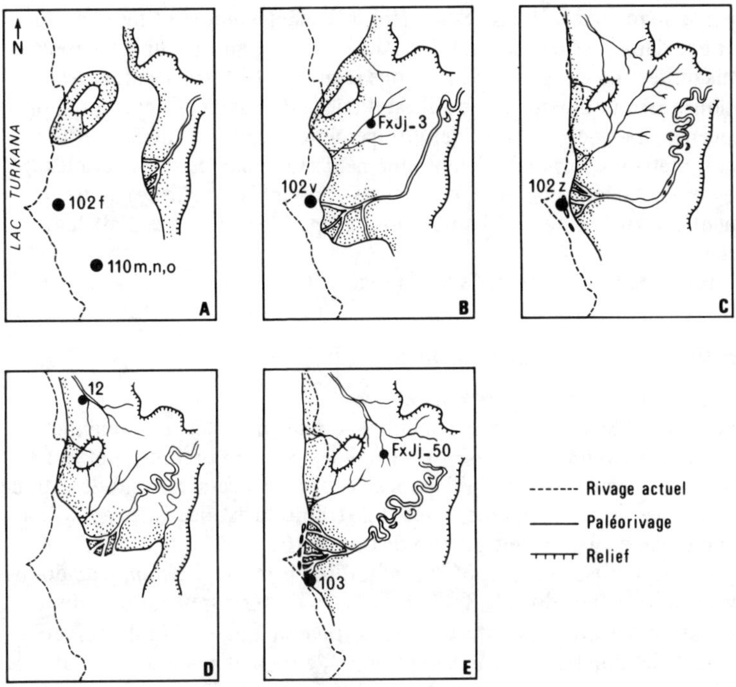

Figure 3. Situation paléogéographique des spectres polliniques plio-pléistocènes d'Est Turkana d'après les reconstitutions de I.C.Findlater (1976):
A – maximum de transgression anté-KBS
B – Dépôt du tuf KBS
C – Maximum d'extension de la surface d'érosion post-KBS
D – Maximum de transgression post-surface d'érosion
E – Dépôt du tuf Okote

tous à la Formation de Koobi Fora et se situent entre 2,0 et 1,4 MA (Fig. 2).

De plus, à partir des travaux de Findlater (1976), la plupart des spectres ont pu être replacés dans un contexte paléogéographique bien défini (Fig. 3).

LES RESULTATS ET INTERPRETATIONS

Un total de 122 taxons ont été différenciés (Vincèns 1982, Bonnefille et al. 1985) et tous les spectres montrent un mélange de taxons que nous avons regroupés par affinités floristiques.

Le premier ensemble groupe 65 taxons à affinités floristiques somalia-masai et soudaniennes. Ils correspondent aux mêmes genres et familles que l'on trouve dans la végétation subdésertique actuelle autour du lac Turkana.

Dans un second ensemble, nous avons regroupé 27 taxons à affinités floristiques afromontagnardes, qu'il est difficile d'imaginer, même au Plio-Pléistocène, en association dans les groupements steppiques de la bordure du lac. En tenant compte de l'environnement établi par les autres études (géologiques, paléontologiques, etc.), il est vraisemblable que ces genres constituaient des groupements à l'amont du bassin, dans des zones d'altitude plus élevée.

Les résultats sont synthétisés sur la Figure 4.

Reconstitution du paléoenvironnement végétal

La bordure est du paléolac Turkana

Entre 2,0 et 1,4 MA, les taxons à affinités somalia-masai et soudaniennes représentant toujours plus de 80% du total des pollens comptés, mettent en évidence la permanence d'un milieu local steppique à caractère subdésertique. Plusieurs types de groupements, dont la distribution est liée à des conditions édaphiques particulières, ont pu être distingués. Ce sont:

– Un tapis herbacé développé sur la bordure littorale, à *Gramineae* dominantes jusqu'au dépôt du tuff KBS (1,88 MA). Postérieurement, on observe une extension des *Chenopodiaceae/Amaranthaceae* qui, du fait du caractère halophile de la plupart des genres déterminés, est à mettre en relation avec la régression lacustre anté-KBS (Findlater 1976) et le développement de sols salés lié à une augmentation de l'alcalinité et de la salinité des eaux du paléolac par ailleurs démontré par des études géochimiques (Cerling 1979) et celle des ostracodes (Cohen 1981). La présence de sols salés est également montrée dans les spectres par l'abondance de pollens de *Salvadoraceae* entre 1,88 et 1,64 MA, arbres à caractère halphile, et dont les peuplements devaient être plus particulièrement développés sur les plaines d'inondation deltaïques comme en témoignent les spectres du site archéologique FxJj-3 et du site à Hominidés de l'*Area* 12.

– En arrière du littoral, l'existence d'une végétation plus arborée est mise en évidence dans presque tous les spectres fossiles. Les taxons les plus abondants, typiquement steppiques, sont *Acacia, Commiphora, Capparidaceae, Anacardiaceae* du genre *Lannea.* Ce sont les mêmes qui dominent actuellement dans la végétation à l'Est du lac Turkana, mais il est difficile d'indiquer d'après les seuls spectres polliniques la physionomie de la végétation, en particulier la densité du couvert arboré. Seuls les spectres situés autour de 1,9 MA, entre le dépôt du tuf Lorenyang et celui du tuf KBS, sont caractérisés par l'absence de taxons arborescents et témoignent de l'extension d'espaces herbacés durant cette période.

– Pour la période postérieure à 1,88 MA, tous les spectres fluviatiles démontrent la présence, le long des chenaux et rivières, d'une végétation riveraine arborée. Les pourcentages des pollens arborescents sont variables suivant les

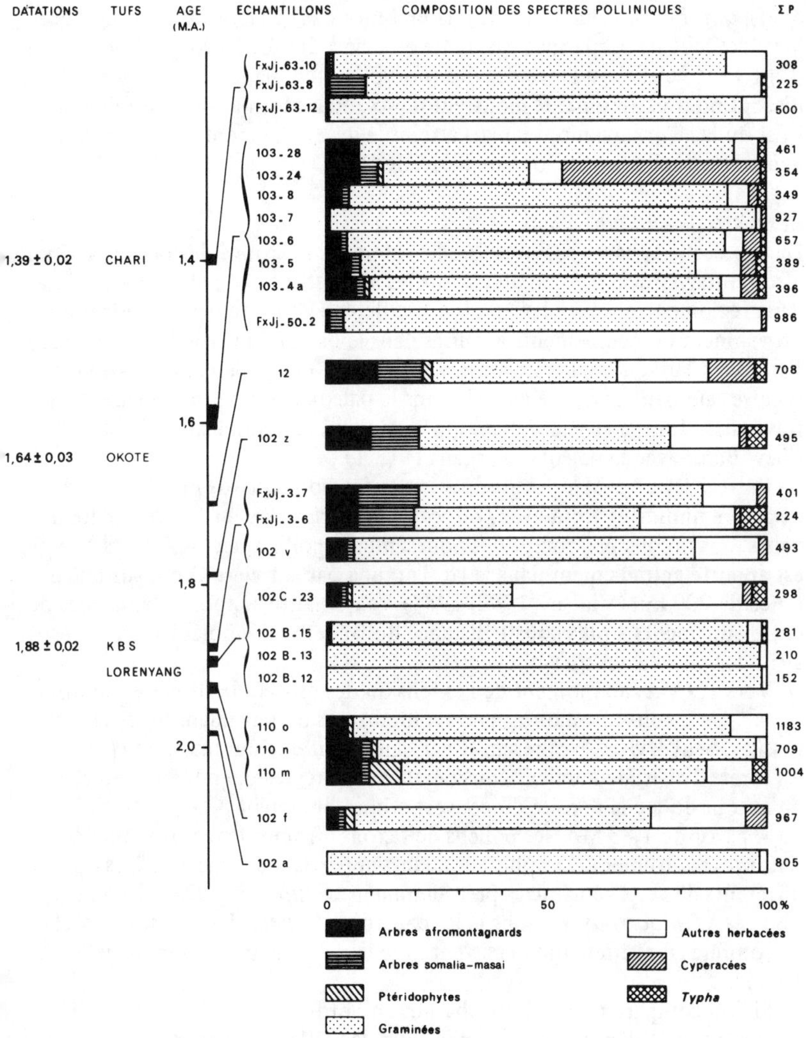

Figure 4. Diagramme pollinique synthétique de la Formation de Koobi Fora, Est Turkana (Kenya)

sites. Ils traduisent des variations de densité du couvert arboré liées soit à des conditions climatiques différentes soit à des conditions édaphiques locales particulières, en relation avec l'importance de la rivière. Dans la composition des spectres fluviatiles trois faits importants sont à souligner. Ce sont: l'absence ou la rareté de taxons mésophiles typiques de forêt riveraine (*Celtis, Ziziphus,*

Syzygium, Ficus, Combretaceae), la présence d'*Hyphaene,* arbre dont l'écologie actuelle ne lui permet pas d'être associé à des formations arborées denses (Carr 1976) et l'abondance des *Gramineae* qui témoigne d'un couvert arboré discontinu. Ces arguments montrent donc que les formations riveraines à l'Est du lac Turkana n'ont jamais atteint le développement des forêts qui existent actuellement le long de la rivière Omo.

Les plateaux sud-éthiopiens

Dans presque tous les spectres fossiles, nous avons trouvé des taxons à affinités afromontagnardes. En s'appuyant sur les reconstitutions paléogéographiques de la région entre 2,0 et 1,4 MA (de Findlater 1976), on peut imaginer qu'ils proviennent de peuplements localisés dans le bassin du Chew Bahir, au Nord-Est du lac Turkana, sur les plateaux éthiopiens. Les pollens trouvés dans les spectres auraient donc été amenés dans le paléolac par les principaux réseaux fluviatiles, dont la rivière Bakate qui mettait en communication le paléolac Chew Bahir avec le paléolac Turkana (Fig. 1).

Entre 2,0 et 1,9 MA, les groupements afromontagnards présentent un caractère himide et froid avec *Podocarpus, Myrica, Ilex* et *Ericaceae* dominants. Ces taxons présentent des pourcentages polliniques supérieurs à ce qui est apporté actuellement dans le lac Turkana par la rivière Omo, sur une distance de 200 km (Vincens 1982, 1984), ce qui laisse supposer l'existence de ces groupements à plus courte distance du lac, et vraisemblablement à des altitudes plus basses qu'actuellement.

Vers 1,9 MA, au moment de l'extension des espaces herbacés en bordure du lac, les taxons afromontagnards sont absents ou rares dans les spectres fossiles. Seul *Myrica* est présent. Ceci signifierait que durant cette période, les forêts d'altitude auraient considérablement régressé, en même temps qu'autour du lac la densité du couvert arboré diminuait.

A partir de 1,88 MA, les pollens de taxons afromontagnards réapparaissent progressivement. Les plus abondants appartiennent à des genres qui sont différents de ceux rencontrés précédemment. *Juniperus* et *Olea* dominent, associés à *Podocarpus.* Ce sont là les constituants essentiels des forêts sèches de conifères d'altitude qui persistent avec la même composition jusqu'à 1,4 MA.

Si l'on compare les résultats obtenus en bordure du paléolac et sur les plateaux sud-éthiopiens, on s'aperçoit que les différents groupements végétaux, steppiques locaux et forestiers régionaux, ont subi entre 2 0 et 1,4 MA, et au même moment, un certain nombre de modifications que nous avons pu mettre en relation avec des fluctuations climatiques.

Mise en évidence des phénomènes climatiques

Une phase aride vers 1,9 MA

Vers 1,9 MA, un changement majeur est mis en évidence dans la séquence palynologique entre le dépôt du tuf Lorenyang et celui du tuf KBS. Il se traduit en bordure du lac par la disparition de la strate arborée et l'extension d'une steppe exclusivement herbacée. Au même moment, sur les plateaux éthiopiens, les forêts régressent considérablement, puis changent de composition; aux peuplements humides succèdent des peuplements secs.

Un tel phénomène enregistré dans un bassin de grande dimension ne peut être expliqué que par une détérioration du climat vers des conditions particulièrement arides. L'existence d'une telle fluctuation vers 1,9 MA est d'ailleurs confirmée par les travaux de Cerling et al. (1977) sur les isotopes de l'oxygène, qui montrent une diminution importante de la pluviosité juste avant le dépôt du tuf KBS, contemporaine de la régression lacustre anté-KBS définiée par Findlater (1976).

Autres fluctuations climatiques
Hormis la phase aride de 1,9 MA, les données palynologiques montrent que la séquence sédimentaire de la Formation de Koobi Fora s'est déposée dans un contexte climatique semi-aride, proche des conditions actuelles. Cependant, si on tient compte des données géologiques en interprétations de Findlater (1976), on s'aperçoit que ce contexte climatique n'a pas été stable. En effet, la séquence sédimentaire est entrecoupée par des surfaces d'érosion majeures ou présente des lacunes de sédimentation. Les plus importantes se situent entre 2,5 et 2,0 MA, entre 1,88 et 1,64 MA (surface d'érosion post-KBS) et postérieurement au dépôt du tuf Chari daté 1,39 MA. Elles ont été interprétées comme des épisodes arides. Seule l'amorce de l'épisode aride le plus récent, postérieur à 1,39 MA, est décelée par la palynologie. Elle se traduit par une régression progressive de la strate arborée locale entre les dépôts des tufs Okote et Chari.

L'ensemble des résultats palynologiques obtenus et leur interprétation en terme de végétation et de climat sont synthétisés dans le Tableau 1.

L'ENVIRONNEMENT DES HOMINIDES DE L'EST TURKANA ENTRE 2,0 ET 1,4 MA

C'est dans les dépôts datés entre 2,0 et 1,4 MA qu'ont été découverts les plus nombreux restes d'Hominidés fossiles ainsi que tous les sites archéologiques de l'Est Turkana (Leakey et al. 1978, Isaac & Harris 1978, Walker & Leakey 1978).

Les informations apportées par la palynologie concernent plus particulièrement l'environnement de trois espèces d'Hominidés parmi les cinq identifiées: *Australopithecus boisei, Homo habilis, H.erectus.* Elles montrent que tous ont vécu entre 2,0 et 1,4 MA au milieu d'une végétation steppique subdésertique, plus ou moins arborée suivant les périodes, et sous des conditions climatiques semi-arides à arides. En ce qui concerne *Australopithecus boisei,* il est vraisemblable que les plus anciens spécimens rapportés à cette espèce sont contemporains ou immédiatement antérieurs à l'extension de la

Tableau 1. Synthèse des résultats palynologiques obtenus dans les dépôts de la Formation de Koobi Fora; paléoenvironnements botaniques et climatiques de la région est du lac Turkana entre 2,0 et 1,4 MA

AGE (M.A.)	HORIZONS MARQUEURS	BORDURE DU LAC — Végétation littorale	Végétation plaine		Conditions climatiques	PLATEAUX SUD-ETHIOPIENS — Végétation		Conditions climatiques	
	Erosion ~				aride				
1.39 / 1.64	CHARI / OKOTE	tapis herbacé graminéen	steppe arborée plus ouverte à *Acacia* + *Commiphora*	VEGETATION A AFFINITES SOMALIA-MASAI ET SOUDANIENNES	semi-aride	groupements secs à *Juniperus*, *Olea*, *Podocarpus* dominants	VEGETATION A AFFINITES AFROMONTAGNARDES	semi-humide	F R O I D
		tapis herbacé graminéen	steppe arborée à *Acacia* + *Commiphora*		semi-aride	groupements secs à *Juniperus*, *Olea*, *Podocarpus* dominants		semi-humide	
	Erosion ~				aride				
1.88	KBS	tapis herbacé abondance d'halophytes (sols salés)	steppe arborée à *Acacia* + *Commiphora*		semi-aride	groupements secs à *Juniperus*, *Olea*, *Podocarpus* dominants		semi-humide	
	(102/TV)	tapis herbacé graminéen	steppe herbacée à *Gramineae*		aride	? (*Myrica*)		?	
	LORENYANG	tapis herbacé graminéen	steppe arborée à *Acacia* + *Commiphora*		semi-aride	groupements humides à *Podocarpus*, *Myrica*, *Ilex*, *Ericaceae* dominants		humide	
2.0	Erosion ≈				aride				

steppe herbacée à *Gramineae,* liée à la phase aride de 1,9 MA. Quant à *Homo erectus,* son apparition vers 1,64 MA semble liée au début de la dernière phase climatique aride mise en évidence au sommet de la Formation de Koobi Fora, au moment où la végétation redevient plus ouverte.

Concernant les industries préhistoriques, il faut noter que les premières trâces de l'activité des Hominidés, l'Industrie du KBS datée 1,88 MA, sont immédiatement postérieures à la phase aride de 1,9 MA et à l'extension des espaces herbacés. L'Industrie du Karari datée 1,64 MA se développe lorsque la végétation recommence à présenter un caractère plus désertique et plus ouvert. Mais plus particulièrement pour les sites FxJj-3, FxJj-50 et FxJj-63, interstratifés dans des dépôts qui ont fourni des pollens, les résultats montrent que ceux-ci, localisés le long de rivières ou de chenaux, correspondent tous à des sites où la végétation est toujours plus dense que sur le reste du territoire.

CONCLUSIONS

Les résultats palynologiques obtenus dans les dépôts fluvio-lacustres plio-pléistocènes de l'Est Turkana témoignent de l'existence en bordure du paléolac d'une végétation steppique subdésertique et de conditions climatiques semi-arides à arides entre 2,0 et 1,4 MA. Il est clair qu'à aucun moment les conditions climatiques n'ont été assez humides pour que s'installent, à basse altitude, des formations forestières du type de celles qui existent actuellement au-delà de 1 500 m et qui se rattachent à la région phytogéographique afromontagnarde. De telles conditions sont probablement à rechercher dans les dépôts antérieurs à 2,0 MA, et plus vraisemblablement antérieurs à 3,0 MA. En effet, l'analyse palynologique de la séquence sédimentaire de Shungura, dans la basse vallée de l'Omo, montre qu'une végétation steppique boisée était déjà installée dans cette région vers 3,0 MA, laissant présager son extension sur toute la bordure du paléolac Turkana (Bonnefille 1980, Bonnefille & Vincens 1985).

REMERCIEMENTS

Mes plus sincères remerciements s'adressent à G.L.Isaac et R.E.F.Leakey, responsables des recherches à l'Est Turkana; A.K.Behrensmeyer, J.W.K.Harris, I.C.Findlater, P.G.Williamson pour leur aide sur le terrain; F.H.Brown et J.M.Harris pour leurs conseils dans l'élaboration de la chronologie des spectres polliniques; R.Bonnefille pour ses conseils et critiques lors de la rédaction de ce travail. Ces recherches ont été financées par le CNRS et la NSF.

BIBLIOGRAPHIE

Behrensmeyer, A.K. 1970. Preliminary geological interpretation of a new hominid site in the Lake Rudolf Basin. *Nature* 226: 225-226.

Bonnefille, R. 1976. Implications of pollen assemblages from the Koobi Fora Formation, East Rudolf, Kenya. *Nature* 264: 403-407.

Bonnefille, R. 1980. Vegetation history of Savanna in East Africa during the Plio-Pleistocene. In *Proc. 4th Intern. Palynol. Conf. (Lucknow, India, 1976-77)* 3: 75-89.

Bonnefille, R., G.Riollet & A.Vincens 1985. Provisional list of Plio-Pleistocene pollen taxa from East Africa. In R.O.Whyte (ed.), *The evolution of the East Asian environment* 2: 613-621. Centre of Asian Studies, Univ. Hong Kong.

Bonnefille, R. & A.Vincens 1985. Apport de la palynologie à l'environnement des Hominidés d'Afrique orientale. In *l'Environnement des Hominidés au Plio-Pléistocène*: 237-278. Coll. int. Singer-Polignac, juin 1981. Paris: Masson.

Bowen, B.E. 1974. The geology of the Upper Cenozoic sediments in the East Rudolf embayment of the Lake Rudolf Basin, Kenya. Ph.D. thesis, Iowa State Univ., 164 pp., 38 fig.

Bowen, B.E. & C.F.Vondra 1973. Stratigraphical relationships of the Plio-Pleistocene deposits, East Rudolf, Kenya. *Nature* 242: 391-393.

Brown, F.H. & T.E.Cerling 1982. The stratigraphic significance of the Tulu Bor tuff of the Koobi Fora Formation, Kenya. *Nature* 299: 212-215.

Brown, F.H. & C.S.Feibel 1985. Stratigraphical notes on the Okote Tuff Complex at Koobi Fora, Kenya. *Nature* 316: 794-797.
Brown, F.H., J.Harris, R.Leakey & A.Walker 1985. Early *Homo erectus* skeleton from west lake Turkana, Kenya. *Nature* 316: 788-792.
Carr, C.J. 1976. Plant ecological variation and pattern in the Lower Omo basin. In Y.Coppens et al. (eds.), *Earliest man and environments in the Lake Rudolf basin*: 432-467. Chicago/London: Univ. Chicago Press.
Cerling, T.E. 1979. Paleochemistry of Plio-Pleistocene lake Turkana, Kenya. *Palaeogeogr., Palaeoclimatol., Palaeoecol.* 27: 245-285.
Cerling, T.E. & F.H.Brown 1982. Additional correlations of tuffs between the Lower Omo Valley and Koobi Fora Region, East Africa. *Nature* 299: 216-221.
Cerling, T.E., R.L.Hay & J.R.O'Neil 1977. Isotopic evidence for dramatic climatic changes in East Africa during the Pleistocene. *Nature* 267: 137-138.
Cohen, A. 1981. Palaeolimnological research at Lake Turkana, Kenya. *Palaeoecology of Africa* 13: 61-82.
Coppens, Y., F.C.Howell, G.L.Isaac & R.E.Leakey (eds.) 1976. *Earliest man and environments in the Lake Rudolf Basin; stratigraphy, paleoecology and evolution.* Chicago/London: Univ. Chicago Press (Prehistoric archaeology and ecology series). 615 pp.
Findlater, I.C. 1976. Stratigraphic analysis and paleoenvironmental interpretation of a Plio-Pleistocene sedimentary basin East of Lake Turkana. Ph.D. thesis, London Univ., 252 pp., 102 fig.
Findlater, I.C. 1978. Isochronous surface within the Plio-Pleistocene sediments East of Lake Turkana, Kenya. In W.W.Bishop (ed.), *Geological background to fossil man:* 415-420. Edinburgh: Scottish Academic Press.
Isaac, G.L. & J.W.K.Harris 1978. Archeology. In M.G.Leakey & R.E.Leakey (eds.), *Koobi Fora Research Project. Vol. I: The fossil hominids and an introduction to their context (1968-1974)*: 64-85. Oxford: Clarendon.
Keay, R.W.J. 1959. *Carte de la végétation de l'Afrique au Sud du Tropique du Cancer.* (Notes explicatives traduites de l'anglais par A.Aubreville.) Oxford: Oxford University Press/AETFAT/Unesco. 24 pp.
Leakey, M.G. & R.E.F.Leakey 1978. *Koobi Fora Research Project. Vol. I: The fossil hominids and an introduction to their context (1968-1974).* Oxford: Clarendon. 191 pp.
Leakey, R.E., M.G.Leakey & A.K.Behrensmeyer 1978. The Hominid catalogue. In M.G.Leakey & R.E.Leakey (eds.), *Koobi Fora Research Project. Vol. I: The fossil hominids and an introduction to their context (1968-1974)*: 86-182. Oxford: Calrendon.
McDougall, I 1981. $^{40}Ar/^{39}Ar$ age spectra from the KBS Tuff, Koobi Fora Formation. *Nature* 294: 120-124.
McDougall, I., T.Davies, R.Maier & R.Rudowski 1985. Age of the Okote Tuff Complex at Koobi Fora, Kenya. *Nature* 316: 792-794.
McDougall, I., R.Maier, P.Sutherland-Hawkes & A.J.W.Gleadow 1980. K-Ar age estimate for the KBS tuff, East Turkana, Kenya. *Nature* 284: 230-234.
Vincens, A. 1982. Palynologie, environnements actuels et plio-pléistocènes à l'Est du lac Turkana (Kenya). Thèse Doct. Etat, Marseille-Luminy, 2 vol., 231 pp., 68 fig., 14 tabl., 7 pl.
Vincens, A. 1984. Environnement végétal et sédimentation pollinique lacustre actuelle dans le bassin du lac Turkana (Kenya). In *Act. 5° Cong. Ass. Palynol. Langue Fr.: 'Palynologie et paléobotanique'. Rev. Paléobiol.*, vol. sp.: 235-242.
Walker, A. & R.E.F.Leakey 1978. The Hominids of East Turkana. *Scientific American* 239 (2): 54-66.

White, F. 1983. *The vegetation of Africa.* A descriptive memoir to accompany the Unesco/AETFAT/UNESCO vegetation map of Africa. Natural Resour. Res. 20. Paris: UNESCO. 356 pp.
White, T.D. & J.M.Harris 1977. Suid evolution and correlation of African hominid localities. *Science* 198: 13-21.
Williamson, P.G. 1982. Molluscan biostratigraphy of the Koobi Fora hominid-bearing deposits. *Nature* 295: 140-142.

FOSSIL OSTRACODES FROM LAKE MOBUTU (LAKE ALBERT): PALAEOECOLOGIC AND TAPHONOMIC IMPLICATIONS

A.S.COHEN
Department of Geosciences, University of Arizona, Tucson, USA

ABSTRACT

Two deep water cores from Lake Mobutu (= Lake Albert) were analyzed for fossil ostracode content to determine palaeohydroclimatic conditions for the lake during the Late Quaternary. Low diversity ostracode assemblages with highly erratic frequency patterns were collected from both cores, spanning a time interval 13 000 BP to the present. Ostracodes dated at approximately 13 000 BP indicate shallow (3-8 m depth), moderately alkaline-saline water conditions at an open lake site currently 47 m deep. Above this horizon, progressive deepening and the establishment of a profundal, fresher water fauna is indicated. Depositional hiatuses, possibly due to deep water erosion, make the dating of this progressive deepening difficult to establish, but it certainly predates 7 500 BP.

Three taphonomic associations are recognized from the ostracode assemblages found. Ostracode sand lag assemblages are life assemblages of unabraded, articulated fossils formed in relatively shallow water conditions, where fines are winnowed out and refractory organics are oxidized, leaving behind a highly concentrated ostracodal sand. Profundal life assemblages form in deep water, where food availability is low. They are characterized by very low density accumulations of unbroken and often articulated valves, reflecting a lack of transport and low initial population densities. Profundal mixed assemblages are transported admixtures of both shallow and deep water ostracode species, which probably form as a result of bottom current transport. They are recognized by the co-occurrence of species with very different habitat requirements. Fossil material in these assemblages is normally broken or abraded, and valves are mostly disarticulated.

INTRODUCTION

Fossil ostracodes are commonly found in African lacustrine sediments. Previous studies have indicated the utility of these fossils for a wide variety of palaeoenvironmental interpretations, including (but not restricted to) palaeochemical, palaeodepth and water temperature analyses (Delorme et al. 1977, De Deckker 1983, Forester & Markgraf 1984). Palaeoecological investigations of fossil ostracodes in Africa have primarily been concerned with the deveopment of palaeochemical interpretations of ancient lake sediments, which can then, in some cases, be used for making palaeoclimatic interpretations. To date, most such studies have concentrated on shallow water lake deposits from East Africa (Cohen 1981, Cohen et al. 1983, Carbonel & Peypouquet 1983, Carbonel et al. 1983, Peypouquet et al. 1983, Nielsen 1984). Ostracode standing crops are normally much higher in shallow water lacustrine settings than in deep water, due to considerable differences in food availability between the littoral and profundal zones (Cohen 1984).

In this study, I present preliminary results from the analysis of fossil ostracodes obtained from two cores taken in Lake Mobutu (= Lake Albert of older literature). This study differs from earlier mentioned investigations of fossil African ostracodes in that the cores were derived from relatively deep water (40 and 47 m). Aside from attempting to document the Late Quaternary ostracode faunal history of the lake, a second goal of this study was to determine the origin of deep water ostracode fossil assemblages in lakes. Specifically, are concentrations of ostracode valves found in profundal settings emplaced through in situ burial as life assemblages, or are they primarily death assemblages which have undergone transport and sorting? Such information will be critical in assessing the utility of these types of assemblages for various palaeoecological and evolutionary studies.

MATERIALS AND METHODS

The two cores analyzed in this study were collected by scientists from the Woods Hole Oceanographic Institution (WHOI) during their 1972 expedition to the African Great Lakes. Detailed descriptions and sedimentological interpretations of these cores may be found in Hecky & Degens (1973). The two piston cores, hereafter referred to as Core # 2 and Core # 3 (Woods Hole Station numbers), were obtained from relatively deep water stations (40 and 47 m respectively) on the Zaire side of the lake (Core # 2: lat. 01°29.2′N, long. 30°42.9′E; Core # 3: lat. 01°31.2′N, long. 30°34.5′E; see Fig. 1). Total recovered core lengths for Cores #2 and # 3 respectively are 532 and 484 cm. The cores are stored and available for subsampling at WHOI.

Ostracode samples were obtained by removing 2-4 g dry mass of sediment from scraped core sample splits. Sample splits were taken every 20 cm for both cores. Sediments were disaggregated by spray washing in warm water. Samples were microscopically examined prior to disaggregation to insure that any breakage of ostracode valves could be ascribed to either original status or damage during preparation.

Samples were sieved in a # 80 US mesh size screen, which insured that all

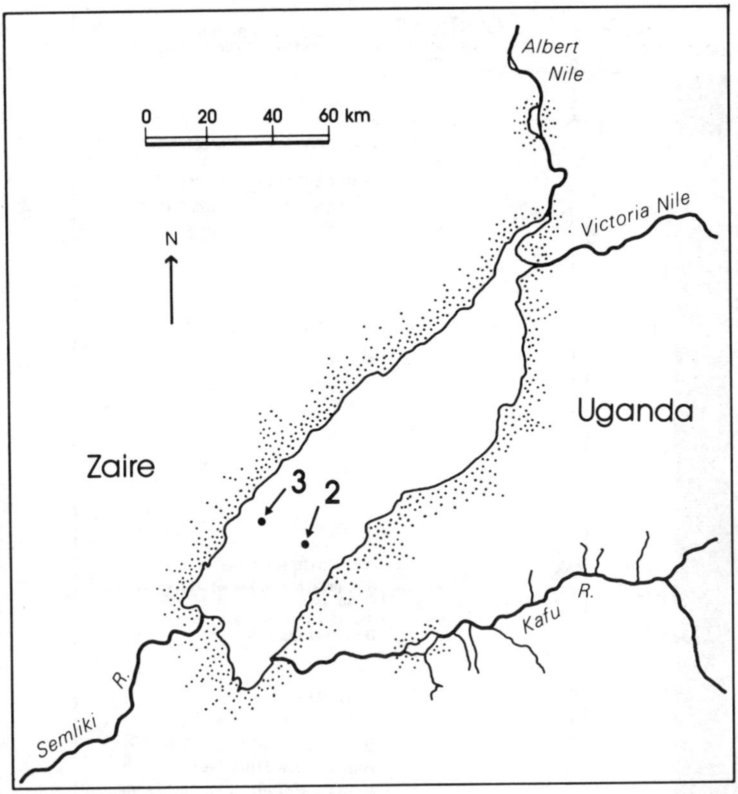

Figure 1. Location map of Lake Mobutu (= Lake Albert). Localities where Cores No. 2 and No. 3 were taken are indicated.

adults and most juvenile ostracodes would be retained. The coarse terrigenous clastic component (as a fraction of total washed sediment) was calculated by weighing the clastic fraction retained on the screen after sieving.

Washed samples were microscopically examined for ostracode content. Percentages were calculated for each species present. No distinction is made here between adult and juvenile valves. Normally 100 valves were counted, but for many samples with very low ostracode concentrations or predominately broken or unidentifiable valves this was not possible. Such samples are indicated in Figure 2 with an asterisk.

Ostracode abundances were estimated by extrapolation from small (0.01-0.05 g) subsamples. For samples with large numbers of broken ostracodes, abundances were estimated by determining the total mass of ostracodal debris and dividing by the average mass of ostracodes present in that core interval.

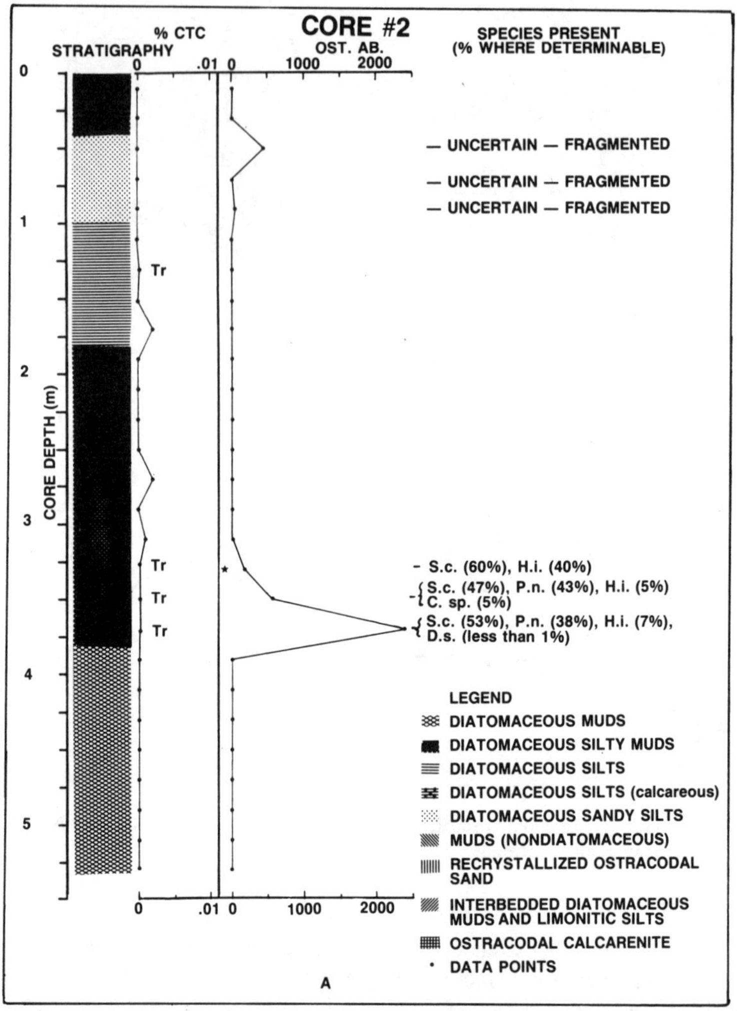

Figure 2. Core stratigraphy and ostracode occurrences discussed in this report. 2a. Core No. 2; and 2b. Core No. 3. % CTC is percent coarse terrigenous clastics which equals the proportion of terrigenous clastic grains larger than 2.5 phi (0.177 mm). Tr means CTC are present in trace amounts, less than .001%. Ost. Ab. is ostracode abundance, in # valves per g sediment (dry weight). Ostracode abundance calculations are discussed in the text.

Abbreviations of species present: S.c. – *Sclerocypris clavularis,* H.i. – *Hemicypris intermedia,* P.n. – *Plesiocypridopsis newtoni,* D.s. – *Darwinula stevensoni,* Cy.sp. – *Cypridopsis* sp., I.sp. – *Ilyocypris* sp., L.sp. – *Limnocythere* sp., C.sp. – *Chrissia* sp.

Percentages refer to total counted valves for that core interval. Asterisks indicate samples for which 100 valves were not available.

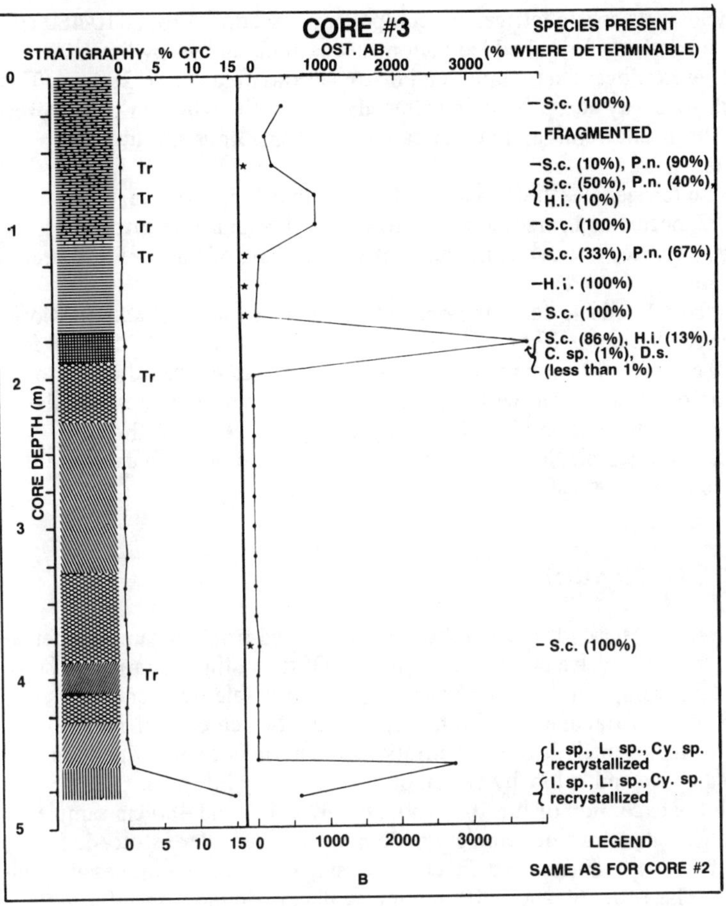

Figure 2 (continued).

CORE STRATIGRAPHY

Stratigraphic columns for the two cores are illustrated in Figure 2. Core # 2 consists of diatomaceous muds in the lower portion of the core which gradually coarsen upcore to diatomaceous silts. Near the top of the core sediments become slightly finer but remain diatom-rich. The diatom stratigraphy is discussed by Hecky & Degens (1973).

Core # 3, deposited somewhat closer to the modern shoreline, is lithologically more complex. The basal 20 cm consists of a recrystallized ostracode sand. The same unit is rich in angular, 0 phi quartz sand at its base, but rapidly loses its coarse terrigenous component upsection. From 410-460 cm the core consists of interbedded diatomaceous muds and wispy limonitic silts. These are overlain by muds and diatomaceous muds up to 200 cm. This interval gives way abruptly to an ostracodal calcarenite which in turn is overlain by both calcareous and non-calcareous, diatomaceous silts at the top of the core.

Coarse terrigenous clastic material (coarser than 2.5 phi) is very rare in Core # 2, occurring in trace amounts to 0.02% at several core intervals. In Core # 3, coarse clastics occur abundantly at the base of the core but rarely elsewhere.

Two correlated age dates and one radiocarbon date are available for Core # 3 (Hecky & Degens 1973). Correlated ages of 13 000 BP and 7 500 BP for the 480 cm and 340 cm levels respectively were suggested, based on diatom and lithologic corrections with other dated cores. A single radiocarbon date of 2 660 ± 90 BP was obtained by Hecky & Degens (1973) for the 140-180 cm core interval. No previous age estimates or radiocarbon dates are available for Core # 2.

FOSSIL OSTRACODES

Low diversity ostracode assemblages were recovered from six sample intervals in Core # 2 and twelve intervals in Core # 3. Of these, three samples in Core # 2 and one sample in Core # 3 consist of unidentifiable fragments only. The occurrence of ostracode material throughout the two cores is extremely erratic with extremely high concentrations at certain horizons which are gounded above and below by barren or low frequency intervals.

With the exception of the 140, 160, 180, 380, 460 and 480 cm samples in Core # 3, all ostracode samples are dominated by angular ostracode fragments. Whole specimens are difficult to obtain, and in some samples are non-existent. The 140, 160 and 380 cm samples all contain monospecific, very low frequency assemblages of unbroken, unabraded *Sclerocypris clavularis* (Sars), with paired valves present in some specimens. Some of these only occur as internal molds. The 180 cm sample from Core # 3 consists of approximately equal proportions of broken and unbroken specimens, although unbroken valves are never attached, even for species in which strong hinges exist. The 460 and 480 cm samples are highly distinctive in that almost all specimens are found intact, and with attached closed valves.

Preservation of ostracode material consists of original shell material or molds in all cases except the 460 and 480 cm samples. Here the material is heavily recrystallized, and in some cases occurs as internal molds, and elsewhere as indeterminate replacement of original shell material.

The 460 and 480 cm horizons in Core # 3 both contain a distinctive assemblage of ostracodes including *Ilyocypris* sp., *Limnocythere* sp. and *Cypridopsis* sp. Specific identification for all three taxa at these horizons is impossible due to the nature of preservation. Charophytes and gastropod shell fragments also occur at these two horizons.

At all other core horizons, the ostracode assemblages consist of five species in variable proportions. *Sclerocypris clavularis* is the most abundant ostracode species, occurring in all but three of the horizons for which identifiable material is available. *Hemicypris intermedia* (Bate) and *Plesiocypridopsis newtoni* (McKenzie) are also quite abundant. Two additional species, *Chrissia* sp. and *Darwinula stevenson* (Brady & Robertson), are much less common in each core, occurring at only a single horizon each. No frequency trends are discernible within the *Sclerocypris clavularis* dominated segments of either core.

PALAEOECOLOGICAL INTERPRETATIONS

The ostracodes recovered in this study, while limited in diversity, are nonetheless highly significant for the understanding of environmental conditions in Lake Mobutu during the Late Quaternary. The occurrence of ostracodal sands composed primarily of *Ilyocypris* sp. and *Limnocythere* sp. indicates moderately alkaline, very shallow, lacustrine conditions at the Core # 3 site at 13 000 BP. Similar associations are commonly encountered in modern Lake Turkana, particularly in semirestricted embayments, at water depths of 3-8 m (Cohen 1982). At these depths, ostracodal sands, consisting primarily of ostracode valves and refractory organics from plant debris, accumulate in thin but extensive sheets. They do not form at wave-swept, littoral depths, accumulating instead in the sublittoral zone, below normal daily wave base. The organics quickly oxidize, as observed from examination of short gravity cores. Like the Mobutu core material, ostracode shell assemblages found in this setting at Lake Turkana tend to have a large proportion of intact and connected valves, reflecting their minimal degree of transport. These ostracode concentrations are primarily the consequence of the winnowing of finer material from the site of accumulation. The interpretation of the 460 and 480 cm horizons as being of shallow water origin is further bolstered by the abundant occurrence of unbroken charophytes and coarse quartz sand at these levels.

Unfortunately, the inability to make a specific diagnosis of the taxa present prevents any precise palaeochemical estimates for the lake at this time. However, the assemblage is clearly representative of more saline and more alkaline lake conditions than the remaining assemblage, found at all other productive horizons of Core # 3 and throughout Core # 2.

The *Sclerocypris clavularis-Hemicypris intermedia* assemblage, found elsewhere in the two cores, is typical of profundal ostracode accumulations in large African lakes. In modern Lake Turkana, these two species form the dominant ostracode association at depths greater than 10 m (Cohen 1982). Thus, the appearance of *S. clavularis* by the 380 cm lvel of Core # 3, even in low concentrations, signifies the onset of profundal conditions in the lake.

Higher levels in Core # 3 and the lower (350-370 cm) ostracode accumulation of Core # 2 contain peculiar admixtures of the *S. clavularis-H. intermedia* assemblages along with common occurrences of *Plesiocypridopsis newtoni* and occasional *Chrissia* sp. Both of these taxa are typical of shallow water and are frequently found around groundwater discharge localities, but rarely occur in the profundal zone of modern African lakes. Thus, their presence with the deep water species is anomalous for a life assemblage. As noted earlier, a very high proportion of valves in both cores are broken and abraded, suggesting that the fossils may have undergone significant post mortem transport. This interpretation is also supported by the very erratic proportions of various species through the upper interval of Core # 3 and the 350-370 cm interval of Core # 2. Variations like this are extremely uncommon in profundal lacustrine ostracode assemblages in the absence of major environmental shifts or the introduction of reworked death assemblages. The sedimentological record does not support an interpretation of major facies changes or sediment accumulation hiatuses at the levels where this sort of change is observed (e.g. Core # 3, between 60 and 100 cm).

The *Sclerocypris-Hemicypris* assemblage indicates water chemistry conditions similar to those of modern Lake Mobutu. The indicated total alkalinity and salinity for this assemblage is 8-20 meg/l and .8-2 ppt respectively. If *Chrissia* specimens found with the deep water assemblage are contemporaneous with that assemblage, the hydrochemical conditions for Lake Mobutu for the 180 cm interval of Core # 3 and the 350 cm level of Core # 2 are further constrained. *Chrissia* is restricted to relatively low alkalinity-salinity water conditions, with maximum alkalinities of approximately 10 meg/l and salinities of 1 ppt, indicating a relatively narrow range of water chemistries (8-10 meg/l alkalinity and .8-1 ppt salinity). Unfortunately, a species specific identification of the *Chrissia* material has not yet been possible. Neither *Plesiocypridopsis newtoni* nor *Darwinula stevensoni* are particularly informative of hydrochemical conditions as they are both highly generalized in their water chemistry requirements.

DISCUSSION

Hecky & Degens (1973) have reviewed the chemical stratigraphic record for Lake Mobutu based upon their analyses of the WHOI cores. They found that the basal interval of Core # 3 was deposited under low-stand conditions for the lake. A pronounced lake level rise is indicated above the 460 cm horizon of Core # 3 by an abrupt decrease in $CaCO_3$, Mg and Sr content and an increase in clay content. As the authors note, however, the precise timing of this event is difficult to ascertain given the likelihood that the facies change was accompanied by a depositional hiatus.

The upper Core # 3 interval (above 200 cm) again represents an environmental shift, with increasing proportions of $CaCO_3$, Sr and organic carbon. Mg increases only slightly, and is believed to be concentrated in the montmorillonite clay fraction. Hecky and Degens attribute these changes to a shift towards higher productivity in the lake, as nutrients from the influent Semliki River increased due to increased volcanic and hydrothermal activity in the region (Hecky & Degens 1973: 84-85).

Ostracode taphonomy reflects these environmental shifts in a significant and predictable fashion. Three important taphonomic associations are represented in the Mobutu core material. Together, they illustrate three of the principal mechanisms by which ostracodes are preserved in large African lakes.

The lowest core interval of Core # 3 contains a winnowed ostracode sand lag assemblage. Such lags have been recovered in modern sediments of Lake Turkana (Cohen 1982) and Lake Tanganyika as well as in Plio-Pleistocene sediments of Lakes Malawi and Turkana. They are notably absent from deposits of smaller Pleistocene African lakes (Cohen et al. 1983, Cohen & Nielsen, in press). Where they have been observed in modern lakes, they accumulate at relatively shallow depths (3-8 m) where wave activity has decreased sufficiently to allow coarser organic material to settle, but is still sufficiently strong to remove most fines. Ostracodes are extremely abundant at such depths, which may be littoral (Tanganyika) or sublitoral (Turkana), depending upon wave intensity and water turbidity (Cohen 1984). As sedimentary particles, ostracode valves reach their greatest proportional frequencies at these depths, in some cases comprising up to over 50% of the surface sediment mass in Lake Turkana. Upon burial, oxidation of intermixed organic particles (the other major sedimentary component at burial) reduces their mass in the sediment, thus increasing the proportion of ostracode valves in the sediment. Some Pliocene, shallow-lacustrine calcarenites from East Turkana, Kenya, consist of over 98% ostracode valves by mass. Fossil ostracodes in these sands, like the modern Lake Turkana accumulations, tend to be unabraded and a large proportion of specimens have paired valves still attached. In both modern Lakes Turkana and Tanganyika, the valve frequencies for species present closely reflect the living ostracode populations of the same areas, indicating that the valves are accumulating as life assemblages.

Ostracode sand lags accumulate as common components of the shallow water deposits of large lakes in Africa. They may be less common in smaller lakes due to the reduced competence of shallow-water wave action in small lakes for winnowing fine particulate matter. Obvious shallow-water life assemblages in the deposits of smaller lakes tend to be associated with abundant clays and silts, preventing the development of calcarenites, even under highly productive situations (Cohen et al. 1983).

Profundal life assemblages appear following the onset of deep-water conditions at the Core # 3 site. They are represented by the monospecific *Sclerocypris clavularis* assemblages from the 140, 160 and 380 cm samples of Core # 3. Extremely low frequencies, coupled with intact, unabraded specimens typify this kind of assemblage. The presence of paired valves in some specimens of *S. clavularis* from these core intervals is particularly noteworthy as this species has an extremely weak hinge structure which normally disarticulates shortly after death. Specimens of *Sclerocypris* cannot undergo even brief post mortem transport without their valves being disarticulated.

Population densities of living ostracodes drop off rapidly from the shallow sublittoral to the profundal zone of Lake Turkana (Cohen 1984), due to a severe limitation of food resources in deep water. For profundal ostracodes, which are mostly detritivores, this reduction is also reflected in a decrease in organic carbon content in the sediment. In Core # 3, not surprisingly, the low density (profundal) life assemblages are associated with the lowest organic carbon contents in the core, often less than 2.5%.

The remaining ostracode samples from both cores are profundal mixed assemblages, incorporating both local profundal species as well as admixed shallow water taxa. Increase in total abundance for these intervals (80, 100 and 180 cm in Core # 3; 50, 350 and 370 cm in Core # 2) probably was caused by the post mortem accumulation of valves due to bottom current activity in the lake, perhaps as turbidity currents. Slight increases in coarse terrigenous clastic material, chiefly angular quartz and feldspar grains, are associated with some of these horizons, and both the 160-180 cm interval of Core # 3 and the 370-33 cm interval of Core # 2 are graded. Hecky & Degens (1973: 23-24) previously noted that the record of depositional hiatuses from Lake Mobutu suggests that vigorous bottom currents have probably been active there during the Holocene. Strong cold water undercurrents, emanating from shallow water regions of the modern lake have been documented by Talling (1969). Such current activity may well be responsible for concentrating the types of mixed shallow-deep water assemblages observed at the previously mentioned core intervals. This would also explain the absence of abundant unbroken and unabraded valves, as well as the erratic species frequency patterns observed at these horizons.

With increasing attention being paid to ostracodes as palaeoecological indicators in African Quaternary strata, it is critical that life assemblages and death assemblages of ostracodes be distinguishable. A better understanding of the mechanisms of transport and burial for these organisms will not only reduce problems of biased palaeoecological and evolutionary interpretations based upon mixed assemblages but may, in the future, provide a new method for evaluating the frequency and nature of subsurface current activity in palaeolakes.

ACKNOWLEDGEMENTS

I would like to thank Rick Forester for many valuable discussions on ostracode ecology, which have served to strengthen this text. James Broda of the WHOI core laboratory, provided me with access to the core material described here.

Debra Gevirtzman and D.van Harten read the manuscript and Carol Erickson typed it, though, of course, all mistakes are my own.

REFERENCES

Carbonel, P., E.Grosdidier, J.P.Peypouquet & J.J.Tiercelin 1983. Les ostracodes, témoins de l'évolution hydrologique d'un lac de rift. Exemple du Lac Bogoria, Rift Gregory, Kenya. *Bull. Centres Rech. Explor,-Prod. Elf-Aquitaine* 7: 301-313.

Carbonel, P. & J.P.Peypouquet 1983. Ostracoda as indicators of ionic concentrations and dynamic variations: methodology (Lake Baringo, Kenya). In R.F.Maddocks (ed.), *Applications of Ostracoda to economic and scientific problems. Proc. 8th Int. Symp. Ostr., Houston*: 264-276.

Cohen, A.S. 1981. Paleolimnological research at Lake Turkana, Kenya. *Palaeoecol. of Africa* 13: 61-82.

Cohen, A.S. 1982. Ecological and paleoecological aspects of the Rift Lakes of East Africa. Thesis Doc., Univ. of California-Davis, 314 pp.

Cohen, A.S. 1984. Effect of zoobenthic standing crop on laminae preservation in tropical lake sediment, Lake Turkana, East Africa. *J. Paleo.* 58: 488-510.

Cohen, A.S., R.Dussinger & J.Richardson 1983. Lacustrine paleochemical interpretations based on eastern and southern African ostracodes. *Palaeogeogr., Palaeoclim., Palaeoecol.* 43: 129-151.

Cohen, A.S. & C.Nielsen (in press). Ostracodes as indicators of paleohydrochemistry in lakes: a Late Quaternary example from Lake Elmenteita, Kenya. *Palaios* 1.

De Deckker, P. 1983. The limnological and climatic environment of modern halobiont ostracodes in Australia. A basis for paleoenvironmental reconstruction. In R.F.Maddocks (ed.), *Applications of Ostracoda to economic and scientific problems. Proc. 8th Int. Symp. Ostr., Houston*: 250-254.

Delorme, D., S.C.Zoltai & L.L.Kalas 1977. Freshwater shelled invertebrate indicators of paleoclimate in northwestern Canada during the late glacial. *Can. J. Earth Sci.* 14: 2029-2046.

Forester, R. & V.Markgraf 1984. Late Pleistocene and Holocene seasonal climatic records from lacustrine ostracode assemblages and regional (pollen) vegetational patterns in southwestern USA. *Amer. Quaternary Assoc., 8th biennial mtg., abst. with programs*: 43-45.

Hecky, R.E. & E.T.Degens 1973. *Late Pleistocene-Holocene chemical stratigraphy and paleoliminology of the Rift Valley Lakes of central Africa.* Woods Hole Oceanographic Inst. Technical Report No. WHOI-73-28, 93 pp.

Nielsen, C. 1984. Ostracodes as paleochemical indicators at Lake Elmenteita, Kenya. *Amer. Quaternary Assoc., 8th biennial mtg., abst. with programs*: 94.

Peypouquet, J.P., P.Carbonel, M.Taieb, J.J.Tiercelin & G.Perinet 1983. Ostracoda and evolution of paleohydrologic environments in the Hadar Formation (the Afar Depression, Ethiopia). In R.F.Maddocks (ed.), *Applications of Ostracoda to economic and scientific problems. Proc. 8th Int. Symp. Ostr., Houston*: 277-285.

Talling, J. 1969. The incidence of vertical mixing and some biological and chemical consequences in tropical African lakes. *Verh. Intern. Verein. Limno.* 17: 998-1012.

VEGETATION AND CLIMATE OF MT ELGON DURING THE LATE PLEISTOCENE AND HOLOCENE

A.HAMILTON
Department of Environmental Studies, Coleraine, Co. Londonderry, UK

SUMMARY AND CONCLUSIONS

The late Pleistocene and Holocene climatic and vegetational history of Mt Elgon were investigated by means of pollen analysis of sediments from four sites. Additional studies were undertaken on modern vegetation, modern pollen deposition and glacial features. During the late Pleistocene (> 23 000 - < 14 000 BP and probably 28 000 - 12 500 BP) the extent of forest and microphyllous thicket on the mountain was much less than now and there is evidence of depression of vegetation belts. The southerly monsoon was greatly reduced in strength. Forest started to spread before 11 000 BP and a series of changes in forest composition are interpreted as due to a moistening climate culminating in the establishment of mesic montane forest by 9 000 BP. The southerly monsoon was moist before 11 000 BP, when temperatures were still depressed. Lower rainfall at about 4 000 BP caused major changes in forest composition. Forest destruction by man during the Holocene has not been extensive.

BACKGROUND TO THE RESEARCH

By 1975 it had become clear that pollen analysis was a useful technique for investigating the Quaternary history of vegetation in the tropics, though the interpretation of pollen diagrams still presented many difficulties. Some of the pioneers of pollen analysis in East Africa were Hedberg (1954), Osmaston (1958, 1965) and Van Zinderen Bakker, the latter with Coetzee publishing pollen diagrams for Mt Kenya and the Cherangani Hills (Coetzee 1967, Van Zinderen Bakker 1962, 1964). Van Zinderen Bakker and Coetzee believed that temperature was the primary factor determining change in vegetation during the Upper Quaternary on the mountains which they studied: high values of grass pollen during the time of the last world glacial maximum were taken as evidence of downward extension of high altitude treeless vegetation. This interpretation of grass pollen was soon questioned by Livingstone (1967), who pointed out that grasses are common in many types of vegetation in tropical Africa. In interpreting his own pollen diagrams for Ruwenzori, where grass pollen is also well-represented during the terminal Pleistocene, he argued that successional development and alterations in climatic moistness rather than temperature depression were the major factors controlling the past composition of vegetation.

A palynological research centre was established at Makerere University in the 1950s, with attention directed at the exciting possibilities offered by the sediments beneath the numerous mires and lakes of the Rukiga Highlands in south-western Uganda. The first published diagram, from Muchoya Swamp, was interpreted as showing major vegetational depression due to lowering in temperature during the last glacial maximum (Morrison 1961, 1968), similar in these respects to the interpretations of the diagrams from Mt Kenya and the Cherangani Hills by Van Zinderen Bakker and Coetzee.

Van Zinderen Bakker encouraged me to publish some of the results of my own research into modern pollen deposition in highland Uganda, carried out as a research student of Michael Morrison (Hamilton 1972). Different types of vegetation were found to vary greatly in pollen productivity and different types of pollen contrasted greatly in their abilities to be carried for long distances in the atmosphere. A method of interpreting pollen diagrams was proposed which gave added weight to pollen types of lower dispersal ability produced by taxa of restricted distribution. Reinterpretation of Livingstone's and Morrison's diagrams suggested that the late Pleistocene in western Uganda was both relatively cool and dry, supporting aspects of previous interpretations. The fact of dryness, which emerged as a logical consequence of the method of interpretation, was a surprise to me in that I was still at that time strongly influenced by the Pluvial Theory correlating world glacial periods with tropical wetness (Nilsson 1931, Wayland 1934). It was only later that I received a copy of Kendall's (1969) study of the ecological history of Lake Victoria with its clear evidence for much greater aridity and massive reduction of lowland forest in southern Uganda before 12 500-12 000 BP. Widespread aridity over much of tropical Africa during the last world glacial maximum has since become widely acknowledged, with evidence from many quarters, including a reassessment of the climatic significance of the pollen diagrams from Mt Kenya and the Cherangani Hills (Van Zinderen Bakker & Coetzee 1972).

In 1971 I was employed at Makerere University to write a field guide to the forest trees of Uganda (Hamilton 1981) and became fascinated by their patterns of distribution (Hamilton 1974). These patterns seemed inexplicable with reference solely to the modern environment, but made more sense if species of lowland forest and moist montane forest had expanded relatively recently across the country from the south-west, many species failing to occupy their potential ranges. The wide occurrences of species of drier and higher altitude montane forests suggested that these forest types had been geographically more persistent (including possibly on Mt Elgon). The spread of lowland forest had already been dated at about 12 500-12 000 BP in both southern and western Uganda (Hamilton 1972, Kendall 1969, Livingstone 1967); the distributional evidence added to our knowledge by suggesting *where* lowland and moist montane forest remained during Pleistocene aridity.

Van Zinderen Bakker then requested me to undertake a general review of the historical significance of patterns of distribution shown by forest organisms in tropical Africa. The information available to me was limited, but, even so, strong evidence emerged for a recent restriction of forest to a number of centres, one of which lay in eastern Zaire, possibly just extending into south-western Uganda and western Rwanda (Hamilton 1976; see also Diamond & Hamilton 1980). The low endemism and relative biotic poverty of the forests of western Kenya (including Mt Elgon) seemed to imply that many forest species had moved there relatively recently, either from eastern Zaire or from refugia near the East African coast. In 1975 I received a grant from the Natural Environment Research Council (UK), enabling me to employ an assistant (Alan Perrott) and undertake a new programme of research into the Quaternary environmental history of East Africa. This programme was designed to test theories of climatic and vegetational change which had emerged from the earlier studies; pollen analysis was to be carried out on sediments from two localities, one believed to have been climatically dry (Mt Elgon) and the other relatively wet (south-western Uganda) during the last world glacial maximum. It was anticipated that calculation of past temperatures from the altitudinal movements of vegetation would be difficult on Mt Elgon, where vegetational history was expected to be dominated by major variations in climatic moistness. Investigations of past glaciation, modern vegetation and modern pollen deposition were to be undertaken on Mt Elgon to back up the palynological work. A Hiller borer was chosen for sediment sampling because of its lightness and versatility.

In the event, permission for field work was not obtained for south-western Uganda and attention was directed instead at nearby Rwanda, where Kamiranzovu Swamp in Rugege Forest was selected as the primary sampling site on the basis of information supplied by Bouxin (1974 & pers. comm.). Time was also found for field work on the Aberdare Range, forming the basis for further studies in central Kenya by Alan Perrott (Perrott 1982; Perrott & Street-Perrott 1982). This chapter deals only with the Elgon component of the research.

PREPARATORY STUDIES AND FIELD WORK

Earlier work on modern pollen deposition on Mt Elgon had shown that large quantities of better dispersed types of pollen are carried up from the montane forest and deposited above the forest limit and also that differences in forest composition between the wetter (western to south-western) and drier (eastern to north-eastern) slopes are reflected in differences in high altitude pollen spectra on different sides of the mountain (Hamilton 1972). It was hoped that

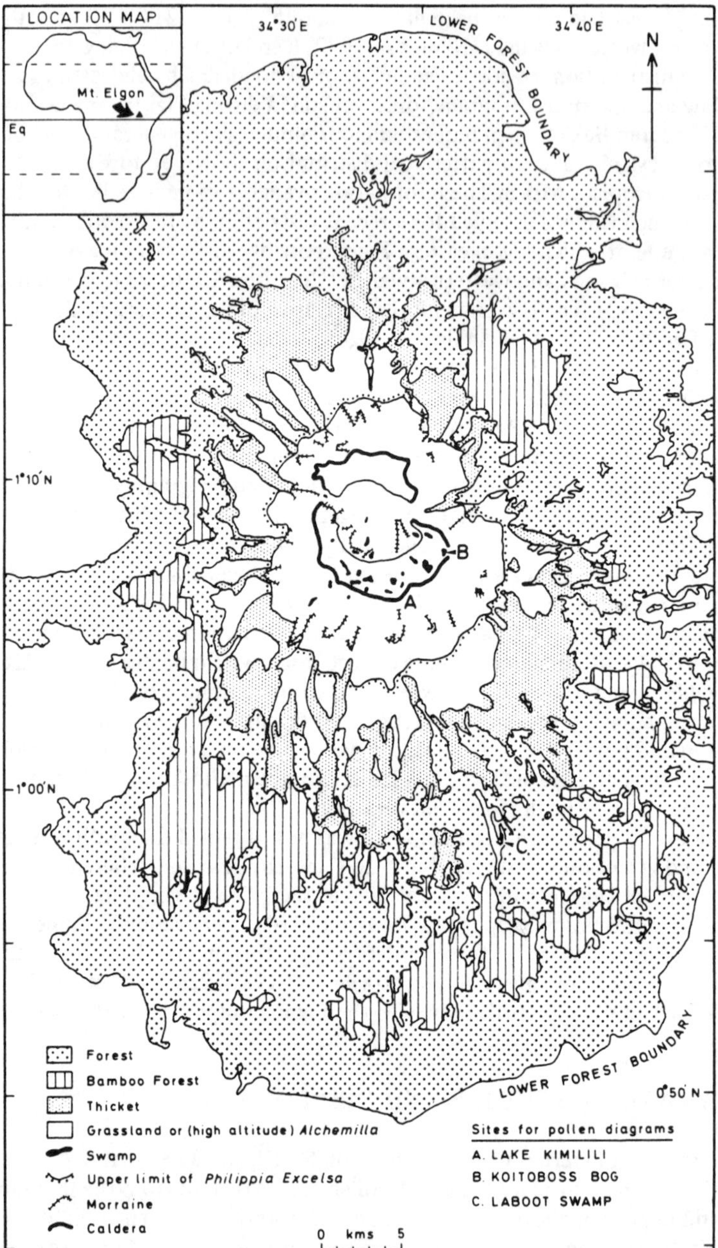

Figure 1. Mt Elgon. Locality map.

a detailed reconstruction of forest history might be possible through comparison of pollen diagrams obtained from high altitude sites on different aspects. A similar approach to the interpretation of pollen diagrams from Ruwenzori (Livingstone 1967) had shown that forest clearance by man was more a feature of the lowlands to the east than to the west of the range (Hamilton 1972).

Aerial photographs were used to help locate sites for sediment sampling. Numerous bogs and a few lakes are present within the formerly glaciated region, but lakes are absent and mires are uncommon at lower altitudes where the only possibilities for Upper Quaternary sediments seen on the photographs were a few long thin features in valleys, interpreted as mires, mainly at altitudes of 2 500-3 000 m on the south and west. Plans for sediment sampling were adjusted after failure to obtain permission to work in Uganda. In the event, only four sites were actually sampled, two at high altitude, Lake Kimilili at 4 150 m on the south and Koitoboss Bog at 3940 m on the east, and two at lower altitude, Scout Hut Mire (3 380 m) and Laboot Swamp (2 880 m), both on the south (Fig. 1). Radiocarbon dates show that sediments at Lake Kimilili extend back to 11 000 BP, those at Koitoboss Bog to 6 500 BP and those at Laboot Swamp to > 23 000 BP. The short undated pollen diagram from Scout Hut Mire is published in Hamilton (1982) and is not discussed here.

Glacial features and vegetation were mapped from the aerial photographs, vegetation being studied also on the ground, along transects running from the lower forest boundary to the summit. One transect (Kimilili) was situated on the relatively wet southern slopes and the other (Koitoboss) on the drier east. Vegetation was sampled by belt transects (trees and lianes) and quadrats (shrubs and herbs). Modern pollen rain was investigated along these transects through the collection and extraction of pollen from epiphytic bryophytes, surface soil samples and pollen traps (Flenley 1973), the latter being emplaced for a five month period. A few pollen traps were positioned in agricultural land around the foot of the mountain.

GLACIATION

There is extensive evidence for a single main glaciation, marked by single or closely spaced multiple termino-lateral moraines in valleys on all sides of the mountain (Hamilton & Perrott 1978, 1979). We failed to find evidence of the three major glacial stages recognized by Nilsson (1931). Temperatures during maximum glaciation were calculated by the morainic method of Osmaston (1975) as being at least 3.5°C lower than at present, assuming no change in precipitation (in fact, it was drier and therefore colder – see later).

There are some interesting glaciological contrasts between the northern and southern sectors of Mt Elgon, the most notable being that the firnline surface during maximum glaciation was tilted strongly downwards towards the north; the implication is that the direction of the main moisture-bearing winds was approximately the opposite to the present. This backs up other evidence for a weakened Southwest Monsoon during the last glacial maximum (Hurni 1981, Nilsson 1931, Osmaston 1975, Prell et al. 1980, Van Campo et al. 1982).

Some valleys facing south-west on Elgon are headed by rock-dammed cirques, suggesting that after the main glacial stage there was a minor glacial stage when the main moisture-bearing winds came from the south-west, as they do now. Radiocarbon dates of 11 012 ± 135 BP (SRR-1118) and 10 708 ± 230 (SRR-1120) for basal sediments at two sites in one of the cirques (now occupied by Lake Kimilili) show that deglaciation from this stage occurred sortly before 11 000 BP (Hamilton & Perrott 1978). The implication of the orientation and dating of this stage is that the Southwest Monsoon became moist before 11 000 BP, at a time when temperatures were still depressed compared with today. It is possible that the Main and Cirque Glacial Stages of Elgon are correlated with the Mahoma and Omurabaho Stages of Ruwenzori (Osmaston 1965, 1975). Retreat of ice from the Mahoma Stage occurred before 14 700 ± 290 BP (Livingstone 1962).

Apart from the pre-11 000 BP date for ice withdrawal from the Lake Kimilili cirque, there is no other firm date for glacial stages on Mt Elgon. However, maximum glaciation here and elsewhere in eastern Africa was probably centred at around 18 000 BP, as in many other parts of the world (Ruddiman & Duplessey 1985).

VEGETATION AND FLORA

Four main physiognomic types of vegetation were identified above the lower forest boundary on Mt Elgon, namely forest (with bamboo a sub-type), microphyllous thicket (dominated by *Philippia* or *Stoebe*), grassland and *Alchemilla* scrub (often with abundant *Dendrosenecio*). The new information on the distribution and characteristics of these types permits revision to earlier schemes of vegetational zonation, such as those of Hedberg (1951) and Langdale-Brown et al. (1964) (Hamilton 1982, Hamilton & Perrott 1981). The floristic composition of montane forest differs markedly on the wetter and drier slopes at lower altitudes.

Andrew Agnew and Christine Kabuye kindly assisted with the identification of plants collected during this study. There were sixteen new records for the mountain. An analysis of the distribution of endemic species by habitat substantiates the view that there are few endemics in the forest (Hamilton 1982); most endemics grow either in *Alchemilla* scrub or on rocky ground at high altitude (Hamilton 1982).

MODERN POLLEN DEPOSITION

The pollen traps showed that the pollen production of montane forest is much higher than that of higher or lower altitude vegetation (Hamilton & Perrott 1980). They also confirmed that considerable quantities of better dispersed pollen types are exported from montane forest. The relative export abilities (Flenley 1973) of forest pollen types were estimated from their representation in the pollen rain above the forest limit compared with the representation of their parent plants in the forest vegetation. Problems in defining source area and in adequately sampling the vegetation make these estimates very approximate, but in general they confirm earlier work (Hamilton 1972), adding an extra element of quantification. The pollen of *Podocarpus* was calculated to be overrepresented about ten times in comparison with *Hagenia* or *Olea* which, in turn, were estimated to be overrepresented about three times in comparison with *Prunus* or *Rapanea.*

The percentage of grass pollen in surface samples or pollen traps is strongly correlated with the amount of grassland near sample sites. Values of grass pollen are always > 30% in samples from grassland and rocky ground communities and always < 30% in samples from other vegetation types. Grass pollen usually constitutes < 10% of total pollen in samples from forest or microphyllous thicket. On Ruwenzori a very large quantity of grass pollen is carried up from below the forest limit and deposited at high altitudes (Hamilton 1972), but this is not the case on Elgon where, it is suggested, the production of grass pollen at lower altitudes is reduced under the drier climate and with the more intensive agriculture.

POLLEN DIAGRAM FROM LABOOT SWAMP

Laboot provides the only direct evidence of the nature of Quaternary vegetation on Mt Elgon before 11 000 BP and it is unfortunate that the site and our approach to studying it were unsatisfactory in some respects. Sedimentation has been irregular, dating control is poor and there are problems of correlation between the two cores used for pollen analysis. Attention is focused here on Pleistocene sediments of clearly established ages: these sediments have well preserved pollen, contrasting with the poor preservation of pollen in many of the younger sediments nearer the surface.

The mean abundance of major pollen has been calculated for samples from three units of sediment (Fig. 2: spectra 13-15). The first unit consists of the depth range 5.75-6.20 m in core A, underlying a level at 5.20-5.40 m, ^{14}C-dated at 13 776 ± 80 BP (SRR-1121). Stratigraphic comparison with core B suggests that these sediments are not older than 23 000 BP. The second unit 7.30-8.30 m in core B: these sediments overlie a level at 8.50-8.85 m which is ^{14}C-dated at 23 073 ± 120 BP (SRR-1122); stratigraphic comparison with core A indicates that this unit is older than 14 000 BP. The final unit is 9.23-

9.68 m in core B, predating the level ^{14}C-dated at 23 073 BP.

Various modern and fossil pollen spectra from East Africa are shown on Figure 2 for comparison. The Pleistocene samples from Laboot contain much less montane forest tree pollen (9.3, 6.3 and 4.6% of total pollen in units **A**, B, C) and much more grass pollen (76.8, 61.5 and 62.9%) than do any modern pollen samples from Mt Elgon and the obvious implication is that the extent

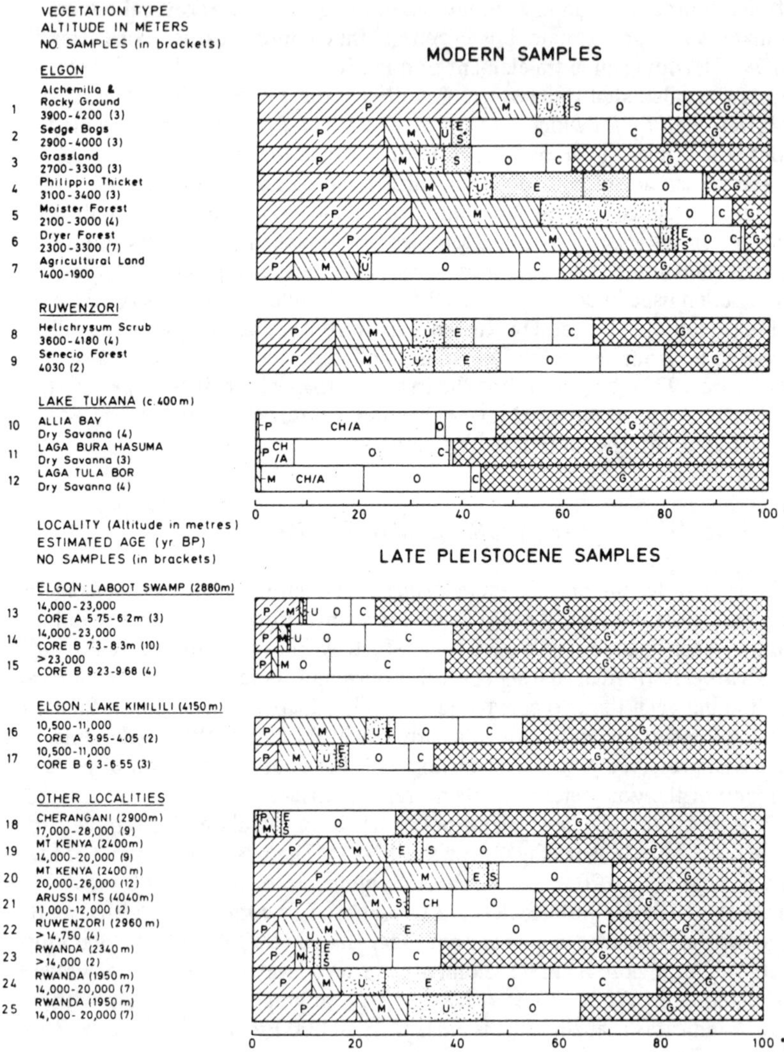

of grassland was greatly increased at the expense of forest. The climate must have been dry. High grass and low tree pollen percentages actually persist in the sediments for several metres above the level in core A dated at 13 776 ± 80 BP, indicating that open vegetation continued for some time after that date.

The percentages of montane forest tree pollen at Laboot can be compared with those in surface samples along transects extending from Lake Turkana towards montane forest. Surface samples from arid savanna east of Lake Turkana, 150 km distant from the nearest montane forest, contain between 0.4 and 0.9% montane forest tree pollen and between 49.3 and 61.7% grass pollen (Bonnefille & Vincens 1977; Fig. 2: spectra 11-12). As the distance to montane forest gets nearer, the percentage of montane forest pollen (mainly *Podocarpus*) increases. Bonnefille (1979) reports that surface samples from north of Lake Turkana, 80-100 km distant from montane forest, contain between 1.3 and 6.2% montane forest tree pollen, while Vincens (1982), working south and south-east of the lake, has found that the representation of montane forest pollen is never over 4% in surface samples from more than 30 km from montane forest. Judging by these findings, patches of montane forest were almost certainly present on or near Mt Elgon during the period of late Pleistocene aridity.

Vegetation on the lower slopes of Mt Elgon during the last glacial period may have borne similarities to that at 2 100-2 600 m on the Nyika Plateau in Malawi today (Meadows 1984) (because of its altitude Laboot Swamp itself is likely to have been situated above the upper montane forest limit). The Nyika is mostly grassland, with patches of montane forest in valley heads covering about 5% of the area. Forest pollen contributes about 5% of modern pollen rain in valley mires, which is a fairly similar figure to Laboot. However,

---

Figure 2. Modern and late Pleistocene pollen spectra from East Africa.
The following pollen types are always shown if sufficiently abundant to display and if included in the pollen sum: *Podocarpus* (P), other montane forest trees (M), Urticaceae (U), Ericaceae (E), *Stoebe* (S), Chenopodiaceae/Amaranthaceae (CH/A), Cyperaceae (C) and Gramineae (G). Others = O. The pollen sum is total pollen, except for Cherangani and Mt Kenya (excluding Cyperaceae and some minor aquatic and swamp taxa), modern Ruwenzori samples (excluding *Dendrosenecio*-type, *Helichrysum*-type, *Lobelia, Rubia*) and the third sample from Rwanda (no. 25) (excluding Cyperaceae, Ericaceae, *Sphagnum, Xyris*). Most spectra are based on the means of several samples: these are standardized to a pollen sum of 100% prior to calculating means. Samples 1-7 from Hamilton & Perrott (1980), samples 8-9 from Hamilton (1972), samples 10-12 from Bonnefille & Vincens (1977), samples 13-17, 21 and 23-25 from Hamilton (1982), samples 18-20 from Coetzee (1967) and sample 22 from Livingstone (1967). Sample 18 is for samples from core B on Kaisungor Swamp, samples 19-20 are from depths of 10.76-12.8 and 12.8-13.66 m in a core from Sacred Lake, sample 21 is from 2.87-3 m in a core from Mt Badda, sample 22 is from Zone 1a (depth 8.9-9.5 m) in a core from Lake Mahoma, samples 23 and 24 are from 1.5-4.55 m in a core from Kamiranzovu Swamp and sample 25 is from a depth of 2.25-2.5 m in a core from Kuwasenkoko Swamp.

Table 1. Montane forest pollen in late Pleistocene sediments of Laboot Swamp

	Number of grains[1]								Conclusions on presence/absence of parent taxa taxa on/near Mt Elgon
Core:	A	B	B	Total	% of montane forest tree pollen	% of total pollen and spore count	Relative export abilities[2]	Relative abundance of parent taxa[3]	
Depth (m):	5.75-6.2	7.3-8.3	9.23-9.68						
Age (kyr BP):	14-23	14-23	>23						
No. of samples:	3	10	4						
Podocarpus	79	145	40	264	77.2	4.93	4.29	100	Present
Juniperus	3	23	11	37	10.8	0.69	1.79	34	
Myrica		11		11	3.2	0.21	(0.5)	36	
Prunus	1	7		8	2.3	0.15	0.11	116	
Hagenia		5	2	7	2.0	0.13	0.47	24	
Olea		4	1	5	1.5	0.09	0.38	22	
Celtis comp.	1	1		2	0.6	0.04	(4)	1	Absent
Macaranga capensis		2		2	0.6	0.04	(1)	3	
Dombeya	1	1		2	0.6	0.04	(0.1)	33	Possibly present
Ilex		1	1	2	0.6	0.04	(0.1)	33	
Ebenaceae		1		1	0.3	0.02	0.11	15	
Rapanea		1		1	0.3	0.02	(0.1)	15	
Total montane forest tree pollen	85	202	55	342					
Total pollen/spores	914	3 230	1 212	5 356					

1. Unstandardized, i.e. the sum of numbers of grains encountered.
2. Figures without brackets from Hamilton & Perrott (1980); figures in brackets are based on the placing of types into the four categories of relative export ability in Hamiton (1972) and assignment of values of 4, 1, 0.5 and 0.1 according to which category they belong. Ebenaceae pollen is assumed to be poorly dispersed.
3. Standardized to *Podocarpus* = 100.

grass pollen production on the Nyika Plateau is probably depressed by burning (Meadows, pers. comm.) and forest cover on Elgon may have been less during the last glacial maximum than a simple comparison suggests.

Comparison of Laboot with other Upper Pleistocene spectra from montane East Africa also demonstrates the unusual rarity of montane forest tree pollen and abundance of grass. Compare, for example, the abundance of these elements in samples dating to between 26 000 and 14 000 BP from Mt Kenya and in samples slightly older than 14 750 BP from Ruwenzori (Fig. 2: spectra 19, 20, 22). As mentioned earlier in this chapter, the abundance of grass pollen in the Pleistocene sediments from Mt Kenya and Ruwenzori has been a focus of interest and interpretational dispute, but at neither of these sites does grass pollen approach the abundance which it achieves at Laboot. Furthermore, the overrepresented *Podocarpus* forms a greater proportion of the montane forest tree pollen at Laboot than at Mt Kenya or Ruwenzori. Thus, the evidence indicates that Mt Elgon was unusually treeless during the Upper Pleistocene.

The most similar Pleistocene pollen spectra to Laboot from montane East Africa are from sediments dating to between 27 750 ± 600 BP and 12 650 ± 100 at Kaisungor Swamp on the Cherangani Hills (Coetzee 1967). For example, montane forest tree pollen and grass pollen average 5% and 73% respectively in samples dating to 28 000 to 17 000 BP from the Cherangani Hills (Fig. 2: spectrum 18), and percentages are similar up to 12 650 BP. A dry climate is indicated. The Cherangani Hills lie only 120 km east of Mt Elgon and share many climatic similarities: if there was low rainfall on the Cherangani Hills, then this is also likely to have been the case on Mt Elgon. Therefore it can be postulated that the period of low forest cover on Mt Elgon may have extended from at least 28 000 to 12 650 BP.

An uncertainty in these comparisons is the extent to which the various sedimentary basins are comparable in terms of pollen deposition and preservation. In particular, the relative contributions of pollen carried in the atmosphere and overland into basins of different types in montane East Africa should be a subject of future research. The good state of preservation of pollen in the Pleistocene sediments at Laboot suggests that overland transport may have been relatively unimportant.

Although montane forest tree pollen is uncommon in the Pleistocene sediments at Laboot, its actual composition is of interest (Table 1). By far the commonest type is *Podocarpus* (77.2% of total montane forest tree pollen), followed by *Juniperus* (10.8%), *Myrica* (3.2%), *Pygeum* (2.3%), *Hagenia* (2.0%) and *Olea* (1.5%); other types are represented by only one or two grains. *Juniperus* and *Myrica* were more important constituents of forest than they are now, suggesting a particularly 'dry' type of montane forest (Hamilton 1982). Considering the relative export abilities of the pollen types, it is probable that *Hagenia, Juniperus, Myrica, Olea, Podocarpus* and *Pygeum* were growing on Elgon, even though confined to small forest patches, but that *Celtis* and *Macaranga capensis* (= *kilimandscharica*: Friis & Gilbert 1986) might not have been present (Table 1). The very few grains of the latter two taxa, both with pollen of moderate or good relative export ability, could well have blown in from a great distance. All tree taxa for which there is palynological evidence of presence on Elgon during the period of late Pleistocene aridity can grow in drier or higher altitude types of montane forest, a finding which is consistent with the model of forest history derived from distribution evidence mentioned earlier.

Among non-arboreal pollen types at Laboot, *Alchemilla* is present in some horizons at abundances which indicate the presence of *Alchemilla* scrub near the sample site (Hamilton 1982). *Alchemilla* scrub is also suggested by occasional grains of *Valeriana,* these being probably derived from *V.kilimandscharica,* a species strongly associated with *Alchemilla* scrub on Mt Elgon today (Hedberg 1964). Laboot is below the modern altitudinal limit of *Alchemilla* crub on Elgon and these signs of its former low altitude occurrence provide the only evidence of temperature depression in any of our pollen diagrams from the mountain. This confirms the general unsuitability of pollen analysis as a technique for establishing past changes in temperature through altitudinal movements of vegetation in regions which have been subject to massive changes in rainfall.

Anthoceros-spores, previously misidentified variously as Pteridium (Hamilton & Perrott 1980) and *Pteris*-type (Hamilton 1982), are common in some of the Pleistocene sediments at Laboot. In surface samples these spores are especially characteristic of montane grassland (Hamilton & Perrott 1980).

It might be supposed that depression of temperatures during the Upper Pleistocene on Mt Elgon would lower the altitude of the Ericaceous Belt, resulting in an abundance of Ericaceae and *Stoebe* pollen at the altitude of Laboot. In fact, Ericaceae and *Stoebe* pollen are very rare at Laboot, with only five grains of Ericaceae and two of *Stoebe* being recorded in a total pollen and spore count (for sediment units A, B, C) of 5 356 grains. There were no grains of *Cliffortia,* another microphyllous shrub of the East African Ericaceous Belt. It is concluded that aridity resulted in a great reduction of microphyllous thicket in addition to montane forest.

In summary, pollen spectra from Laboot contain many indications of aridity on Mt Elgon during a period extending from before 23 000 BP to after 14 000 BP (28 000-12 500 BP, considering also the evidence from the Cherangani Hills). Montane grassland was extensive and the cover of montane forest and microphyllous thicket greatly reduced.

POLLEN DIAGRAM FROM LAKE KIMILILI

Samples were collected from two cores in the lake sediments and pollen counts made by Alan Perrott on the whole of one core (A) and the basal part of the other (B). Complete pollen diagrams have been published previously in Hamilton (1982) and here montane forest tree pollen is presented

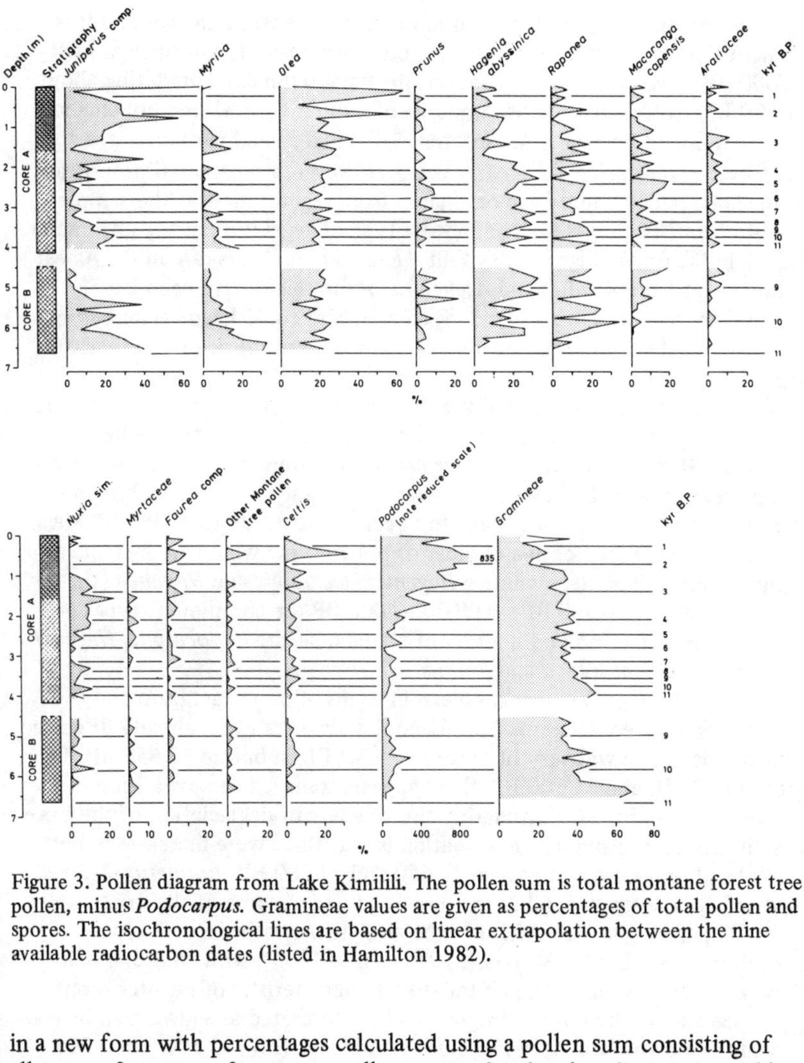

Figure 3. Pollen diagram from Lake Kimilili. The pollen sum is total montane forest tree pollen, minus *Podocarpus.* Gramineae values are given as percentages of total pollen and spores. The isochronological lines are based on linear extrapolation between the nine available radiocarbon dates (listed in Hamilton 1982).

in a new form with percentages calculated using a pollen sum consisting of all types of montane forest tree pollen except for the abundant and variable *Podocarpus* (Fig. 3). Exclusion of *Podocarpus* makes it easier to appreciate changes in the abundance of other pollen types. Nine ^{14}C dates show that sedimentation has been more or less continuous at both coring sites since 11 000 BP. For earlier times, attention is directed here particularly at the pollen diagram from core B which is believed to give the clearer picture of vegetation; the core B sediments accumulated in deeper water, at a faster rate of deposition and are better stratified palynologically. Unfortunately, this early period is rather inadequately dated.

Montane forest tree pollen is common in sediments of all ages and it is thought that montane forest has been extensive on Mt Elgon since at least 11 000 BP. Given this overall context, the basal sediments, predating about 11 500 BP, have relatively low amounts of montane forest tree pollen and high amounts of grass (Fig. 2: spectra 16-17; Fig. 3) and perhaps there was less forest cover during these early years. However, a substantial proportion of the grass pollen could have originated from vegetation near the sample site. *Alchemilla* scrub is the commonest type of vegetation at the altitude of Lake Kimilili on Mt Elgon today, but *Alchemilla* pollen is rare in the oldest sediments at Lake Kimilili and *Alchemilla* scrub might have been largely replaced by Afro-alpine grassland. Such grassland could have been a response to a dry climate or have constituted a seral vegetation type on the recently deglaciated land.

The basal sediments, especially in the better stratified core B, show major changes in the abundance of some types of montane forest tree pollen. *Juniperus, Myrica, Olea* and *Podocarpus* are the commonest montane forest tree pollen types at the base of the diagram (roughly 11 000 BP), but thereafter *Juniperus* and *Myrica* decline to become rare by about 8 000 BP. Meanwhile, there are increases in a number of pollen types which are infrequent in the lowest sediments, including *Hagenia, Macaranga* and *Rapanea* (these three from about 10 500 BP). At 9 000-7 000 BP the commonest types of montane forest tree pollen, in order of abundance, are *Podocarpus, Hagenia, Olea, Rapanea, Macaranga* and *Prunus.*

What do these pollen changes mean in terms of past vegetation and climate? The pollen diagrams from Laboot and Cherangani, already discussed, indicate that there was very little forest on Mt Elgon before 12 650 BP. Thus, between 12 650 and 11 500 BP, there appears to have been expansion in the area of montane forest. Considering also the date of deglaciation of the Lake Kimilili cirque, the simplest explanation is that there were increases in both rainfall and temperature between 12 650 and 11 000 BP. *Juniperus, Myrica, Olea* and *Podocarpus* are all characteristic genera of some of the driest types of montane forest in East Africa (some also grow in other forest types) (Hamilton 1982; Lind & Morrison 1974; White 1983), while *Hagenia* and *Macaranga,* at least on Mt Elgon today, are characteristic of moister forest types (see later). The pollen changes can be interpreted as showing an increase in rainfall over the late Pleistocene/early Holocene (say, 12 500-9 000 BP). Problems in dating and in disentangling the effects of other influences such as succession on plant populations inhibit a more detailed reconstruction of climatic history, though there is an indication of a temporary increase in the dry forest indicators *Juniperus* and *Myrica* at about 10 000 BP interrupting their general fall.

The evidence thus suggests major climatic and vegetational change on Mt Elgon between 12 500 and 9 000 BP. This is but one contribution to a whole body of information arguing for major environmental change in many parts of tropical Africa at this time (Livingstone 1975). For example, there were major increases in lake levels at about 12 500-10 000 BP in Kenya and Ethiopia, as well as in the Bosumtwi, Kivu and Victoria basins (Butzer et al. 1972, Gasse et al. 1980, Hecky & Degens 1973, Kendall 1969, Talbot et al. 1984, Tiercelin et al. 1981). The climate became much warmer and wetter on Ruwenzori at about 12 700 BP (Hamilton 1972, Livingstone 1967) and there was a rise in temperature on Mt Kenya at about 11 000 BP (Coetzee 1967). Vegetation in Somalia became more mesic after about 12 500 BP (Van Campo et al. 1982). Forest became established around Lake Victoria at 12 000 BP (Kendall 1969) and around Lake Bosumtwi soon after 9 000 BP (Talbot et al. 1984). The overall picture is of a warming and moistening climate throughout tropical Africa between 12 500 and 9 000 BP, with the spread of more mesic types of vegetation.

The Lake Kimilili diagram contains a very large increase in *Podocarpus* commencing soon after 4 000 BP. Two species of *Podocarpus* grow on Mt Elgon today, *P.falcatus* (= *gracilior*) in direr forests and *P.latifolius* (= *milanjianus*) in moister forest types (Hamilton & Perrott 1980). Generally, the genus is much commoner on the drier sides of the mountain, partly because extensive stands of bamboo at 2 450-3 000 m on the wetter slopes severely restricts the occurrence of trees such as *P.latifolius.*

Podocarpus pollen has a particularly high relative export ability and the rises in *Podocarpus* seen in pollen diagrams from East Africa could be but the most visible manifestations of climatically induced changes in forest composition involving many other taxa. The next section of this chapter contains an analysis of changes in abundance of montane forest pollen types at around 4 000 BP for both the Kimilili and Koitoboss diagrams. It is shown that there were indeed many other changes in the abundance of pollen types and that many of these could have been caused by the onset of a drier climate.

Studies of Holocene fluctuations in lake levels in tropical Africa have sometimes revealed complex stories, but an important fall in water level is reported from many basins at about 4 000 BP, including in Ethiopia (Gasse et al. 1980), along the Western Rift (Hecky & Degens 1973) and in Ghana (Talbot & Delibrias 1980, Talbot et al. 1984). Palynology suggests that increased dryness at this time was a climatic alteration of great importance to East African montane ecosystems.

The Kimilili diagram, in common with others from Mt Elgon and Mt Kenya (Coetzee 1967), shows few changes attributable to human influence, contrasting in this respect with pollen diagrams from western Uganda and Ethiopia (Hamilton 1972, 1982, Hamilton et al. 1986, Livingstone 1967, Morrison & Hamilton 1974). Reasons are believed to include the low palynological visibility of vegetation below the Montane Forest Belt due to low pollen production (Hamilton & Perrott 1980) and the maintenance of a substantial montane forest cover on Mts Elgon and Kenya into historic times.

POLLEN DIAGRAM FROM KOITOBOSS BOG

This bog lies at 3 940 m inside the eastern wall of the caldera. When selected for coring, its sediments were expected to contain a larger proportion of pollen derived from the drier eastern forests than those at Lake Kimilili. It was therefore a surprise that pollen analysis of the 2 m core of fibrous peat (basal date 6 505 ± 65 BP) actually revealed a relatively large quantity of pollen derived from the more mesic forests on the west. The pattern of air movements on the upper slopes of Elgon had clearly not been appreciated.

The pollen diagram is divided into two zones (upper A and lower B) on the basis of a marked increase in *Podocarpus.* This rise is assumed to be of a similar age to that at Lake Kimilili, enabling comparison of contemporaneous pollen spectra from the two sites. For the purpose of analysis the mean percentages of long-distance pollen types (those originating from taxa which do not grow in the Ericaceous or Afro-alpine Belts) have been calculated for Zones A and B at each site. The samples used for Zone B at Lake Kimilili are restricted to those estimated to postdate 6 500 BP. *Podocarpus* and Urticaceae pollen are very abundant at both sites (averaging 47 and 21% of the long-distance pollen) and have high relative export abilities; they have been excluded from the pollen sum in the belief that differences in the abundance of pollen types between zones will then more accurately reflect differences in the actual sizes of plant populations. This is a coarse analysis, lumping together pollen spectra covering long periods of time and hiding interesting details, but it is a first step in attempting to reconstruct forest history according to aspect using differential analysis of pollen diagrams.

Macaranga capensis and Urticaceae decline in abundance from Zones B to A at both sites (Table 2). *Macaranga* is a species confined to the Moist Lower Montane Forest Zone on Mt Elgon and the decline in abundance of its pollen could be due to decrease in the area of this forest type as a result of a drier climate. Increased dryness is also indicated by the Urticaceae fall, since Urticaceae pollen is better represented in surface samples from moist as compared with dry montane forest on Mt Elgon (Hamilton 1972, Hamilton & Perrott 1980). Both *Macaranga* and Urticaceae are more abundant in both zones at Koitoboss than Kimilili, which is part of the evidence that Koitoboss has received relatively more pollen from the moister forests.

Juniperus and *Podocarpus* behave in an opposite manner to *Macaranga* and Urticaceae, increasing in abundance from Zone B to Zone A at both localities and being more abundant at Kimilili than Koitoboss. *Juniperus* is a species confined to the drier slopes of Mt Elgon today and its expansion in Zone A and greater abundance at Kimilili are evidence of expansion in the area of Dry Montane Forest. As already mentioned, the genus *Podocarpus* is widely distributed in the modern Elgon forests, but achieves its greatest abundance on the drier slopes. It is therefore not surprising that *Podocarpus* pollen is most abundant at the site which has received most pollen from the drier forests.

Table 2. Koitoboss Bog and Lake Kimilili: comparative abundance of long-distance pollen in Zones A and B

Pollen type	Mean abundance (%)	More (+) or less (−) abundant at Koitoboss comparied with Kimilili		More (+) or less (−) abundant in Zone A compared with Zone B	
		Zone B	Zone A	Koitoboss	Kimilili
a. Pollen types originating predominantly from the west in both zones					
Urticaceae	21.5	+++	+++	−−	−−−
Macaranga capensis	3.0	+++	+++	−−−	−−−
Prunus	1.4	++	+++	++	−−−
Celtis comp.	1.3	++	++	+++	++
b. Pollen types originating predominantly from the east in both zones					
Podocarpus	47.4	−−	−−	+++	+++
Juniperus	1.8	−−−	−−−	+++	+++
Rapanea	1.4	−−−	−−−	+++	−−
Combretaceae comp.	0.5	−−	−−−	−−−	−−
c. Others					
Olea	6.0	−−	+	+++	+
Hagenia	4.6	−	+	−−	−−−
Monolete ferns	3.8	++	−−−	−−−	−
Acalypha	2.1	−	+	++	+
Myrica	0.6	+	−−−	−−−	+
Others	4.6	+	−	+	+

Samples from 0-187 cm at Kimilili (17 samples) and 0-67.5 cm at Koitoboss (7 samples) are assigned to Zone A; samples from 187-300 cm at Kimilili (9 samples) and 76.5-205 cm at Koitoboss (14 samples) are assigned to Zone B. The top (0-1 cm) sample from Kimilili contains abundant pollen of planted trees and has been omitted. The total long-distance pollen in each sample has been standardized to give a sum of 100, excluding *Podocarpus* and Urticaceae. The mean of these standardized figures has then been calculated for each pollen type for each zone and it is these means which are used for comparisons between sites and zones. Three symbols indicate that the larger of the two figures in a comparison exceeds the smaller by $>100\%$; two by 50-100%; one by $<50\%$. Mean abundance (first column) are the means of the means for the four zones (pollen sum including *Podocarpus* and Urticaceae).

Three pollen types (*Celtis, Acalypha,* Combretaceae comp.) are believed to have blown in entirely or mainly from vegetation growing at exceptionally low altitudes. *Celtis* pollen is more abundant at Koitoboss than at Kimilili, implying that it has been derived mainly from the west, which at first sight is surprising in that *Celtis* is unrecorded from montane forest on this side of the mountain, but is present in the lowest altitude montane forests on the east. The explanation could be that the pollen has come predominantly not from montane but rather from lowland forest, such as still survives in places in southern Uganda and which before agricultural clearance probably extended right up to the western slopes of Mt Elgon (Langdale-Brown et al. 1964). *Celtis* is a characteristic and abundant genus of lowland semi-deciduous forest in Uganda. The increase in *Celtis* pollen from Zone B to Zone A which occurs at both sites could thus have been due to an increase in the area of this forest type in southern Uganda, perhaps because it replaced evergreen forest under a drier climate. In fact, there is previous palynological evidence for such displacement (Kendall 1969).

An increase in abundance of *Acalypha* pollen from Zone B to Zone A at both Koitoboss and Kimilili could also have been due at least partly to expansion of drier types of lowland forest. The genus can be abundant today in semi-deciduous forest around Lake Victoria. Like *Celtis, Acalypha* pollen also increases in abundance during the Upper Holocene in the Lake Victoria pollen diagram (Kendall 1969).

The family Combretaceae (the likely source of the Combretaceae comp. pollen) is a characteristic component of moist woodland in Uganda and western Kenya (Lind & Morrison 1974). Combretaceae comp. pollen is more abundant in both zones at Kimilili than at Koitoboss, suggesting a source mainly from the east. The decline in abundance of Combretaceae pollen from Zone B to Zone A at both sites could have been due in part to replacement of moist by drier woodland.

Today, *Hagenia* has a greater altitudinal range and is generally more abundant on the wetter western slopes of Mt Elgon, where it is both more common in the Upper Montane Forest Zone than on the east and where it is also the most frequent tree growing among the very extensive groves of bamboo (Hamilton & Perrott 1981). Declines in abundance of *Hagenia* pollen from zones B to A at both sites can be attributed to reduction in the extent of moister types of forest on the mountain. The greater abundance of *Hagenia* pollen in Zone A at Koitoboss compared with Kimilili matches the greater abundance of the tree on the western slopes. Less clear is why *Hagenia* values in Zone B are higher at Kimilili than at Koitoboss. A partial explanation could be that extensive stands of bamboo with *Hagenia* once occurred on the eastern slopes prior to expansion of dry montane forest.

The greater abundance in both zones of *Prunus* at Koitoboss and of *Rapanea* at Kimilili broadly reflects the relative abundance of these genera on the different aspects of Mt Elgon today. Both species are common in the Upper Montane Forest Zone, most markedly so on the east where *Hagenia* is less frequent. *Prunus* is also a conspicuous element of the Moist Lower Montane Forest Zone. Explanation of differences in abundance of these pollen types between zones requires some special pleading, although the pollen data can be interpreted as indicating an increase in Zone A of *Prunus* and *Rapanea* at the expense of *Hagenia* in the Upper Montane Forest Zone on the west and a decline in the extent of this zone as a whole on the east. The latter could have been due to altitudinal restriction caused by expansion of such trees as *Juniperus, Olea* and *Podocarpus.*

Two species of *Olea* grow on Elgon today, complicating interpretation of the pollen diagrams. However, the pollen of one species, *O.africana,* was distinguished during counting of the Koitoboss samples and this was found to

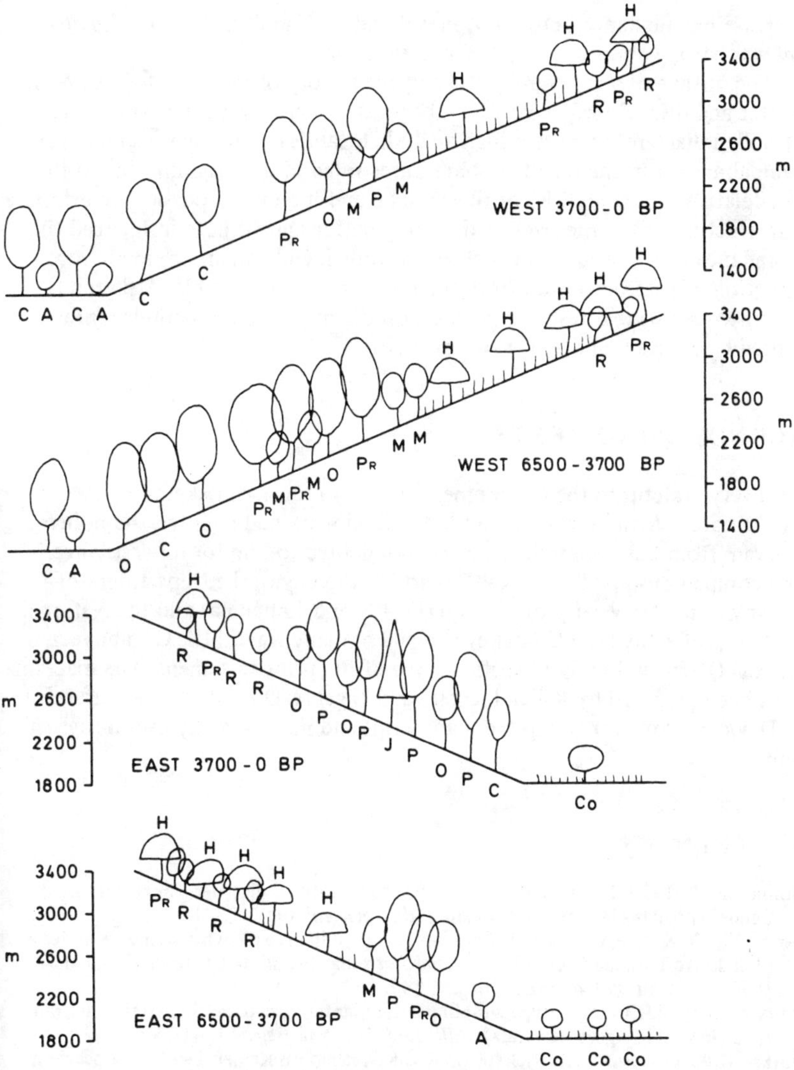

Figure 4. Reconstructed forest and lower altitude vegetation on the wetter western and drier eastern sides of Mt Elgon at 6 500-3 700 and 3 700-0 BP. The idea is to give an impression of how the sizes and distributions of species populations may have changed. Altitudes in metres. Trees unrepresented in the pollen record would have been abundant in some forest types, but are not shown.

A = *Acalypha*
C = *Celtis*
Co = *Combretaceae*
H = *Hagenia*
J = *Juniperus*
M = *Macaranga*
O = *Olea*
P = *Podocarpus*
Pr = *Prunus*
R = *Rapanea*
vertical lines = bamboo

increase in abundance between Zones B and A (Hamilton 1982). *Olea africana* shows a strong preference for dry montane forest.

The outline of a coherent picture of the history of montane forest on the wetter and drier slopes of Mt Elgon begins to emerge from this comparison of pollen diagrams from two high altitude localities. Some species must have been abundant in the forests but are unrecorded in the diagrams due to the low relative export abilities of their pollen and it seems impossible to estimate how fluctuations in the sizes of their populations might have influenced the abundance of those species which are recorded. Only an approximate reconstruction of past vegetation from the pollen record is possible. A sketch is presented showing forest composition on Elgon during the periods represented by Zones B and A at Koitoboss (Fig.4).

ACKNOWLEDGEMENTS

I am very grateful to the Government of Kenya for permission to carry out this research. Alan Perrott shared in the field work and counted the pollen diagram from Lake Kimilili: I am much indebted to him for his assistance and companionship. The Forest Department (Kenya), the Department of Botany at the University of Nairobi (Professor S.Imbamba) and the National Museums of Kenya (Dr C.Leakey) helped in many ways. Miss C.Kabuye and Dr A.D.Q.Agnew kindly identified some of the plant specimens. The diagrams have been prepared by Kilian McDaid and Nigel McDowell and I am grateful to David Taylor for reading the manuscript and making many useful suggestions.

REFERENCES

Bonnefille, R. 1979. Méthode palynologique et reconstitutions paléoclimatiques au Cénozoique dans le Rift Est Africain. *Bull. Soc. géol. Fr.* 21: 331-342.

Bonnefille, R. & A.Vincens 1977. Représentation pollinique d'environnements arides à l'Est du lac Turkana (Kenya). In *Recherches françaises sur le Quaternaire.* INQUA 1977, suppl. au *Bull. d'AFEQ* 50: 235-247.

Bouxin, G. 1974. Etude phytogéographique des plantes vasculaires du marais Kamiranzovu (forêt de Rugege, Rwanda). *Bull. Jard. bot. Nat. Belg.* 44: 41-159.

Butzer, K.W., G.L.Isaac, J.L.Richardson & C.K.Washbourn-Kamau 1972. Radiocarbon dating of East African lake levels. *Science (N.Y.)* 175: 1069-1076.

Coetzee, J.A. 1967. Pollen analytical studies in East and Southern Africa. *Palaeoecol. of Africa* 3: 1-146.

Diamond, A.W. & A.C.Hamilton 1980. The distribution of forest passerine birds and Quaternary climatic change in tropical Africa. *J. Zool. (Lond.)* 191: 379-402.

Flenley, J.R. 1973. The use of modern pollen rain samples in the study of the vegetational history of tropical regions. In H.J.B.Birks & R.G.West (eds.), *Quantitative plant ecology:* 131-141. Oxford: Blackwell.

Friis, I. & M.G.Gilbert 1986. Variation in *Macaranga capensis* (Euphorbiaceae). *Kew Bull.* 41: 68.

Gasse, F., R.Rognon & F.A.Street 1980. Quaternary history of the Afar and Ethiopian Rift lakes. In M.A.J.Williams & H.Faure (eds.), *The Sahara and the Nile*: 361-400. Rotterdam: Balkema.

Hamilton, A.C. 1972. The interpretation of pollen diagrams from highland Uganda. *Palaeoecol. of Africa* 7: 45-179.

Hamilton, A.C. 1974. Distribution patterns of forest trees in Uganda and their historical significance. *Vegetatio* 29: 21-35.

Hamilton, A.C. 1976. The significance of patterns of distribution shown by forest plants and animals in tropical Africa for the reconstruction of upper Pleistocene palaeo-environments: a review. *Palaeoecol. of Africa* 9: 63-97.

Hamilton, A.C. 1981. *A field guide to Uganda forest trees.* Kampala: Priv. publ., 95 pp.

Hamilton, A.C. 1982. *Environmental history of East Africa.* London: Academic Press. 328 pp.

Hamilton, A.C. & R.A.Perrott 1978. Date of deglacierisation of Mount Elgon. *Nature (Lond.)* 273: 49.

Hamilton, A.C. & R.A.Perrott 1979. Aspects of the glaciation of Mt Elgon, East Africa. *Palaeoecol. of Africa* 11: 153-161.

Hamilton, A.C. & R.A.Perrott 1980. Modern pollen deposition on a tropical African mountain. *Pollen Spores* 22: 437-468.

Hamilton, A.C. & R.A.Perrott 1981. A study of altitudinal zonation in the montane forest belt of Mt Elgon, Kenya/Uganda. *Vegetatio* 45: 107-125.

Hamilton, A.C., D.Taylor & J.C.Vogel 1986. Early forest clearance and environmental degradation in south-west Uganda. *Nature (Lond.)* 320: 164-167.

Hecky, R.E. & E.T.Degens 1973. *Late Pleistocene-Holocene chemical stratigraphy and paleolimnology of the Rift Valley lakes of Central Africa.* Woods Hole Oceanogr. Inst., Mass., 93 pp.

Hedberg, O. 1951. Vegetation belts of the East African mountains. *Svenk. bot. Tidskr.* 45: 140-202.

Hedberg, O. 1954. A pollen-analytical reconnaissance in tropical East Africa. *Oikos* 5: 137-166.

Hedberg, O. 1964. Features of Afroalpine plant ecology. *Acta phytogeogr. sued.* 49: 1-144.

Hurni, H. 1981. Simen Mountains – Ethiopia: palaeoclimate of the last cold period (Late Wurm). *Palaeoecol. of Africa* 3: 127-137.

Kendall, R.L. 1969. An ecological history of the Lake Victoria basin. *Ecol. Monogr.* 39: 121-176.

Langdale-Brown, I., H.A.Osmaston & J.G.Wilson 1964. *The vegetation of Uganda and its bearing on landuse.* Govt. Printer, Uganda.

Lind, E.M. & M.E.S.Morrison 1974. *East African vegetation.* London: Longman. 257 pp.

Livingstone, D.A. 1962. Age of deglaciation in the Ruwenzori Range, Uganda. *Nature (Lond.)* 194: 859-860.

Livingstone, D.A. 1967. Postglacial vegetation of the Ruwenzori Mountains in Equatorial Africa. *Ecol. Monogr.* 37: 25-52.

Livingstone, D.A. 1975. Late Quaternary climatic change in Africa. *Ann. Rev. Ecol. Syst.* 6: 249-280.

Meadows, M.E. 1984. Past and present environments of the Nyika Plateau, Malawi. *Palaeoecol. of Africa* 16: 353-390.

Morrison, M.E.S. 1961. Pollen analysis in Uganda. *Nature (Lond.)* 190: 483-486.

Morrison, M.E.S. 1968. Vegetation and climate in the uplands of south-western Uganda during the Later Pleistocene Period, I. Muchoya Swamp, Kigezi District. *J. Ecol.* 56: 363-384.

Morrison, M.E.S. & A.C.Hamilton 1974. Vegetation and climate in the uplands of south-western Uganda during the Later Pleistocene Period, II. Forest clearance and other vegetational changes in the Rukiga Highlands during the past 8000 years. *J.Ecol.* 62: 1-31.

Nilsson, E. 1931. Quaternary glaciations and pluvial lakes in British East Africa. *Geogr. Annlr.* 13: 249-349.
Osmaston, H.A. 1958. *Pollen analysis in the study of the past vegetation and climate of the Ruwenzori and its neighbourhood.* Uganda Forest Dept., 32 pp.
Osmaston, H.A. 1965. The past and present climate and vegetation of Ruwenzori and its neighbourhood. D.Phil. thesis, Oxford Univ.
Osmaston, H.A. 1975. Models for the estimation of firnlines of present and Pleistocene glaciers. In R.F.Peel et al. (eds.), *Processes in physical and human geography* (Bristol Essays): 218-245. London: Heinemann.
Perrott, R.A. 1982. A high altitude pollen diagram from Mount Kenya: its implications for the history of glaciation. *Palaeoecol. of Africa* 4: 77-83.
Perrott, R.A. & F.A.Street-Perrott 1982. New evidence for a late Pleistocene wet phase in northern intertropical Africa. *Palaeoecol. of Africa* 14: 57-83.
Prell, W.L., W.H.Hutson, D.F.Williams, A.W.H.Be, K.Geitzenauer & B.Molfino 1980. Surface circulation of the Indian Ocean during the Last Glacial Maximum, approximately 18,000 yr BP. *Quat. Res.* 14: 309-336.
Ruddiman, W.F. & J.-C.Duplessey 1985. Conference on the last deglaciation: timing and mechanism. *Quat. Res.* 23: 1-17.
Talbot, M.R. & G.Delibrias 1980. A new late Pleistocene-Holocene water-level survey for Lake Bosumtwi, Ghana. *Earth and Planetary Science Letters* 47: 336-344.
Talbot, M.R., D.A.Livingstone, P.G.Palmer, J.Maley, J.M.Melack, G.Delibrias & S.Gulliksen 1984. Preliminary results from sediment cores from Lake Bosumtwi, Ghana. *Palaeoecol. of Africa* 16: 173-192.
Tiercelin, J.J., R.W.Renaut, G.Delibrias, J.le Fournier & S.Bieda 1981. Late Pleistocene and Holocene lake level fluctuations in the Lake Bogoria basin, northern Kenya rift valley. *Palaeoecol. of Africa* 13: 105-120.
Van Campo, E., J.C.Duplessy & M.Rossignol-Strick 1982. Climatic conditions deduced from 150-kyr oxygen isotope-pollen record from the Arabian Sea. *Nature (Lond.)* 296: 56-59.
Van Zinderen Bakker, E.M. 1962. A late-glacial and post-glacial climatic correlation between East Africa and Europe. *Nature (Lond.)* 194: 201-203.
Van Zinderen Bakker, E.M. 1964. A pollen diagram from equatorial Africa, Cherangani, Kenya. *Geologie & Mijnb.* 43: 123-128.
Van Zinderen Bakker, E.M. & J.A.Coetzee 1972. A re-appraisal of late Quaternary climatic evidence from tropical Africa. *Palaeoecol. of Africa* 7: 151-181.
Wayland, E.J. 1934. Rifts, rivers, rains and early man in Uganda. *J. Roy. anthrop. Inst.* 64: 333-352.
White, F. 1983. *The vegetation of Africa.* UNESCO. 356 pp.

CENTRAL AND WEST AFRICAN PALAEOENVIRONMENTS

FRAGMENTATION DE LA FORET DENSE HUMIDE AFRICAINE ET EXTENSION DES BIOTOPES MONTAGNARDS AU QUATERNAIRE RECENT: NOUVELLES DONNEES POLLINIQUES ET CHRONOLOGIQUES. IMPLICATIONS PALEOCLIMATIQUES ET BIOGEOGRAPHIQUES

J.MALEY
ORSTOM, UR 103 & CNRS, UA 327, Institut des Sciences de l'Evolution, Laboratoire de Palynologie, Université des Sciences et Techniques du Languedoc, Montpellier, France

RESUME

Des données polliniques nouvelles obtenues dans les forêts denses humides ('rain forests') du Ghana et de l'Ouest Cameroun, permettent de renouveler la question des refuges forestiers durant les périodes arides et froides du Quaternaire récent. Durant les périodes froides ou fraîches et suivant les variations de l'humidité, les biotopes montagnards soit forestiers, soit de milieu ouvert, ont pu s'étendre à basse altitude, remplaçant plus ou moins les forêts de plaine, réfugiées alors dans trois secteurs isolés, constituant ainsi un refuge dit de Haute-Guinée, une série de refuges Camerouno-Gabonais et enfin un refuge du Zaïre oriental (Fig. 1). Cette vaste extension montagnarde a probablement maintenu un écran entre les savanes soudaniennes au nord et zambéziennes au sud, expliquant en particulier la différence au niveau spécifique de leurs flores.

D'un point de vue paléoclimatique, jusqu'à maintenant il a été difficile de réconcilier les données de température de surface des océans tropicaux obtenues par CLIMAP avec les données obtenues, par exemple, sur l'Afrique tropicale. Ces dernières montrent que lors du dernier maximum glaciaire, l'aridité a été alors très forte et l'abaissement de la température compris en moyenne entre 5 et 8°C. Toutefois, lorsqu'on prend en compte la grande intensification des upwellings à cette époque, avec par exemple pour la bande équatoriale de l'Atlantique un abaissement maximum de 8 à 9°C en été, il est possible d'expliquer les changements importants survenus sur l'Afrique tropicale par l'action refroidissante et aridifiante des couvertures nuageuses stratiformes qui sont générées par les eaux froides de l'Atlantique tropical. Ces nuages, qui se présentent souvent sous forme de manteaux nuageux persistants et généralement pas ou peu précipitants, interceptent le rayonnement solaire et provoquent un abaissement des températures de surface.

Des exemples actuels d'extension localisée de biotopes montagnards à basse altitude sont décrits. Ces modèles actuels montrent comment des couvertures nuageuses persistantes peuvent expliquer l'extension des conditions montagnardes en plaine durant les périodes froides du Quaternaire. Les upwellings, actuellement synchrones à travers le Golfe de Guinée, amplifient donc le signal des alizés qui seraient ainsi à l'origine des changements climatiques et biogéographiques survenus sur l'Afrique tropicale.

ABSTRACT

The fragmentation of the African rain forest and the extension of montane biotopes during the late Quaternary: new pollen and chronological data. Palaeoclimatic and biogeographic inferences.

New pollen data from the rain forests of Ghana and West Cameroun permit us to re-examine the question of forest refugia during late Quaternary arid and cold periods. During the cold or cool periods and in response to changes in humidity, the montane biotopes, forested or open, spread to low altitude, more or less replacing the lowland forests. These forests occurred in three isolated refugia, the Upper Guinea, the Cameroun-Gabon series, and the eastern Zaïre (Fig. 1). This extensive montane complex probably served as a screen between the sudanian savanna to the north and the zambezian to the south, particularly explaining the difference at the specific level of their floras.

The palaeoclimatic implications of, on the one hand, the CLIMAP sea surface temperature data, and on the other, the temperature and rain data from continental tropical Africa, are difficult to reconcile. The continental data for the last glacial maximum indicate increased aridity and temperatures lowered by 5 to 8°C. However, it is possible to explain the major changes in continental temperatures in terms of an interaction between tropical Atlantic upwellings and cloud formation. In particular, stratiform cloud cover produced by the cold Atlantic waters (maximum lowering during the last glacial maximum, 8 to 9°C), caused a cooling and drying of the adjacent land areas. These clouds formed persistent covers, producing little or no rain, but intercepting solar radiation and causing lowered temperatures.

Examples are described of present-day extensions to low altitudes of montane biotopes. These models illustrate that persistent cloud covers could have caused the spread of montane conditions to lowlands during cold Quaternary periods. The upwellings, presently synchronous throughout the Guinea Gulf, amplify the trade winds which could account for the climatic and biogeographic changes in tropical Africa.

1 INTRODUCTION: LES PROBLEMES BIOGEOGRAPHIQUES

1.1 *La forêt dense humide de basse altitude*

Les botanistes et les zoologistes travaillant en Afrique tropicale et en particulier dans la zone occupée par la forêt dense humide de basse altitude ont été souvent intrigués par la répartition assez curieuse des plantes ou des animaux qu'ils étudiaient. En effet, vu la grande homogénéité climatique actuelle du bloc forestier, on pourrait s'attendre à ce que la répartition des espèces soit homogène, ou tout au moins qu'elle soit surtout affectée par la grande 'coupure du Dahomey', qui voit les savanes de type soudanien s'étendre jusqu'à la mer, interrompant le bloc forestier sur environ 200 km au niveau du Togo et du Bénin (Fig. 1). Or en fait, la plupart des auteurs ont constaté que cette coupure avait un rôle assez mineur, c'est à dire que peu d'espèces inféodées au milieu forestier y terminent leur répartition. Par contre, ces auteurs ont abouti à la conclusion que la coupure majeure se situe plus à l'est, entre le fleuve Niger et la Cross River à l'est du Nigéria. De plus, lorsqu'on peut disposer de nombreux points de prélèvements dans ce secteur, on constate que la coupure se fait surtout au niveau de la Cross River (pour les plantes, cf. par ex. Léonard 1965, Clayton & Hepper 1974, Brenan 1978, Soladoye 1985; pour les vertébrés, cf. par ex. Rand 1951, Marchant 1954, Moreau 1966, 1969, Eisentraut 1963, 1970, Bernardi 1966, Hall & Moreau 1970, Mayr & O'Hara 1986; pour les insectes, cf. Carcasson 1964, Bernardi 1975, Pierre 1979, Lachaise et al. 1986). La plupart de ces auteurs ont ainsi défini des flores ou des faunes dites de Haute Guinée pour celles à l'ouest de la Cross River et de Basse Guinée pour celles à l'est, qui comprennent en particulier les biotopes forestiers du bassin Congolais. Les biotopes les plus riches sont de loin ceux du Cameroun et du Gabon (ibid.). La coupure géographique occasionnée par le cours de la Cross River est infime par rapport à l'ampleur du phénomène biogéographique constaté, c'est pourquoi quelques auteurs considèrent que cette coupure majeure du bloc forestier, comme d'autres coupures moins importantes repérées en d'autres points, seraient, d'une manière ou d'une autre, la résultante des fluctuations climatiques sur l'Afrique équatoriale (cf. Marchant 1954, Moreau 1966, 1969, Eisentraut 1963, 1970, Hall & Moreau 1970, Richards 1963, Hamilton 1976, 1982; etc.).

1.2 *L'extension des conditions montagnardes*

En Afrique tropicale, les études sur la répartition des plantes et des animaux entre des montagnes souvent très éloignées, comme par exemple entre les montagnes de la Dorsale Camerounaise et celles du Rift de l'Afrique orientale, ont miss en évidence une grande similitude, soit par des espèces communes, soit par des sous-espèces ou des espèces très proches (espèces soeurs) (cf. les synthèses de Moreau 1963, 1966, Hamilton 1976, 1982, Schnell 1977, Bernardi 1979, White 1981). Ainsi, pour la végétation, les espèces communes entre le Mont Cameroun et l'Afrique orientale, sont de 57% pour la forêt montagnarde et de 49% pour la prairie montagnarde (Hall 1973). Ces flores et ces faunes typiquement montagnardes sont strictement inféodées à l'étage afromontagnard qui se situe généralement entre 1 000 et 3 000 à 3 500 m d'altitude; au-dessus on passe à l'étage afroalpin (cf. les synthèses citées ci-dessus).

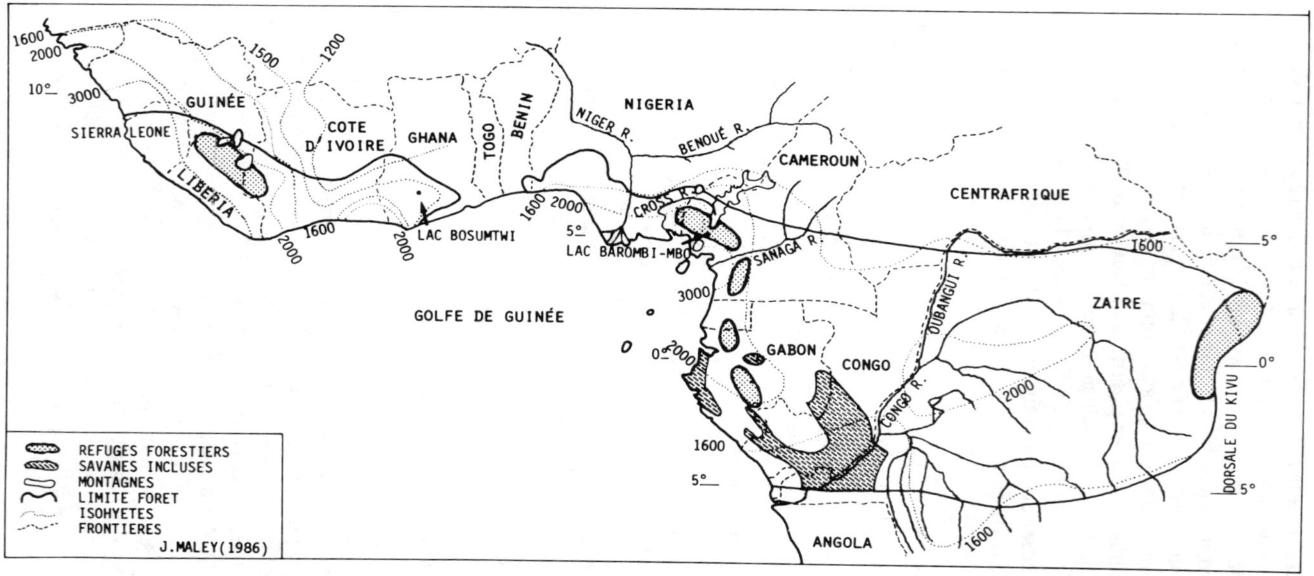

Figure 1. Schéma des refuges de forêts denses humides planitiaires durant le dernier maximum glaciaire. Les refuges de Haute Guinée et du Zaïre oriental sont d'après Endler (1985) et Mayr & O'Hara (1986), à partir des données de Moreau (1966, 1969) et Hall & Moreau (1970). Les conditions actuelles (limites de la forêt, savanes incluses et isohyètes principaux) sont adaptées de White (1983, Fig. 5 et map).

Les milieux montagnards sont actuellement des îles écologiques en position de refuge par rapport aux biotopes de plaine qui les isolent sur les montagnes. Afin d'expliquer les grandes similitudes floristiques et faunistiques rapportées ci-dessus, la plupart des auteurs ont considéré qu'à l'occasion de changements climatiques et particulièrement durant des périodes plus fraîches du Quaternaire, ces faunes et ces flores avaient pu s'étendre en plaine et ainsi migrer entre les massifs montagneux (Schnell 1952: 437, Morton 1961; et les synthèses citées, ibid.). Ces similitudes qui s'appliquent aussi à des êtres vivants aussi fragiles que des oiseaux, des batraciens ou divers groupes d'insectes (papillons, mouches) rendent improbable une explication par des transports à longue distance (cf. les discussions dans les synthèses citées, ibid., et Morton 1961, 1972, Hedberg 1969, Livingstone 1975).

En Afrique orientale, les travaux des géologues (Flint 1959, Livingstone 1962, Osmaston 1965, Downie & Wilkinson 1972, Messerli & Winiger 1980) sur l'extension maximum des glaciers au Quaternaire récent ont abouti à un abaissement de leur partie frontale de 650 m à plus de 900 m en certains points. Au Kilimandjaro, les névés se sont abaissés de 750 à 900 m environ d'après Hamilton (1982). Avec un abaissement de la pluviosité que Livingstone (1980) estime à environ 25%, la température aurait ainsi diminué de 6 à 9°C durant la dernière grande phase glaciaire. Les recherches des palynologues ayant travaillé en Afrique orientale avaient déjà apporté des données importantes (Van Zinderen Bakker 1962, 1964, Coetzee 1964, 1967, Morrison 1966, 1968, Livingstone 1967, Van Zinderen Bakker & Coetzee 1972, Hamilton 1972, 1982, Roche & Van Grunderbeek 1985, Vincens 1986) qui, pour la plupart de ces auteurs, et en premier lieu Coetzee (ibid.), montrent clairement l'extension des végétations montagnardes à des altitudes nettement plus basses qu'actuellement. Ainsi au lac Sacré sur le Mont Kenya, les données polliniques montrent que la forêt montagnarde s'est abaissée de 1 000 à 1 100 m vers 18 000 ans NP, conduisant à une diminution de la température de 5,1 à 8,8°C (Coetzee, ibid, Van Zinderen Bakker & Coetzee 1972). En conclusion, Hamilton (1973, 1982) et Flenley (1979) montrent dans des études synthétiques qu'une bonne convergence émerge de ces diverses données, indiquant clairement pour cette époque un abaissement de la température compris entre 5 et 9,5°C.

2 LES DONNEES POLLINIQUES OBTENUES EN FORET DENSE HUMIDE

Les différents problèmes biogéographiques exposés ci-dessus ont été renouvelés par les recherches palynologiques qui sont actuellement menées sur les dépôts quaternaires récents de deux lacs situés à basse altitude en forêt dense équatoriale; le lac Bosumtwi au Ghana et le lac Barombi-Mbo dans l'Ouest Cameroun (Fig. 1).

2.1 *Le lac Bosumtwi au Ghana*

Pour le lac Bosumtwi (Maley & Livingstone 1983, Talbot et al. 1984, Maley 1986 et en préparation) (Fig. 2) les données principales concernent tout d'abord l'extension à basse altitude d'un élément montagnard typique, caractérisé par l'Olivier montagnard, *Olea hochstetteri* – actuellement présent 700 km plus à l'ouest, vers 1 200 m d'altitude – dont les pollens se rencontrent parfois en abondance entre environ 9 000 ans BP et la base de la carotte vers 28 000 ans BP, avec des pourcentages pouvant atteindre 17% avant 19 000 ans BP et 25% entre 15 000 et 9 000 ans BP environ (Maley, en préparation). Grâce à l'examen de fragments de cuticules de Gramineae, Palmer (1982) (Fig. 2C) a corroboré cette donnée en mettant en évidence dans les mêmes unités de la carotte une flore graminéeenne de Pooideae qui ne se rencontrent actuellement en zone tropicale que sur les hautes montagnes (Livingstone & Clayton 1980), comme par exemple au Mont Cameroun où elles n'apparaissent qu'au-dessus de 2 000 m (Letouzey 1968, 1985) (cf. Section 7.3). Les données polliniques impliquent une réduction de la température moyenne de 3°C au minimum. La présence d'une flore de Pooideae pourrait impliquer un abaissement de la température supérieur à 6°C.

En même temps que ces éléments montagnards s'étendaient à basse altitude, la forêt disparaissait pour laisser la place à une prairie de type montagnard avec des bouquets d'arbres épars. La base de la carotte, vers 27-28 000 ans BP, avec de nombreux pollens d'arbres, supérieurs à 50%, a vu l'extension d'un faciès montagnard nettement forestier. Entre environ 25 000 ans et 9 000 ans BP les pollens de Gramineae étaient en mouenne supérieurs à 50% (actuellement moins de 10%). Une phase d'aridité maximum s'est produite entre environ 19 000 et 15 000 ans BP. Les pollens de Gramineae et de Cyperaceae ont atteint alors de 91 à 94%. Les pollens d'arbres recensés ne représentaient plus que 4 à 5% – alors qu'actuellement ils sont de 76 à 85% – et, fait remarquable, ils appartenaient alors dans leur grande majorité à la flore forestière semi-caducifoliée et non à la flore des savanes soudano-guinéennes. Seules actuellement les prairies montagnardes guinéennes de moyenne altitude (étage submontagnard, cf. Section 7.3) comportent des bouquets d'arbres isolés constitués, en plus des espèces montagnardes typiques, de tout un cortège d'espèces dont l'aire s'étend aussi aux basses altitudes (Schnell 1977).

La forêt est réapparue assez brusquement entre 9 000 et 8 500 ans BP, en même temps que les pollens de l'élément montagnard disparaissaient presque complètement des spectres. Cette forêt a perduré jusqu'à l'actuel, malgré une diminution marquée de la pluviosité au cours de l'Holocène récent, comme le montre la baisse des niveaux lacustres après 4 000 ans BP. Le maintien de la forêt indique que, si le total des pluies a diminué, sa répartition annuelle n'a pas dû varier.

Le Lac BOSUMTWI au Ghana.

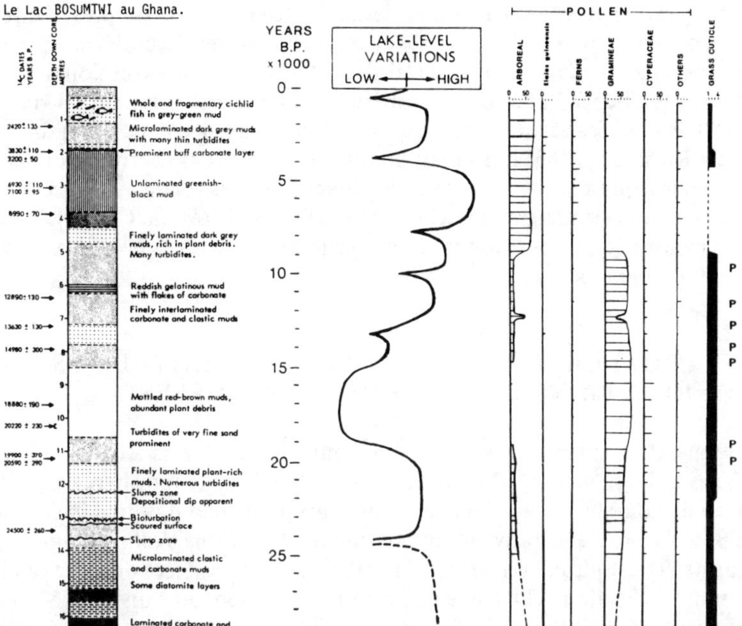

A - Stratigraphie de la carotte B-7

B - Schéma des variations du niveau du lac Bosumtwi.

C - Données polliniques synthétiques. P, Cuticules de Pooideae (Gramineae afro-montagnardes)

Figure 2. Le lac Bosumtwi au Ghana depuis 28 000 ans.
A: d'après Talbot et al. (1984); B: d'après M.R.Talbot & K.Kelts, figure à paraître in *Geology*, la partie antérieure à 15 000 ans BP a été modifiée par référence à des données polliniques inédites de J.Maley; C: d'après Talbot et al. (1984), adapté avec des données polliniques inédites de J.Maley. Cuticules de Pooideae d'après P.G.Palmer.

2.2 *Le lac Barombi-Mbo dans l'Ouest Cameroun*

Au lac Barombi-Mbo, des analyses polliniques préliminaires ont été effectuées sur une trentaine d'échantillons allant de l'actuel jusque vers 25 000 ans BP (Maley & Brenac 1985, Maley, Brenac & Giresse, à paraître, Brenac, recherches en cours). Les phénomènes ont été différents dans l'Ouest Cameroun, car d'une part, l'extension d'un élément montagnard, caractérisé surtout par *Olea hochstetteri* et apparu dès la base des couches étudiées vers – 24 m, s'est achievé au cours du Pléistocène terminal, peu après 15 000 ans BP, plus tôt qu'au lac Bosumtwi. D'autre part, la grande importance des pollens d'arbres montre que les faciès forestiers ont toujours persisté depuis la base jusqu'à l'Actuel. La période située approximativement entre 20 000 et 15 000 ans BP est caractérisée par des pourcentages de pollens de Gramineae compris entre

20 et 30%, valeur plus élevée que pour les périodes antérieures ou postérieures (Brenac, commun. pers.). Il faut probablement interpréter l'accroissement de ces pourcentages par la présence de clairières et aussi par une extension des faciès ouverts au-delà d'une zone forestière subsistant dans le secteur du lac (Fig. 1). Ces données et interprétations s'accordent bien à la conclusion du botaniste Richards (1963) qui situait un refuge forestier dans cette région, en se basant sur la présence de plantes endémiques à aire restreinte à l'Ouest Cameroun et aussi sur la grande richesse floristique de la région. Cette richesse est bien confirmée par les récoltes effectuées récemment par le botaniste D.W. Thomas (commun. pers.).

3 LES PROBLEMES PALEOCLIMATIQUES: RAPPEL DES DONNEES PRINCIPALES ET DES HYPOTHESES

L'abaissement de la température de 5 à 8°C mis en évidence en Afrique orientale (Section 1.2), qui a pu atteindre aussi 6°C dans la zone forestière de l'Afrique de l'ouest (Section 2.1), pose plusieurs problèmes d'ordre climatique.

Le premier problème provient du fait que ce phénomène s'est poursuivi en Afrique de l'ouest jusque vers 9 000 ans BP, d'après les données obtenues au lac Bosumtwi (Section 2.1), à une époque qui a correspondu à un fort réchauffement à des latitudes plus hautes des deux hémisphères. Le second problème provient du fait que l'abaissement des conditions montagnardes à basse altitude s'est achievé plus tôt dans l'Ouest Cameroun, probablement peu après 15 000 ans BP (Section 2.2). Tout ceci semble bien montrer qu'il s'agit d'un phénomène intéressant, suivant les époques, toutes ou certaines parties de l'Afrique tropicale – pour ne parler que de ce continent. Cette constatation devrait donc aider à trouver une solution au problème posé.

De plus, quelques auteurs (cf. Rind & Peteet 1985), à la suite des résultats de l'opération Climap (1981) sur des estimations climatiques pour le dernier maximum glaciaire, ont constaté une inadéquation entre les températures estimées pour la surface des océans intertropicaux et les données climatiques déduites dans les régions tropicales (continents ou îles). En effet, concernant l'Afrique tropicale, CLIMAP (ibid.) a estimé que vers 18 000 ans BP les températures moyennes de surface de l'Océan Atlantique tropical avaient été en moyenne comparables aux actuelles et, d'une manière très localisée, inférieures de 1 à 3°C. Les simulations effectuées avec ces valeurs aboutissent à un abaissement de la température de ca. 3°C sur l'Afrique orientale (cf. Rind & Peteet, ibid.), ce qui est plus faible de moitié environ à celui obtenu par les données in situ (Section 1.2) et à un résultat totalement irréaliste pour les précipitations qui auraient ainsi doublé par rapport aux valeurs actuelles, alors que toutes les données in situ indiquent une forte aridification que Livingstone (1980) estime correspondre à une diminution de la pluviosité d'environ 25% (Section 1.2). Rind & Peteet (ibid.) montrent aussi qu'il n'est pas possible de concevoir entre 15 000 et 20 000 ans BP un gradient de température avec l'altitude différent de l'actuel, c'est à dire de 5 à 6°C pour 1 000 m. Ils concluent finalement (ibid.) que les estimations de CLIMAP (ibid.) pour les températures de surface des océans tropicaux ne sont pas adéquates. Or effectivement des estimations de température basées sur les Foraminifères (Prell et al. 1976) montrent en particulier un renforcement considérable de l'upwelling équatorial, avec des abaissements de température atteignant 8 à 9°C en été vers 18 000 ans BP (Fig. 5). Les températures de surface en dehors des zones de remontées d'eaux froides auraient assez peu changées. Lorsqu'on aura examiné le rôle climatique actuel des upwellings, on verra que ces constatations devraient permettre de réconcilier les données océaniques avec les données continentales (Section 6).

4 LE ROLE CLIMATIQUE ACTUEL DES UPWELLINGS ET DES NUAGES STRATIFORMES

La plupart des climatologues travaillant sur l'Afrique tropicale ont bien montré la relation étroite qui existe entre les remontées d'eaux froides dans le Golfe de Guinée (s. lato) – essentiellement le courant de Benguela, le sous-courant équatorial et la zone côtière allant de la Côte d'Ivoire au Nigéria (Fig. 3) – et la génération des nuages stratiformes ou des phénomènes climatiques associés à ces nuages (Hoflich 1972, Drochon 1976, Saint-Vil 1977, 1979, Bakun 1978, Herman et al. 1980, Slingo 1980, etc.) (Fig. 4). L'apparition en surface de la mer, d'eaux à une température inférieure à celle du flux de mousson, entraîne sa stabilisation en supprimant les potentialités convectives de cette masse d'air humide. Dans ces conditions, seuls des nuages stratiformes peuvent apparaître. Ensuite, sur le continent voisin, tant qu'il n'y aura pas de réchauffement par la base ou de convection dynamique, ces nuages n'évolueront pas ou peu et ils pourront ainsi s'étendre très loin à l'intérieur des terres. Les nuages stratiformes sont généralement peu ou pas précipitants; lorsqu'ils donnent des pluies, il s'agit de pluies fines, de type crachin, non-orageuses (cf. Maley 1981, 1982). De plus, la persistance de manteaux nuageux intercepte le rayonnement solaire, entraînant ainsi un fort rafraîchissement de la température des basses couches de l'atmosphère (cf. Herman et al. 1980, Crane & Barry 1984, Le Treut 1985), limitant aussi l'évaporation (Fig. 4). Ce refroidissement ne peut que renforcer la stabilité de ces couvertures nuageuses.

Actuellement, l'influence des remontées d'eaux froides et des nuages stratiformes a un impact assez réduit sur la végétation forestière au nord du Golfe de Guinée, car la petite saison sèche estivale qui en résulte ne dure habituellement que deux mois (août et septembre) (Fig. 4). Par contre, vers l'est, cette influence est beaucoup plus marquée, avec d'importantes conséquences pour la végétation. En effet, durant l'été boréal, de juin à septembre, la forêt Gabonaise est recouverte en permanence par une couverture nuageuse stratiforme qui peut s'étendre à plus de 800 km des côtes (Saint-Vil

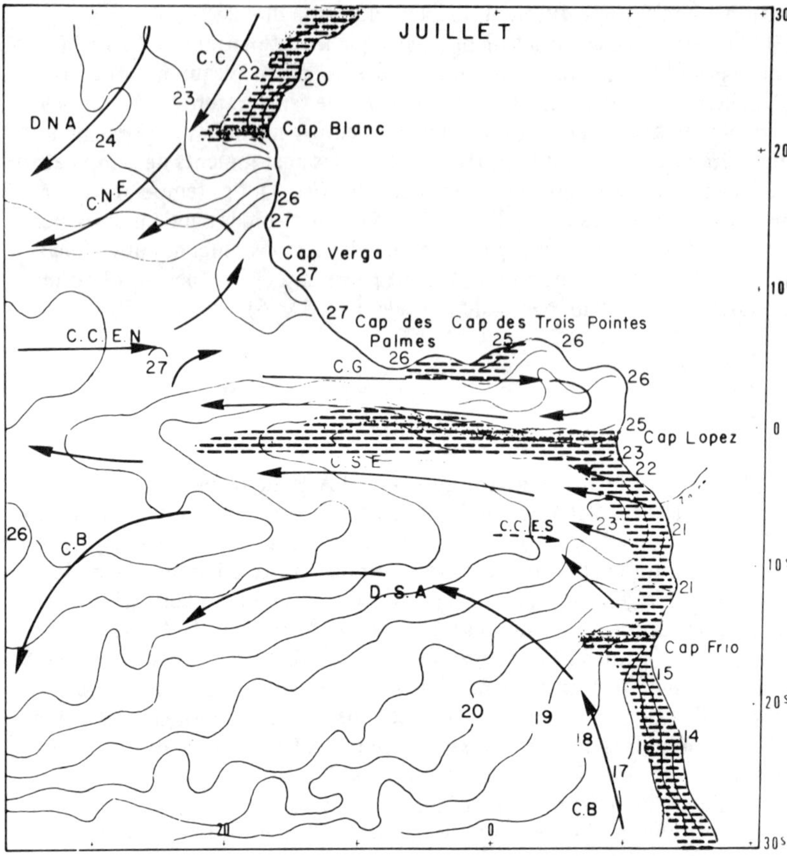

Figure 3. Température, circulation de surface et remontées d'eaux froides (upwellings) sur l'Atlantique tropical. *Les courants* (du nord au sud):
CC: Courant des Canaries; DNA: Dérive nord-Atlantique; CNE: C. nord-équatorial; CCEN: Contre courant équatorial nord; CG: C. de Guinée; CSE: C. sud équatorial; CCES: Contre courant équatorial sud; DSA: Dérive sud Atlantique; CB: C. de Benguela. Extrait de Wauthy (1983, Fig. 20A).

1977, 1979). Ces nuages, quasiment non précipitants, déterminent la grande saison sèche, toutefois l'abaissement de la température qui diminue beaucoup les pertes par évaporation, est responsable d'un faible déficit de saturation. C'est ce phénomène qui permet à la forêt Gabonaise de subsister, car avec une quasi absence de pluie pendant quatre mois – même avec des précipitatons très importantes durant les autres mois – la plus grande partie du Gabon devrait être couverte de savanes. Cependant, une trop longue persistance des

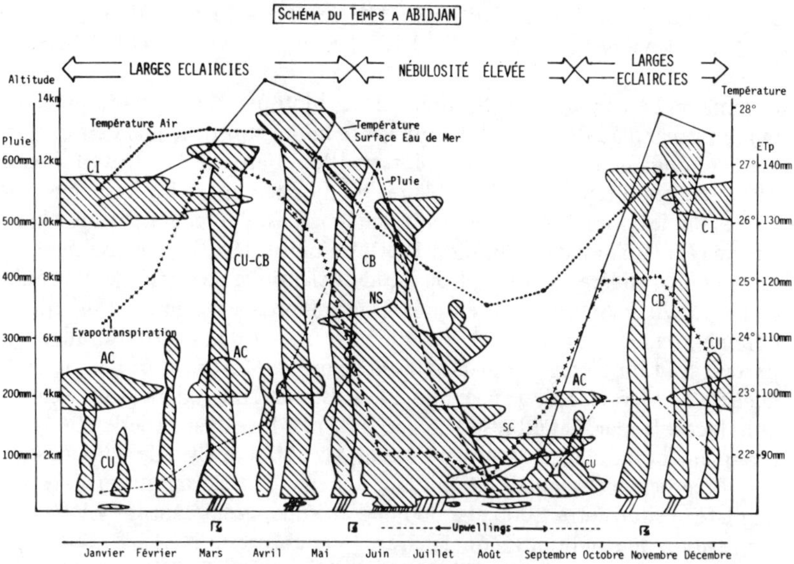

Figure 4. Schéma de l'évolution annuelle du temps à Abidjan, montrant la succession des principaux types de nuages (d'après Drochon 1976, ASECNA) et des éléments principaux du climat (température de surface de l'eau de mer, température de l'air, pluie et évapotranspiration: données ASECNA et ORSTOM). La station d'Abidjan est représentative du climat de la zone forestière occidentale.

couvertures nuageuses non précipitantes peut finalement entraîner l'élimination de la forêt, car l'extension des savanes de la zone côtière (Fig. 1) dépend de cette saison sèche estivale qui dure cinq mois dans ce secteur (cf. Saint-Vil ibid.). Il est probable aussi que les savanes incluses du Congo (Fig. 1) ont, pour l'essentiel, la même origine (Samba-Kimbata 1978).

En conclusion, la relation entre la température des eaux de surface de l'océan et les facteurs climatiques est donc évidente (cf. Flohn 1983, Hastenrath 1984). En particulier, la relation avec la présence ou l'absence de pluies est très nette car, a contrario, les années où les upwellings sont faibles ou nuls, et donc avec persistance des eaux chaudes de surface, des pluies exceptionnelles sont enregistrées sur la zone forestière (Hisard 1980), et parfois aussi sur une grande partie de la zone tropicale (Hooke 1970). L'année 1984 qui a vu une suppression presque complète des upwellings, s'est traduite au Gabon par des précipitations considérables de juin à septembre (Guillot 1985).

5 QUELQUES MODELES ACTUELS D'EXTENSIONS MONTAGNARDES LOCALISEES A BASSE ALTITUDE

Comme on l'a vu plus haut (Section 1.2) les biotopes montagnards apparaissent en général au-dessus de 1 000 m. Cette limite constitue une barrière biologique fondamentale (cf. par ex. Moreau 1966, Bernardi 1979, etc.). Toutefois quelques auteurs ont critiqué la réalité du phénomène car localement des faunes ou des flores habituellement montagnardes, s'étendent parfois à basse altitude (cf. Diamond & Hamilton 1980, Hamilton 1982). Effectivement, lorsqu'on considère par exemple la façade Atlantique de l'Afrique, les collines de Freetown en Sierre Leone en présentent un premier exemple. Ces collines dominent la mer en culminant vers 900 m et, dès 500 m d'altitude, elles supportent tout un cortège de plantes montagnardes (Morton 1968), dont *Olea hochstetteri* (cf. Hepper 1963). Plus à l'est, sur le flanc sud du Mont Cameroun, face à la mer, des arbres (Thomas 1985) et des oiseaux (Serle 1964) typiquement montagnards se manifestent dès 500 m d'altitude. Plus au sud, sur l'Escarpement Angolais qui fait face à la mer en s'élevant jusqu'au rebord du Plateau situé entre 1 000 et 2 000 m d'altitude, des botanistes y indiquent l'apparition de conditions montagnardes ('cloud forest', etc.) dès 200 à 300 m d'altitude (Exell & Mendonca 1937, Airy-Shaw 1947), en même temps que des oiseaux de type montagnard (Hall & Moreau 1962, Moreau 1966). D'autres exemples existent sur la côte de l'Océan Indien (cf. les Monts Usambaras; Moreau 1935, 1938, 1966).

Ces exemples montrent qu'on est en présence d'un phénomène écologique non fortuit qui étend à basse altitude les conditions climatiques – essentiellement la température – qu'on trouve habituellement sur les montagnes à partir de 1 000 à 1 500 m. La plupart des auteurs qui ont observé ces phénomènes localisés, les ont rattachés à la persistance de couvertures nuageuses et de brouillard qui sont particulièrement fréquents sur les reliefs face à la mer (Moreau 1966, Serle ibid., Airy-Shaw ibid.). Sur ces montagnes, la zonation des biotopes est donc atypique lorsqu'on considère seulement les altitudes, mais normale d'après les conditions écologiques. Ainsi, ce qui semblait au départ un point faible dans la démonstration de l'isolement actuel des taxons montagnards, se révèle finalement un point fort.

Ces exemples, qui correspondent en fait à des modèles actuels d'extension à basse altitude de biotopes montagnards, apparaissent très importants pour tenter d'expliquer les abaissements de température qui se sont produits sur l'Afrique tropicale au cours du Quaternaire récent (Sections 1.2 et 2). Le modèle de l'Escarpement Angolais est particulièrement précieux ici car les botanistes cités ci-dessus (ibid.) ont montré que l'abondance de la couverture nuageuse dépendait directement des nuages bas qui viennent de la mer où circule le courant froid de Benguela. Hoflich (1972) a démontré comment ce courant froid est à l'origine des épais manteaux de nuages stratiformes qui ont une influence majeure sur le climat du continent voisin en y réduisant les pluies et en y abaissant la température.

6 LE ROLE MAJEUR DES UPWELLINGS DANS LES CHANGEMENTS CLIMATIQUES SUR LA ZONE TROPICALE AFRICAINE

Les divers exemples actuels rapportés plus haut (Sections 4 et 5) de l'action climatique soit refroidissante, soit aridifiante des couvertures nuageuses stratiformes, montrent que les upwellings qui les génèrent ont un rôle climatique primordial. Quelques auteurs ont déjà montré comment les upwellings pouvaient avoir un impact important sur les paléoclimats de l'Afrique (cf. par ex. Van Zinderen Bakker 1982). Le rôle paléoclimatique des nuages stratiformes, comme relai des upwellings sur l'Afrique tropicale, a déjà été présenté dans ses grandes lignes (Maley 1985, 1986), en reprenant des idées exposées précédemment sur le rôle des nuages et des types de pluie associés aux processus de la sédimentation (Maley 1981, 1982, 1984).

La reconstitution des températures océaniques de surface par les assemblages de Foraminifères et de Radiolaires, montre que vers 18 000 ans BP les upwellings de l'Atlantique tropical et en particulier ceux de la bande équatoriale (sous-courant équatorial), à peu près sans changement dans leur position, étaient très intenses en été avec des températures inférieures aux actuelles de 4 à 9°C (Prell et al. 1976, Morley & Hays 1979) (Fig. 5). De plus, Prell et al. (ibid.) ont montré que même durant l'hiver boréal (février), qui est actuellement la saison aux eaux marines les plus chaudes, la température des eaux était encore de 3°C inférieure à l'actuelle. De ce fait, il apparaît que les remontées d'eaux froides devaient alors durer une grande partie de l'année, entraînant sur le continent voisin un abaissement important de la température ainsi qu'une forte diminution des pluies. Actuellement, les fluctuations interannuelles des différents upwellings (équatoriaux, côtiers, courant de Benguela, etc.) sont synchrones d'un point à l'autre du Golfe de Guinée (s. lato) (Merle 1980), ce qui conduit à penser que l'ampleur du phénomène durant les phases les plus froides du Quaternaire a dû avoir un impact renforcé sur le continent africain.

Les données de Prell et al. (1976) semblent montrer que dans la bande de l'upwelling équatorial les températures de surface sont revenues à un niveau voisin de l'Actuel vers 9 ou 10 000 ans BP, ce qui correspondrait assez bien au moment où la forêt équatoriale est réapparue sur l'Afrique de l'ouest (Fig. 2; cf. Section 2.1). Les océanographies ayant montré récemment (Servain et al. 1982, Picaut 1983) que c'est le renforcement des alizés sur l'Atlantique équatorial, particulièrement dans son secteur ouest, qui déclenche les upwellings dans le Golfe de Guinée, on peut donc relier la variation des alizés au cours du Quaternaire aux phénomènes climatiques discutés ici (cf. Flohn 1982). Or effectivement, diverses études ont montré un renforcement considérable des vents vers 18 000 ans BP et en particulier des alizés (Newell et al. 1981).

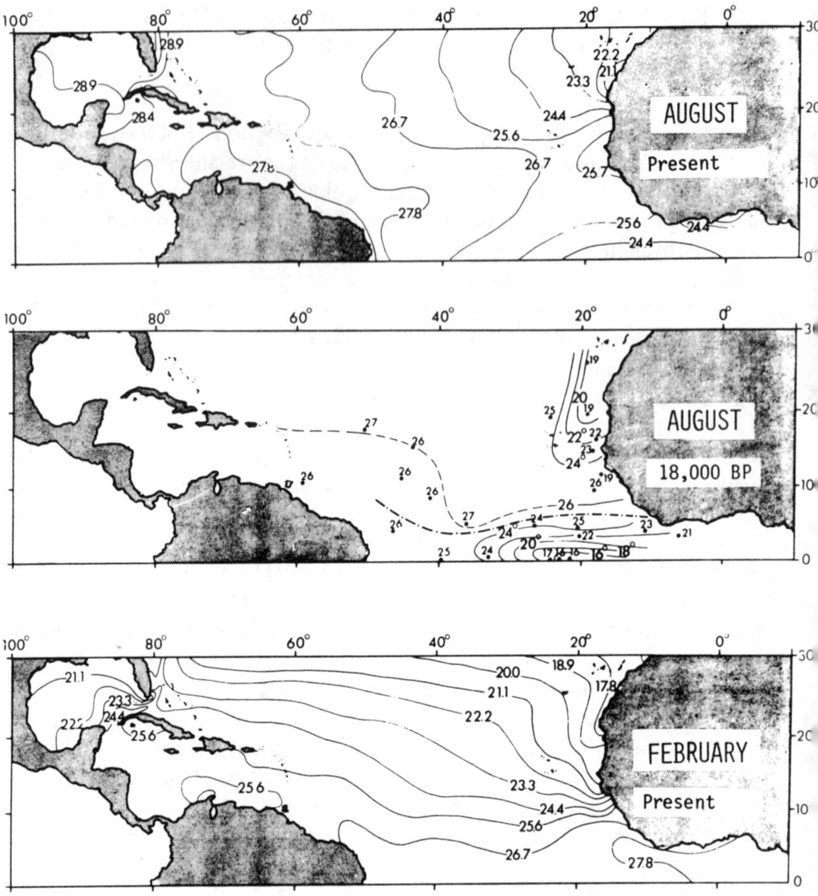

Figure 5. Reconstitution par les Foraminifères des températures de surface (août et février) sur l'Atlantique nord tropical durant le dernier maximum glaciaire, vers 18 000 ans BP. Valeurs actuelles en comparaison. Extrait de Prell et al. (1976, Fig. 11 et 12).

7 LES IMPLICATIONS BIOGEOGRAPHIQUES

7.1 *Les refuges forestiers durant la dernière grande phase aride et froide*

Les données polliniques présentées ci-dessus (Section 2) apportent des preuves directes, d'une part de la fragmentation de la forêt dense humide au cours du dernier maximum glaciaire, d'autre part de la présence effective d'un refuge forestier dans l'Ouest Cameroun. Ensuite, ces données ayant été acquises sur des sédiments déposés en continu, elles apportent une chronologie relativement précise des phénomènes grâce aux datations par le radiocarbone. Les lumières ainsi jetées sur l'histoire du domaine forestier équatorial africain au cours du Quaternaire récent, d'une part permettent d'établir une nouvelle approximation sur la localisation des refuges forestiers, particulièrement ceux de la zone centrale Camerouno-Gabonaise, d'autre part apportent une explication paléoclimatique de la coupure biogéographique mise en évidence au niveau de la Cross River.

Les biogéographes qui ont cherché à établir la position approximative des refuges forestiers, se sont basés essentiellement sur la grande richesse floristique et faunistique actuelle de certains secteurs forestiers, y compris l'endémisme spécifique et infra-spécifique. Pour le présent auteur, l'importance de la pluviométrie actuelle ne peut être qu'un critère secondaire car dans ce domaine, ce qui compte surtout est la longueur de la saison des pluies, plus que le total annuel, pourvu que celui-ci soit au moins supérieur à 1 200 m (cf. le nord de la forêt en Côte d'Ivoire et Ghana, Fig. 1). Il semble exister actuellement un certain consensus pour reconnaître trois régions principales ayant pu fonctionner comme refuges forestiers durant des périodes arides du Quaternaire (cf. Myers 1982).

1. Un secteur situé sur le nord du Libéria et l'ouest de la Côte d'Ivoire, correspondant au refuge dit de Haute Guinée (cf. Section 1.1).

2. Un secteur assez large situé sur le sud-ouest du Cameroun et l'ouest du Gabon, correspondant à la série de refuges Camerouno-Gabonais.

3. Un secteur assez étendu situé vers l'extrémité nord-est du Zaïre et correspondant au refuge du Zaïre oriental (parfois appelé refuge central en considération des bastions forestiers actuels de la Côte de l'Océan Indien).

La carte schématique des refuges présentée sur la figure 1, prend comme point de départ le schéma de Endler (1985), établi à partir des données de Moreau (1966, 1969); ce schéma a été récemment repris et commenté par Mayr & O'Hara (1986). Sur cette carte et sur d'autres un peu différentes des refuges en position côtière situés sur les principaux saillants vers le Golfe de Guinée, en particulier sur le saillant Gabonais, avaient été proposés d'abord par Aubreville (1949, 1962 pro parte) puis repris par Hamilton (1976) et Van Zinderen Bakker (1976). Pour des raisons climatiques qui ont été discutées plus haut (Section 4), une telle position côtière ne peut pas être retenue. En effet ces divers saillants présentent actuellement des savanes rélictuelles, parfois importantes comme au Gabon (Fig. 1) (pour la Côte d'Ivoire, cf. Portères 1950, Aubreville 1962: 28-30, Bellier et al. 1969; Gabon: cf. Aubreville

1962: 31-37). Le renforcement considérable des upwellings lors du dernier maximum glaciaire ne peut qu'avoir entraîné l'extension de ces savanes (Section 6). Ces données actuelles et ces déductions font donc préférer des refuges situés à quelque distance des côtes, tel le refuge de Haute Guinée de Endler (ibid.) et Mayr & O'Hara (ibid.).

A part cette critique des refuges côtiers, on ne discutera ici que des refuges du secteur Camerouno-Gabonais. Les données polliniques du lac Barombi-Mbo (Section 2.2) apportent en premier lieu une confirmation de la présence effective d'un refuge sur l'Ouest Cameroun. Ensuite ce sont des diverses données floristiques et faunistiques actuelles, telles la richesse taxonomique et la présence de taxons endémiques ou isolés par rapport à d'autres refuges, qui permettent d'essayer de cerner davantage leurs contours. Les limites proposées sur la Figure 1 sont une approximation qui devra être amendée lorsque de nouvelles données seront disponibles. De plus, tous ces refuges ne devaient pas comporter une forêt d'un seul bloc, mais plutôt de nombreux petits ensembles isolés ou faiblement rattachés entre eux.

L'extension du refuge de l'Ouest Cameroun en direction des collines d'Oban vers l'ouest, se base sur toutes les données floristiques et faunistiques rapportées en Section 1, plaçant une coupure majeure au niveau de la Cross River. Sa limite vers l'est, près du fleuve Sanaga, se base tout d'abord sur la grande richesse floristique des forêts de l'Ouest Cameroun qui d'après Richards (1963) et D.W.Thomas (commun. pers.) seraient les plus riches de toute l'Afrique forestière. Sa limitation avant la Sanaga se base aussi, par exemple, sur la découverte dans l'Ouest Cameroun, entre le nord des Monts Roumpi, le secteur de Kumba et à l'est vers Ngambé, d'un genre endémique, *Medusandra*, comportant deux espèces (Brenan 1952, Letouzey & Satabie 1974), pour lequel fût créée une Famille et même un Ordre nouveau. Une coupure très nette des biotopes forestiers de plaine au niveau de la Sanaga a été mise en évidence par Amiet (1985) dans ses études sur les Anoures du Cameroun, Bernardi (1966) a fait aussi la même constatation à partir de l'étude de certains groupes de papillons. La coupure matérialisée par le fleuve Sanaga se situe dans le secteur où les savanes du nord de la forêt s'approchent le plus de la côte. Un parallèle avec le 'V Baoulé' de Côte d'Ivoire a été fair par Amiet (ibid.).

Vers le sud-ouest, ce refuge ne devait pas atteindre le pied nord du Mont Cameroun, car dans ce secteur il existe encore actuellement quelques petites savanes rélictuelles entourées de taches de forêts semi-caducifoliées (Letouzey 1978), subsistant à l'heure actuelle par un effet d'abri sous le vent du Mont Cameroun: de Barombi-Kotto à Muyuka la pluviométrie tombe en-dessous de 2 000 mm/an (Eisentraut 1963, Suchel 1980). Autour du Mont Cameroun se rencontrent fréquemment des taxons de la forêt semi-caducifoliée (Ulmaceae et Sterculiaceae) (Vivien, Faure & Letouzey 1985), alors que la zone du refuge est dominée surtout actuellement par la forêt sempervirente très riche en Caesalpiniaceae (Letouzey 1968, 1985). L'isolement du Mont Cameroun par rapport à la Dorsale Camerounaise plus au nord, apparaîtra aussi plus loin lorsqu'on discutera des peuplements montagnards (Section 7.2).

Le refuge du sud-Cameroun, au sud de la Sanaga et le long du versant occidental du Plateau sud-Camerounais, a été délimité approximativement d'après les conclusions présentées par Amiet (1985) à la suite de ses recherches sur les Anoures. Ce refuge est confirmé par Myers (1982) qui rapporte qu'au sud de la Sanaga se situe une forte concentration de Mammifères endémiques, particulièrement des Primates (cf. Grubb 1982, Eisentraut 1963).

Les deux refuges du Gabon ont été délimités par référence à Aubreville (1962) qu'était un des botanistes connaissant le mieux ce pays. Discutant du problème des refuges forestiers, cet auteur écrivait (ibid., p. 63): 'C'est au Gabon qu'il nous semble que la forêt a conservé un domaine assez grand, non pas dans la région littorale sablonneuse où elle a dû disparaître rapidement, mais sur les massifs montagneux intérieurs, Monts de Cristal au nord de l'Ogooué, Monts de Chaillu au sud de l'Ogooué. Aujourd'hui c'est dans ces régions que la flore est la plus riche en genres et espèces de tout le massif forestier guinéo-congolais, parce qu'elle a résisté sur place à la période de dessèchement.' Deux critiques minimes sont à faire: d'une part des recherches ultérieures semblent avoir montré que la plus grande richesse floristique se trouve dans l'Ouest Cameroun (cf. supra), d'autre part les refuges de forêts de plaine n'ont pas pu exister vers le sommet des montagnes désignés-ci-dessus, mais probablement à leur pied et périphérie.

La présence actuelle de savanes incluses dans la boucle de l'Ogooué, vers Booué (Aubreville 1967) (Fig. 1), implique probablement qu'en période aride une jonction s'est effectuée avec les savanes côtières – comme le pense aussi Aubreville (1962 cf. supra) – et donc qu'une coupure a existé entre les deux refuges. Par contre, il serait possible que le refuge des Monts de Cristal ait eu quelques relations directes avec celui du sud-Cameroun, à travers la Guinée équatoriale qui est assez mal connue.

7.2 *La reconstitution de la forêt planitiaire vers 9 000 ans BP*

Les données polliniques et chronologiques obtenues au lac Bosumtwi montrent que la reconstitution de la forêt entre les refuges s'est produite assez rapidement aux alentours de 9 000 ans BP, en moins d'un millénaire. Cette donnée concerne plus spécialement la forêt ouest-africaine, mais quelques données polliniques (Roche 1979, Caratini & Giresse 1979, M'Benza-Muaka & Roche 1980) et d'autres d'ordre géologique (De Ploey 1969, Giresse & Lanfranchi 1984, Peyrot & Lanfranchi 1984) obtenues dans le bassin du Congo et au Congo occidental montrent une histoire comparable.

De part et d'autre de la 'coupure du Dahomey' (Section 1.1), la similitude relativement grande des flores et des faunes implique qu'à l'Holocène inférieur la forêt a aussi recolonisé cette zone. Plus à l'est, il ne semble pas faire de doute que la coupure biogéographique majeure de la Cross River soit la résultante de la longue disparition de la forêt à l'ouest de cette rivière (cf. Section 1.1). La 'coupure du Dahomey' est probablement un phénomène survenu plus récemment et résultant peut-être de la diminution de la pluviosité vers 4 000 ans BP (cf. Livingstone 1982, Robbins 1978), comme cela apparaît par exemple dans la chute du niveau du lac Bosumtwi à cette époque (Fig. 2B).

7.3 *Les migrations des taxons afromontagnards*

Une brève présentation de l'état de la question a été donnée en Section 1.2, puis, grâce à des données polliniques et d'autres sur les cuticules de Gramineae (Section 2.1), on a montré que des taxons montagnards ont existé à basse altitude, loin des montagnes, au cours du Quaternaire récent. De plus, des exemples actuels d'extensions localisées de biotopes montagnards (Section 5) montrent comment ces migrations ont pu s'effectuer. A ce propos on doit noter que des écologistes étudiant les conditions montagnardes ont bien mis en évidence le rôle primordial joué par les couvertures nuageuses et le brouillards (Grubb & Whitmore 1966, Grubb 1971, 1974, 1977).

Le problème de ces migrations apparaît toutefois assez complexe, car, si on a insisté jusqu'ici sur la coupure fondamentale entre flores ou faunes planitiaires et montagnardes, des distinctions toutes aussi importantes séparent physionomiquement les formations afromontagnardes de forêts de celles de prairies. La distinction écologique résulte essentiellement de la répartition des pluies. En effet, les prairies montagnardes apparaissent lorsque la pluviosité diminue et surtout lorsque la saison sèche augmente. Ce fait est bien illustré sur le Mont Cameroun où la limite forêt montagnarde – prairie correspond à une forte diminution du total annuel et surtout à un racourcissement de la saison des pluies qui ne dure plus en prairie que six à sept mois vers 2 500 m, juste au-dessus de la limite supérieure de la forêt montagnarde, alors que dans cette formation vers 1 000 m, il n'y a que zéro ou un mois sec chaque année (Fontes & Olivry 1976).

Sur les montagnes du Cameroun et particulièrement sur le Mont Cameroun, Letouzey (1968, 1985) a introduit une subdivision plus détaillée. Il distingue vers le bas un étage submontagnard s'étendant en moyenne de 1 000 à 2 000 m et comportant surtout des forêts. Leur caractéristique est de posséder, outre des espèces typiquement montagnardes comme *Olea hochstetteri,* tout un cortège d'espèces de forêt semi-caducifoliée guinéo-congolaise (Letouzey, ibid.). Plus haut, se trouve l'étage proprement montagnard entre 2 000 et 3 000 m environ. Suivant les conditions climatiques (cf. supra) et aussi actuellement les influences anthropiques, cet étage correspond soit à la forêt avec un cortège d'arbres typiquement montagnards, soit à des paysages ouverts de type prairie avec des bouquets d'arbres ou des arbres isolés. La prairie renferme de nombreuses Gramineae à affinités tempérées (Pooideae) (cf. Section 2.1).

Dans son étude de la vicariance chez les Anoures du Cameroun, Amiet (1985) a recensé chez les orobiontes (taxons montagnards) une répartition qui se calque sur les subdivisions établies par Letouzey (ibid.), avec un groupement particulier en forêt submontagnarde, un autre en forêt montagnarde et un troisième largement réparti dans ces deux étages. Il faut rappeler ici que les Anoures sont rares en milieu franchement ouvert. De cette répartition on pourrait déduire une histoire paléoclimatique assez complexe avec des phases de refroidissement distinctes et d'intensités différentes (Amiet 1975, 1980, 1985), ce qui est très plausible au vu des résultats présentés ici.

Lorsqu'on considère le peuplement des montagnes de l'Ouest Cameroun, on constate pour certains groupes un net appauvrissement du Mont Cameroun, ou la présence de formes parfois différentes, par rapport aux montagnes plus au nord de la Dorsale Camerounaise (pour les oiseaux, cf. Serle 1950, 1954; pour les Anoures, cf. Amiet 1980, 1985). Une coupure écologique persistante semble donc avoir existé dans le passé (cf. par exemple les savanes résiduelles du pied nord du Mont Cameroun; Section 7.1) entre cette montagne et le secteur des Monts Rumpi et Koupé.

Durant les périodes froides du Quaternaire le niveau de la mer s'est abaissé, parfois de plus de 100 m. C'est ainsi que l'île de Fernando Po a été rattachée au continent voisin par un isthme dont le point le plus bas se trouve actuellement vers −60 m. La communication a dû s'interrompre pour la dernière fois au cours du Pléistocène terminal (15 à 10 000 ans BP). Or il est intéressant de remarquer avec Eisentraut (1970) que le pourcentage des petits mammifères et des oiseaux vivant uniquement en plaine est, par rapport aux espèces pouvant vivre en altitude, nettement plus faible sur l'île (35 et 39% respectivement) que sur le continent voisin (60% pour les deux groupes). Ceci semble indiquer que le peuplement insulaire se serait fait à partir du continent préférentiellement durant les périodes froides et, on le constate encore ici, à un moment où les formes montagnardes s'étendaient à basse altitude (ibid.).

Les analyses polliniques effectuées tant au Bosumtwi qu'au Barombi-Mbo montrent que, d'une part des périodes froides et arides, comme par exemple celle survenue entre environ 20 000 et 15 000 ans BP, auraient permis la migration des taxons de biotopes très ouverts, d'autre part des périodes fraîches et plus humides, caractérisée par un environnement plus forestier comme vers 27-28 000 ans BP au Bosumtwi, auraient autorisé la migration des taxons de milieux forestiers. A ce propos il est intéressant de remarquer que Van Zinderen Bakker & Coetzee (1972) ont montré qu'en Afrique orientale une phase fraîche et relativement humide s'était produite vers 27 000 ans BP au Mont Kenya. Par ailleurs, dans son étude des oiseaux des montagnes africaines. Moreau (1966) a pu mettre en évidence deux phases de migration, l'une ayant concerné des oiseaux vivant en forêt montagnarde (s. lato), l'autre des oiseaux vivant en prairie montagnarde (moorland).

Une autre question beaucoup débattue a été la voie préférentielle de communication entre les montagnes d'Afrique orientale et celles du Cameroun (cf. Bernardi 1979). Certains auteurs, comme par exemple Moreau (ibid.), ont privilégié la voix la plus directe par le nord du bassin du Congo à travers la Centrafrique. Toutefois, s'il est probable que certains biotopes montagnards, préférentiellement de milieux ouverts, ont pu s'étendre à travers le bassin du Congo (cf. M'Benza-Muaka & Roche 1980), il apparaît que le flux principal des biotopes forestiers a dû emprunter surtout le bord sud du bassin du Congo, qui forme une large dorsale située en général au-dessus de 1 000 m et qui comporte actuellement de nombreuses populations satellites de taxons afromontagnards, isolés parmi des végétations typiquement planitiaires (cf. les cartes de White 1981, cf. aussi par ex. Lawton 1972). Après la 'plaque tournante' de l'Angola, qui a ici une position clé, cet axe de migration s'est poursuivi le long de l'Atlantique par la chaîne des collines du Mayombe, puis les Monts du Chaillu et les Monts de Cristal. Bien que les sommets ne soient pas très élevés – de 700 à 1 000 m – des plantes montagnardes ont été parfois collectées dans des stations rélictuelles à des altitudes assez basses. On peut ainsi noter deux stations de *Podocarpus latifolius (P. milanjianus),* l'une vers 700 m sur le flanc Congolais du Mont Chaillu (collection P.Sita 1977), l'autre vers 900 m dans le sud-Cameroun, près d'Ebolowa (Letouzey 1968). En ajoutant ces données à celles présentées par White (ibid.), on peut conclure que cette chaîne de collines, encore actuellement fréquemment baignée dans les nuages, a pu servir par intermittence au cours du Quaternaire de voie de communication préférentielle entre les peuplements montagnards d'Afrique orientale et ceux de l'Angola et du Cameroun.

Des confirmations directes de l'extension de peuplements montagnards sur ces régions durant des périodes plus fraîches sont apportées par quelques analyses polliniques effectuées en Angola et au Congo. Au nord-est de l'Angola, dans une région comprise entre 700 et 900 m, Van Zinderen Bakker & Clark (1962) et Clark & Van Zinderen Bakker (1963) présentent des analyses polliniques de sédiments datés entre environ 30 000 ans BP et le Pléistocène terminal, qui contiennent des pollens de plantes afro-montagnardes, *Ilex, Myrica, Erica* et *Podocarpus*; ce dernier genre est présent avec des valeurs atteignant de 18 à 25%, alors qu'actuellement les plus proches stations sont éloignées de 600 à 800 km et à des altitudes comprises entre 1 800 et 2 000 m (ibid.). Plus au nord, près de Pointe Noire au Congo, au large de l'estuaire de la rivière Kouilou qui draîne les collines du Mayombe, Caratini & Giresse (1979) ont trouvé dans des niveaux situés vers la fin du Pléistocène, des pourcentages élevés de pollens de *Podocarpus*, près de 20%. Cette donnée est la preuve de la grande extension prise à cette époque par les peuplements montagnards sur le Mayombe.

8 LE PROBLEME DES SAVANES SOUDANO-ZAMBEZIENNES

Tous les auteurs qui avaient traité jusqu'à présent de la question des refuges forestiers durant les périodes arides (Aubreville 1949, 1962, Moreau 1966, 1969, Hamilton 1976, Van Zinderen Bakker 1976) ainsi que le présent auteur (Maley 1976), avaient envisagé un envahissement des savanes de type soudano-zambézien entre les refuges forestiers. Pour cette hypothèse, la carte en couleur de Van Zinderen Bakker (ibid.) était très suggestive, indiquant un contact vers le centre du bassin du Congo entre les savanes soudanienne et zambézienne. Or il y a un empêchement majeur à cette hypothèse: les flores soudanienne et zambézienne sont composées en grande partie d'espèces différentes, même si de nombreux genres sont souvent communs (White 1965). On doit donc conclure qu'un écran s'est depuis longtemps interposé entre la flore soudanienne au nord et la flore zambézienne au sud.

Il apparaît bien maintenant que cet écran a été composé par des végétations montagnardes. Les données polliniques obtenues au lac Bosumtwi (Section 2.1), situé à environ 70 km au sud de la limite méridionale des savanes, en apportent une confirmation importante. En effet, même durant la phase aride intense survenue entre 19 000 et 15 000 ans BP, à un moment où le rapport entre les pourcentages de pollens de Gramineae et Cyperaceae et ceux des arbres était comparable à celui qu'on peut rencontrer actuellement au Sahel (cf. la station de Tjéri au Tchad, Maley 1981: 389), même à cette époque, quasiment aucun pollen de taxons typiquement savanicoles n'a été rencontré, sauf dans un échantillon quelques pollens d' *cacia* cf. *hockii* qui vit actuellement juste au nord de la forêt. Cela est très remarquable, surtout lorsqu'on sait qu'à un peu plus de 125 km à l'est, les savanes soudaniennes s'avancent jusqu'à la mer en formant la 'coupure du Dahomey' (cf Section 1.1) (Fig. 1).

9 CONCLUSIONS ET HYPOTHESES

De nombreux problèmes importants restent cependant sans réponse précise. Parmi ces problèmes, il y a en premier lieu celui de la relation spatiale entre les refuges de forêts planitiaires et les végétations montagnardes. En effet, on a montré que l'emplacement approximatif des refuges forestiers semblait se situer au pied et à la périphérie de certaines montagnes (Fig. 1). Par ailleurs, les diverses données exposées ou rappelées dans ce chapitre, concernant les migrations des végétations montagnardes, montrent clairement que ces végétations, soit forestières, soit de milieu ouvert, ont existé en plaine. Toutefois, on ne connaît pas l'extension réelle de ces végétations. On peut cependant raisonnablement supposer que durant les phases froides et arides, les formations de milieu ouvert se sont largement répandues et pendant les phases fraîches et plus humides, celles de milieux forestiers les ont plus ou moins remplacées.

Au cours de ces diverses périodes, il faut donc envisager une certaine imbrication des types de végétations montagnardes et planitiaires. Dans les secteurs occupés par les refuges de forêts de plaine, on peut penser que les végétations de plaine et de montagne ont pu se cotoyer suivant les conditions climatiques locales. Les refuges de forêts planitiaires étaient probablement constitués d'une série de petits refuges éparpillés au gré des conditions topographiques qui ont dû jouer un grand rôle pour exacerber les potentialités des facteurs climatiques.

Pour essayer de comprendre comment une telle mosaïque aurait pu exister, on peut rappeler que la petite saison sèche estivale – qui est sous la dépendance des upwellings et qui affecte la majeure partie du bloc forestier africain, mais avec des durées variables (un à deux mois en général, sauf au Gabon et dans certaines parties du Congo où elle peut atteindre quatre mois; cf. Section 7.1 et Leroux 1983) – n'affecte pas la zone forestière de l'Ouest Cameroun où, au contraire, cette période de l'année correspond aux précipitations maximales, à ce que Suchel (1972) appelle le 'paroxisme pluvial' de l'Ouest Cameroun. Ainsi on pourrait émettre l'hypothèse que durant les périodes plus froides, les reliefs de l'Ouest Cameroun auraient pu favoriser, peut-être comme actuellement, par des variations localisées d'ascendance surtout à la périphérie de ces reliefs, par effet orographique, la transformation des nuages stratiformes non précipitants en nuages précipitants, de type nimbo-stratus ou même de type cumuliforme (cf. Ludlam 1966). Ainsi localement, il aurait pu exister des températures plus élevées et des chutes de pluie nécessaires à la survie de végétations forestières de plaine.

Des études détaillées de climatologie sur l'Ouest Cameroun et aussi dans les secteurs localisés où s'observent actuellement des phénomènes d'extension à basse altitude de biotopes montagnards, pourraient peut-être apporter de nouveaux éléments concernant la paléoclimatologie du bloc forestier guinéo-congolais.

REMERCIEMENTS

Ce chapitre est un des résultats des recherches débutées il y a quelques années au Zoology Department, Duke University, USA, dirigé par D.A.Livingstone, dans le cadre du programme 'Deep Drilling in African Lakes'. Ces recherches ont été poursuivies ensuite au Laboratoire de Palynologie (Dir. M.Van Campo) de l'Institut des Sciences de l'Evolution de Montpellier (UA 327 du CNRS) (Dir. L.Thaler). Les carottages des lacs Bosumtwi et Barombi-Mbo ont été effectués par les équipes de D.A.Livingstone. Les recherches au Cameroun sont effectuées dans le cadre d'un programme sous la co-responsabilité de D.A.Livingstone et de l'auteur (ORSTOM, Unité de Recherches no. 103), en collaboration avec plusieurs organismes scientifiques du Cameroun (MESRES, ISH, CGN, CRH), avec l'Institut Polytechnique de Zurich en Suisse (K.Kelts) et avec les moyens logistiques et en personnel de l'ORSTOM au Cameroun (Dir. Ph.Mathieu). Ce programme, outre les aspects traités dans ce chapitre, comporte particulièrement des recherches en Limnologie, Stratigraphie, Sédimentologie, Paléomagnétisme, sur les Isotopes de l'Oxygène, du Carbone et du Beryllium. Le financement des opérations de carottage au Cameroun provient pour l'essentiel de la France (ORSTOM et CNRS) et des Etats-Unis (NSF et Duke University); pour le lac Bosumtwi au Ghana, le financement a été fourni par la NSF.

BIBLIOGRAPHIE

Airy-Shaw, H.K. 1947. The vegetation of Angola. *J. Ecology* 35: 23-48.

Amiet, J.L. 1975. Ecologie et distribution des Amphibiens Anoures de la région de Nkongsamba (Cameroun). *Ann. de la Fac. Sc. Yaoundé* 20: 33-107.

Amiet, J.L. 1980. Révision du genre *Leptodactylodon* Anderson (*Amphibia, Anura, Astylosterminae*). *Ann. de la Fac. Sc. Yaoundé* 27: 69-224.

Amiet, J.L. 1985. Aires disjointes et taxons vicariants chez les Anoures du Cameroun: implications paléoclimatiques. Manuscrit, Yaoundé, 18 p.

Aubreville, A. 1949. *Contribution à la paléohistoire des forêts de l'Afrique tropicale.* Paris: Soc. Edit. Géogr., Marit. et Colon., 99 pp.

Aubreville, A. 1962. Savanisation tropical et glaciations quaternaires. *Adansonia* 2 (1): 16-84.

Aubreville, A. 1967. Les étranges mosaïques forêt-savane du sommet de la boucle de l'Ogooué au Gabon. *Adansonia* 7 (1): 13-22.

Bakun, A. 1978. Guinea current upwelling. *Nature* 271: 147-150.

Bellier, L., D.Gillon, Y.Gillon, J.L.Guillaumet & A.Perraud 1969. Recherches sur l'origine d'une savane incluse dans le bloc forestier du Bas-Cavally (Côte d'Ivoire) par l'étude des sols et de la biocénose. *Cahier ORSTOM, série Biologie* 4: 65-94.

Bernardi, G. 1966. Aréotypes et chorologie de l'ouest africain principalement d'après les *Pieridae* (Lepidopt.). *J. W.Afric. Sc. Assoc.* 11: 49-67.

Bernardi, G. 1975. Contribution à la biogéographie de la Région Guinéenne: I. Le genre *Hypolimnas* Hübner (Lepidopt.-Nymphalidae). *C.R. Soc. Biogéogr. (Paris)* 452: 6-10.

Bernardi, G. 1979. Contribution à la biogéographie des montagnes africaines. I. Généralités. *C.R. Soc. Biogéogr. (Paris)* 479: 5-28.

Brenan, J.P.M. 1952. Plants of the Cambridge expedition, 1947-1948: II. A new order of flowering plants from the British Cameroons. *Kew Bull.* 2: 227-244.

Brenan, J.P.M. 1978. Some aspects of the phytogeography of Tropical Africa. *Ann. Missouri Bot. Gard.* 65: 437-478.

Caratini, C. & P.Giresse 1979. Contribution palynologique à la connaissance des environnements continentaux et marins du Congo à la fin du Quaternaire. *C.R. Acad. Sc. (Paris)* série D, 288: 379-382.

Carcasson, R.H. 1964. A preliminary survey of the zoogeography of African butterlies. *East Afr. Wild Life J.* 2: 122-157.

Clark, J.D. & E.M.Van Zinderen Bakker 1963. Prehistoric cultures of north-east Angola and their significance in tropical Africa. *Comp. Diamantes Angola* 62, 222 pp.

Clayton, W.D. & F.N.Hepper 1974. Computer-aided chorology of West African grasses. *Kew Bull.* 29: 213-234.

CLIMAP Project Members 1981. Seasonal reconstruction of the earth's surface at the Last Glacial Maximum. *Geol. Soc. America Map & Chart* Ser. MC-36.

Coetzee, J.A. 1964. Evidence for a considerable depression of the vegetation belts during the Upper Pleistocene on the East African mountains. *Nature* 204: 564-566.

Coetzee, J.A. 1967. Pollen analytical studies in East and Southern Africa. *Palaeoecol. of Africa* 3, 146 pp.

Crane, R.G. & R.G.Barry 1984. The influence of clouds on climate with a focus on high latitude interactions. *J. Climatology* 4: 71-93.

De Ploey, J. 1969. Report on the Quaternary of the western Congo. *Palaeoecol. of Africa* 4: 65-68.

Diamond, A.W. & A.C.Hamilton 1980. The distribution of forest passerine birds and Quaternary climatic change in tropical Africa. *J. Zoology (London)* 191: 379-402.

Downie, C. & P.Wilkinson 1972. *The geology of Kilimanjaro.* Dept. Geology, Univ. of Sheffied. 253 pp.

Drochon, A. 1976. Données climatologiques au sol et en altitude pour la station d'Abidjan. *Note ASECNA (Dakar)* 55, 14 pp.

Eisentraut, M. 1963. *Die Wirbeltiere des Kamerungebirges.* Hamburg: Parey. 353 pp.
Eisentraut, M. 1970. Eiszeitklima und heutige Tierverbreitung im tropischen Westafrika. *Umschau Wiss & Techn. (Frankfurt a.M.)* 3: 70-75.
Endler, J.A. 1985. Pleistocene forest refuges: fact or fancy. In G.T.Prance (ed.), *Biological diversification in the tropics*: 641-657. New York: Columbia Univ. Press.
Exell, A.W. & F.A.Mendonca 1937. *onspectus Florae Angolensis. I. Ranunculaceae-Aquifoliaceae.* Lisboa.
Flenley, J.R. 1979. *The equatorial rain forest: a geological history.* London: Butterworths. 162 pp.
Flint, R.F. 1959. Pleistocene climates in Eastern and Southern Africa. *Bull. Geol. Soc. America* 70: 343-374.
Flohn, H. 1982. Oceanic upwelling as a key for abrupt climatic change. *J. Meteo. Soc. Japan* 60 (1): 268-273.
Flohn, H. 1983. A climate feedback mechanism involving oceanic upwelling, atmospheric CO_2 and water vapour. In A.Street-Perrott et al. (eds.), *Variations in the global budget*: 403-418. Dordrecht: Reidel.
Fontes, J.C. & J.C.Olivry 1976. Premiers résultats sur la composition isotopiques des précipitations de la région du Mont Cameroun. *Cah. ORSTOM, série Hydrol.* 13 (3): 179-194.
Giresse, P. & R.Lanfranchi 1984. Les climats et les océans de la région congolaise pendant l'Holocène – bilans selon les échelles et les méthodes de l'observation. *Palaeoecol. of Africa* 16: 77-88.
Grubb, P.J. 1971. Interpretation of the 'Massenerhebung' effect on tropical mountains. *Nature* 229: 44-45.
Grubb, P.J. 1974. Factors controlling the distribution of forest-types on tropical mountains: new facts and a new perspective. In J.R.Flenley (ed.), *Altitudinal zonation in Malesia: Trans. 3rd Aberdeen-Hull Symp. on Malesian Ecology*: 13-45.
Grubb, P.J. 1977. Control of forest growth and distribution on wet tropical mountains, with special reference to mineral nutrition. *Ann. Rev. Ecol. Syst.* 8: 83-107.
Grubb, P.J. 1982. Refuges and dispersal in the speciation of African mammals. In G.T. Prance (ed.), *Biological diversification in the tropics*: 537-553. New York: Columbia Univ. Press.
Grubb, P.J. & T.C.Whitmore 1966. A comparison of montane and lowland rain forest in Ecuador. II. The climate and its effects on the distribution and physiognomy of the forests. *J. Ecology* 54: 303-333.
Guillot, B. 1985. Température de surface de la mer et pluviosité autour du Golfe de Guinée. *Veille Climatiq. Satellitaire Lannion, Mars* (6): 13-14.
Hall, J.B. 1973. Vegetational zones on the southern slopes of Mount Cameroon. *Vegetatio* 27: 49-69.
Hall, B.P. & R.E.Moreau 1962. A study of the rare birds of Africa. *Bull. of British Museum, Nat. Hist., Zoology* 8 (7): 313-378.
Hall, B.P. & R.E.Moreau 1970. *An atlas of speciation in African passerine birds.* London: British Museum Nat. History.
Hamilton, A.C. 1972. The interpretation of pollen diagrams from highland Uganda. *Palaeoecol. of Africa* 7: 46-149.
Hamilton, A.C. 1973. The history of the vegetation. In E.M.Lind & M.E.S.Morrison (eds.), *The vegetation of East Africa.* Longman: 188-209.
Hamilton, A.C. 1976. The significance of patterns of distribution shown by forest plants and animals in tropical Africa for the reconstruction of upper Pleistocene palaeoenvironments: a review. *Palaeoecol. of Africa* 9: 63-97.
Hamilton, A.C. 1982. *Environmental history of East Africa. A study of the Quaternary.* London: Academic Press. 311 pp.
Hastenrath, S. 1984. Interannual variability and annual cycle: mechanisms of circulation and climate in thr tropical Atlantic sector. *Mont. Weather Rev.* 112: 1097-1107.

Hedberg, O. 1969. Evolution and speciation in a tropical high mountain flora. *Biol. J. Linn. Soc.* 1: 135-148.
Hepper, F.N. 1963. Flora of West Tropical Africa. In J.Hutchinson & J.M.Dalziel (eds.), Pub. Crown Agents for Oversea Govern. Vol. 2. London.
Herman, G.F., M.C.Wu & W.T.Johnson 1980. The effect of clouds on the earth's solar and infrared radiation budgets. *J. Atmos. Sc.* 37: 1251-1261.
Hisard, P. 1980. Observation de réponses de type 'El Nino' dans l'Atlantique tropical oriental, Golfe de Guinée. *Oceanologica Acta* 3: 69-78.
Hoflich, O. 1972. Die meteorologischen Wirkungen kalter Auftriebswassergebiete. *Geoforum* 11: 35-46.
Hooke, P. 1970. Revenge of the gods. *Weather* 25 (9): 425-428.
Lachaise, D., M.L.Cariou, J.R.David, F.Lemeunier & L.Tsacas 1986. The origin and dispersal of the *Drosophila melanogaster* species subgroup: a speculative paleobiogeographic essay. Manuscrit, Gif/Yvette, 37 pp.
Lawton, R.M. 1972. A vegetation survey of northern Zambia. *Palaeoecol. of Africa* 6: 253-256.
Léonard, J. 1965. Contribution à la subdivision phytogéographique de la région guinéo-congolaise d'après la répartition géographique d'Euphorbiacées d'Afrique tropicale. *Webbia* 19 (2): 627-649.
Leroux, M. 1983. *Le climat de l'Afrique tropicale.* Paris: Champion. 633 pp. et Atlas (250 cartes).
Letouzey, R. 1968. *Etude phytogéographique du Cameroun.* Paris: Lechevalier. 508 pp.
Letouzey, R. 1978. Notes phytogéographiques sur les Palmiers du Cameroun. *Adansonia* 18: 293-325.
Letouzey, R. 1985. *Notice de la carte phytogéographique du Cameroun au 1:500.000.* Yaoundé: IRA/Toulouse: Inst. Carte Intertn. Végét.
Letouzey, R. & B.Satabie 1974. Une seconde espèce du genre *Medusandra* Brenan (Médusandracées). *Adansonia* 14: 63-68.
Le Treut, H. 1985. *Modélisation du système climatique: paramétrisation de la couverture nuageuse et étude de l'interaction entre les nuages et le rayonnement dans un modèle de circulation générale de l'atmosphère; variabilité du système climatique à l'échelle de temps des âges glaciaires.* Thèse Doct. Etat, Univ. Paris VI. 343 pp.
Livingstone, D.A. 1962. Age of deglaciation in the Ruwenzori range, Uganda. *Nature* 194: 859-860.
Livingstone, D.A. 1967. Postglacial vegetation of the Ruwenzori Mountains in Equatorial Africa. *Ecolog. Mongr.* 37: 25-52.
Livingstone, D.A. 1975. Late Quaternary climatic change in Africa. *Ann. Rev. Ecol. & Syst.* 6: 249-280.
Livingstone, D.A. 1980. Environmental changes in the Nile headwaters. In M.A.J.Williams & H.Faure (eds.), *The Sahara and the Nile. Quaternary environments and prehistoric occupation in northern Africa*: 339-359. Rotterdam: Balkema.
Livingstone, D.A. 1982. Quaternary geography of Africa and the Refuge Theory. G.T. Prance (ed.), *Biological diversification in the Tropics*: 523-536. New York: Columbia Univ. Press.
Livingstone, D.A. & W.D.Clayton 1980. An altitudinal cline in Tropical African grass, floras and its paleoecological significance. *Quat. Res.* 13: 392-402.
Ludlam, F.H. 1966. Cumulus and Cumulonimbus convection. *Tellus* 18: 687-698.
Maley, J. 1976. Essai sur le rôle de la zone tropicale dans les changements climatiques: l'exemple africain. *C.R. Acad. Sci. (Paris)* série D, 283: 337-340.
Maley, J. 1981. Etudes palynologiques dans le bassin du Tchad et paléoclimatologie de l'Afrique nord-tropicale de 30.000 ans à l'époque actuelle. *Travaux et Documents de l'ORSTOM* 129: 586 pp. (Thèse Sc. Montpellier, juin 1980).
Maley, J. 1982. Dust, clouds, rain types and climatic variations in tropical North Africa. *Quat. Res.* 18: 1-16.
Maley, J. 1984. Influence des nuages sur les paléoenvironnements: quelques exemples

pris dans le Quaternaire africain. *10ème Réunion Ann. Sc. Terre (Bordeaux)*: 373.
Maley, J. 1985. Histoire de la forêt dense au Quaternaire récent et reconstitution des paléoclimats en Afrique de l'Ouest par les analyses polliniques. *Colloque Prog. Nat. Et. Dyn. Climat.*: 15-16. Meudon: CNRS.
Maley, J. 1986. Fragmentation et reconstitution de la forêt dense humide ouest-africaine au cours du Quaternaire récent: Hypothèse sur le rôle des upwellings. In H.Faure, L.Faure & E.S.Diop (eds.), *Changements globaux durant le Quaternaire. Passé-présent-futur (Symp. INQUA, Dakar). Travaux & Docu. ORSTOM* 97: 281-282.
Maley, J. & P.Brenac 1985. Analyses polliniques préliminaires du Quaternaire récent de l'ouest Cameroun. *Symp. 'Palynologie et Milieux Tropicaux'. 9ème Congr. Assoc. Palyn. Langue Franç. (Montpellier)*, Résumé: 22.
Maley, J., P.Brenac & P.Giresse 1986. Histoire de la végétation et du climat de l'Ouest Cameroun au Quaternaire récent: résultats préliminaires. A paraître in *Revue de Géographie du Cameroun.*
Maley, J. & D.A.Livingstone 1983. Extension d'un élément montagnard dans le sud du Ghana (Afrique de l'ouest) au Pléistocène supérieur et à l'Holocène inférieur: premières données polliniques. *C.R. Acad. Sci. (Paris)* série 2, 296: 1287-1292.
Marchant, S. 1954. The relationship of the southern Nigerian avifauna to those of Upper and Lower Guinea. *The Ibis* 96: 371-379.
Mayr. E. & R.J.O'Hara 1986. The biogeographic evidence supporting the Pleistocene forest refuge hypothesis. *Evolution* 40: 55-67.
M'Benza-Muaka & E.Roche 1980. Exemple d'évolution paléoclimatique au Pléistocène terminal et à l'Holocène au Shaba (Zaïre). *Symp. 'Palynologie et Climats' (Paris). Mém. Museum Nat. Hist. Nat., série Bota.* 27: 137-148.
M'Benza-Muaka, E.Roche & H.Doutrelepont 1984. Note sur les apports de la palynologie et de l'étude des bois fossiles aux recherches géomorphologiques sur la vallée de la Lupembashi. *Revue Paléobiologie,* vol. sp., 149-154.
Merle, J. 1980. Variabilité thermique annuelle et interannuelle de l'océan Atlantique équatorial est. L'hypothèse d'un 'El Nino' Atlantique. *Oceanologica Acta* 3: 209-220.
Messerli, B., M.Winiger & P.Rognon 1980. The Saharan and East African uplands during the Quaternary. In M.A.J.Williams & H.Faure (eds.), *The Sahara and the Nile. Quaternary environments and prehistoric occupation in northern Africa*: 87-118. Rotterdam: Balkema.
Moreau, R.E. 1935. A synecological study of Usambara, Tanganyika Territory, with particular reference to birds. *J. Ecology* 23: 1-43.
Moreau, R.E. 1938. Climatic classification from the standpoint of East African biology. *J. Ecology* 26: 467-496.
Moreau, R.E. 1963. The distribution of tropical African birds as an indicator of past climatic changes. In F.C.Howell & F.Bourlière (eds.), *African ecology and human evolution*: 28-42. Chicago: Aldine.
Moreau, R.E. 1966. *The bird faunas of Africa and its islands.* New York: Academic Press. 424 pp.
Moreau, R.E. 1969. Climatic changes and the distribution of forest vertebrates in West Africa. *J. Zool. (London)* 158: 39-61.
Morley, J.J. & J.D.Hays 1979. Comparison of glacial and interglacial oceanographic conditions in the south Atlantic from variations in calcium carbonate and Radiolarian distributions. *Quat. Res.* 12: 396-408.
Morrison, M.E.S. 1966. Low-latitude vegetation history with special reference to Africa. *Proc. Int. Symp. World Climate 8000-0 BC (Royal Meteo. Soc. London)*: 142-148.
Morrison, M.E.S. 1968. Vegetation and climate in the Upslands of South-Western Uganda during the later Pleistocene period. *J. Ecology* 56: 363-384.
Morton, J.K. 1961. The upland floras of West Africa – their composition, distribution and significance in relation to climate changes. *C.R. 4ème Réunion Plénière AETFAT (Lisbonne)*: 391-409.

Morton, J.K. 1968. Sierre Leone. In I. & O.Hedberg (eds.), *Conservation of vegetation in Africa south of the Sahara. Acta Phytogeographica Suec.* 54: 72-74.
Morton, J.K. 1972. Phytogeography of the West African mountains. In D.H.Valentine (ed.), *Taxonomy, phytogeography and evolution*: 221-236. London: Academic Press.
Myers, N. 1982. Forest refuges and conservation in Africa, with some appraisal of survival prospects for tropical moist forests throughout the biome. In G.T.Prance (ed.), *Biological diversification in the Tropics*: 658-672. New York: Columbia Univ. Press.
Newell, R.E., S.Gould-Stewart & J.C.Chung 1981. A possible interpretation of paleoclimatic reconstructions for 18 000 BP for the region 60°N to 60°S, 60°W to 100°E. *Palaeoecol. of Africa* 13: 1-19.
Osmaston, H.A. 1965. The past and present climate and vegetation of Ruwenzori and its neighbourhood. Ph.D. thesis, Oxford.
Palmer, P.G. 1982. A stratigraphic analysis of fossil grass cuticles from Lake Bosumtwi, Ghana, West Africa and its paleoecological significance. *XIème Intern. Quat. Congress (Moscou)*. Vol. I, Abstract: 244.
Peyrot, B. & R.Lanfranchi 1984. Les oscillations morphoclimatiques récentes dans la vallée du Niari. *Palaeoecol. of Africa* 16: 265-281.
Picaut, J. 1983. Propagation of the seasonal upwelling in the eastern equatorial Atlantic. *J. Phys. Oceanogr.* 13: 18-37.
Pierre, J. 1979. Contribution à la biogéographie de la région guinéenne. III. Le genre *Acraea Fabricius* (Lépidoptère). *C.R. Soc. Biogéogr.* 481: 73-79.
Portères, R. 1950. Problème sur la végétation de la Basse Côte d'Ivoire. *Bull. Soc. Bota. France* 97: 153-156.
Prell, W.L., J.V.Gardner, A.W.H.Be & J.D.Hays 1976. Equatorial Atlantic and Caribbean foraminiferal assemblages, temperatures and circulation: interglacial and glacial comparisons. *Geological Soc. of America Memoir* 145: 247-266.
Rand, A.L. 1951. Birds from Liberia with a discussion for barriers between Upper and Lower Guinea subspecies. *Fieldiana Zool.* 32: 561-653.
Richards, P.W. 1963. Ecological notes on West African vegetation. II. Lowland forest of the southern Bakundu Forest Reserve. *J. Ecology* 51: 123-149.
Rind, D. & D.Peteet 1985. Terrestrial conditions at the Last Glacial Maximum and CLIMAP sea-surface temperature estimates: are they consistent? *Quat. Res.* 24: 1-22.
Robbins, C.B. 1978. The Dahomey Gap. A reevaluation of its significance as a faunal barrier to West African high forest mammals. *Bull. Carnegie Mus. Nat. Hist.* 6: 168-174.
Roche, E. 1979. Végétation ancienne et actuelle de l'Afrique centrale. *African Economic History* 7: 30-37.
Roche, E. & M.C.van Grunderbeek 1985. Apports de la palynologie à l'étude du Quaternaire supérieur au Rwanda. *Symposium 'Palynologie et Milieux tropicaux, IXème Congrès Assoc. Palyn. Langue Franç. (Montpellier)*. A paraître.
Saint-Vil, J. 1977. Les climats du Gabon. *Ann. Univ. Nat. Gabon* 1: 101-125.
Saint-Vil, J. 1979. *La grande saison sèche au Gabon.* Inst. Géogr. Trop., Univ. Abidjan. 17 pp.
Samba-Kimbata, M.J. 1978. Le climat du Bas-Congo. Thèse 3ème cycle. Dijon.
Schnell, R. 1952. Végétation et flore de la région montagneuse du Nimba (Côte d'Ivoire). *Mém. Inst. Franç. Afriq. Noire* 22: 604 pp.
Schnell, R. 1977. *Introduction à la phytogéographie des pays tropicaux. 4: La flore et la végétation de l'Afrique tropicale.* Paris: Gauthier-Villars. 378 pp.
Serle, W. 1950. A contribution to the ornithology of the British Cameroons. *The Ibis* 92 (3): 343-376.
Serle, W. 1954. A second contribution to the ornithology of the British Cameroons. *The Ibis* 96: 47-80.
Serle, W. 1964. The lower altitudinal limit of the montane forest birds of the Cameroon mountains, West Africa. *Bull. British Ornitho. Club* 84: 87-91.

Servain, J., J.Picaut & J.Merle 1982. Mise en évidence d'un couplage à longue distance entre le vent et la température de surface dans l'Atlantique Equatorial. *C.R. Acad. Sci. (Paris)* série 2, 294: 789-792.

Slingo, J.M. 1980. A cloud parametrization scheme derived from GATE data for use with a numerical model. *Quart. J. R. Meteo. Soc.* 106: 747-770.

Soladoye, M.O. 1985. A revision of *Baphia (Leguminosae-Papilionoideae). Kew Bull.* 40 (2): 291-386.

Suchel, J.B. 1972. La répartition des pluies et les régimes pluviométriques au Cameroun. *Trav. & Docu. Géogr. Trop., CEGET (Bordeaux)* 5: 287 pp.

Suchel, J.B. 1980. La répartition des pluies dans la région du Mont Cameroun. *Actes 16ème Congr. Intern. Météo Alpine, Soc. Météo France*: 191-198.

Talbot, M.R., D.A.Livingstone, P.G.Palmer, J.Maley, J.M.Melack, G.Delibrias & S.Gulliksen 1984. Preliminary results from sediment cores from Lake Bosumtwi, Ghana. *Palaeoecol. of Africa* 16: 173-192.

Thomas, D.W. 1985. Montane forests of Cameroon. Manuscrit, Missouri Bota. Gard., 9 pp.

Van Zinderen Bakker, E.M. 1962. A late-glacial and post-glacial climatic correlation between East Africa and Europe. *Nature* 194: 201-203.

Van Zinderen Bakker, E.M. 1964. A pollen diagram from equatorial Africa, Cherangani, Kenya. *Geologie & Mijnb.* 43: 123-128.

Van Zinderen Bakker, E.M. 1976. Tentative vegetation maps of Africa south of the Sahara during a glacial and an interglacial maximum. *Palaeoecol. of Africa* 9: colour map in frontispiece and p. IV.

Van Zinderen Bakker, E.M. 1982. African palaeoenvironments 18 000 yrs BP. *Palaeoecol. of Africa* 15: 77-99.

Van Zinderen Bakker, E.M. & J.D.Clark 1962. Pleistocene climates and cultures in north-eastern Angola. *Nature* 196: 639-642.

Van Zinderen Bakker, E.M. & J.A.Coetzee 1972. A re-appraisal of late-Quaternary climatic evidence from tropical Africa. *Palaeoecol. of Africa* 7: 151-181.

Vincens, A. 1986. Diagramme pollinique d'un sondage Pléistocène supérieur-Holocène du lac Bogoria (Kenya). *Rev. Palaeobota. & Palyno.* 7: 169-192.

Vivien, J., J.J.Faure & collab. R.Letouzey 1985. *Arbres des forêts denses d'Afrique centrale* (avec carte de distribution au Cameroun de 312 espèces, par R.Letouzey). Minist. Rel. Ext., Coop. & Dévelop., et Agence Coop. Cult. & Techn. 565 pp.

Wauthy, B. 1983. Introduction à la climatologie du Golfe de Guinée. *Océanographie Tropicale, ORSTOM* 19 (2): 103-138.

White, F. 1965. The savanna woodlands of the Zambezian and Sudanian domains. An ecological and phytogeographical comparison. *Webbia* 19 (2): 651-681.

White, F. 1981. The history of the Afromontane archipelago and the scientific need for its conservation. *Afr. J. Ecol.* 19: 33-54.

White, F. 1983. *The vegetation of Africa.* UNESCO/AETFAT/UNSO, Maps and Memoirs: 356 pp.

SOUTHERN AFRICAN PALAEOENVIRONMENTS

REGIONAL GEOMORPHIC EVIDENCE FOR CLIMATIC CHANGE IN SOUTHERN AFRICA SINCE THE MESOZOIC

R.R.MAUD
437 Smith Street, Durban, South Africa

T.C.PARTRIDGE
University of the Witwatersrand, Johannesburg, South Africa

ABSTRACT

Regional geomorphic evidence for climatic change in Southern Africa since the Mesozoic is reviewed. Since it was initiated at the breakup of Gondwanaland the Great Escarpment has had a marked effect on regional and local climates. The African erosion cycle initiated at the breakup carries laterites and silcretes on a characteristically deeply weathered kaolinized profile indicative of warm moist and semi-arid conditions respectively. Two periods of Tertiary uplift and tilting of the margins of the subcontinent, the first in the early Miocene and the second and greater episode in the late Pliocene, have had marked geomorphic and climatic effects. In the Quaternary, fluctuating climatic conditions are reflected in a variety of geomorphic effects.

FOREWORD

The contribution of Eduard van Zinderen Bakker to an understanding of African palaeoclimates is unrivalled and will remain a source of inspiration to workers in this field for many years to come. Although based primarily on his own speciality of palynology, his analyses and reconstructions have drawn on data from a variety of disciplines, including that of geomorphology. It is, therefore, a pleasure and a privilege for us to contribute the present chapter on geomorphic evidence for climatic change to a volume in celebration of Eduard van Zinderen Bakker's eightieth birthday.

INTRODUCTION

The elevated plateau of Africa is bounded by escarpments and punctuated by basins and rift valleys which bear testimony to the prolonged interplay of internal tectonic forces and external processes of denudation. It is possible to descry, within the gross physiography of the continent, considerable evidence of major climatic trends of regional significance. This is particularly true of southern Africa, whose geomorphic history, although by no means perfectly comprehended, has nonetheless been unravelled in broad outline by three generations of workers, several of whom have, in the process, achieved international recognition and distinction. Equally significant is the realisation that has grown in recent years of the implications which major uplift of the subcontinent must have had for local atmospheric circulations and temperature regimes. In this chapter we hope to highlight the more significant geomorphic evidence for changing climates in southern Africa over the last 150 million years.

LATE MESOZOIC AND PALAEOGENE CONDITIONS

Geomorphically it is not possible to differentiate to any significant extent in respect of erosional and physiographic conditions prevailing over southern Africa in the late Mesozoic and Palaeocene. The offshore sedimentary history of this period, however, is reasonably well preserved, and detailed information on this aspect is now for the most part available.

Recent reconstructions of Gondwanaland (e.g. Dingle et al. 1983) indicate that at the time of its breakup in the late Jurassic, Africa occupied a central position therein. Absolute elevations over southern Africa at this time, after the outporings of the Stormberg volcanicity prior to rifting, were relatively great, varying between about 1 800 and 2 350 m (Partridge & Maud, in press). Rifting breakup of Gondwanaland initiated the Great Escarpment which separates the high-standing interior plains of southern Africa from the lower-standing contiguous marginal areas to the east, south and west, in which vigorous erosion at various times has given rise to the characteristically deeply dissected nature of these areas. The Great Escarpment has had a marked effect on both regional and local climatic conditions and of note is the fact that the central plains of the subcontinent have always stood relatively high, with the ensuing climatic implications of such an elevation.

The breakup of Gondwanaland initiated a long, continent-wide cycle of erosion, which probably comprised a number of subphases or episodes, but definite evidence for which is lacking. This cycle, known as the African, lasted until it was terminated by uplift and tilting of the marginal areas in the early Miocene. The African erosion surface (King 1951, Partridge & Maud, in press) is characterised by a deeply weathered and kaolinized soil profile which carries on it, in places where it survives on the Transvaal highveld and in Natal and the north-eastern Cape, laterite, and in the eastern, southern and western Cape and Namaqualand, silcrete. Deep laterite profile formation is generally characteristic of hot humid tropical regions (McFarlane 1976). The origin of silcrete remains a matter of debate to the extent that while it is agreed that it formed under warm to hot conditions, whether the climate was also humid or semi-arid has not as yet been decided (Summerfield 1983, Twidale & Hutton 1986). The occurrence in South Africa of silcretized 'B' horizons of former solonetzic soil profiles characteristic of a semi-arid environment, as at Kentani in the Transkei, and the widespread occurrence in the arid portions of South Africa of 'dorbank', an incipient form of silcrete, would seem to support the existence of the latter type of environment.

In that pebbles of silcrete are found at the base of the higher portions of the marine Alexandria Formation in the south-eastern Cape, which is here of early Eocene age (Siesser & Miles 1979; Siesser, Maud & Partridge, in prep.), the indications are that both the laterite and silcrete are mainly of Cretaceous or earlier Tertiary age. In Namibia near Luderitz, the Chalcedon-Tafelberg Silcrete Formation (formerly Pomona Beds) of probable Palaeocene age carry a silcrete capping, but these are overlain by sediments of the Buntfeldschuh Formation of late Palaeocene or early Eocene age (Siesser & Salmon 1979). This time of major laterite and silcrete formation is in keeping with the inferred ages of major laterite accumulations in East and West Africa and in Australia, and with the age of the major widespread silcrete occurrences in Australia.

Thus the geomorphic evidence is that during the Cretaceous in southern Africa, climatic conditions were generally hot, varying from humid in the east to semi-arid in the south and west. To what extent these conditions prevailed into the Palaeogene is uncertain.

Corroborative evidence for the general warmth of the Cretaceous period in southern Africa is afforded by the abundant marine faunas, together with wood and plant remains, found at places in the fringing marine sediments of this age, and the occurrence of dinosaur remains of this age in Namaqualand (Haughton 1915). The occurrence of a 'drier sub-tropical type of forest flora' also from this time from the infillings of kimberlite pipes in Namaqualand (Scholtz 1985) and Botswana (Tyson 1986) indicates however the beginning of a trend towards cooler and drier climates in the late Tertiary.

The subcontinent having been reduced, apart from the vicinity of the Great Escarpment, to a region of relatively low relief by end-Cretaceous and Eocene times, and as a result there being no fringing sedimentary deposits of significance of Late Eocene and Oligocene age from which the climatic conditions prevailing in the middle to late Palaeogene may be inferred, the next geomorphic indications of climatic conditions are found in the Miocene.

NEOGENE CONDITIONS

Miocene

A general trend towards cooler and drier climates evident during the Tertiary was significantly accelerated during the Miocene. Partridge & Maud (in press) present evidence for the termination of the prolonged African cycle of erosion by uplift of the subcontinent about 10 MY ago towards the end of the early Miocene. By the end of the African cycle, the interior of southern Africa had, except in a few elevated areas such as the Lesotho Highlands, being reduced to a low relief pediplain at altitudes ranging from about 500 to 700 m above sea-level. Deformation of the African surface and the divergence between this and the succeeding Post-African I surface indicate that this movement was asymmetrical, ranging from about 300 m in the east to about 150 m inland of the west coast. Larger movements occurred locally along inland axes of warping. Whereas these movements were probably of insufficient magnitude to cause major climatic shifts on a regional scale, they appear to have complemented, locally, the effects of changing oceanic circulations. The influence of the latter differed markedly along the east and west coasts of the subcontinent.

Clear palaeoclimatic evidence is forthcoming largely from coastal sediments, although isolated sequences preserved in incised valleys of the interior have also yielded important data. At Uloa on the coast of Zululand, a near-shore sequence interpreted as of late Miocene or Mio-Pliocene age (Maud & Orr 1975, Siesser & Miles 1979) has yielded a warm water marine fauna, including the giant shark *Carcharodon megalodon,* molluscs and coral fragments (King 1953). Along the southern Cape coast, the early Miocene uplift resulted in the cutting of a coastal platform, partly by marine agencies, below the level of the African surface, which is here represented by eroded silcrete cappings overlying a deeply kaolinized residual mantle. On this coastal platform, at elevations ranging from about 250 m to 330 m, are preserved several lignite occurrences which have yielded wood fragments, leaves and pollen representing a Tertiary forest community. Although ages as early as Eocene or Oligocene have been suggested for these remains, their presence on a surface cut in response to the late Lower Miocene uplift of the subcontinent precludes their accumulation earlier than the mid-Miocene. Recent palynological analyses of core samples from these deposits by J.A.Coetzee (pers. comm.) tend to confirm this conclusion.

At Noordhoek in the Cape peninsula, core drilling to below sea-level has revealed that a silcrete capped surface overlies kaolinized granite at –50 m (Rogers 1980); the silcrete which represents a downwarped remnant of the African surface, is buried by polleniferous sediments of the Elandsfontyn Formation which, according to Coetzee (1980), date to the early Middle Miocene. The pollen spectra yielded by these deposits are regarded as indicative of a warm, temperate, broad-leafed rain forest with abundant palms (Coetzee 1978, Coetzee et al. 1983). Molluscan assemblages of similar age from Langebaanweg, some 120 km to the north, contain subtropical taxa confirming the presence of a warm ocean along the west coast at this time (Tankard & Rogers 1978). Upper Miocene peats from the same area include pollen spectra with a much higher proportion of shrubland (fynbos) elements than at Noordhoek (Coetzee & Rogers 1982). This suggests that, by the end of the Miocene, considerable climatic change involving both cooling and, perhaps, the establishment of a more seasonal rainfall regime, had occurred.

This trend along the west coast, from warm, humid conditions in the earlier part of the Miocene, to cooler and drier environments towards its close, is confirmed by the evidence from several other sites. The early to mid-Tertiary faunas recovered from Arris Drift and Elizabeth Bay on the seaboard of southern Namibia, and at Bosluispan in the now arid interior of Bushmanland (Stromer 1926, Hopwood 1929, Corvinus 1978, Hendey 1978) suggest that a warm, mesic environment prevailed until the mid-Miocene; the assemblage unearthed at Bosluispan is of particular interest in view of its preservation in the filling of a fossil drainage channel cut below the African surface of the Bushmanland Plateau in the ensuing Post-African I cycle.

Perhaps the most spectacular evidence of climatic deterioration at the end of the Miocene comes from the Namib desert to the north of the Orange river. The main Namib surface is, like the coastal platform of the southern Cape, cut below kaolinized and silcrete capped remnants of the African surface. In the Luderitz area the silcreted surface of the Chalcedon-Tafelberg Silcrete Formation is overlain by Eocene sediments (Siesser & Salmon 1979), but the far more extensive lower platform carries Miocene deposits. There is thus little doubt that, as is the case elsewhere around the coast of southern Africa, the lower surface was cut in the Post-African I cycle in response to uplift and tilting at the end of the early Miocene. This platform cannot, therefore, be older than mid-Miocene, and this provides a firm *terminus-post-quem* the overlying Tsondab Sandstone Formation, which is generally accepted as reflecting arid conditions. It is highly likely that this aridification was, in fact, contemporaneous with the conspicuous change evident across the Miocene-Pliocene boundary on the evidence of offshore sedimentological data obtained by Siesser (1978) from DSDP Site 362 on the Walvis Ridge. Siesser's results indicate that, while weak upwelling of cool waters, associated with the incipient development of the Benguela Current, began as early as the middle or late Oligocene, upwelling was greatly intensified during the late Miocene.

That change in oceanic circulations, in particular the initiation of cold upwelling, was a major contributing factor in the Miocene climatic deterioration along the western margin of southern Africa, is widely acknowledged. Less certain is the effect of fluctuating oceanic levels. Vail et al. (1977) have cited evidence for major global eustatic fluctuations during the Cainozoic, and similar sea-level curves have been reconstructed for southern Africa (Siesser & Dingle 1981, Hendey 1981, 1983a). Hendey (1983a) considered that the sequence of land surfaces and benches present throughout the coastal hinterlands of the southern and western Cape could, in fact, be explained solely in terms of eustatic fluctuations, and discounted entirely the effects of tectonic factors. This view is challenged by Partridge & Maud (in press), who cite abundant evidence from the south-eastern Cape and elsewhere to show that marine deposits of similar age occur at a wide range of elevations on an inclined coastal platform, indicating that their present altitudinal distribution is predominantly the product of tectonic movements rather than eustatic effects. This does not imply that sea-level fluctuations were absent during the Cainozoic, but rather that their amplitude and significance have been exaggerated. There is, in fact, a body of evidence in favour of an overall reduction in sea-level since the upper Cretaceous of no more than 150 m (e.g. Kerr 1984), in contrast wit the estimate of 350 m (Pitman 1978) used to calibrate the Vail curve. Within this reduced range, the fluctuations which did occur were of significance, particularly during the Neogene, in producing a number of marine terraces. Of special significance was the reduction of global oceanic levels of some 70 m during Messinian times at the end of the Miocene; this has been correlated with an increase in global ice volumes (Hsü 1974), and it is noteworthy that the marked climatic deterioration evident along the western margin of the country was generally contemporaneous with this event. We do not, however, subscribe to Hendey's (1983a) view, based on the arguments of Cooper (1977), that marine transgressions and regressions were major instruments of climatic change in southern Africa. Indeed, since the extent of some of these fluctuations is now in question (Partridge & Maud, ibid.), and since their geomorphic significance in initiating onshore denudational cycles has been challenged (Summerfield 1985), it is likely that their effects were of no more than local significance in the vicinity of the coastal margins. In this context the occurrence of two successive transgressions, one in the Middle Miocene and the other in the early to mid-Pliocene, within a period characterized by general climatic deterioration in the western areas of southern Africa, presents no further difficulties.

Geomorphic evidence for climatic change during the Miocene is, on the whole, less satisfactory in the continental interior than in the coastal areas. Data from the southern Kalahari suggest that the Molopo-Nossob drainage became re-integrated on top of the sedimentary filling of the Kalahari Basin during Miocene times (Partridge & Maud, in press). This would have been possible only under more mesic climatic circumstances than prevail in this area today: at present that channel of the Molopo carries flow, even in times of flood, for no more than a fraction of its previous length.

Pliocene

The general climatic trends established during the Miocene appear to have continued into the Pliocene. Until almost the end of this period very warm conditions appear to have prevailed along the east and south coast on the evidence of marine faunas (Maud 1986, McMillan 1986). On the basis of similar evidence, the cold Benguela circulation appears to have exerted a progressively stronger influence on the western margin of the subcontinent. This influence is clearly evident in biological remains from the estuarine Varswater Formation at Langebaanweg, which is considered to be of Lower Pliocene age. Peats interbedded in this succession contain pollen in which fynbos and grassland elements are strongly represented (Coetzee & Rogers 1982), indicating marked cooling and, in all probability, a more strongly seasonal rainfall pattern. Analyses of the associated vertebrate fauna by Hendey

(1983b) provide broad confirmation of these results and indicate the presence of a shrubland-grassland-forest mosaic with local intergrades during this period. Hence, by the Pliocene, the pronounced east-west temperature and rainfall gradient across southern Africa was well established.

This situation was significantly modified by disturbance towards the end of the Pliocene, which contributed more to the present distribution of elevations within southern Africa than did any previous tectonic event since the fragmentation of Gondwanaland. Deformation of the Post-African I surface indicates that uplift was concentrated along much the same axes as in the Miocene, but was of vastly greater amplitude in the eastern part of the subcontinent. Hence uplifts in the south-eastern hinterland ranged from 600 to 900 m; movements declined to about 200 m in the southern hinterland and to 100 m or less inland of the west coast. The resulting westward tilt caused significant rejuvenation along the major inland drainage lines. As was the case in the Miocene, uplift in the Kalahari Basin lagged behind that of surrounding areas, increasing the significance of this feature as a sedimentary depository.

The effects of this uplift, particularly on the climatic regime of the eastern part of the interior plateau of southern Africa, must have been considerable, and have, in fact, been alluded to by Van Zinderen Bakker (1986). A temperature decline, at least as great as that which characterised the Last Glacial Maximum in southern Africa, must have been felt in these elevated eastern areas, which included the interior of Natal and the eastern Cape, the Highveld of the Transvaal and the Orange Free State, and the Lesotho Highlands. Marine evidence suggests that this uplift occurred about 2,5 MY ago near the end of the Pliocene (Partridge & Maud, in press). A simultaneous global decline in temperature, reflected by the onset of ice rafting in the North Sea and major growth of the East Antarctic ice sheet (Denton 1985), would have greatly extended the significance of this event.

That the local climatic response to these late Pliocene changes was in the direction of greater aridity is clear from a number of lines of evidence. The aeolian Kalahari sand, which constitutes the uppermost unit of the Kalahari Group, extends into areas far beyond limits of the Kalahari Basin, including the western Transvaal, western Orange Free State, the northern Cape and parts of the Karoo extending as far south as the Cape Fold Mountains. Although dating evidence is tenuous, this spread of desertic conditions probably dates from this time. The upper terraces of the Vaal river, which, on limited faunal evidence, are probably of late Pliocene age, reflect extensive coarse alluviation; this is essentially a semi-arid phenomenon (Partridge & Brink 1967). In the cave depositories of Makapansgat and Sterkfontein in the Transvaal, which have yielded important hominid and faunal remains, a change in style of sedimentation, from predominantly fine materials, deposited at least in part under subaqueous conditions, to more gravelly debris and sheetflood accumulations, occurs at levels which, from broad biostratigraphic evidence and an incomplete palaeomagnetic record, date to around 2,5 MY (Partridge 1985a). This transition argues for the onset of more arid conditions than had prevailed previously; a cyclic pattern of deposition suggests that marked fluctuations of rainfall did, in fact, occur after this time. More or less concurrent faunal changes indicating a shift to a more open vegetation have been recognised by Vrba (1974, 1975), and Tobias (1985, 1986) has pointed to a significant nodal event in hominid speciation at about this time. Of significance in this regard is the fact that uplift on a similar scale occurred concurrently in East Africa (Baker et al. 1972).

Pleistocene and Holocene

Abundant geomorphic evidence for climatic change and associated glacio-eustatic sea-level movements during this time is forthcoming both from the marginal areas of the subcontinent, and in the interior.

In the eastern coastal areas a sequence of dune cordons of varying ages and degrees of weathering (Davies 1976) attest to a sequence of climatic changes varying from warm and moist during interglacial periods to cooler and somewhat drier during the glacial periods, particularly during the last two glacial maxima which were major periods of dune formation (Maud 1968). Weathering of the older dunes has given rise to the red clayey sands and sands of the Berea Formation of the east coast, the younger dunes being brown or grey in colour and less weathered (Maud 1968). Related to the dunes are a series of river terraces and gravels of varying degrees of weathering, the older terraces being red weathered and more clayey than the younger brown soil less weathered terraces.

There is much evidence for a warm Eemian period of high sea-level on the east coast as shown by the occurrence of corals of this age at False Bay in the Lake St Lucia system (Hobday 1976) and elevated marine bench oyster beds as at Mdloti on the Natal coast (Maud 1968). In the eastern and southern Cape, the warm water 'Swartkops fauna' of Davies (1980) is of the same age. On the west coast, the evidence of marine faunas preserved on lower coastal terraces is that cool conditions prevailed during the Pleistocene (McMillan 1986, Tankard 1976). The glacio-eustatic lowering of sea-levels around the South African coastlines to depths as great as 70 m and more below present sea-level during the last two glacial maxima is indicated both by the incision of the lowermost courses of major rivers into the bedrock beneath their beds, and the occurrence to these depths of subaerially deposited calcareous dune aeolianites of this age both off- and onshore, as for example beneath the Durban Bluff. The lower courses of the rivers have been filled by highly variable alluvial and estuarine sediments ranging from sands to clays (Maud 1968).

Of widespread occurrence in the eastern portions of the country is a superficial colluvium which evidently was deposited during the cooler and drier times leading up to the last glacial maxima (Price Williams et al. 1982).

Of note is the complete absence of true glacial features such as cirques, striated pavements, etc., on the Lesotho Highlands, which must have carried an icecap at this time in view of the periglacial features now developed there. This suggests that conditions there were so dry that insufficient snow accretion to the icecap occurred such that significant glacial erosional movements could take place.

Some coastal features, such as lagoonal beach bars and marine cut benches on the Natal coast, indicate a warmer somewhat wetter period during the early to mid-Holocene in this region (Maud 1968).

Geomorphic features of the interior of the subcontinent generally reflect the influence of the frequent climatic cycles which characterised the Pleistocene, although much of the evidence is of a strictly local nature and is hence beyond the scope of this chapter. It is widely agreed that the maximum extensions of the aeolian sands of the Kalahari and the Namib Sand Sea were achieved during the course of repeated arid cycles which, on rather limited local evidence, appear, at least towards the end of the Pleistocene, to have coincided with glacial maxima in higher latitudes. Based on the distribution of linear dunes Lancaster (1979) has deduced that, at the time of maximum dune formation in the Kalahari region, the 150 mm rainfall isohyet must have been some 1 200 km to the north-east of its present position, and the southern African anticyclone must have been substantially larger than today. The lower terraces of the Vaal river, the oldest of which have yielded Acheulian artifacts and a Middle Pleistocene fauna, contain cut-and-fill sequences which point to an increasing cyclicity in flow patterns. Successive tufa carapaces along the Ghaap escarpment of Griqualand West likewise reflect alternating humid and semi-arid conditions. Although older accumulations are present at Ulco (J.C.Vogel, pers. comm.), the first major tufa deposition appears to have begun at Taung around 1MY ago; later carapaces date to approximately 760 000 years, 340 000-230 000 years and 103 000-30 000 years (Vogel & Partridge 1984). Analysis of the local springflow regime suggests that tufa deposition was at a maximum during semi-arid interludes, while tufa erosion occurred during the more humid episodes (Partridge 1985b).

Important evidence for Pleistocene climatic fluctuations is forthcoming from the many pans which dot the interior of southern Africa. There is considerable evidence that cycles of pan deflation and lunette formation alternated regionally with a rise in the phreatic surface, accompanied by pan flooding and the development of beach ridges and marginal terraces. Although detailed analyses have yet to be undertaken at many key pan sites, there is evidence that several such cycles occurred between 250 000 years and the beginning of the Holocene at Haaskraal Pan in the central Karoo (Partridge & Dalbey 1986). Several authors have attempted to correlate lake levels in the Kalahari during this period; this and other local evidence for Pleistocene climatic change in southern Africa is summarised in Deacon et al. (1984) and will not be repeated here. It is however, true to say that an enormous wealth of geomorphic, sedimentological and isotopic evidence remains to be analysed before an unambiguous history of Pleistocene climatic fluctuations, constrained by a reliable dating framework, can be written.

REFERENCES

Baker, B.H., P.A.Mohr & L.A.J.Williams 1972. *Geology of the eastern rift system of Africa.* Geol. Soc. Amer., Spec. Publ. 136, 67 pp.

Coetzee, J.A. 1978. Climatic and biological changes in south-western Africa during the late Cainozoic. *Palaeoecol. of Africa* 10: 13-29.

Coetzee, J.A. 1980. Tertiary environmental changes along the south-western African coast. *Palaeontologia Africana* 23: 197-203.

Coetzee, J.A. & J.Rogers 1982. Palynological and lithological evidence for the Miocene palaeoenvironment in the Saldanha region (South Africa). *Palaeogeogr., Palaeoclimatol., Palaeoecol.* 39: 71-85.

Coetzee, J.A., A.Scholtz & H.J.Deacon 1983. Palynological studies and the vegetation history of the fynbos. In H.J.Deacon et al. (eds.), *Fynbos palaeoecology. A preliminary synthesis:* 156-173. S.Afr. Natn. Prog. Rep. 75, CSIR, Pretoria.

Cooper, M.R. 1977. Eustasy during the Cretaceous: its implications and importance. *Palaeoclimatol. Palaeoecol.* 22: 1-60.

Corvinus, G. 1978. Palaeontological and archaeological investigations in the lower Orange valley from Arisdrift to Obib. *Palaeoecol. of Africa* 10: 75-92.

Davies, O. 1976. The older coastal dunes in Natal and Zululand and their relation to former shorelines. *Ann. S.Afr. Mus.* 71: 19-32.

Davies, O. 1980. Last interglacial shorelines in the south Cape. *Palaeont. Afr.* 23: 153-171.

Deacon, J., N.Lancaster & L.Scott 1984. Evidence for late Quaternary climatic change in southern Africa. In J.C.Vogel (ed.), *Late Cainozoic palaeoclimates of the southern hemisphere*: 391-404. Rotterdam: Balkema.

Denton, G.H. 1985. Did the Antarctic ice sheet influence Late Cainozoic climate and evolution in the southern hemisphere? *S.Afr. J. Sci.* 81: 224-229.

Dingle, R.V., W.G.Siesser & A.R.Newton 1983. *Mesozoic and Tertiary geology of southern Africa.* Rotterdam: Balkema. 375 pp.

Haughton, S.H. 1915. On some dinosaur remains from Bushmanland. *Trans. Roy. Soc. S.Afr.* 5: 259-164.

Hendey, Q.B. 1978. Preliminary report on the Miocene vertebrates from Arisdrift, South West Africa. *Ann. S.Afr. Mus.* 76: 1-141.

Hendey, Q.B. 1981. Palaeoecology of the late Tertiary fossil occurrences in 'E' quarry, Langebaanweg, South Africa, and a reinterpretation of their geological context. *Ann. S.Afr. Mus.* 84: 1-104.

Hendey, Q.B. 1983a. Cenozoic geology and palaeogeography of the fynbos region. In H.J.Deacon et al. (eds.), *Fynbos palaeoecology. A preliminary synthesis*: 35-60. S.Afr. Natn. Prog. Rep. 75, CSIR, Pretoria.

Hendey, Q.B. 1983b. Palaeoenvironmental implications of the late Tertiary vertebrate fauna of the Fynbos region. In *Fynbos palaeoecology. A preliminary synthesis*: 100-115.

Hobday, D.K. 1976. Quaternary sedimentation and development of the lagoonal complex, Lake St Lucia, Zululand. *Ann. S.Afr. Mus.* 71: 93-113.

Hopwood, A.J. 1929. New and little known mammals from the Miocene of Africa. *Ann. Mus. Novit* 344: 1-9.

Hsü, K.J. 1974. The Miocene desiccation of the Mediterranean and its climatical and zoogeographical implications. *Naturwissenschaften* 61: 137-142.

Kerr, R.A. 1984. Sea-floor spreading is not so variable. *Science* 223: 472-473.

King, L.C. 1951. *South African scenery* (2nd ed.). Edinburgh: Oliver & Boyd. 379 pp.

King, L.C. 1953. Miocene marine fauna from Zululand. *Trans. geol. Soc. S.Afr.* 56: 59-91.
Lancaster, I.N. 1979. Quaternary environments in the arid zone of southern Africa. Dept. of Geogr. and Envtl. Studies, Univ. of Witwatersrand, Occ. Paper 22, 77 pp.
Maud, R.R. 1968. Quaternary geomorphology and soil formation in coastal Natal. *Z. geomorph. N. F.* Suppl-Bd. 7: 155-199.
Maud, R.R. 1986. Neogene palaeoclimates in the eastern and south-eastern coastal areas of South Africa. *S.Afr. J. Sci.* 82: 65-66.
Maud, R.R. & W.M.Orr 1975. Aspects of post-Karoo geology in the Richards Bay area. *Trans. Geol. Soc. S.Afr.* 78: 101-109.
McFarlane, M.J. 1976. *Laterite and landscape.* London: Academic Press.
McMillan, I.K. 1986. Cainozoic planktonic and larger foraminifera distributions around southern Africa and their implications for past changes of oceanic water temperatures. *S.Afr. J. Sci.* 82: 66-69.
Partridge, T.C. 1985a. The palaeoclimatic significance of Cainozoic terrestrial stratigraphic and tectonic evidence from southern Africa: a review. *S.Afr. J. Sci.* 81: 245-247.
Partridge, T.C. 1985b. Spring flow and tufa accretion at Taung. In P.V.Tobias (ed.), *Hominid evolution: Past, present and future.* New York: Alan R.Liss.
Partridge, T.C. & A.B.A.Brink 1967. Gravels and terraces of the lower Vaal basin. *S.Afr. Geogr. J.* 49: 21-38.
Partridge, T.C. & T.S.Dalbey 1986. Geoarchaeology of Haaskraal Pan: a preliminary palaeoenvironmental model. *Palaeoecol. of Africa* 17: 69-78.
Partridge, T.C. & R.R.Maud (in press). Geomorphic evolution of southern Africa since the Mesozoic. *Trans. Geol. Soc. S.Afr.* 89.
Pitman, W.C. 1978. Relationship between eustacy and stratigraphic sequences of passive margins. *Geol. Soc. Amer. Bull.* 89: 1389-1403.
Price Williams, D., A.Watson & A.S.Goudie 1982. Quaternary colluvial stratigraphy, archaeological sequences and palaeoenvironments in Swaziland, southern Africa. *Geogr. J.* 148: 50-67.
Rogers, J. 1980. *First report on Cenozoic sediments between Cape Town and Elands Bay.* Geol. Surv. Rpt. 165, Pretoria. 65 pp.
Scholtz, A. 1985. The palynology of the upper lacustrine sediments of the Arnot pipe, Banke, Namaqualand. *Ann. S.Afr. Mus.* 95: 1-109.
Siesser, W.G. 1978. Aridification of the Namib Desert: evidence from oceanic cores. In *Antarctic glacial history and world palaeoenvironments*: 105-113. Rotterdam: Balkema.
Siesser, W.G. & R.V.Dingle 1981. Tertiary sea-level movements around southern Africa. *J. Geol.* 89: 523-536.
Siesser, W.G. & G.A.Miles 1979. Calcareous nannofossils and planktic foraminifera in Tertiary limestones, Natal and Eastern Cape, South Africa. *Ann. S.Afr. Mus.* 79: 139-158.
Siesser, W.G. & D.Salmon 1979. Eocene marine sediments in the Sperrgebiet, South West Africa. *Ann. S.Afr. Mus.* 79: 9-34.
Stromer, E. 1926. Festland- und Süsswasser-bewohnende Wirbeltiere aus den Diamantfeldern Deutsch, Südwestafrikas. In E.Kaiser (ed.), *Die Diamantwüste Südwestafrikas* 2: 107-153. Berlin: Dietrich Reiner.
Summerfield, M.A. 1983. Silcrete as a palaeoclimatic indicator: evidence from southern Africa. *Palaeogeogr., Palaeoclimatol., Palaeoecol.* 41: 65-79.
Summerfield, M.A. 1985. Plate tectonics and landscape development on the African continent. In M.Marisawa & J.T.Hack (eds.), *Tectonic geomorphology.* Boston: Allen & Unwin. 390 pp.
Tankard, A. 1976. Pleistocene history and coastal morphology of the Ysterfontein-Elands Bay area, Cape Province. *Ann. S.Afr. Mus.* 69: 73-119.

Tankard, A. & J.Rogers 1978. Late Cenozoic palaeoenvironments on the west coast of southern Africa. *J. Biogeogr.* 5: 319-337.

Tobias, P.V. 1985. Ten climacteric events in hominid evolution. *S.Afr. J. Sci.* 81: 271-272.

Tobias, P.V. 1986. Delineation and dating of some major phases in hominidisation and homonization since the Middle Miocene. *S.Afr. J. Sci.* 82: 92-94.

Twidale, C.R. & J.T.Hutton 1986. Silcrete as a climatic indicator: discussion. *Palaeogeogr., Palaeoclimatol., Palaeoecol.* 52: 351-360.

Tyson, P.D. 1986. *Climatic change and variability in southern Africa.* Cape Town: Oxford University Press. 220 pp.

Vail, P.R., R.M.Mitchum & S.Thompson 1977. Seismic stratigraphy and global changes in seal level. Part 4: Global cycles of relative changes of sea level. *Mem. Amer. Ass. Petrol. Geol.* 26: 83-97.

Van Zinderen Bakker, E.M. 1986. African climates and palaeoenvironments since Messinian times. *S.Afr. J. Sci.* 82: 70-71.

Vogel, J.C. & T.C.Partridge 1984. Preliminary radiometric ages for the Taung tufas. In J.C.Vogel (ed.), *Late Cainozoic palaeoclimates of the southern hemisphere*: 507-514. Rotterdam: Balkema.

Vrba, E.S. 1974. Chronological and ecological implications of the fossil Bovidae at the Sterkfontein Australopithecine site. *Nature* 250: 19-23.

Vrba, E.S. 1975. Some evidence of chronology and palaeoecology of Sterkfontein, Swartkrans and Kromdraai from the fossil Bovidae. *Nature* 254: 301-304.

ISLANDS: PALAEOENVIRONMENTS

PALEOCLIMATS DE L'ILE DE FUERTEVENTURA (ARCHIPEL CANARIEN)

N.PETIT-MAIRE / J.C.ROSSO
Laboratoire de Géologie du Quaternaire, CNRS, Marseille, France

G.DELIBRIAS
Centre des Faibles Radioactivités, Domaines du CNRS, Gif s/Yvette, France

J.MECO
Escuela Universitaria del Profesorado de EGB, Universidad de la Laguna, Las Palmas, Spain

S.POMEL
Institut de Géographie, Faculté des Lettres, Aix-en-Provence, France

ABSTRACT

Fuerteventura, the most arid island in the Canary archipelago, due to its location about 100 km off the Saharan coast of Morocco, holds important Quaternary deposits indicating recent climatic change. Three levels with Hymenoptera nests and terrestrial mollusc shells were observed in the palaeodunes, testifying to a cessation of aeolian processes and to the fixation of the dunes by vegetation: they are radiocarbon dated at 23 600 ± 550 BP, 15 000 ± 200 BP and 9 800 ± 140 BP. These ages correspond to the climatic chronology known for the Sahara (humid maxima at 24 000-22 000 BP, during the early Holocene and (which is logical considering the oceanic position of Fuerteventura) to the first phase of deglaciation, which occurred after the glacial maximum at c. 18 000 BP. This phase induced a decrease in aridity in the Saharan mountainous areas, even if not in its basins.

It would be interesting to observe the chronological differences in climatic sensitivity between the eastern and western Canary Islands.

l'île de Fuerteventura, la plus occidentale de l'archipel canarien, est située par 28°45'-28°02'N et 13°49'-14°30'W, à moins de 100 km de la côte marocaine. Elle couvre une surface de 1 725 km^2 et se caractérise, par rapport aux autres îles, par son climat aride et par l'importance des formations sédimentaires quaternaires: dunes, encroûtements calcaires, cônes de déjection, paléosols. Elle se prête donc parfaitement à l'étude des variations climatiques récentes d'une région mixte: déjà atlantique mais encore saharienne.

Le climat de Fuerteventura est conditionné par le courant froid des Canaries qui réduit les précipitations et règle la température, situation comparable à celle du Sahara occidental, du Namib et de l'Atacama, induisant une aridité importante.

Les précipitations varient de 53,1 mm sur la côte est (Entallada) à 220,7 mm à l'intérieur (Betancuria), pour un nombre de jours de pluie, dont 60% sont

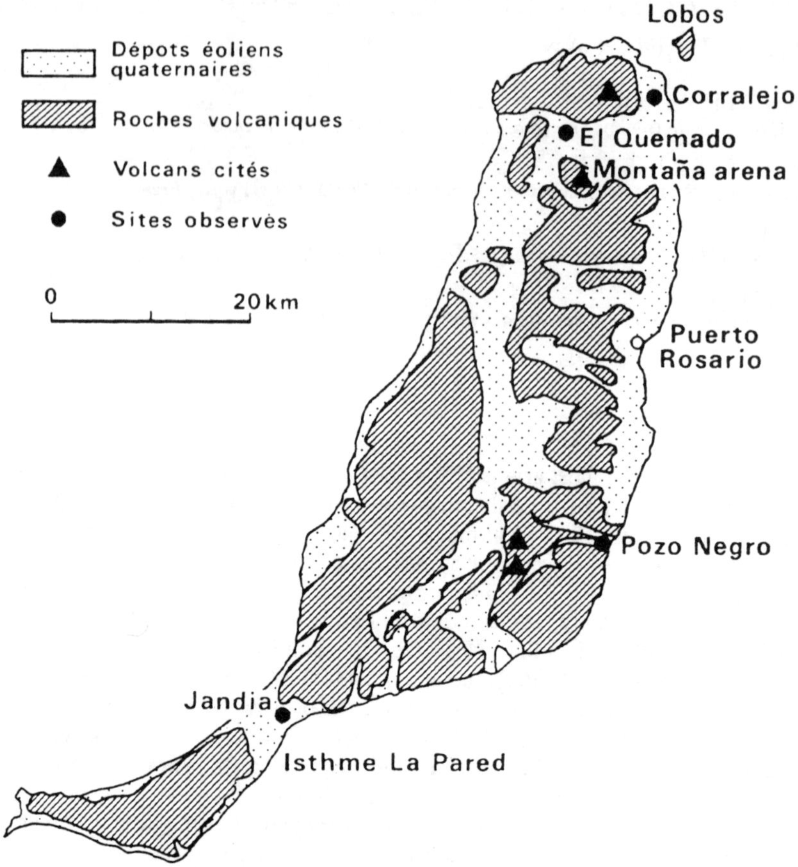

Figure 1. L'île de Fuerteventura.

répartis en hiver, allant respectivement de 13,1 à 28,3 jours; la moyenne annuelle pour l'ensemble de l'île est de 105 mm. Les températures maximales et minimales absolues varient de 44° à 5°4 C; à Los Estancos, point situé à 230 m d'altitude, la moyenne annuelle est de 18°8 C. La radiation solaire est forte à cette latitude subtropicale, l'insolation moyenne étant de 7.9 heures de soleil par jour à Los Estancos. Tout au long de l'année, mais surtout durant l'été et l'automne, des invasions d'air saharien touchent Fuerteventura; les températures y peuvent alors atteindre plus de 40° C; les masses d'air très sèches sont chargées de poussières en suspension, véritables 'vents de sable' sahariens (Marzol 1984). La flore est caractérisée par la présence de quelques espèces endémiques psammophiles (*Argyranthemum winteri, Onomis christii*), associées à des espèces sahariennes, Chenopodiacées pour la plupart: *Suaeda* sp., *Salsola* sp., *Traganum moquini, Zygophyllum fontanesii, Euphorbia balsaminifera* (Santos 1984). Parmi les faunes actuelles, nous ne mentionnerons que la présence, dans les zones où la végétation est présente, d'un insecte Hyménoptère de grande taille, à rayures noires et jaunes, *Anthophora*, qui construit ses nids individuels, groupés en colonies, dans les sols les plus humides (Bacallado Aranega et al. 1984) et celle d'*Helicidae* indéterminés.

Ces conditions bioclimatiques sont proches de celles du Sud marocain: à Tiznit (29°42′N), la température annuelle moyenne est de 19°3 C et à Goulimine (29°59′N), les précipitations atteignent 119.5 mm (Dubief 1968).

Depuis longtemps, on a reconnu la signification paléoclimatique des dépôts récents de Fuerteventura: sables éoliens ('jables') compactés par un ciment calcaire, vastes cônes alluviaux et dépôts de fonds de vallées, 'caliches' épaisses atteignant plusieurs mètres d'épaisseur au centre de l'île (Fuster et al. 1968) mais aussi grandes extensions de paléosols (Pomel et al. 1983, 1985, Pomel 1986). Des essais de corrélation avec les niveaux marins du Pléistocène supérieur et de l'Holocène ont également été proposés (Meco & Pomel 1985). Cependant, les niveaux à Mollusques terrestres et à nids d'*Anthophora* observés dans les jables (Fuster et al. 1968) n'avaient, à ce jour, pas été étudiés, datés, ni considérés en relation avec la chronologie climatique saharienne. A cet effet, une courte mission de terrain a été financée par le Gouvernement des Iles Canaries (projet no. 33/3.9.84) et quatre coupes ont été observées et échantillonnées dans les formations quaternaires de Fuerteventura, deux d'entre elles dans des formations dunaires, deux autres dans des entailles d'oueds (Fig. 1):

1. 'Jables' de l'isthme de Jandia (Fig. 2a) à équidistance des côtes Ouest et Est.
2. Dunes de Corralejo (Fig. 2b), à environ 2 km de la côte.
3. Oued de El Quemado (Fig. 2c), à environ 4 km du volcan Bayuyo.
4. Oued de Pozo Negro (Fig. 2d), à environ 500 m de la côte.

Toutes ces coupes révèlent l'existence de plusieurs niveaux alternés de type 'aride' (dépôts éoliens de forte puissance: 0,50 à > 3 m) et de type 'humide' (sols argileux, graviers, coquilles de Mollusques, colonies de nids d'Hyménoptères), l'importance des couches variant selon la position géographique et topographique des sites.

La malacofaune est représentée par de nombreux Pulmonés stylommatohpores appartenant à la famille des Subulinidae (= Stenogyridae, subfam. Rumininae) et à celle des Helicidae (subfam. Helicellinae et Helicinae). *Rumina decollata* (Linné) est une espèce circum-méditerranéenne largement répandue en Europe méridionale et au Maghreb; cette forme xérophile peut coloniser les milieux subdésertiques; une 'variété' de petite taille, cylindroïde, est connue dans le Sud algérien et dans le Sahara (f. *saharica* Debeaux) à l'état subfossile (Pallary 1901, 1924, 1934); elle a été récoltée vivante, dans l'Atakor, jusqu'à 2 000 m d'altitude (Pallary 1934). Chevallier (1969)

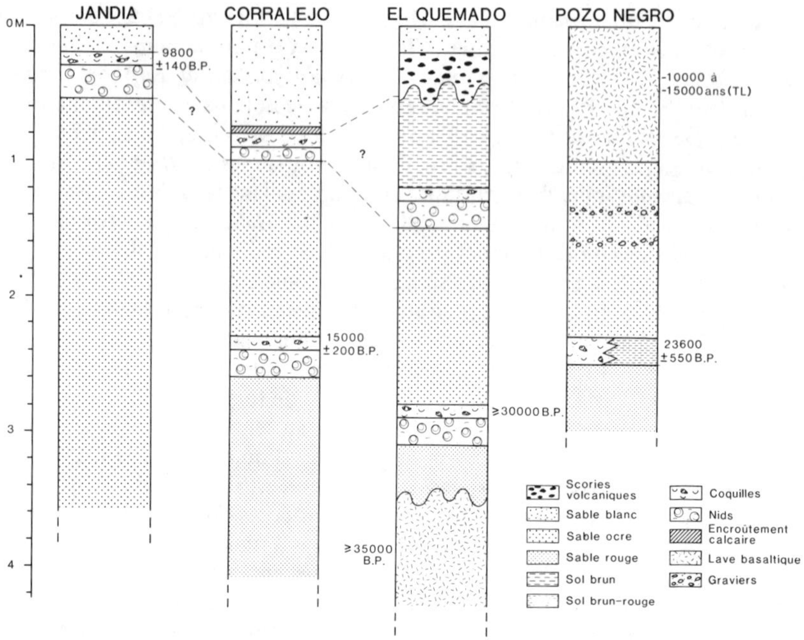

Figure 2. Les quatre coupes étudié.

signale sa présence fréquente en Algérie; elle a été maintes fois signalée aux Canaries. Parmi les Helicidae, *Hemicycla glasiana* (Shuttleworth) est une espèce peu exigeante, fortement affectée par le polymorphisme, récoltée à Gran Canaria à 300 m d'altitude (Dautzenberg 1890). Plusieurs espèces référables au genre *Helicella* Férussac 1821 sont présentes dans nos coupes: l'une d'elles évoque *H. lineata* (Olivier) (= *H. maritima* Draparnaud) qui colonise d'ordinaire les herbes rases fixant les dunes. Ainsi, il s'agit dans l'ensemble d'une faune homogène, xérophile, abondante; elle est correctement développée avec des individus nombreux et de taille normale. En stratigraphie, une disposition est constante: chaque niveau à Mollusques est toujours précédé par un niveau à nids d'*Anthophora.* La signification climatique de cette séquence paraît nette: bien qu'il s'agisse d'espèces xérophiles, leur présence implique cependant un arrêt durable des processus éoliens, l'humidification des dunes vives et leur colonisation par une végétation du type de celle observée actuellement dans les parties les plus humides de l'île, encore fréquentées par ces espèces.

Le niveau à Mollusques le plus superficiel est daté à 9 800 ± 140 BP à Jandia (Gif 7033); il correspond en stratigraphie au niveau à Mollusques supérieur de la coupe de Corralejo et au sol brun argileux de El Quemado;

celui-ci est fossilisé par les scories du volcan Bayuyo dont l'activité est peu antérieure aux plages de Corralejo datées sur coquilles marines de 3 640 ± 100 BP (Gif 5346; Pomel 1986).

Le niveau à Mollusques séparé du précédent, à Corralejo, par le dépôt de 1,30 m de sable éolien, est daté de 15 000 ± 200 BP (Gif 7032). Il correspond peut-être aux niveaux de ruissellement à graviers de l'oued de Pozo Negro, situés sous une coulée de vallée dont un âge préliminaire par thermoluminescence se situe vers –10 000 à –15 000 ans (Pomel et al. 1985, Pomel 1986).

Les niveaux à Mollusques les plus inférieurs des coupes de El Quemado et Pozo Negro sont datés respectivement de 23 600 ± 550 BP (Gif 7034) et ⩾ 30 000 BP (Gif 7031). A Pozo Negro, les couches à Mollusques et à nids correspondent parfois à un niveau de sol brun rouge calcifère; ils surmontent tous la dune rouge. A El Quemado, les couches fossilifères surmontent aussi cette même dune et/ou une coulée du volcan de Montaña Arena, donné par thermoluminescence comme postérieur à –51 000 ans (Pomel et al. 1985), une plage sous-coulée du même appareil étant donnée sur coquilles, par ^{14}C, comme ⩾ 35 000 BP (Gif 5347). Ces niveaux à Mollusques peuvent être contemporains des encroûtements calcaires, datés (Lab. Inst. Royal Patr. part. Bruxelles) de 25 847 ± 560 BP et 19 406 ± 540 BP (Pomel 1986) qui affectent cette même coulée.

Ces âges des niveaux 'humides' de Fuerteventura sont parfaitement cohérents avec les données actuellement admises sur l'évolution climatique du Sahara et de ses marges, qui peut se résumer ainsi, d'après les travaux les plus récents:

– 35 000 (?)-29 000 BP: période aride, dunes rouges (Durand et al. 1983, Durand & Lang 1986). Les dunes rouges de Fuerteventura sont probablement corrélables à ces dépôts.

– 29 000-20 000 BP: période humide avec maximum entre 24 000 et 22 000 BP (Durand et al. 1983, Petit-Maire 1986, Durand & Lang 1986).

– 20 000-17 000 à 9 500 BP (selon la situation géographique: latitude, altitude, proximité océanique): période aride (Durand et al. 1983, Petit-Maire 1986, Rognon 1980, Talbot 1984).

– 17 000/9 500 BP-4 500/3 000 BP (selon la situation géographique, les massifs montagneux et les bordures océaniques ou sahéliennes réagissant à la première phase courte de déglaciation, les bassins seulement à la seconde (Duplessy 1984, Berger et al. 1985); période humide avec maximum lacustre entre 9 000 BP et 7 000 (Petit-Maire 1986, Durand & Lang 1986, Fontes et al. 1985, Petit-Maire & Riser 1983).

Il est tout à fait logique que Fuerteventura, aux marges occidentales du Sahara, ait connu des régimes climatiques alignés sur les durées maximales des humides sahariens. Il serait fort intéressant de déterminer maintenant l'influence croissante du régime atlantique et l'influence décroissante des conditions et des masses d'air sahariennes, en observant les éventuels dépôts du Quaternaire récent dans les Iles Canaries plus occidentales.

REFERENCES

Bacallado Aranega, J.J. et al. 1984. Fauna del Archipielago Canario. *Geografia de Canarias* 1: 242.

Berger, W.H., J.S.Killingley, C.V.Metzler & E.Vincent 1985. Two-stop deglaciation. *Quaternary Research* 23: 258-271.

Chevallier, H. 1969. Mollusques subfossiles récoltés par Mr Henri Lhote dans le Sud-Oranais et le Sahara. *Bull. Mus. natn. Hist. nat. Paris* 41: 266-294.

Dautzenberg, Ph. 1890. Récoltes malacologiques de Mr l'Abbé Cullièret aux îles Canaries et au Sénégal. *Mém. Soc. géol. Fr.* 3: 147-168.

Dubief, J. 1959-1963. *Le climat du Sahara* (2 tomes). Alger: Inst. Rech. Sahar. (Mém. hs).

Duplessy, J.C. 1984. Les calottes glaciaires. *La Recherche* 156: 806-818.

Durand, A., J.Lang, A.Morel & J.P.Roset 1983. Evolution géomorphologique, stratigraphique et paléoclimatique au Pléistocène supérieur et à l'Holocène de l'Aïr oriental (Sahara méridional, Niger). *Rev. Géol. dyn. Géogr. phys.* 24 (1): 47-59.

Durand, A. & J.Lang 1986. Approche critique des méthodes de reconstitution paléoclimatique: le Sahel nigéro-tchadien depuis 40 000 ans. *Bull. Soc. géol. Fr.* (8) 2/2: 267-278.

Fontes, J.C., F.Gasse, Y.Callot, J.C.Plaziat, P.Carbonel, P.A.Dupeuble & I.Kaczmarska 1985. Freshwater to marine-like environments from Holocene lakes in northern Sahara. *Nature* 317: 608.

Fuster, J.M. et al. 1968. *Fuerteventura.* Madrid: Instituto Lucas Mallada, 239 pp.

Geografia de Canarias I: *Geografia fisica.* Interinsular Canaria, 1984.

Klug, H. 1968. *Morphologische Studien auf den Kanarischen Inseln.* Kiel: Schmidt & Klaunig, 184 pp.

Marzol, M.V. 1984. El Clima. *Geografia de Canarias* 1: 158-202.

Meco, J. & R.Pomel 1985. Les formations marines et continentales des îles Canaries orientales (Grande Canarie, Fuerteventura et Lanzarote): stratigraphie et signification paléoclimatique. *Estudios geologicos* (Madrid) 41: 223-227.

Pallary, P. 1901. *Mémoire sur les Mollusques fossiles terrestres, fluviatiles et saumâtres de l'Algérie.* Paris: Soc. géol. Fr. (Mém. sér. Paléonto., 22).

Pallary, P. 1924. Notes sur quelques Mollusques d'un dépôt pléistocène de Colomb-Béchar. *Bull. Soc. Hist. nat. Afr. N.* 5: 111-113.

Pallary, P. 1934. Mollusques du Sahara central. *Mém. Soc. Hist. nat. Afr. N.* 4: 58-72.

Petit-Maire, N. 1986. Paleoclimates in the Sahara of Mali: a multidisciplinary study. *Episodes* 9 (1): 7-16.

Petit-Maire, N. & J.Riser (eds.) 1983. *Sahara ou Sahel? Quaternaire récent du bassin de Taoudenni (Mali).* Marseille: Imprimerie Lamy, 473 pp.

Pomel, R.S. 1986. *Morphologie volcanique et paléoclimatologie des îles Canaries: comparaison avec d'autres milieux volcaniques insulaires (îles de la Mer Tyrrhénienne et de la Mer Egée, île de la Réunion).* Univ. Aix-Marseille: Thèse Doct. Et., 3 tomes: 1083 pp. (Publ. Inst. géogr.).

Pomel, R.S. et al. 1983. Les îles Canaries: Fuerteventura, Lanzarote et Tenerife. *Excursion Com. natn. Géogr. – Commission Géomorphologie des Reliefs volcaniques, 27 mars - 3 avril 1983 – Livret guide.* Clermont-Ferrand: Univ./CNRS-ERA 054. 358 pp.

Pomel, R.S., D.Miallier, J.Fain & S.Sanzelle 1985. Datation d'un sol brun-rouge calcifère par une coulée d'âge Würm ancien (51 000 ans) à Fuerteventura (Iles Canaries). *Méditerranée* 4: 59-68.

Rognon, P. 1980. The Saharan and East African uplands during the Quaternary: fluviatile piedmont deposits. In M.A.J.Williams & H.Faure (eds.), *The Sahara and the Nile*: 118-132. Rotterdam: Balkema.

Santos, A. 1984. Flora y vegetacion. *Geografia de Canarias* 1: 258-293.

Talbot, M.R. 1984. Late pleistocene rainfall and dune building in the Sahel. *Palaeoecology of Africa* 16: 203-214. Rotterdam: Balkema.

PRE-SETTLEMENT VEGETATION CHANGES AT LAKE TRITRIVAKELY, MADAGASCAR

D.A.BURNEY
Department of Zoology, Duke University, Durham, USA

ABSTRACT

A 5 m sediment core from Lake Tritrivakely in central Madagascar provides a record of changes in the crater's vicinity over the last c. 11 000 years. Pollen, charcoal and sedimentary data indicate that changes occurred in the vegetation even before the presumed arrival of human settlers c. 1 500-2 000 years BP, and that fire ecology and pyrogenic vegetation types occurred in Madagascar in the pre-settlement Holocene. Although some forest refugia may have persisted in the region throughout the Holocene, the prevailing early Holocene vegetation at the site was composed of graminoids and ericoids, and these alternate as the prevailing pollen type. By the middle Holocene, pollen of Gramineae has attained long-term dominance, with Ericaceae in a secondary role, and pollen of dry-adapted trees of wooded savannas forms the primary non-ericoid arboreal component of the spectrum. At c. 4 000 years BP, the previously high values for sedimentary charcoal drop to consistently low levels that persist for the next 2 000 years. Pollen of *Celtis* and many other types of woody vegetation, especially small trees and shrubs associated with forest edge, open woodland, and riparian environments, increase steadily in importance, while Gramineae pollen ranges between 25 and 38%. At c. 1 500-2 000 years BP, charcoal values begin to increase again, although they never attain the high values of the early and middle Holocene, and arboreal pollen percentages decrease as Gramineae and Compositae increase. The apparent existence of open vegetation types, pyrogenic communities, and shifting vegetation dominance in pre-settlement Madagascar is in direct contradiction to key aspects of the Perrier/Humbert hypothesis that pre-settlement Madagascar was characterized by closed 'climax' vegetation types that have changed little throughout the late Cenozoic. If a more dynamic vegetation history is more plausible, then these early authors' corollary that human-induced vegetation change was the primary cause of the late Holocene megafaunal extinctions in Madagascar may be an overly simplified version of actual occurrences leading up to the mass extinctions.

Key-words: Madagascar; extinctions; pollen analysis; fire ecology.

INTRODUCTION

Madagascar is perhaps unique among large tropical land masses in that, for most of the Holocene, no people lived on the island. The earliest estimates for human settlement, based on glottochronological evidence, reach back only to c. 2 000 years BP (Dahl 1951, Vérin et al. 1970). The earliest ^{14}C date for an archaeological site (Sarodrano) is only 1 460 ± 90 years BP (Battistini & Vérin 1971). The island is thus an interesting case study in tropical vegetation change, since Holocene changes elsewhere have generally been subject to dual causation. Nearly everywhere else in the tropics, natural causes have been difficult to separate from the effects of the rapidly evolving technologies of Holocene man.

Almost nothing is known concerning climate and vegetation changes in pre-settlement Madagascar. Perrier de la Bathie (1921, 1936) and Humbert (1927, 1949, 1955) believed that the island was essentially covered by dense 'climax' formations for the entire Cenozoic, until settlers introduced fire to these virgin environments with devastating consequences. They further assert that the extinction of the Malagasy megafauna, including the giant lemurs, elephantbirds, and the pigmy hippopotamus, was a direct consequence of this drastic modification of the formerly pristine environment by man.

This hypothesis has been repeated uncritically by many secondary sources for decades but was, until very recently, essentially untested. While ecologists have been forced to abandon notions of unchanging climate and vegetation in most of the rest of the tropics (Livingstone & Van der Hammen 1978), authors have generally overlooked climatic change as a factor in evolutionary and ecological patterns in Madagascar. Mahé & Sourdat (1972), however, have pointed to geomorphological and pedological evidence for relatively drastic climatic change in the arid southwest of the island. Very recently, doubts about the Perrier/Humbert hypothesis have been expressed on grounds of the inconsistency of such a simple explanation with the views of modern climatologists and ecologists regarding the long-term instability of tropical climates and vegetation (Dewar 1984, Burney et al. 1986). In addition, explicit tests of the notion of continuous, unchanging forest cover in the interior of Madagascar during the Holocene have been possible using diachronic data from stratigraphic pollen analysis (MacPhee et al. 1985, Burney et al. 1985). Stratigraphic charcoal analysis from late Quaternary lake sediments has also shown that fire has in actuality been a factor in some Malagasy environments for at least 34 000 years prior to the presumed time of human arrival (Burney 1986).

The new synthesis emerging from this work is that the original hypothesis is too extreme in its contrast between pre- and post-settlement ecological conditions. Degradation of wooded environments by man has occurred since the arrival of the first settlers, and in fact continues today at an alarming rate. However, the hypothesis as originally stated is almost certainly an overly simplified version of circumstances leading up to the mass extinctions. Open environments, pyrogenic communities, and vegetation changes were a fact of life in Madagascar long before human arrival, which suggests that changes of this type alone would hardly be expected to wipe out the entire megafauna in a matter of a few centuries.

This study presents the results from pollen analysis of a sediment core from Lake Tritrivakely in central Madagascar. The core is thought to provide an essentially uninterrupted record of the entire Holocene, and fills some gaps in the chronology of vegetation dynamics prior to the time of human arrival and the megafaunal extinctions.

METHODS

The general location of the coring site is shown in Figure 1. Lake Tritrivakely (S 19°47′E 46°55′) lies at an elevation of 1 800 m in rolling volcanic terrain. Grassland, terraced agriculture and rice paddies dominate the modern vegetation nearby. Within a 10 km radius of the lake, the vegetation was classified from LANDSAT imagery, stereo airphotos, and ground survey (Burney 1986) to consist of 47% cropland, 37% rangeland and 16% wooded (mostly plantations of *Pinus* and *Eucalyptus*).

The lake lies in the bottom of a volcanic crater. When visited in July, 1985, in the dry season, the lake had a surface area of 4.5 ha and a maximum depth of 1.4 m. Local people said that the lake was nearly dry for a short time during droughts in 1959 and 1965. *Cyperus* and *Juncus* stems are emergent from even the deepest parts of the lake. The lake has no visible outlet, and, except for one small man-made diversion from a rice paddy near the crater rim, run-off into the lake is restricted to the interior of the crater.

Sounding transects indicate that the lake bottom is essentially flat and featureless, dropping off gradually from shore. Coring was done from a platform supported by two rubber rafts, anchored near the geographic center of the lake, in water 115 cm deep. A piston-corer of the Livingstone-sampler type, mounted on stiff aluminium rods, was used to push 1.2 m thin-walled aluminium pipes into the sediment. The pipes had a 6 cm OD, and were sealed with corks for transport to the US. Recovery was essentially 100% on all sections but the first. The top sediments, which were laden with live roots, were sampled by first collecting the sediment-water interface with a scoop (surface sample H in the analyses to follow), then beginning the coring in a hole c. 10 cm deep cut into the root mat. Subsequent coring proceeded normally, so the core probably begins about 10 cm below the modern surface.

The resulting 5 m core was opened and subsampled in the palynological laboratory of the Duke University Zoology Department. Five samples were submitted to Beta Analytic, Inc., for ^{14}C dating. The results are shown in Figure 2. The core was described and photographed in fresh condition. Samples were analyzed for organic matter and charcoal by nitric acid digestion and loss-on-ignition as described by Winkler (1985). These results are shown in Figure 3.

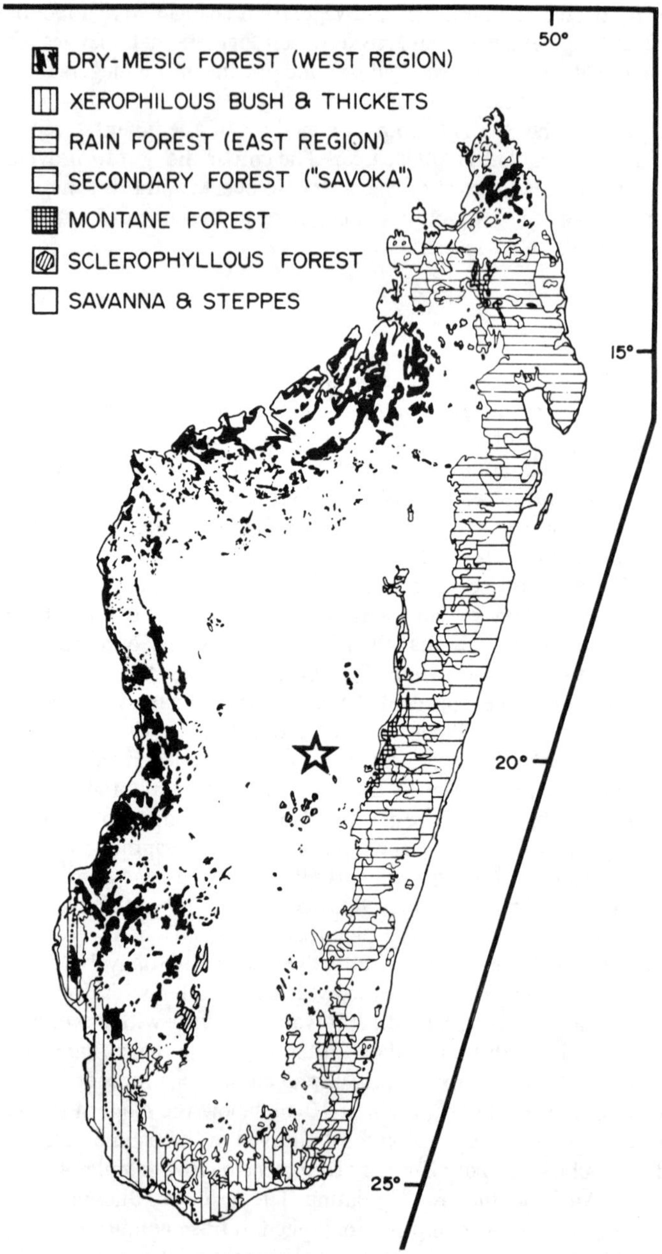

Figure 1. Existing vegetation of Madagascar, modified from Humbert & Cours Darne (1965) and MacPhee et al. (1985). Star indicates location of Lake Tritrivakely.

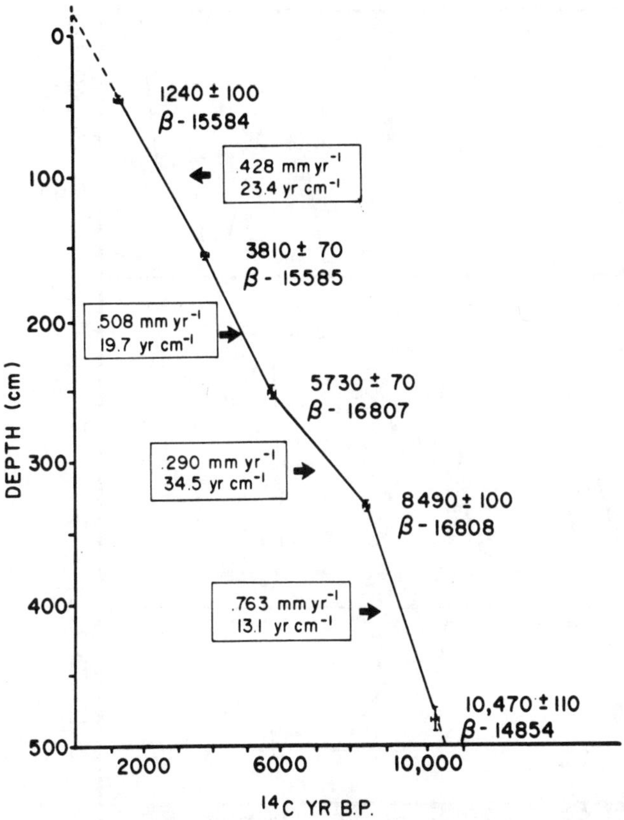

Figure 2. Radiocarbon chronology for the Lake Tritrivakely core

Fossil pollen was concentrated from the sediment samples according to the methods described by Faegri & Iversen (1975), with modificaions developed in D.A.Livingstone's laboratory at Duke University. Pollen residues were treated volumetrically as described by Kendall (1969). Pollen concentrations thus calculated were converted to influx values using the sedimentation rates interpolated from the five radiocarbon dates (Fig. 3).

Pollen grains were identified using the pollen reference collection in the Duke Zoology Department and the pollen morphological literature. Especially pertinent to identifications were the treatments of the pollen morphology of key families in the 'Palynologia de Madagassica et Mascarenica' series (Senesse 1980, Straka 1966, 1980, Straka & Friedrich 1983, Straka & Simon 1967, Straka et al. 1967). Photomicrographs of many of the pollen types identified, together with information on identification criteria and ecological significance,

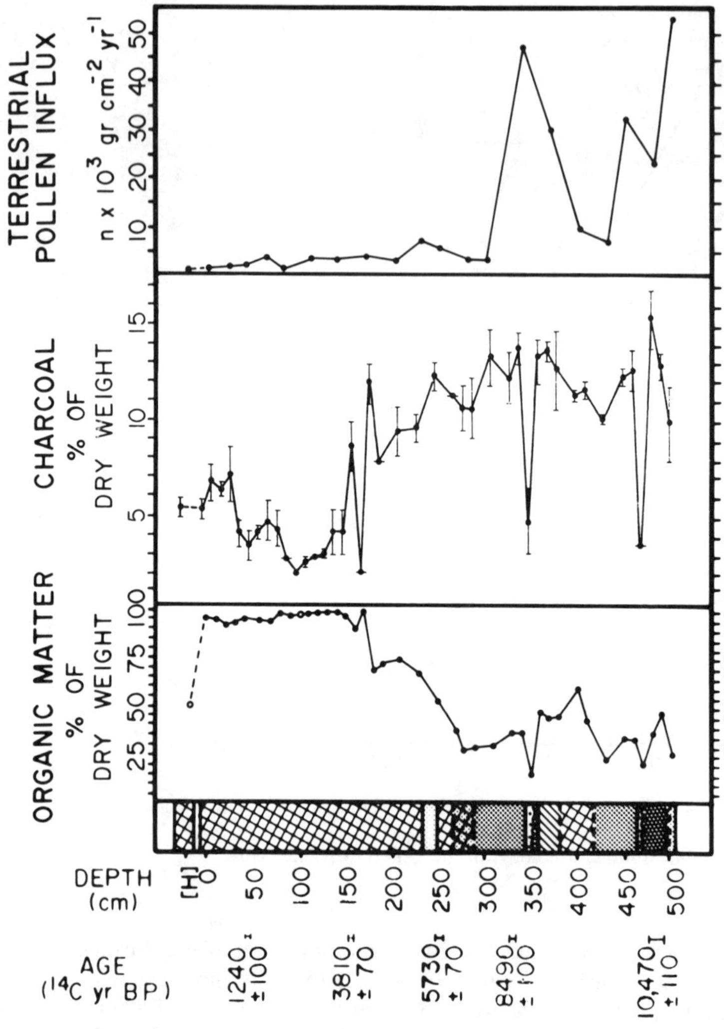

Figure 3. Lithology, percent organic matter, percent charcoal and pollen influx for the Lake Tritrivakely core. 'H' denotes a surface sample collected from the sediment-water interface c. 10 cm above the top of the core. Sediment conventions follow Faegri & Iversen (1975).

are included in a paper by the author concerning modern pollen-vegetation relationships in Madagascar which is in review elsewhere.

Between 500 and 1 000 pollen grains were counted at each level, using a Leitz Orthoplan microscope with Nomarsky interference and polarization, at 400 X and 1000 X oil immersion. Pollen samples were mounted in permanent liquid glycerine preparations which permit manipulation of the pollen grains for viewing on all sides. Pollen diagrams were plotted on a Tandy TRS-80 Model II microcomputer using COREPLOT software (Tucker & Tucker 1985).

RESULTS

Sediments

Sediment types represented in the 5 m core from Lake Tritrivakely (Fig. 3) showed considerable variation in the early and middle Holocene, but the Late Holocene sediments were much more uniform. For convenience, the sedimentary facies have been divided into eleven units, beginning at the bottom of the core.

Coring was terminated at a depth of 5 m because a hard, gritty layer was encountered that prevented extraction of a sixth section. The interval 503-496 cm in the core (Unit 1) contained a silty brown gyttja (10 YR 3/2) with many sub-angular to rounded clastics and some clay. The dense refractory material beneath that was not recovered may have been a bed of volcanic ash or sand. The second unit in the core was a fine detrital gyttja (same color as Unit 1), similar to the sediments recovered from deeper lakes in the region such as Lakes Andraikiba and Kavitaha (Burney et al. 1985). This material, which extends from c. 496-470 cm, includes a ^{14}C date, centered on 481.5 cm, of 10 470 ± 110 years BP (Beta-14854). On top of this sedimentary unit lies a laterally variable, 1 cm band of grayish-brown clay (2.5 Y 5/2) followed by 5 cm of dark yellowish-brown sand (10 YR 4/4). The sand contains both angular and subrounded material of apparently volcanic origin, suggesting that it may be a mixture of air-borne and water-borne ash.

From about 465-420 cm (Unit 4) there is a dark grayish brown (10 YR 3/2) humic clay layer, with volcanic cinders up to 3 mm at the bottom of the unit. This sediment contains about half as much organic matter per dry weight as Unit 2, and is in this respect more similar to Unit 1. Around 420 cm there is a gradual transition to a coarse fibrous gyttja (Unit 5), becoming darker toward the top. This type of material continues up to c. 383 cm.

From 383-360 cm (Unit 6) a crumbly peat occurs, very dark greyish brown (2.5 Y 3/2) and c. 45 % organic matter. On top of the peat is a single clay-sand-clay intercalation (Unit 7) similar to Unit 3). The bottom clay unit is 4 cm of light and dark brown mottled clay, followed by 5 cm of olive (5 Y 5/4) sand which contains angular to subrounded clastics, probably of volcanic origin. Above the sand is a 7 cm band of mottled clay.

Unit 8 (344-290 cm) is a humic clay similar to Unit 4. There are a few yellowish cinders and reddish flecks in the lowest 5 cm of this unit, but the material is generally uniform above. A sample centered on 330.5 cm dated to 8 490 ± 100 years BP (Beta-16808). This material gives way gradually to a gritty coarse gyttja (Unit 9), olive in colour (5 Y 4/2), becoming less gritty upwards.

At 250 cm (dated 5 370 ± 70 years BP, Beta-16807) a few light olive-gray (5 Y 6/2) basaltic fragments occur, angular and up to 8 mm in length. These are followed by Unit 10 (251-238 cm), a silty humic clay (5 Y 3/1).

From 238 cm to the top of the core (Unit 11), no further sedimentary changes are noted. Organic matter values rise from about 70% at the bottom of the unit to nearly 100% at 160 cm, and values remain high. The material is a coarse fibrous gyttja, nearly black (5 Y 2/2) and containing some very fine clastics but generally little else except plant fibres, pollen and algal skeletons.

Radiocarbon dates were obtained of 3 810 ± 70 years BP (Beta-15885) from material centered on 153 cm, and 1 240 ± 100 years BP (Beta-15884) from material centered at 43 cm below the top of the core. The core was unconsolidated from the top down to about 20 cm, and a few live roots penetrated down to ca. 50 cm. Rootlets were picked from the samples prior to dating, and repeated rinsing removed soluble humic acids from them. Nevertheless, the 43 cm date in particular should be regarded with some suspicion, as the presence of a few pollen grains of plants introduced in the twentieth century AD were observed in a pollen sample from 3 cm above the level dated (see 'Pollen analysis' for details), suggesting that some sediment mixing or surface contamination has probably occurred.

Figure 3 also presents charcoal percent of dry weight and pollen influx data for comparison. The charcoal values are discussed in detail elsewhere (Burney 1986), with comparisons to results from other methods of measuring charcoal in the same core. For the present study, it is sufficient to note that all the types of charcoal results obtained for this core consistently showed the same pattern: charcoal values were high in the early and mid-Holocene, but a drastic decline in charcoal concentration and charcoal influx occurred at about 3 800 years BP. These low levels were maintained for about 2 000 years, followed by a rise beginning around 1 500-2 000 years BP. In the present millennium, values never rise to the maxima attained during several periods of the early and middle Holocene.

Pollen influx values are also shown in Figure 3. It should be noted that this parameter is heavily dependent on the calculated sedimentation rate. Although sedimentation within the more organic units (such as the late Holocene Unit 11) might be expected to have been roughly constant, there is every reason to believe that this is not true for the earlier periods, since at least ten distinct units are covered in this period by only three radiocarbon dates. The influx results for the early and middle Holocene are therefore extremely variable and too suspect to be very useful. It is, however, interesting to note that pollen influx was generally high, even in the humic clay units, suggesting that material (except volcanic ash) probably accumulated gradually, rather than in episodic fashion. Also, the graminoid/ericoid-dominated pollen spectra of the early Holocene (presented in the next section) appear, as might be expected, to have produced much more pollen than later environments, when such heavy local pollen producers as the graminoids and ericoids (Hamilton 1982) have declined to some extent.

Pollen analysis

A total of 107 different pollen types were identified in this core. Although firm identifications to genus or family were possible in most cases, less certain identifications were suffixed with 'comp.' or 'sim.' according to whether the degree of uncertainty was low or higher, in accordance with the system devised by Benninghoff & Kapp (1962). Botanical nomenclature used follows Koechlin et al. (1974). In addition, 57 different unknown pollen types were photographed and described. Only four unknown types were observed with frequent recurrence, and total unknowns ranged between 0 and 9% for all levels. Crumpled indeterminates ranged between 1.8% and 7.6% of total pollen grains at all levels.

Figure 4 shows percentages for Gramineae and the major woody taxa, and Figure 5 shows miscellaneous terrestrial types, including plants with variable life forms, herbs, pteridophytes and introductions. In Figure 6, the aquatic and paludal types, which have been excluded from the pollen sum for calculating terrestrial pollen percentages, are shown. Aquatics and paludals are calculated as percentages of the raw sum (terrestrial + aquatics and paludals). Table 1 lists the counts and percentages for trace pollen types that are not shown in any of the diagrams, but have been included in the terrestrial sum.

The relations between percentages of Gramineae, Ericaceae, and *Celtis* pollen are informative. In a study of modern pollen/vegetation relationships in Madagascar (Burney 1986), it was shown that percentages of Gramineae pollen over c. 30-40% were characteristic of sediments from areas predominantly covered with grasslands, whereas Ericaceae pollen (probably mostly *Philippia*, by far the most common ericoid genus in Madagascar) dominated montane shrub grasslands and shrublands, and *Celtis* pollen percentages were highest in woodland/grassland mosaics. Modern spectra from closed forest areas were low in these three pollen types. Instead the forest spectra were composed of pollen of a large number of tree taxa, as well as abundant spores from Cyatheaceae and other pteridophytes.

The percentages of Gramineae, Ericaceae, and *Celtis* in the Tritrivakely core fall into four spectral categories. There are some troublesome intermediate spectra, gradual transitions and alternating spectral types that make it undesirable to define 'pollen zones' in the classic sense. Instead of zones, the author proposes the use of the terms 'early', 'middle' and 'late Holocene' to define the pre-settlement pôrtion of the core, and 'post-settlement' for the period beginning about 1 500 years BP.

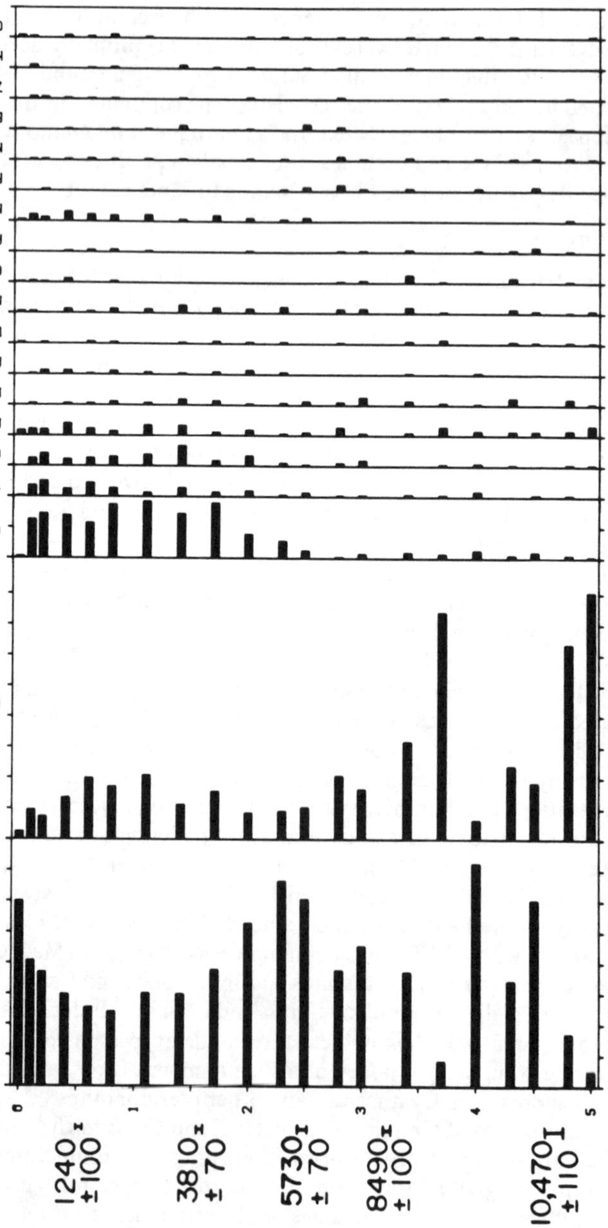

Figure 4. Pollen diagram showing percentages of Gramineae and major woody taxa in the Lake Tritrivakely core. Surface sample H is denoted by 'O' and the top of the core is at the 10 cm mark. All other samples are denoted as in Figure 3.

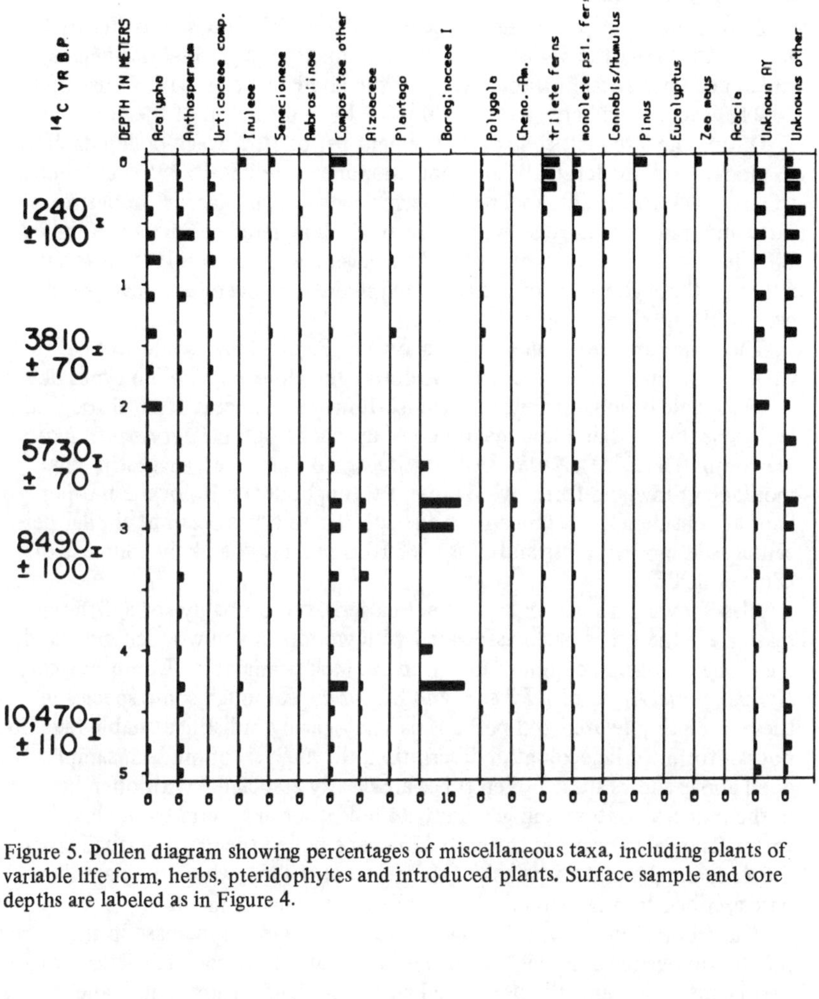

Figure 5. Pollen diagram showing percentages of miscellaneous taxa, including plants of variable life form, herbs, pteridophytes and introduced plants. Surface sample and core depths are labeled as in Figure 4.

In the early Holocene, two major types of spectra occur, sometimes in alternation. In the earliest spectrum, extrapolated to c. 10 750 years BP, the Gramineae percentage is the lowest and Ericaceae is the highest of the entire core (4% and 81% respectively). For the next four levels, the two gradually switch positions, with grass pollen reaching 73% and Ericaceae declining to 6% at a level interpolated to an age of 9 400 years BP. At the very next level (c. 9 000 years BP), the two have again switched position, to a configuration similar to the level at the bottom of the core. Over the next five levels, grass pollen increases and Ericaceae decreases again, to 67% and 9%, respectively. This level is 30 cm above a ^{14}C date of 5 730 ± 70 years BP and interpolates to c. 5 400 years BP.

During these early and mid-Holocene times, *Celtis* shows consistently low values (< 5%). At the 5 400 years BP level, however, a gradual rise begins which becomes an important feature of the late Holocene, with values of 13-20% for every level between c. 4 000 years BP and the top of the core.

During the late Holocene pre-settlement period, Ericaceae pollen declines to moderately low levels (9-20%) and Gramineae are 25-38%. It is instructive to refer back to the charcoal results, which show high charcoal values for the early and middle Holocene, with a drastic decline dated to 3 810 ± 70 years BP. Charcoal values remain low until the level dated 1 240 ± 100 years BP, although the high values of the early and middle Holocene are never reached, even in the modern sample.

The major arboreal pollen types shown in Figure 4 give some clues as to what might have occurred at the site during the Holocene. Pollen types likely to be capable of long-distance transport show no distinct pattern through the Holocene, but rather a low, nearly-constant background (*Podocarpus, Noronhia* comp. (Oleaceae), *Antidesma* and *Maesa,* for instance), suggesting that montane or lowland forest were present throughout the Holocene but perhaps at considerable distance from the site. These types occur at similar percentages in modern grassland sites over 100 km from the likely source areas (Burney 1986).

Also showing no particular distinct pattern, but probably for a different reason, are the types often associated with woodland and wooded grassland, including pyrogenic communities: *Uapaca,* Melastomataceae-Combretaceae, *Gymnosporia, Ephippiandra* sim. and *Zizyphus.* Although some species in these or closely related and perhaps palynologically indistinguishable taxa do not conform to the ecological description, the analysis of modern samples cited above shows these pollen types are clearly associated with open habitats in the modern context and generally do not occur in spectra from closed forest. They also occur in the early Holocene pollen spectra described from the Ampasambazimba fossil site (McPhee et al. 1985), where they are likewise ascribed to mosaic habitats including open environments.

The arboreal types which show a pattern of a sudden increase in the middle Holocene to the highest values in the late Holocene pre-settlement period are types generally associated with woodlands, forest edges and riparian habitats. These include *Trema, Chaetacme* (these two types are in the same family as *Celtis,* the Ulmaceae, and can be distinguished only with great care, but the combined values for the three show the same trend), *Alchornea,* Moraceae comp., *Hymenocardia,* and Cyatheaceae. These together might be taken as indicators of somewhat wetter or less intensely seasonal conditions than earlier in the Holocene, although decreasing fire frequency (also possibly a result of wetter or less intensely seasonal conditions) and landscape modification by volcanism cannot be ruled out as primary causes. Regarding volcanism, it should be noted that, whereas the core contains evidence of three or more volcanic episodes in the early and middle Holocene, late Holocene sediments are almost entirely free of both angular and reworked clastics. Lava-dammed valleys are a common feature in this area, and the increase in woody vegetation could reflect trees growing on a local lava-dammed valley.

Some of the pollen types shown in Figure 5 may provide information concerning the character of pre-settlement vegetation. The notion that early and mid-Holocene vegetation at the site was more open, with an increase in forest vegetation in the late (pre-settlement) Holocene, is supported by the relatively higher values for all categories of Compositae prior to c. 4 000 years BP. Total composites attain c. 10% in the millennia prior to this date, drop to a trace, and do not attain 10% again until the modern period (surface sample H, shown as 'O' on the pollen diagrams). The only composite which perhaps fails to conform to the pattern is *Brachylaena* (Table 1), a fire-resistant tree of dry woodlands and wooded grasslands which occurs as a trace at one level each in the late, middle, and early Holocene and in the post-settlement zone.

Acalypha, a physiognomically and ecologically diverse genus, shows the

Table 1. Minor taxa not shown on plots of pollen diagrams, but included in pollen sum. Depth in meters, pollen count/% of total for depth.

Index and taxon	Index and taxon
1 Acanthaceae	28 Liliaceae comp.
2 Anacardiaceae	29 *Linociera* sim.
3 Anthocerotae	30 Loranthaceae
4 Bignoniaceae	31 *Lycopodium*
5 *Brachylaena*	32 *Macaranga* comp.
6 *Bridelia*	33 Malvaceae
7 Capparidaceae	34 Meliaceae
8 Celastraceae	35 Mutiseae
9 Cichorieae	36 *Myrica*
10 *Croton* comp.	37 Myrsinaceae
11 *Cupressus*	38 *Pandanus*
12 Davalliaceae	39 Periplocaceae
13 *Dombeya*	40 *Phoenix*
14 *Drypetes*	41 *Phyllanthus*
15 *Elaeocarpus*	42 Polypodiaceae
16 *Eugenia*	43 Proteaceae II
17 *Euphorbia*	44 Proteaceae IV
18 Euphorbiac. prolate	45 Rhanmaceae
19 *Faurea*	46 *Rhus*
20 *Ficus*	47 Sapotaceae
21 *Gagnebina* sim.	48 Sapot. retic.
22 *Harungana*	49 *Selaginella*
23 *Hibiscus*	50 Solanaceae
24 *Ilex*	51 *Stereospermum*
25 *Indigofera*	52 *Strychnos*
26 *Isoglossa*	53 *Syzygium*
27 Labiatae	54 *Tragia* sim.

Depth	Acanthac	Anacardi.	Anthocer.	Bignonia.	Brachyla.	Bridelia.	Capparid.	Celastra.
0.0	2/00.6	0/00.0	0/00.0	0/00.0	2/00.6	0/00.0	0/00.0	0/00.0
0.1	0/00.0	0/00.0	0/00.0	0/00.0	0/00.0	0/00.0	0/00.0	0/00.0
0.2	0/00.0	2/00.7	1/00.3	0/00.0	0/00.0	0/00.0	0/00.0	0/00.0
0.4	0/00.0	0/00.0	0/00.0	0/00.0	0/00.0	0/00.0	0/00.0	0/00.0
0.6	0/00.0	0/00.0	0/00.0	0/00.0	0/00.0	0/00.0	0/00.0	0/00.0
0.8	0/00.0	1/00.3	0/00.0	0/00.0	0/00.0	0/00.0	0/00.0	0/00.0
1.1	0/00.0	0/00.0	0/00.0	0/00.0	1/00.3	1/00.3	0/00.0	0/00.0
1.4	1/00.4	1/00.4	0/00.0	0/00.0	0/00.0	0/00.0	0/00.0	1/00.4
1.7	0/00.0	0/00.0	0/00.0	0/00.0	0/00.0	0/00.0	0/00.0	0/00.0
2.0	0/00.0	0/00.0	0/00.0	0/00.0	0/00.0	0/00.0	0/00.0	0/00.0
2.3	0/00.0	1/00.3	0/00.0	0/00.0	0/00.0	0/00.0	1/00.3	0/00.0
2.5	0/00.0	1/00.3	1/00.3	0/00.0	1/00.3	0/00.0	0/00.0	0/00.0
2.8	0/00.0	0/00.0	0/00.0	0/00.0	0/00.0	0/00.0	1/00.3	0/00.0
3.0	1/00.4	0/00.0	0/00.0	1/00.4	0/00.0	0/00.0	0/00.0	0/00.0
3.4	1/00.2	0/00.0	0/00.0	0/00.0	0/00.0	0/00.0	1/00.2	0/00.0
3.7	0/00.0	0/00.0	0/00.0	1/00.2	1/00.2	0/00.0	0/00.0	0/00.0
4.0	0/00.0	1/00.3	0/00.0	0/00.0	0/00.0	0/00.0	0/00.0	0/00.0
4.3	1/00.3	0/00.0	0/00.0	0/00.0	0/00.0	0/00.0	2/00.7	0/00.0
4.5	0/00.0	0/00.0	0/00.0	0/00.0	0/00.0	0/00.0	0/00.0	0/00.0
4.8	0/00.0	0/00.0	0/00.0	0/00.0	0/00.0	0/00.0	0/00.0	0/00.0
5.0	0/00.0	0/00.0	0/00.0	0/00.0	0/00.0	0/00.0	0/00.0	0/00.0

Depth	Cichorie.	Croton.c.	Cupressu.	Davallia.	Dombeya.	Drypetes.	Elaeocar.	Eugenia.
0.0	1/00.3	0/00.0	1/00.3	0/00.0	1/00.3	0/00.0	0/00.0	0/00.0
0.1	0/00.0	0/00.0	0/00.0	0/00.0	0/00.0	0/00.0	0/00.0	0/00.0
0.2	0/00.0	1/00.3	0/00.0	1/00.3	0/00.0	0/00.0	0/00.0	2/00.7
0.4	0/00.0	0/00.0	0/00.0	0/00.0	0/00.0	0/00.0	0/00.0	0/00.0
0.6	0/00.0	1/00.3	0/00.0	0/00.0	0/00.0	0/00.0	1/00.3	0/00.0
0.8	0/00.0	0/00.0	0/00.0	0/00.0	2/00.6	0/00.0	1/00.3	2/00.6
1.1	0/00.0	0/00.0	0/00.0	0/00.0	0/00.0	1/00.3	1/00.3	0/00.0
1.4	0/00.0	0/00.0	0/00.0	0/00.0	0/00.0	0/00.0	3/01.1	0/00.0
1.7	0/00.0	0/00.0	0/00.0	0/00.0	0/00.0	0/00.0	1/00.3	0/00.0
2.0	0/00.0	0/00.0	0/00.0	0/00.0	0/00.0	0/00.0	0/00.0	0/00.0
2.3	0/00.0	0/00.0	0/00.0	0/00.0	0/00.0	0/00.0	0/00.0	0/00.0
2.5	0/00.0	0/00.0	0/00.0	0/00.0	0/00.0	0/00.0	0/00.0	0/00.0
2.8	0/00.0	0/00.0	0/00.0	0/00.0	0/00.0	0/00.0	0/00.0	0/00.0
3.0	0/00.0	0/00.0	0/00.0	0/00.0	0/00.0	0/00.0	0/00.0	0/00.0
3.4	0/00.0	1/00.2	0/00.0	0/00.0	0/00.0	0/00.0	0/00.0	0/00.0
3.7	0/00.0	0/00.0	0/00.0	1/00.2	0/00.0	0/00.0	0/00.0	0/00.0
4.0	0/00.0	0/00.0	0/00.0	0/00.0	0/00.0	0/00.0	0/00.0	0/00.0
4.3	0/00.0	0/00.0	0/00.0	0/00.0	0/00.0	0/00.0	0/00.0	0/00.0
4.5	0/00.0	1/00.3	0/00.0	0/00.0	0/00.0	0/00.0	0/00.0	0/00.0
4.8	0/00.0	1/00.2	0/00.0	0/00.0	0/00.0	0/00.0	0/00.0	0/00.0
5.0	0/00.0	0/00.0	0/00.0	0/00.0	0/00.0	0/00.0	1/00.2	0/00.0

Depth	Euphorbic..	Euphorb.p.	Faurea.	Ficus.	Gagnebin.	Harungan.	Hibiscus.	Ilex
0.0	1/00.3	0/00.0	0/00.0	0/00.0	0/00.0	0/00.0	0/00.0	0/00.0
0.1	0/00.0	0/00.0	0/00.0	0/00.0	0/00.0	1/00.3	0/00.0	0/00.0
0.2	0/00.0	0/00.0	1/00.3	0/00.0	1/00.3	1/00.3	0/00.0	0/00.0
0.4	0/00.0	0/00.0	0/00.0	0/00.0	0/00.0	1/00.3	0/00.0	0/00.0
0.6	0/00.0	0/00.0	0/00.0	1/00.3	0/00.0	0/00.0	0/00.0	0/00.0
0.8	1/00.3	0/00.0	0/00.0	0/00.0	1/00.3	0/00.0	0/00.0	0/00.0
1.1	0/00.0	0/00.0	0/00.0	0/00.0	0/00.0	1/00.3	0/00.0	1/00.3
1.4	1/00.4	0/00.0	0/00.0	2/00.7	0/00.0	0/00.0	0/00.0	0/00.0
1.7	1/00.3	1/00.3	0/00.0	4/01.2	0/00.0	0/00.0	1/00.3	0/00.0
2.0	0/00.0	0/00.0	0/00.0	0/00.0	0/00.0	0/00.0	0/00.0	1/00.3
2.3	2/00.5	0/00.0	0/00.0	0/00.0	0/00.0	0/00.0	0/00.0	0/00.0
2.5	0/00.0	0/00.0	0/00.0	0/00.0	0/00.0	0/00.0	0/00.0	0/00.0
2.8	0/00.0	0/00.0	0/00.0	0/00.0	0/00.0	0/00.0	0/00.0	0/00.0
3.0	0/00.0	0/00.0	0/00.0	0/00.0	1/00.4	0/00.0	0/00.0	0/00.0
3.4	0/00.0	0/00.0	0/00.0	1/00.2	0/00.0	0/00.0	0/00.0	1/00.2
3.7	0/00.0	0/00.0	1/00.2	0/00.0	0/00.0	0/00.0	0/00.0	0/00.0
4.0	0/00.0	0/00.0	0/00.0	0/00.0	0/00.0	0/00.0	0/00.0	0/00.0
4.3	0/00.0	0/00.0	0/00.0	0/00.0	0/00.0	0/00.0	0/00.0	0/00.0
4.5	2/00.7	0/00.0	0/00.0	0/00.0	0/00.0	0/00.0	0/00.0	0/00.0
4.8	1/00.2	0/00.0	0/00.0	0/00.0	1/00.2	0/00.0	0/00.0	0/00.0
5.0	0/00.0	0/00.0	0/00.0	0/00.0	0/00.0	0/00.0	0/00.0	0/00.0

Depth	Indigofe.	Isogloss.	Labiatae.	Liliacea.	Linocier.	Lorantha.	Lycopodi.	Macarang.
0.0	0/00.0	0/00.0	0/00.0	0/00.0	0/00.0	0/00.0	1/00.3	1/00.3
0.1	0/00.0	0/00.0	0/00.0	0/00.0	0/00.0	0/00.0	0/00.0	0/00.0
0.2	0/00.0	0/00.0	0/00.0	0/00.0	0/00.0	0/00.0	0/00.0	0/00.0
0.4	0/00.0	0/00.0	0/00.0	0/00.0	0/00.0	0/00.0	0/00.0	0/00.0
0.6	0/00.0	1/00.3	2/00.5	0/00.0	0/00.0	0/00.0	0/00.0	0/00.0
0.8	0/00.0	0/00.0	0/00.0	0/00.0	0/00.0	0/00.0	0/00.0	1/00.3
1.1	1/00.3	0/00.0	0/00.0	0/00.0	1/00.3	0/00.0	0/00.0	1/00.3
1.4	0/00.0	0/00.0	0/00.0	0/00.0	0/00.0	1/00.4	0/00.0	0/00.0
1.7	1/00.3	0/00.0	0/00.0	0/00.0	1/00.3	0/00.0	1/00.3	2/00.6
2.0	0/00.0	0/00.0	0/00.0	0/00.0	0/00.0	0/00.0	0/00.0	0/00.0
2.3	2/00.5	1/00.3	0/00.0	0/00.0	0/00.0	0/00.0	0/00.0	0/00.0
2.5	0/00.0	1/00.3	2/00.6	1/00.3	0/00.0	0/00.0	0/00.0	0/00.0
2.8	0/00.0	0/00.0	0/00.0	0/00.0	1/00.3	0/00.0	0/00.0	0/00.0
3.0	0/00.0	0/00.0	1/00.4	0/00.0	0/00.0	0/00.0	0/00.0	0/00.0
3.4	0/00.0	1/00.2	0/00.0	0/00.0	0/00.0	0/00.0	0/00.0	0/00.0
3.7	0/00.0	0/00.0	0/00.0	0/00.0	1/00.2	0/00.0	0/00.0	0/00.0
4.0	0/00.0	0/00.0	0/00.0	0/00.0	1/00.3	0/00.0	0/00.0	0/00.0
4.3	2/00.7	0/00.0	0/00.0	0/00.0	1/00.3	0/00.0	0/00.0	0/00.0
4.5	0/00.0	1/00.3	0/00.0	0/00.0	0/00.0	0/00.0	0/00.0	0/00.0
4.8	1/00.2	0/00.0	0/00.0	0/00.0	0/00.0	0/00.0	0/00.0	0/00.0
5.0	0/00.0	0/00.0	0/00.0	0/00.0	0/00.0	0/00.0	1/00.2	0/00.0

Depth	Malvacea.	Meliacea.	Mutiseae	Myrica	Myrsinac	Pandanus	Periploc	Phoenix
0.0	1/00.3	0/00.0	0/00.0	0/00.0	0/00.0	0/00.0	1/00.3	2/00.6
0.1	0/00.0	3/00.9	0/00.0	0/00.0	1/00.3	0/00.0	0/00.0	0/00.0
0.2	0/00.0	0/00.0	0/00.0	0/00.0	1/00.3	0/00.0	0/00.0	0/00.0
0.4	0/00.0	1/00.3	0/00.0	0/00.0	0/00.0	0/00.0	0/00.0	0/00.0
0.6	0/00.0	0/00.0	0/00.0	0/00.0	0/00.0	0/00.0	0/00.0	0/00.0
0.8	1/00.3	2/00.6	0/00.0	0/00.0	0/00.0	1/00.3	0/00.0	0/00.0
1.1	0/00.0	0/00.0	0/00.0	1/00.3	2/00.5	0/00.0	0/00.0	0/00.0
1.4	0/00.0	1/00.4	0/00.0	0/00.0	0/00.0	0/00.0	0/00.0	0/00.0
1.7	0/00.0	0/00.0	0/00.0	0/00.0	0/00.0	0/00.0	0/00.0	0/00.0
2.0	0/00.0	1/00.3	0/00.0	0/00.0	3/00.8	0/00.0	0/00.0	0/00.0
2.3	0/00.0	0/00.0	0/00.0	0/00.0	2/00.5	0/00.0	0/00.0	0/00.0
2.5	0/00.0	0/00.0	0/00.0	0/00.0	2/00.6	1/00.3	0/00.0	0/00.0
2.8	0/00.0	0/00.0	0/00.0	0/00.0	2/00.7	0/00.0	0/00.0	0/00.0
3.0	0/00.0	0/00.0	0/00.0	1/00.4	5/01.8	0/00.0	0/00.0	0/00.0
3.4	0/00.0	0/00.0	0/00.0	0/00.0	5/01.1	0/00.0	0/00.0	0/00.0
3.7	0/00.0	0/00.0	0/00.0	0/00.0	1/00.2	0/00.0	0/00.0	0/00.0
4.0	0/00.0	0/00.0	0/00.0	0/00.0	4/01.1	0/00.0	0/00.0	0/00.0
4.3	0/00.0	0/00.0	1/00.3	1/00.3	0/00.0	0/00.0	0/00.0	0/00.0
4.5	0/00.0	1/00.3	0/00.0	0/00.0	2/00.7	0/00.0	0/00.0	0/00.0
4.8	0/00.0	0/00.0	0/00.0	0/00.0	0/00.0	0/00.0	0/00.0	0/00.0
5.0	0/00.0	0/00.0	0/00.0	0/00.0	3/00.6	0/00.0	0/00.0	0/00.0

Depth	Phyllant	Polypodi	Proteace	Proteace	Rhamnace	Rhus	Sapotace	Sapot.r.
0.0	1/00.3	0/00.0	0/00.0	0/00.0	1/00.3	0/00.0	0/00.0	0/00.0
0.1	0/00.0	1/00.3	0/00.0	0/00.0	0/00.0	0/00.0	0/00.0	0/00.0
0.2	0/00.0	1/00.3	0/00.0	0/00.0	0/00.0	2/00.7	0/00.0	0/00.0
0.4	0/00.0	0/00.0	0/00.0	0/00.0	0/00.0	1/00.3	0/00.0	0/00.0
0.6	0/00.0	0/00.0	0/00.0	1/00.3	0/00.0	1/00.3	0/00.0	0/00.0
0.8	2/00.6	2/00.6	1/00.3	0/00.0	0/00.0	0/00.0	0/00.0	0/00.0
1.1	0/00.0	0/00.0	0/00.0	0/00.0	0/00.0	0/00.0	0/00.0	0/00.0
1.4	1/00.4	0/00.0	0/00.0	0/00.0	0/00.0	0/00.0	1/00.4	3/01.1
1.7	1/00.3	0/00.0	0/00.0	0/00.0	1/00.3	1/00.3	0/00.0	0/00.0
2.0	4/01.1	0/00.0	0/00.0	0/00.0	0/00.0	0/00.0	2/00.5	0/00.0
2.3	0/00.0	0/00.0	0/00.0	1/00.3	0/00.0	0/00.0	0/00.0	0/00.0
2.5	2/00.6	0/00.0	0/00.0	0/00.0	0/00.0	0/00.0	0/00.0	0/00.0
2.8	0/00.0	0/00.0	0/00.0	0/00.0	0/00.0	0/00.0	0/00.0	0/00.0
3.0	1/00.4	0/00.0	0/00.0	1/00.4	0/00.0	0/00.0	0/00.0	0/00.0
3.4	1/00.2	0/00.0	0/00.0	0/00.0	0/00.0	1/00.2	0/00.0	0/00.0
3.7	0/00.0	0/00.0	0/00.0	0/00.0	0/00.0	0/00.0	1/00.2	0/00.0
4.0	0/00.0	0/00.0	0/00.0	0/00.0	0/00.0	0/00.0	0/00.0	0/00.0
4.3	2/00.7	0/00.0	0/00.0	0/00.0	0/00.0	0/00.0	0/00.0	0/00.0
4.5	0/00.0	0/00.0	0/00.0	0/00.0	0/00.0	0/00.0	1/00.3	0/00.0
4.8	0/00.0	0/00.0	0/00.0	0/00.0	0/00.0	0/00.0	1/00.2	0/00.0
5.0	0/00.0	1/00.2	0/00.0	0/00.0	0/00.0	0/00.0	0/00.0	0/00.0

Depth	Selagine	Solanace	Stereosp	Strychno	Syzygium	Tragia s.
0.0	2/00.6	0/00.0	0/00.0	0/00.0	0/00.0	0/00.0
0.1	1/00.3	0/00.0	0/00.0	0/00.0	0/00.0	0/00.0
0.2	0/00.0	0/00.0	0/00.0	0/00.0	0/00.0	1/00.3
0.4	0/00.0	0/00.0	0/00.0	0/00.0	0/00.0	4/01.2
0.6	0/00.0	1/00.3	0/00.0	0/00.0	1/00.3	0/00.0
0.8	0/00.0	0/00.0	0/00.0	0/00.0	0/00.0	0/00.0
1.1	0/00.0	0/00.0	0/00.0	0/00.0	0/00.0	0/00.0
1.4	0/00.0	0/00.0	0/00.0	0/00.0	0/00.0	0/00.0
1.7	0/00.0	1/00.3	0/00.0	0/00.0	0/00.0	1/00.3
2.0	0/00.0	0/00.0	0/00.0	1/00.3	2/00.5	0/00.0
2.3	0/00.0	0/00.0	0/00.0	0/00.0	0/00.0	0/00.0
2.5	1/00.3	0/00.0	0/00.0	1/00.3	1/00.3	0/00.0
2.8	0/00.0	0/00.0	0/00.0	0/00.0	0/00.0	0/00.0
3.0	0/00.0	0/00.0	1/00.4	1/00.4	0/00.0	0/00.0
3.4	0/00.0	0/00.0	0/00.0	3/00.7	0/00.0	0/00.0
3.7	0/00.0	0/00.0	1/00.2	1/00.2	0/00.0	1/00.2
4.0	0/00.0	0/00.0	0/00.0	0/00.0	0/00.0	0/00.0
4.3	0/00.0	0/00.0	0/00.0	2/00.7	0/00.0	0/00.0
4.5	0/00.0	0/00.0	0/00.0	2/00.7	0/00.0	0/00.0
4.8	0/00.0	0/00.0	0/00.0	0/00.0	0/00.0	0/00.0
5.0	0/00.0	0/00.0	0/00.0	2/00.4	0/00.0	0/00.0

same pattern as *Celtis,* increasing from a low background value in the early and mid-Holocene to its highest values (up to c. 5%) in the late Holocene. Although a difficult genus to characterize palynologically, being variously ascribed in East Africa to forest decline associated with human disturbance (Livingstone 1967) and drier forest types (Hamilton 1982), the majority of *Acalypha* species grow on forest edges or in openings, rather than under dense cover (Dale & Greenway 1961). Thus, when the late Holocene vegetation of the site was yielding the highest percentages of arboreal pollen types, *Acalypha* seems to have been indicating that the vegetation was never entirely closed. (Recall that Gramineae and Ericaceae percentages were still moderately high in this period as well.)

Anthospermum, various species of which may indicate either cool dry conditions (ericoid bush habitats for instance) or moist disturbed habitats (such as forest edges), is present at 1-5% in all pre-settlement levels except in some mid-Holocene spectra.

Urticaceae comp. shows the same pattern in this core as in the modern samples, as well as in the late Holocene core from Lake Kavitaha in the Itasy region c. 150 km to the north. This type seems associated with the prevalence of riparian woodlands. It appears in the Tritrivakely core at the same time that the *Celtis* rise begins and declines when *Celtis* declines.

Pollen of plant families that are not restricted to, but are primarily found in xeric habitats occur primarily in two distinct periods of the early and mid-Holocene, with a noticeable gap between. Pollen of Chenopodiaceae-Amaranthaceae, Boraginaceae, and Aizoaceae are generally absent from the core, except in the periods c. 5 500-9 000 and 9 500-10 000 years BP. These are also periods of very high charcoal and grass pollen values.

Pteridophyte spores are generally low at all pre-settlement levels. Both triletes and monoletes occur as traces at most levels throughout, and only very small traces of Davalliaceae and Polypodiaceae, which are generally associated with forests, occur at widely-scattered intervals in the core (Table 1). Pteridophytes reach their highest values (up to 6% triletes in the modern sample) during the post-settlement period. The triletes in this period appear to be *Pteridium,* although the very limited selection of Malagasy fern spores in the reference collection precludes certainty. *Pteridium* is favoured by burning, and hunters and gatherers in some parts of the world intentionally set fires in order to encourage a growth of bracken, which has highly edible rhizomes (McGlone 1983).

The post-settlement zone of this core contains interesting features that are comparable to the results from the same time period in the Lake Kavitaha core (Burney et al. 1985). *Cannabis/Humulus* pollen, probably derived from the early introduction of hemp by mariners (Vavilov 1949), appears just prior to the beginning of the presumed post-settlement zone (the first appearance of this pollen type is interpolated as 2 300 years BP in this core – hemp pollen was present in the earliest level in the Kavitaha core, which was c. 1 500 years BP).

Pollen evidence from introductions of the European period suggests that the top of this core, as one might expect for sediments collected in a mat of live emergent vegetation, is somewhat mixed stratigraphically. *Pinus, Eucalyptus* and *Cupressus* are believed to have first arrived in Madagascar about 1 900 AD (Koechlin et al. 1974), but a few grains of the first two occur down to 40 cm, a level only 3 cm above the level dated 1 240 ± 100 years BP. Clearly this core is less suitable for discriminating trends of the present millennium than was the Kavitaha core, which was taken from a deeper lake that had a sedimentation rate over four times as fast, a much better situation for discriminating events of recent centuries.

Two other introduced plants contribute pollen to the modern sample from the site, but do not occur in the core. *Zea mays* (7 grains – 2%) and *Acacia* (2 grains – 0.6%) reflect the fact that both are presently growing inside the crater. Maize was probably introduced to Madagascar, as in East Africa, in the early nineteenth century AD or shortly before (Kokwaro 1979). Although there are a few indigenous *Acacia* species, especially in southern and western arid regions of the island, they are uncommon in the central region. However, an Australian species, *Acacia dealbata* Link., is widely planted as a source of firewood and charcoal, and this fast-growing species is a very common escape. Inside the Tritrivakely crater and along the rim, it is quite abundant and is practically the only tree present, except for a few small pines.

Figure 6 shows the aquatic and paludal plants as a percentage of the raw sum. Cyperaceae are very common at most levels. They decline in the early Holocene at the same levels as Gramineae. *Potamogeton* does not conform to

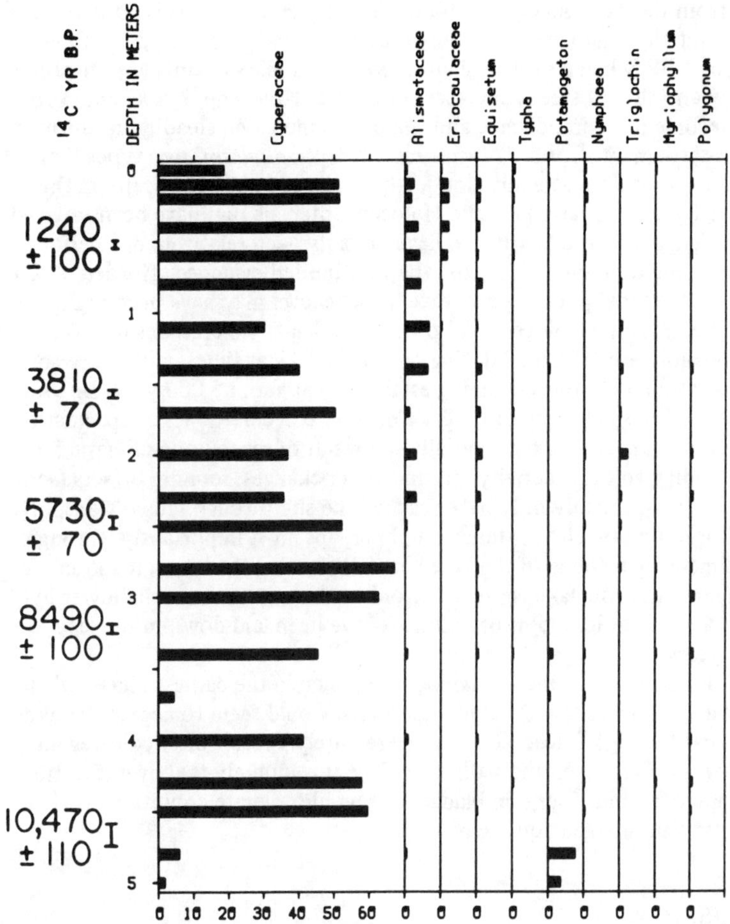

Figure 6. Pollen diagram showing percentages of aquatic and paludal taxa. Pollen counts for these have been excluded from the sum used for calculating percentages of terrestrial types. Surface sample and core depths are labeled as in Figure 4.

the pattern exhibited by the paludals and emergent aquatics. In the earliest two spectra (c. 10 400 and 10 700 years BP) it occurs at 7 and 4% respectively, and only sporadically as a trace in the rest of the core. Deep crater lakes in the area, such as Lake Andraikiba near Antsirabe, contain few aquatic plants except *Potamogenton pectinatus* L., which forms a dense mat around the steep shoreline of this lake (personal observation). Other aquatics are virtually absent from the Tritrivakely spectra of the early Holocene, as is *Potamogeton* after its initially high levels, and they only very slowly increase until about 5 000 years BP, when a host of shallow-water and marsh plants appear rather suddenly and show a steady increase. In the late Holocene, *Typha* and *Nymphaea* become more important, and these have increased steadily up through the post-settlement period. *Polygonum* and *Myriophyllum,* two types that frequently colonize seasonally flooded habitats, appear sporadically in the core, particularly in early and mid-Holocene intervals that have been inferred from evidence presented earlier to have possibly been relatively dry periods.

Thus it might be suggested from the very limited evidence afforded by the aquatic and paludal pollen types, that the lake level may have been highest at the beginning of the Holocene, followed by at least two episodes of lower levels, possibly even seasonal drying out of the lake at times, with more stable shallow-water conditions becoming established at about 5 000 years BP. Since that time, some shallowing of the lake may have occurred. These speculations, if they in fact are true, are potentially the result of several non-climatic factors, including volcanic activity, ground-water changes, founder effects (some aquatics could probably only have reached the site through faunal transport, but this is not a rare circumstance), and perhaps most importantly, through the sedimentary infilling of the lake basin. Regarding the latter, it should be remembered that the lake is presently only about 1.4 m deep, shallower in some years, and at least 5 m of sediment have been laid down in the last 11 000 years.

The presence of so much *Potamogeton* pollen in the earliest phase, when Ericaceae pollen was also at its highest levels, would seem to negate the likelihood that the high Ericaceae values were purely a result of ericoid vegetation growing directly on the sediment, since it is unlikely that even the most wet-adapted ericoids found in Madagascar would tolerate conditions that supported a submerged aquatic plant.

DISCUSSION

The Perrier/Humbert hypothesis, that Madagascar's environment prior to human arrival was composed of closed, static 'climax' communities, does not hold up well in the face of the evidence presented here. While pre-settlement Madagascar may in some sense have been less disturbed than mainland environments, the sedimentary record from Lake Tritrivakely points to dynamic features in the environment qualitatively similar to those of other tropical lands in the late Quaternary: pyrogenic communities, open vegetation types in mosaic with more closed types, and, most likely, forest core areas or refugia that have shrunk and expanded with the vicissitudes of climatic, geological or hydrological processes.

Although the possibility of pre-settlement changes in Malagasy environments seems fairly likely, ideas about the nature and extent of the changes are very tentative, pending analysis of sedimentary microfossils from other sites for comparison. It would be unwise to assume that the events recorded at Tritrivakely were all regional in extent, when some may in fact be related to local volcanic activity, pedogenesis, hydrology and so forth. However, as a 'first guess' at what happened in the region around Tritrivakely during the Holocene, the following scenario is proposed:

1. At the beginning of the Holocene, c. 11 000-10 000 years BP, conditions were wetter than today, with a deeper lake in the crater and montane bushland around the site. During this time, Antsifotra or some other nearby volcano rained ash on the site at least once, and fires occurred frequently enough to leave much charcoal in the sedimentary record.

2. For the rest of the early Holocene, a variable climate may have prevailed. Limited evidence suggests that there may have been two or more episodes of warmer and/or drier climate (for the same level of precipitation, a higher temperature would yield a lower precipitation/evaporation ratio). These hypothetical dry periods were possibly separated by a brief cooler or moister episode between perhaps 9 000-9 500 years BP. A short, possibly very dry episode occurred c. 9 500 years BP, and the other, of longer duration, was from c. 9 000-5 500, perhaps with maximum dryness or highest temperatures at about 7 000 years BP. Volcanic eruptions may have occurred about 8 800 and 5 700 years BP. During the early Holocene, grassland and ericoid bushland shared (and sometimes alternated) dominance at the site, with some forest vegetation in the region, although possibly at considerable distances.

3. By mid-Holocene times (c. 5 000 years BP) the ericoid community seems to have given way to a more savanna-like assemblage, dominated by grassland but with acomponent of dry-adapted trees. Some forest vegetation persisted, although it was most likely not near the site, since only well-dispersed types are represented. Between c. 6 000-4 000 years BP, the organic matter content in the sediment rises consistently (except for a thin silty band around 5 500 years BP).

4. By c. 4 000 years BP, a change of uncertain type seems to have occurred. Since there has been no evidence of nearby volcanism for almost two millennia, it is more likely that the change was of a climatic nature. The sediment changes gradually from the muds containing some clays in the middle Holocene to a more purely organic fibrous gyttja. Pollen of *Celtis* and a host of other trees rises from the near-constant low background earlier in the Holocene to increase gradually over the next two millennia to the highest values of the entire Holocene. Charcoal in the sediments drops to very low levels, although it never entirely disappears, suggesting a decline in the frequency or intensity of fires. The meaning of these changes awaits additional data and analyses, but one likelihood is that the nature of the precipitation may have changed to a more equable (less seasonal) distribution, so that rainy seasons were not necessarily wetter, but dry seasons may have been less dry than previously. Whatever the reason, charcoal dropped to consistently low values

for c. 2 000 years, and pollen of graminoids, composites, ericoids and other indicators of open habitats declined to moderate percentages but never disappeared from the spectra.

5. The arrival of man is heralded in the late Holocene by the appearance of pollen of introduced plants and a rise of charcoal content in the sediments. This charcoal level is still well below those of the early Holocene, both in terms of percent dry weight and influx, but it cannot be determined from the data presented whether the nature or frequency of the fires was in some way critically different. At any rate, the arboreal vegetation shows a decline following the arrival of man. It is possible that man's effects are superimposed on a natural decline, since several major shifts in the vegetation have been documented for earlier parts of the Holocene.

Palaeoclimatic results from other areas of the world should be compared with these new results from Madagascar only with great caution. Sites from similar latitudes in southern Africa, Australia and Easter Island have shown vaguely similar trends in the Holocene record, with generally less extreme climatic variation in the Holocene than in the late Pleistocene. Patterns for these areas have been variable from one site to another, with localized and sometimes short-term trends. Most of the sites studied in these latitudes have been from much lower elevations, which makes comparison difficult. Scott & Vogel (1983), however, describe pollen spectra from a site in the Transvaal Highveld (26°S, c. 1 500 m) with cool subhumid conditions in the early Holocene, followed by a mid-Holocene warming and cooler, moist conditions from about 3 000 years BP to present. At a site in Soutpansberg, Northern Transvaal (Scott 1982), there was little change in bushveld from 5 000 years BP, although slight cooling may have occurred between 3 500-2 460 years BP. Butzer et al. (1978) found a warmer or drier period during the middle Holocene at a site on the southwestern margin of the Kalahari.

In northern Australia (Bowler et al. 1976) Holocene climates have generally been stable, although warmer and moister conditions may have occurred between 8 000 and 5 000 years BP. Likewise, on Easter Island (27°S), Flenley & King (1984) found that Holocene climatic changes have been of lower magnitude than those of the late Pleistocene. As at Tritrivakely, no severe late Holocene drying was detected.

The data from Madagascar, as well as these other southern sites, suggest a less extreme and more localized set of trends than those described for the Holocene of equatorial East Africa. In this relatively well-studied area, a general trend of a warm, wet early and middle Holocene, followed by considerably drier conditions beginning around 4 000 years BP, seems to apply consistently over a fairly large area within about 10° of the equator (Livingstone 1975, Hamilton 1982, Stager 1985).

All of the scenario outlined above for Madagascar is highly tentative and comparisons with other areas may not be very reliable. The fragmentary results from Ampasambazimba (MacPhee et al. 1985) and Kavitaha (Burney et al. 1985) are consistent with these interpretations, however. The particulars of this interpretation are nevertheless not essential for evaluating the validity of the Perrier/Humbert hypothesis. For that purpose, it might be more instructive to simply note that there is a high degree of similarity in many details between the pollen spectra and charcoal values of the early and middle Holocene at Tritrivakely and the post-settlement spectra throughout the interior. Likewise, charcoal data (Burney 1986) and pollen spectra in preparation that reach back as early as c. 36 000 years BP at several sites in the centre and north of the island confirm that pyrogenic conditions and open vegetation types were characteristic of the late Pleistocene in some parts of Madagascar as well. Every microfossil analysis published so far, as well as those in preparation, points to the same unavoidable impression: all of Madagascar has not been covered with climax forest for the entire pre-settlement portion of the late Quaternary. Open habitats, fire ecology and vegetation change have been a factor here, in the presumed absence of man, just as they have elsewhere in man's presence.

The Perrier/Humbert hypothesis postulated that a major cause of the late Holocene megafaunal extinctions was that the long-intact climax communities of the island were radically disrupted by man, and that the megafauna were unable to cope with such unfamiliar changes. While it remains true that man's drastic modification of the Malagasy environment began in the few centuries of overlap between human arrival and the megafaunal extinctions, it now seems less certain that the 'introduction' of fire and the opening of some habitats would have been sufficient disturbance in itself to wipe out so suddenly a diverse and presumably (at least in the case of the elephantbirds and hippopotamus) abundant megafauna. If, on the other hand, competition from introduced herbivores and hunting of large animals unfamiliar with the danger imposed by man, an unlikely-looking carnivore, were added to the environmental change, these causes might (or equally well might not) add up to enough to drive a host of animals to extinction in a relatively short time over a vast and diverse area.

The cause of Holocene mass extinctions in Madagascar is an issue believed by most authors to have been settled for more than half a century. In the light of new types of data previously unavailable, the classic 'solution' to the mystery looks less convincing than before.

ACKNOWLEDGEMENTS

Fieldwork in Madagascar was carried out with the permission of the Service de la Conservation du Patrimoine National (Ministère de la Culture et de l'Art Revolutionnaires) according to the terms of a collaborative agreement with several Malagasy institutions (Laboratoire d'Anthropobiologie and Laboratoire de Palynologie of the Université de Madagascar, and Musée d'Art et d'Archéologie). Many thanks to B.Rakotosamimanana, M.Villaume-Randriamanantena, H.Rakotoarivelo and J.-A.Rakotoarisoa for their assistance with logistics. Thanks also to the staff and students of the Service de Paléontologie, and to R.Dewar, N.Wells and P.Johnson for assistance in the field. D.Livingstone of the Duke University Department of Zoology, E. Simons of the Duke University Primate Center, and J.Kokwaro, F.Muthuri and S.Njuguna of the University of Nairobi Botany Department furnished laboratory facilities and field equipment. R.MacPhee provided useful comments on early drafts of this chapter, and rendered valuable assistance in all phases of the work. L.Roth, D.Billings, H.Wilbur, S.Khorram and D.Livingstone read the manuscript and provided helpful suggestions. D.Shaw provided technical assistance in the laboratory. L.Burney and M.Frost collected pollen specimens and prepared reference slides. The research was supported by National Geographic Society grant 2826-84 and National Science Foundation grants BSR 8501079, BSR 83-08221 and BSR 8516925.

REFERENCES

Benninghoff, W.J. & R.O.Kapp 1962. Suggested notations to indicate identification status of fossil pollen. *Pollen et Spores* 4 (1): 332.

Bowler, J.M., G.S.Hope, J.N.Jennings, G.Singh & D.Walker 1976. Late Quaternary climates of Australia and New Guinea. *Quat. Res.* 6: 359-394.

Burney, D.A. 1986. Late Quaternary environmental dynamics of Madagascar. Ph.D. dissertation, Duke University, Durham.

Burney, D.A., R.D.E.MacPhee, R.E.Dewar & N.A.Wells 1985. Cause, effect and megafaunal extinction in Holocene Madagascar: Pollen and charcoal influx at Lake Kavitaha (abstract). *Am. J. Phys. Anthrop.* 69: 182-183.

Burney, D.A., R.D.E.MacPhee, R.E.Dewar, N.A.Wells, E.H.Andriamanantena & M.Villaume-Randriamanantena 1986. L'Environnement au cours de l'Holocène et la disparition de la mégafaune à Madagascar: Quel rapport avec la conservation de la nature? In *Proceedings of the IUCN/SSC Symposium on Species Conservation Priorities in Madagascar (Antananarivo, November 1985).*

Butzer, K.W., R.Stuckenrath, A.J.Bruzewicz & D.M.Helgren 1978. Late Cenozoic palaeoclimates of the Gaap Escarpment, Kalahari margin, South Africa. *Quat. Res.* 10: 310-339.

Dahl, O. 1951. *Malgache et Maanyan.* Oslo: Egede Institutett.

Dale, I.R. & P.J.Greenway 1961. *Kenya trees and shrubs.* Nairobi: Buchanan's Kenya Estates.

Dewar, R.E. 1984. Recent extinctions in Madagascar: The loss of the subfossil fauna. In P.S.Martin & R.G.Klein (eds.), *Quaternary extinctions: a prehistoric revolution*: 574-593. Phoenix: Univ. of Arizona Press.

Faegri, K. & J.Iversen 1975. *Textbook of pollen analysis* (rev. 3rd ed.). New York: Hafner.

Hamilton, A.C. 1982. *Environmental history of East Africa: a study of the Quaternary.* New York: Academic Press.

Humbert, H. 1927. Destruction d'une flore insulaire par le feu. Principaux aspects de la végétation à Madagascar. *Mém. Acad. Malgache* 5: 1-80.

Humbert, H. 1949. La dégradation de sols à Madagascar. *Mém. Inst. Rech. Sci. Madagascar* D1: 33-52.

Humbert, H. 1955. Les territoires phytogéographiques de Madagascar: leur cartographie. *Ann. Biol.* 31: 439-448.

Humbert, H. & G.Cours Darne 1965. Carte internationale du tapis végétal et des con-

ditions écologiques à 1/1,000,000: Notice de la carte Madagascar. *Trav. Sect. Sci. Tech. Inst. Fr. Pondichéry (hors. sèr.)* 6: 1-162.
Kendall, R.A. 1969. An ecological history of the Lake Victoria basin. *Ecol. Monogr.* 39: 121-176.
Koechlin, J., J.-L.Guillaumet & P.Morat 1974. *Flore et végétation de Madagascar.* Vaduz: Cramer.
Kokwaro, J.O. 1979. *Classification of East African crops.* Nairobi: Kenya Literature Bureau.
Livingstone, D.A. 1967. Postglacial vegetation of the Ruwenzori mountains in Equatorial Africa. *Ecol. Monogr.* 37: 25-52.
Livingstone, D.A. 1975. Late Quaternary climatic change in Africa. *Ann. Rev. Ecol. Syst.* 6: 249-280.
Livingstone, D.A. & T.van der Hammen 1978. Palaeogeography and palaeoclimatology. In *Tropical forest ecosystems.* National Resources Research XIV: 61-90. Paris: UNESCO.
MacPhee, R.D.E., D.A.Burney & N.A.Wells 1985. Early Holocene chronology and environment of Ampasambazimba, a Malagasy subfossil lemur site. *Int. J. Primatol.* 6 (5): 463-489.
Mahé, J. & M.Sourdat 1972. Sur l'Extinction des vertébrés subfossiles et l'aridification du climat dans le Sud-ouest de Madagascar. *Bull. Soc. Géol. Fr.* 14: 295-309.
McGlone, M.S. 1983. Polynesian deforestation of New Zealand: a preliminary synthesis. *Arch. Oceania* 18 (1983): 11-25.
Perrier de la Bathie, H. 1921. La végétation malgache. *Ann. Mus. Col. Marseille* 9: 1-266.
Perrier de la Bathie, H. 1936. *Biogéographie des plantes de Madagascar.* Paris: Société d'Editions Géographiques, Maritimes et Coloniales.
Scott, L. 1982. A late Quaternary pollen record from the Transvaal Bushveld, South Africa. *Quat. Res.* 17: 339-370.
Scott, L. & J.C.Vogel 1983. Late Quaternary pollen profile from the Transvaal Highveld, South Africa. *S.Afr. J. Sci.* 79: 266-272.
Senesse, S. 1980. Palynologia de Madagassica et Mascarenica. Fam. 98 bis Caesalpinaceae. *Pollen et Spores* 22 (3-4): 355-423.
Stager, J.C. 1985. Environmental changes at Lake Cheshi, Zambia, since 40,000 years B.P. Ph.D. dissertation, Duke University, Durham.
Straka, H. 1966. Palynologia de Madagassica et Mascarenica. Fam. 50-59 bis. *Pollen et Spores* 8: 241-264.
Straka, H. 1980. PMM. Supplement. *Pollen et Spores* 22 (3-4): 343-354.
Straka, H. & B.Friedrich 1983. PMM. Fam. 121-127. *Pollen et Spores* 25 (1): 49-73.
Strake, H. & A.Simon 1967. PMM. Fam. 122-125. *Pollen et Spores* 9 (1): 59-70.
Straka, H., A.Simon & M.-T.Cerceau-Larrival 1967. PMM. Fam. 155-166. *Pollen et Spores* 9 (3): 427-466.
Tucker, V.A. & A.E.Tucker 1985. A microcomputer data management program for plotting pollen diagrams. *Pollen et Spores* 27 (2): 277-288.
Vavilov, N.I. 1949. The origin, variation, immunity and breeding of cultivated plants. *Chronica Botanica* 13: 1-6.
Vérin, P., C.Kottak & P.Gorlin 1970. The glottochronology of Malagasy speech communities. *Oceanic Linguistics* 8: 26-83.
Winkler, M.G. 1985. Charcoal analysis for paleoenvironmental interpretation: a chemical assay. *Quat. Res.* 23 (3): 313-326.

Photo 1. Top: Sorted stripes on Marion Island, bottom: Sorted polygons on Kerguelen

K.Hall: Periglacial landforms and processes of the sub-Antarctic and Antarctic islands, pp. 383-392.

Photo 2. Top: Stripes-within-stripes on Kerguelen, bottom: An exposure of frost shattered quartz-micaschist on Signy Island

PERIGLACIAL LANDFORMS AND PROCESSES OF THE SUB-ANTARCTIC AND ANTARCTIC ISLANDS

K.HALL
Geography Department, University of Natal, Pietermaritzburg, South Africa

ABSTRACT

Periglacial research in the Antarctic and sub-Antarctic has not yet achieved the level of sophistication found for Arctic and Alpine regions. Nevertheless, there is a wide assemblage of periglacial landforms on the southern islands and a number of valuable studies have been undertaken, most of which are not that well known by workers in the northern hemisphere. In this species-poor environment a strong relationship is often found between plant life and the patterned ground. As land areas in the vast Southern Ocean are very limited, the palaeoenvironmental and palaeoclimatological evidence provided by fossil and active periglacial features and processes is of the utmost importance. A brief summary of much of the available periglacial literature is presented in an attempt to generate interest in this field of study on the southern islands.

INTRODUCTION

Recent volumes on Antarctica and its surrounding regions (e.g. Bonner & Walton 1985, Walton 1986) have covered such topics as climate, soils, geology and glaciers in some detail but the periglacial system has been relatively ignored. Sugden (1982), in his geographic synthesis of both the Arctic and Antarctic, does show examples of periglacial activity from the Maritime Antarctic, and Walton (in Laws 1984) briefly notes periglacial processes but, compared to the extensive studies in the northern hemisphere (e.g. Washburn 1979, Harris 1986), this is a field comparatively unexplored in the South. However, the periglacial system is important as it is here that many cold region geomorphic processes interact intimately with the biological world. In fact, the relationship between periglacial features and botanical activity is frequently so strong that a study of either requires an intimate knowledge of the other. The periglacial processes operative around the margins of the ice cover take over as the glaciers recede. Several features of the periglacial sys-

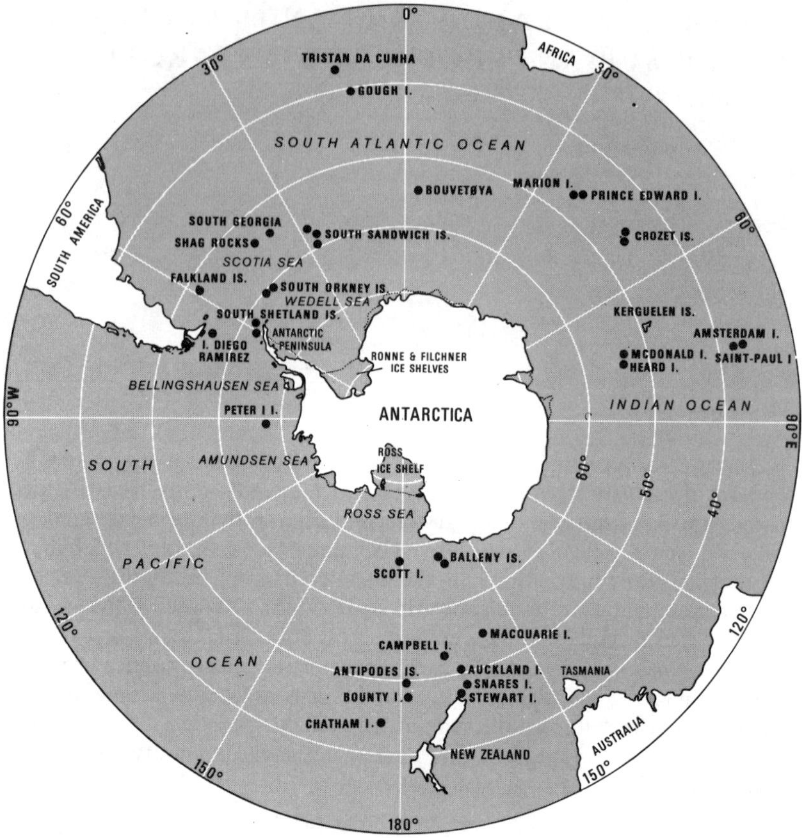

Figure 1. Map of the Southern Ocean showing the location of the islands mentioned in the text.

tem are also very useful as palaeoclimatological/palaeoenvironmental indicators and, as such, can be useful tools in our study of the southern islands.

The number of islands, and their areal extent, is very small in the context of the great Southern Ocean (Fig. 1). However, they have great importance for they not only provide accessible land sites, bridging the gap between Antarctica and the surrounding continents, but they also constitute important locations for the acquisition of data that can be compared with those derived from ocean cores. Some of the islands are so extensively covered by ice (e.g. Bouvet Island, the Balleny Islands, Scott Island and Peter I Øy) that they do not support any periglacial activity except perhaps for frost weathering on the few exposed cliffs. Thus the number of islands to be considered is relatively small.

The information which is available on periglacial topics is published in a number of languages and widely spread through various books, journals and reports. The only concise source of information presently available is the annotated bibliography of Walton (1980) which deals with both pedology and periglacial processes in the Antarctic and sub-Antarctic. However, this is probably not as widely known as it should be and much more work has been undertaken in the six years since Walton's volume was completed. Thus, an attempt will be made here to briefly present much of the available information on periglacial features and processes from the southern islands and to comment on their importance.

MARION ISLAND AND PRINCE EDWARD ISLAND

Marion Island and Prince Edward Island (latitudes 46°48′-46°59′S, longitudes 37°35′-37°55′E) are the twin peaks of a submerged volcano (King 1969). The islands are small in size (290 km^2 and 40 km^2 respectively). Extensive studies were initiated here by Van Zinderen Bakker in 1965 (1971) and a bibliography of the research on the islands has been compiled by Siegfried et al. (1979).

Van Zinderen Bakker (1978) observed the presence of fossil and active sorted patterned ground (Washburn 1978) on both Marion and Prince Edward Islands. On Price Edward he recorded exceptionally long sorted stripes (270 m) above 500 m altitude, whilst on Marion it was suggested that permafrost might be present in the central mountain area. The occurrence of needle ice (pipkrake) was thought to account for the origin of moss balls and of some deflated areas. Hall (1979), following the above work, made a number of observations on the patterned ground, particularly that of sorted stripes (Photo 1). It was found that sorted stripes had a preferred alignment parallel to the dominant westerly winds, and that the effect of the wind was so strong that stripes even occurred on horizontal surfaces. The mode of stripe formation was not determined but a strong association with pipkrake activity was observed. Stripe width was found to increase with altitude but the fine stripes were consistently broader than the coarse.

Stone banked lobes (Hall 1981) and *Azorella*-banked steps (Hall 1983a) were recorded but both are thought to be largely inactive at the present time as the depth of lenticular freezing is so limited (c. 0.1 m). Both of these features owe their formation, in part, to gelifluction and, like similar features observed on Macquarie Island (see below), are considered to be related to a more severe climate. It is suggested that *Azorella* associated with the steps grew in that location due to the sheltered position offered by the step riser (Löffler et al. 1983). During this more 'severe climate' it is highly likely that gelifraction was extremely active (Hall 1983a) as there would have been numerous freeze-thaw cycles in what is a relatively wet environment. Gelifraction is probably an active process at higher elevations at the present time but no research has been undertaken there.

ILES CROZET

This group of five main islands resides in similar latitudes (lat. 46°-46°30′S, long. 50°30′-52°30′E) to those of the Marion and Prince Edward Islands and experiences a similar climate, particularly the incessant westerly winds.

Although these volcanic islands are mountainous, have been glaciated, and are sparsely vegetated at higher elevations (Aubert de la Rue 1967), there appears to be very little available information on periglacial activity. Philippi (1908) describes sorted stripes and polygons, as does Bellair (1969) who also noted solifluction activity. Bellair suggests that it is wetting and drying of the ground that is the formative process for stripes and polygons, but the recent work from Marion and Kerguelen (Hall 1979, 1983b) would indicate some freeze-thaw action.

ARCHIPEL DE KERGUELEN

This is an extensive archipelago (lat. 48°27′-49°58′S, long. 68°25′-70°35′E) of some 300 islands, with the main island comprising an area of approximately 6 000 km^2 of which about one-tenth is currently ice covered (King 1969). Situated just to the north of the Antarctic Convergence, the island experiences low temperatures ($\bar{x}$ = 4.6° C), frequent frosts, strong westerly winds and extensive cloud cover (Weyant 1967).

Small-scale sorted patterned ground, particularly sorted stripes, has been reported from numerous parts of the main island (Troll 1958, Aubert de la Rue 1969, Bellair 1959, Markov 1971, Nougier 1964, 1970). Recent work on Peninsule Loranchet (Hall 1983a and b) has detailed sorted nets, polygons (Photo 1) and stripes. Due to the more rugged terrain, stripes do not show a particularly strong relationship to the westerly winds, but some were found (Hall 1983b) orientated across a slope in alignment with a katabatic wind. Stripe width increased with altitude, but below about 120 m coarse stripes were wider than fine stripes whilst at the higher elevations fine stripes were broader than the coarse ones. Several areas of well developed sorted nets and circles were observed (Hall 1983a), and at one location stripes-within-stripes (Photo 2) were found (Hall 1983b; Photos 1 and 2). Surprisingly, gelifluction features, although not absent (Aubert de la Rue 1967), appeared to be relatively rare (Hall 1983a).

The sorting mechanism was said by Bellair (1969) to be wetting and drying of the soil but, although this process is certainly active, the role of ground freezing must be considered. The change of tripe width with altitude and the change from a dominant coarse stripe to a dominant fine stripe with increase in height both intimate frost action (Hall 1983b). In addition, the observation of clasts in the coarse borders of sorted nets and stripes with their a/b planes vertical and in the fine centres with the a/b planes upward facing, argues for ground freezing (Hall 1983b). Numerous weathering features indicative of both chemical and mechanical action are present (Hall, pers. obs.) but no data on the actual processes or their rates are available.

MACQUARIE ISLAND

Situated approximately half-way between Australia and Antarctica (lat. 54° 37′S, long. 158°54′E) this small island is currently ice-free but exposed to the full force of the dominant westerly winds. The island mostly comprises a plateau surface with little land above 350 m (Löffler et al. 1983).

Sorted polygons and stripes, and solifluction features, particularly terraces, have been recorded (Mawson 1943, Bunt 1954, Taylor 1955, Colhoun & Goede 1974, Löffler et al. 1983). Bunt (1954) ascribed the sorted stripes to the action of needle ice and water action and noted that wind currents may well influence the freezing pattern of the ice needles. Taylor (1955), discussing the formation of terraces, observed that different forms occurred on windward and leeward slopes but that both were associated with the establishment of *Azorella selago* which acts as a movement inhibitor with respect to continued downslope transport of material. Löffler et al. (1983) point out, however, that periglacial solifluction was also of great importance in the formation of the terraces and that the large terraces are relicts of a former, colder climate. A very strong relationship between vegetation, slope processes, and wind exposure was noted (Löffler et al. 1983: 234-235).

SOUTH GEORGIA

This long, narrow island (160 km by 36 km), with its central chain of mountains dissected by glaciers, experiences a cold, damp climate (King 1969). The islands position (lat. 54°20′S, long. 36°40′W) locates it as sub-Antarctic but it does, nevertheless, experience a dynamic periglacial environment around the margins of the extensive ice cover.

Sorted stripes, nets, and polygons, non-sorted circles, stripe hummocks, solifluction lobes, stone-banked lobes and frost boils have all been reported (Smith 1960, Clapperton 1971, Stone 1974 and 1975, Clayton 1977, Thom 1981, Walton in Laws 1984). In addition, extensive weathering, particularly by frost action, has also been recognised as a major process below c. 700 m (Stone 1974, Gordon 1985). Weathering is said to have been active throughout the Holocene, to have been abetted by the structural controls of rock bedding and jointing, and to have possibly been aided by dilatation subsequent to deglaciation (Gordon 1985: 49). The debris resulting from weathering is seen as a major factor in landform development and a prime source of material for transport, and subsequent deposition, by glaciers (Gordon 1985; Gordon & Thom, in press).

A strong relationship between patterned ground and vegetation has been recognised (Heilbronn & Walton 1984a and b, Walton & Heilbron 1984). Certain grasses and herbs colonise sorted stripes first and their survival is related to both tiller and root morphology (Heilbronn & Walton 1984a). The rate of downslope plant movement is a function of plant size, and certain plants, notably *Phleum alpinum*, stabilise the soil sufficiently to enable other species to begin colonisation. Differences in microtopography resulting from periglacial activity also generate floristic differences between stripe crest and trough (Walton in Laws 1984). There is a correlation between the effects of frost heaving and plant mortality, and needle ice also plays a role (Heilbronn & Walton 1984a, Walton & Heilbronn 1984, Walton in Laws 1984). In fact, except for the work undertaken on Signy Island in the South Orkneys (see below), the greatest depth of multidisciplinary periglacial studies undertaken on any of the southern islands has probably been achieved on South Georgia.

SOUTH ORKNEY ISLANDS

Situated in the Maritime Antarctic (lat. 60°30′S, long 44°25′-46°10′W) this group consists of two large islands, Coronation and Laurie, and two smaller ones, Powell and Signy. With the exception of Signy Island the islands are almost totally ice-covered. Unlike the sub-Antarctic islands discussed above, the flora of the South Orkneys is very limited and predominantly comprises lichens.

Chambers (1966a and b, 1967, 1970) detailed the patterned ground of Signy Island, describing sorted and unsorted circles, sorted stripes, and stone streams. Solifluction was considered the main agent of downslope movement on Signy Island (Chambers 1967) and forms suggesting solifluction activity were observed on Coronation Island (Hall, pers. obs.). Chambers (1970) found that a disturbed miniature sorted pattern would re-establish itself within three years. In Moraine Valley, on Signy Island, large sorted circles were observed (Hall, pers. obs.) that indicate downslope movement subsequent to their original formation such that they now have an almost 'barchanoid' form.

Weathering of the rock (Photo 2) on these islands was first reported by Grange (in Dumont D'Urville 1841-54) and recently a programme to consider the role of weathering in the Maritime Antarctic environment of Signy Island were initiated (Walton & Hall, in press). Walton (1985) has described the mechanical disruption of quartz-micaschist by lichens, whilst Hall (1986a, b and c, subm. a, b, c and d, and Hall et al. 1986) has detailed the rock water content and chemistry, rock engineering properties affecting weathering, rock fracture mechanics, a series of laboratory experiments on the freeze-thaw weathering of the local rock, and the results of ultrasonic investigations of the freezing mechanism. This programme probably constitutes the largest long-term study of mechanical, chemical and biological weathering in any periglacial environment (Walton & Hall, in press).

SOUTH SHETLAND ISLANDS

This group of mountainous, extensively-glaciated islands (lat. 61°-63°30′S, long. 53°30′-62°45′W) includes Elephant and Clarence Islands, Livingston Island, Deception Island, King George Island, as well as numerous small islets (King 1969). Plant life is scarce with lichens and mosses predominating.

Sorted polygons and stripes together with solifluction lobes have been reported from Elephant Island (Allison & Smith 1973) and from Deception Island (Corte & Somoza 1957) whilst Simonov (1977) describes King George Island as comprising a 'periglacial landscape'. Olsacher et al. (1956), Araya & Herve (1966) and Corte & Somoza (1957) all report the role of frost weathering on rocks. Other periglacial features described include mud streams and palses (Araya & Herve 1972), erosion by water resulting from snowbank-melt and erosion pavements due to wind and water action (Corte & Somoza 1957), permafrost, nivation and 'biogenous' weathering (Simonov 1977). The role of plant life is stressed, with Simonov (1977) noting its effects as a weathering agent and Corte & Somoza (1957) and Araya & Herve (1966) describing the relationship between sorted structures and vegetation cover plus the role solifluction plays in affecting plant growth on slopes.

SUMMARY

Overall it seems that, compared to the northern hemisphere, much of the work on the southern islands is still in its infancy with little emphasis on highly detailed or long-term studies. Nevertheless, there does appear to be a wealth of information from this very important area, much of which still remains relatively unknown to many researchers working in Arctic regions. The islands are extremely important for they comprise the only land areas within the vast Southern Ocean from which detailed information on the relationship between climatic change, geomorphic processes, plant succession and distribution, and biological activity can be obtained. Due to their isolated and frequently exposed positions, the large distances between island groups (affecting plant dispersal), and their particular climatic conditions (notably tne prevailing westerly winds and the overwhelming oceanic influence), the islands might be considered to experience an almost 'special' type of periglacial environment. This may be all the more so in terms of the relationship between periglacial processes and botanical activity as the islands are notably species-poor. In turn, this leads to the realisation of the role periglacial activity plays in the plant colonisation that takes place immediately upon deglaciation, something which botanical studies may not have fully recognised. Thus, the periglacial realm is one which cannot be ignored as it is so intimately part of the past and present environment of this region.

Overall, these wonderful islands suggest an exciting prospect for periglacial research. Sadly, though, this is not a field of study undertaken to any large extent by most Antarctic Treaty nations and, unlike in much of the Arctic and alpine regions, logistical constraints necessitate working through a national Antarctic programme. Until some detailed work on these widely separated locations is initiated, there remains a gap in our knowledge of the periglacial environment of an important part of our world.

ACKNOWLEDGEMENTS

I wish to express my sincere gratitude to Professor Van Zinderen Bakker for giving me the opportunity to move from the Arctic to the Antarctic and for the encouragement he has continued to give me over the years. Numerous people have been responsible for allowing me to visit a large part of the sub-Antarctic and Antarctic, but I would particularly like to thank Dr R.M.Laws, Director of the British Antarctic Survey, and Dr David Walton for their trust and help.

REFERENCES

Allison, J.S. & R.I.L.Smith 1973. The vegetation of Elephant Island, South Shetland Islands. *British Antarctic Survey Bull.* 33/34: 185-212.

Araya, R. & F.Herve 1966. *Estudio geomorfologico y geologico en las Islas Shetland der Sur, Antarctica.* Instituto Antartico Chileno, Publ. 8, 76 pp.

Araya, R. & F.Herve 1972. Periglacial phenomena in the South Shetland Islands. In R.J. Adie (ed.), *Antarctic geology and geophysics*: 105-109. Oslo: Universitetsforlaget.

Aubert de la Rue, E. 1959. Phénomènes périglaciales et actions éoliennes aux Iles de Kerguelen. *Mém. de l'Institut Scientifique de Madagascar* (Sér. D) 9: 1-21.

Aubert de la Rue, E. 1967. *Les terres australes.* Paris: Presses Universitaires de France, 126 pp.

Bellair, P. 1969. Soil stripes and polygonal ground in the subantarctic islands of Crozet and Kerguelen. In T.L.Pewe (ed.), *The periglacial environment, past and present*: 217-222. Montreal: McGill-Queens Univ. Press.

Bonner, W.N. & D.W.H.Walton (eds.) 1985. *Key environments: Antarctica.* Oxford: Pergamon Press, 381 pp.

Bunt, J.S. 1954. The effect of freezing and thawing on the surface structure of certain soils on Macquarie Island. *Australian J. Science* 17: 36.

Chambers, M.J.G. 1966a. Investigations of patterned ground at Signy Island, South Orkney Islands: I. Interpretation of mechanical analyses. *British Antarctic Survey Bull.* 9: 21-40.

Chambers, M.J.G. 1966b. Investigations of patterned ground at Signy Island, South Orkney Islands: II. Temperature regimes in the active layer. *British Antarctic Survey Bull.* 10: 71-83.

Chambers, M.J.G. 1967. Investigations of patterned ground at Signy Island, South Orkney Islands: III. Miniature patterns, frost heaving and general conclusions. *British Antarctic Survey Bull.* 12: 1-22.

Chambers, M.J.G. 1970. Investigations of patterned ground at Signy Island, South Orkney Islands: IV. Long-term experiments. *British Antarctic Survey Bull.* 23: 93-100.

Clapperton, C.M. 1971. Geomorphology of the Stromness Bay-Cumberland Bay area, South Georgia. *British Antarctic Survey Sci. Reports* 70: 1-25.

Clayton, R.A.S. 1977. The geology of north-western South Georgia I. Physiography. *British Antarctic Survey Bull.* 46: 85-98.
Colhoun, E. & A.Goede 1974. A reconnaissance survey of the glaciation of Macquarie Island. *Papers and Proc. R. Soc. of Tasmania* 108: 1-19.
Corte, A.E. & A.L.Somoza 1957. Observaciones criopedologicas y glaciologicas en las Islas Decepcion, Media Luna y Melchior. *Publicacion del Inst. Antarctico Argentino* 4: 65-131.
D'Urville, D. 1841. *Voyage au Pôle Sud et dans l'Océanie sur les corvettes l'Astrolabe et la Zelée.* Paris, 32 pp.
Gordon, J.E. 1985. Glaciology and geomorphology. *South Georgia Joint Services Expedition 1981-1982*: 42-49. Nature Conservancy Council, Berksh., UK.
Gordon, J.E. & G.Thom (in press). Debris production and transfer on South Georgia. *British Antarctic Survey Bull.*
Hall, K.J. 1979. Sorted stripes orientated by wind action: some observations from sub-Antarctic Marion Island. *Earth Surface Processes* 4: 281-289.
Hall, K.J. 1981. Observations on the stone-banked lobes of Marion Island. *S. Afr. J. Sci.* 77: 129-131.
Hall, K.J. 1983a. Observations of some periglacial features and their palaeoenvironmental implications on sub-Antarctic islands Marion and Kerguelen. *S. Afr. J. Antarctic Res.* 13: 35-40.
Hall, K.J. 1983b. Sorted stripes on sub-Antarctic Kerguelen Island. *Earth Surface Processes and Landforms* 8: 115-124.
Hall, K.J. 1986a. Rock moisture content in the field and the laboratory and its relationship to mechanical weathering studies. *Earth Surface Processes and Landforms* 11: 131-142.
Hall, K.J. 1986b. The utilisation of the stress intensity factor (K_{IC}) in a model for rock fracture during freezing: an example from the maritime Antarctic. *British Antarctic Survey Bull.* 73.
Hall, K.J. 1986c. Freeze-thaw simulations on quartz-micaschist and their implications on weathering studies on Signy Island, Antarctica. *British Antarctic Survey Bull.* 73.
Hall, K.J. (in press). The physical properties of quartz-micaschist and their relationship to mechanical weathering studies. *Earth Surface Processes and Landforms.*
Hall, K.J. (Subm. a.) A laboratory simulation of rock breakdown due to freeze-thaw in a maritime Antarctic environment: I. Large blocks. *Earth Surface Processes and Landforms.*
Hall, K.J. (Subm. b.) A laboratory simulation of rock breakdown due to freeze-thaw in a maritime Antarctic environment: II. Small, saturated samples. *Earth Surface Processes and Landforms.*
Hall, K.J. (Subm. c.) A laboratory simulation of rock breakdown due to freeze-thaw in a maritime Antarctic environment: III. Ultrasonic results. *Earth Surface Processes and Landforms.*
Hall, K.J. (Subm. d.) Daily monitoring of a rock tablet as an aid to the investigation of frost weathering: an example from the maritime Antarctic. *British Antarctic Survey Bull.*
Hall, K.J., A.Verbeek & K.Meiklejohn 1986. The extraction and analysis of solutes from rock samples and their implication for weathering studies: an example from the maritime Antarctic. *British Antarctic Survey Bull.* 70: 79-84.
Harris, S.A. 1986. *The permafrost environment.* London: Croom Helm, 276 pp.
Heilbronn, T.D. & D.W.H.Walton 1984a. Plant colonisation of actively sorted stone stripes in the subantarctic. *Arctic and Alpine Res.* 16: 161-172.
Heilbronn, T.D. & D.W.H.Walton 1984b. The morphology of some periglacial features on South Georgia and their relationship to local environment. *British Antarctic Survey Bull.* 64: 21-36.
King, H.G.R. 1969. *The Antarctic.* London: Blandford Press, 276 pp.
Laws, R. (ed.) 1984. *Antarctic ecology.* London: Academic Press. 60 pp.

Löffler, E., M.E.Sullivan & A.N.Gillison 1983. Periglacial landforms on Macquarie Island, Subantarctic region. *Zeitschr. für Geomorphologie* 27: 223-236.
Markov, K.K. 1971. Kerguelen Islands. *Antarctica Commission Reports* 1968: 43-58.
Mawson, D. 1943. Macquarie Island: its geography and geology. *Australian Antarctic Exp. 1911-14, Sci. Rep.* (Ser. A) 5: 1-194.
Nougier, J. 1964. Observations periglaciaires dans l'Archipel de Kerguelen, Terres Australes et Antarctiques françaises. *Zeitschr. für Geomorphologie* 8: 143-149.
Nougier, J. 1970. *Contribution à l'étude géologique et géomorphologique des Iles Kerguelen.* Comité National Français des Recherches Antarctiques, 27, Vol. 1, 440 pp.
Olsacher, J., H.Diaz & M.Teruggi 1956. *Contribucion a la geologia de la Antartica Occidental.* Inst. Antartico Argentino, Publ. 2, 152 pp.
Philippi, E. 1908. Geologische Beobachtungen auf der Possession-Insel (Crozet-Gruppe). *Deutsche Südpolar-Expedition, 1901-1903* 2: 315-323.
Siegfried, W.R., P.F.Forbes & P.R.Condy 1979. Scientific research at the Prince Edward Islands, 1847-1979: a bibliography. *S. Afr. J. Antarctic Res.* 9: 35-41.
Simonov, I.M. 1977. Physical-geographic description of the Fildes Peninsula (South Shetland Islands). *Polar Geogr.* 1: 223-242.
Smith, J. 1960. Cryoturbation data from South Georgia. *Biuletyn Perglacjalny* 8: 72-76.
Stone, P. 1974. Physiography of north-east South Georgia. *British Antarctic Survey Bull.* 8: 17-36.
Stone, P. 1975. An unusual sort of patterned ground, Cooper Bay, South Georgia. *British Antarctic Survey Bull.* 41/42: 195-197.
Sugden, D. 1982. *Arctic and Antarctic: a modern geographical synthesis.* Oxford: Blackwell, 472 pp.
Taylor, B.W. 1955. Terrace formation on Macquarie Island. *J. of Ecology* 43: 133-137.
Thom, G. 1981. Patterned ground in South Georgia, Antarctica. Ph.D. thesis, University of Aberdeen, 178 pp.
Troll, C. 1958. *Structure soils, solifluction and frost climates of the earth.* US Army Snow Ice and Permafrost Research Establishment Translation 43, 121 pp.
Van Zinderen Bakker, E.M. 1971. Introduction. In E.M.van Zinderen Bakker, J.M. Winterbottom & R.A.Dyer (eds.), *Marion and Prince Edward Islands*: 1-15. Cape Town: Balkema.
Van Zinderen Bakker, E.M. 1978. Geoecology of the Marion and Prince Edward Islands (sub-Antarctic). In C.Troll & W.Lauer (eds.), *Geoecological relations between the Southern temperate zones and the tropical mountains.* Erdwissenschaftliche Forschung 11: 495-515.
Walton, D.W.H. 1980. *An annotated bibliography of Antarctic and sub-Antarctic pedology and periglacial processes.* British Antarctic Survey Data 5, 75 pp.
Walton, D.W.H. 1985. A preliminary study of the action of crustose lichens on rock surfaces in Antarctica. In W.R.Siegfriend, P.R.Condy & R.M.Laws (eds.), *Antarctic nutrient cycles and food webs*: 180-185. Berlin: Springer.
Walton, D.W.H. 1986. *Antarctic science.* Cambridge: Cambridge Univ. Press, 288 pp.
Walton, D.W.H. & K.J.Hall (in press). Rock weathering and soil formation in the Maritime Antarctic: an integral approach for Signy Island. *British Antarctic Survey Bull.*
Walton, D.W.H. & T.D.Heilbronn 1984. Periglacial activity on the sub-antarctic island of South Georgia. In *Proc. Fourth Intern. Conf. on Permafrost, Fairbanks, Alaska*: 1356-1361.
Washburn, A.L. 1979. *Geocryology.* London: Arnold, 406 pp.
Weyant, W.S. 1967. *The Antarctic atmosphere: climatology of the surface environment.* American Geographical Society, Antarctic Map Folio Series 8.

ANTHROPOLOGY

PRELIMINARY REPORT ON NEW HOMINID DENTAL TAPHO-DATA, A NEW TAPHONOMIC HYPOTHESIS AND TWO NEW TAPHONOMIC RATIOS

P.V.TOBIAS
Palaeo-anthropology Research Unit, Department of Anatomy, University of the Witwatersrand Medical School, Johannesburg, South Africa

SUMMARY

The numbers of maxillary and of mandibular teeth represented in early hominid fossil assemblages, considered with data on maxillae and mandibles, suggest the hypothesis that, under bland depositional regimen, conditions of accumulation and survival of remains, upper and lower jaws and teeth stand a more or less equal chance of featuring in fossil collections. In a more exacting palaeo-ecological setting and unprotected potentially destructive taphonomic milieu, mandibular teeth and the sturdier lower jaws may be expected to be favoured over maxillary teeth and the more fragile maxillae. The ratio of maxillae to mandibles (Mx/Mn jaw ratio) and the ratio of maxillary to mandibular teeth (Mx/Mn dental ratio) are proposed as two new measures of the blandness or severity of the depositional and taphonomic milieux.

AVANT-PROPOS

It is a pleasure to offer this little contribution on taphonomy and palaeoecology in honour of my old and dear friend and admired colleague, Emeritus Professor Eduard van Zinderen Bakker. I salute you, Eduard, and thank you for all that you have taught us.

INTRODUCTION

In my studies on the Olduvai hominid deciduous teeth (Tobias 1986), I was struck by the marked discrepancy between the numbers of maxillary and of mandibular teeth available (respectively, fifteen and one). While on so small a sample the results may reflect no more than a sampling factor, the finding provoked me to compare the relative numbers of upper and lower teeth, both deciduous and permanent, from a number of open, East African sites as well as from several sealed cave deposits in South Africa. The overall pic-

Table 1. Numbers and percentages of maxillary and mandibular hominid permanent teeth from Southern and East African sites*

	Maxillary teeth No.	%	Mandibular teeth No.	%	Total
Sterkfontein	263	56.0	207	44.0	470
Makapansgat	26	44.1	33	55.9	59
Taung	2	50.0	2	50.0	4
Swartkrans	219	54.75	181	45.25	400
Kromdraai	13	41.9	18	58.1	31
S.African cave deposits	523	54.25	441	45.75	964
Omo	88	44.0	112	56.0	200
Koobi Fora	62	33.3	124	66.7	186
Hadar	56	32.9	114	67.1	170
Olduvai	65	51.6	61	48.4	126
Laetoli	34	40.5	50	59.5	84
Nariokotome III	8		9		17
Peninj	0		16		16
Chesowanja	6		0		6
Garusi	3		0		3
Lothagam	0		1		1
E.African open sites	322	39.8	487	60.2	809

*Since we are concerned here with the likelihood of preservation of individual teeth, the numbers given in this table are numbers of actual teeth, not of specimens or individuals.

Table 2. Numbers and percentages of maxillary and mandibular hominid teeth, deciduous and permanent combined, from nine Southern and East African site-samples*

	Maxillary teeth No.	%	Mandibular teeth No.	%	Total
Sterkfontein	280	54.1	238	45.9	518
Makapansgat	26	42.6	35	57.4	61
Swartkrans	234	52.2	214	47.8	448
Kromdraai	13	33.3	26	66.7	39
S.African cave deposits	553	51.9	513	48.1	1066
Omo	92	44.0	117	56.0	209
Koobi Fora	67	33.0	136	67.0	203
Hadar	66	34.7	124	65.3	190
Olduvai	80	56.3	62	43.7	142
Laetoli	42	40.8	61	59.2	103
E.African open sites	347	41.0	500	59.0	847

*In contrast with Table 1, this table does not include the teeth from Taung, Nariokotome III, Peninj, Chesowanja, Garusi and Lothagam, comprising 29 maxillary and 38 mandibular teeth in all.

ture that emerged from an analysis of 1773 early hominid permanent teeth revealed an interesting contrast between the assemblages from the former and from the latter sites.

RELATIVE NUMBERS OF MAXILLARY AND MANDIBULAR PERMANENT TEETH

In an earlier analysis, the author found that, from fourteen African sites, the hominid permanent teeth comprised 469 maxillary and 425 mandibular teeth (Tobias 1972: 71). These figures refer to numbers of individual teeth, not of specimens or individuals. Of the total 894 teeth, 52.5% were maxillary and 47.5% mandibular. To these totals must now be added at least 140 more from Omo (based on Howell & Coppens 1974); 178 more from Koobi Fora (after R.E.Leakey et al. 1978); 170 from Hadar (after Johanson, White & Coppens 1982); 84 from Laetoli (from White 1977, 1980); 273 more from Sterkfontein (based on Tobias et al. 1977, and on the subsequent annual reports of the Palaeo-anthropology Research Group of the University of the Witwatersrand from 1977 to 1985); 17 from Nariokotome III (Brown et al. 1985); 13 more from Swartkrans (Grine 1981); 4 more from Kromdraai (based on check-list, Table 1 of Vrba 1981); and a few additional items from Makapansgat, Sterkfontein, Swartkrans and East African sites. This gives a total of at least 879 additional, African, early hominid, permanent teeth, since I completed my last tally early in 1971 (Tobias 1972). In other words, the stockpile of early hominid permanent teeth from the African continent has doubled in the fifteen years from 1971 to 1985!

The overall total of early hominid permanent teeth currently available is thus 1773, of which 964 have stemmed from the five southern African sites and 809 from ten East African sites (Table 1).

When we examine the distribution of all fifteen South and East African site-samples between maxillary and mandibular permanent teeth, we see first that 845 or 47.7% are maxillary and 928 or 52.3% mandibular. These figures compare with values for deciduous teeth of 40.6% maxillary and 59.4% mandibular.

Thus, for both deciduous and permanent teeth of African early hominids, there is a slightly greater chance of mandibular teeth being preserved, than of maxillary teeth.

When the samples of permanent teeth from the five South African cave deposits are examined, then irrespective of the taxa involved, there is a small majority (54.25%) of maxillary teeth. Although the departure from a 50-50 distribution is small and probably not significant, under the conditions of bone accumulation prevailing in these caves, there is a suggestion of a slightly greater likelihood of maxillary permanent teeth being accumulated in these cave deposits, within which bone preservation appears to have been excellent (C.K.Brain, pers. comm.).

The relative frequencies of permanent teeth in the East African open sites are quite different from those in the southern African cave deposits: approximately 60% are mandibular and about 40% maxillary. In four of the five East African sites which have yielded appreciable numbers of hominid permanent teeth – Omo, Koobi Fora, Hadar and Laetoli – the frequencies of mandibular teeth vary from 56 to 67%. The Olduvai sample of permanent teeth is exceptional among the East African site samples in showing a majority – albeit a small one – of maxillary teeth. Thus the conditions of tooth accumulation and survival at Olduvai have slightly favoured permanent maxillary teeth.

RELATIVE NUMBERS OF MAXILLARY AND MANDIBULAR TEETH (DECIDUOUS AND PERMANENT COMBINED)

Table 2 summarises the total numbers of maxillary and of mandibular teeth for the nine dentally richest African sites, irrespective of whether the teeth are deciduous or permanent and irrespective of taxon. As in Table 1, the totals include teeth in jaws and isolated teeth.

DISCUSSION

It seems clear that, under conditions prevailing at the fluvial, deltaic and lake margin deposits of Laetoli, Omo and Koobi Fora, mandibular teeth – both deciduous and permanent – seem to have stood a better chance of accumulating, being preserved and recovered than maxillary teeth. From these three sites have come 201 maxillary teeth (deciduous plus permanent), but 314 mandibular teeth (combined), that is 39.0 and 61.0% respectively. Hadar is a similar case in respect of permanent teeth – its value of 67.1% mandibular permanent teeth is the highest of all those for which percentages are cited in Table 1; on the other hand, its modest sample of deciduous teeth (n = 20) is equally divided between maxillary and mandibular dentitions. The overall totals for Hadar are thus 66 maxillary teeth (deciduous plus permanent) and 124 mandibular teeth (combined), that is, 34.7 and 65.3% respectively. With these figures, the Olduvai combined samples contrast sharply: 80 maxillary teeth (deciduous plus permanent) and 62 mandibular teeth (combined), the corresponding percentages of the total site-samples being 56.3 and 43.7. At Olduvai the collection and preservation of mandibular teeth is slightly less favoured, it would seem, than that of maxillary teeth, as is true also of the total samples (deciduous plus permanent) from Sterkfontein (54.1 and 45.9%), Swartkrans (52.2 and 47.8%) and combined southern African sites (51.9 and 48.1%).

The overall southern African sample of 1066 deciduous-plus-permanent teeth, of which 51.9% are maxillary and 48.1% mandibular, departs so slightly from a 50-50 distribution that it would be reasonable to infer that maxillary and mandibular teeth have had a virtually equal chance of finding their way into the dolomitic limestone cave deposits, of being preserved therein and recovered therefrom. On the other hand, the East African sample of 847 deciduous-plus-permanent teeth, comprising 41.0% maxillary and 59.0% mandibular teeth, departs more strikingly from a 50-50 distribution: it would be a fair inference that mandibular teeth have had a better chance of accumulating and surviving in, and being recovered from, the open, fluvial, lacustrine and deltaic sedimentary environments of these sites.

It is interesting in this connection to see that the 67-33% of mandibular to maxillary teeth from Koobi Fora finds a parallel in the large predominance of mandibles (78.4%) over maxillae (21.6%) among the hominid fossils from Koobi Fora (after Behrensmeyer 1978).

As might be expected from the more fragile bony structure of the maxilla and the much sturdier construction of the mandible (which was well-known, it seems, to long-haired Samson), the survival of the maxilla under destructive or unprotected conditions is relatively infrequent. The maxillary teeth seem to be affected by this, since they are found in most of these open East African sites far less frequently than are the mandibular teeth, especially at Koobi Fora, Hadar and Laetoli.

Though Olduvai Beds I and II represent a lake margin situation similar in some respects to that of Koobi Fora (Behrensmeyer 1978), hominid maxillary teeth, curiously enough, have had as good a chance of being accumulated, surviving and being found at Olduvai as at Sterkfontein. It is not immediately clear what taphonomic and palaeo-ecological factors at Olduvai have seemingly operated to protect the maxillary teeth and, presumably, the maxillae from destruction and loss. Of course it remains possible that the 56.3-43.7% figures for Olduvai maxillary and mandibular teeth are an artefact of sampling and that more Olduvai teeth may tip the scales into line with the ratios in the larger samples from Omo, Koobi Fora and Hadar. Moreover, as Shipman (1981) has pointed out: 'Comparing the representation of different body parts in different assemblages is a major problem. Equally problematical is assessing the significance of observed similarities or differences in skeletal representation' (Shipman 1981: 137).

As far as the author has been able to ascertain, Behrensmeyer's (1978) study and the earlier one of Tobias (1972) are the only previous investigations to have analysed the occurrences of the skeletal parts of the very ancient hominids themselves, while the present study is the first known to the author which examines the differential distribution of maxillary and mandibular teeth of the Plio-Pleistocene hominids.

CONCLUSIONS

A new taphonomic hypothesis and two new taphonomic ratios are proposed.

It may be predicted that, where taphonomic conditions have been non-destructive, non-dispersive, relatively mild and protective, maxillae and mandibles may be expected to have been conserved and recovered in approximately equal proportions; moreover, if the hypothesized interrelatedness of teeth and their respective socketing host organs is correct (Tobias 1986), one would expect the maxillary and mandibular teeth also to be available in nearly equal proportions. On the other hand, the more brutal and destructive the sedimentary environment and other taphonomic influences have been, the more likely it is that the maxillary and mandibular remains of jaws and teeth will deviate from equality of proportions, generally at the expense of the maxillae and upper teeth and in favour of the mandibles and lower teeth.

Hence, it is suggested that the upper jaw/lower jaw ratio (Mx/Mn jaw ratio) and the maxillary teeth/mandibular teeth ratio (Mx/Mn dental ratio) may serve as useful gauges of the rigour of the taphonomic environment, that is the rigour of all processes which can differentially affect the concentration and survival of maxillae and mandibles. In much the same way, Behrensmeyer (1975, 1978) has used the relative numbers of teeth and vertebrae (the T/V ratio) in fossil assemblages, to assess whether or not a bone assemblage had been selectively sorted prior to burial.

The Mx/Mn dental ratio proposed here suggests that the modest hominid assemblages from Kromdraai (dental ratio 0.50) and Makapansgat (0.74) had been more rigorously sorted prior to inhumation than the large ones of Sterkfontein (1.18) and Swartkrans (1.09). Similarly, harsher sorting agencies would seem to have pertained in the hominid assemblages of Koobi Fora (0.49), Hadar (0.53), Laetoli (0.69) and Omo (0.79), whereas a blander taphonomic regimen would appear to have been the lot of the Olduvai hominids (1.29). The Mx/Mn dental ratio of the Olduvai assemblage suggests slightly less destructive sorting and milder, more protective conditions than prevailed even within the seclusion of the dolomitic cave repositories of the Transvaal.

These hypotheses and these new survival ratios (Mx/Mn dental ratio and Mx/Mn jaw ratio), require to be tested rigorously on the fossil hominid caches, the associated comparative faunae, and analogous recent bone assemblages.

ACKNOWLEDGEMENTS

The author thanks the Foundation for Research Development and the Council of the University of the Witwatersrand for their generous subvention. He expresses his appreciation to Mrs Heather White.

REFERENCES

Behrensmeyer, A.K. 1975. The taphonomy and paleoecology of Plio-Pleistocene vertebrate assemblages of Lake Rudolf, Kenya. *Bull. Mus. Comp. Zool.* 146: 473-578.

Behrensmeyer, A.K. 1978. The habitat of Plio-Pleistocene hominids in East Africa: taphonomic and micro-stratigraphic evidence. In C.Jolly (ed.), *Early hominids of Africa*: 165-189. London: Duckworth.

Brown, F., J.Harris, R.Leakey & A.Walker 1985. Early *Homo erectus* skeleton from west Lake Turkana. *Nature* (Lond.) 316: 788-792.

Grine, F.E. 1981. Description of some juvenile hominid specimens from Swartkrans, Transvaal. *Ann. Transvaal Mus.* 86: 43-71.

Howell, F.C. & Y.Coppens 1974. Les faunes de mammifères fossiles des formations Plio-Pleistocéne de l'Omo en Ethiopie (Tubulidentata, Hyracoidea, Lagomorpha, Rodentia, Chiroptera, Insectivora, Carnivora, Primates). *C.R. Acad. Sci. Paris* 278: 2421-2424.

Johanson, D.C., Y.Coppens & T.D.White 1982. Dental remains from the Hadar Formation, Ethiopia: 1974-1977 collections. *Am. J. Phys. Anthropol.* 57: 545-603.

Leakey, R.E.F., M.G.Leakey & A.K.Behrensmeyer 1978. The hominid catalogue. In M.G.Leakey & R.E.F.Leakey (eds.), *Koobi Fora Research Project* Vol. 1: 86-182. Oxford: Clarendon Press.

Shipman, P. 1981. *Life history of a fossil: an introduction to taphonomy and paleoecology.* Cambridge, Mass.: Harvard Univ. Press, 222 pp.

Tobias, P.V. 1972. Progress and problems in the study of early man in Sub-Saharan Africa. In R.Tuttle (ed.), *The functional and evolutionary biology of primates*: 63-93. Chicago: Aldine-Atherton.

Tobias, P.V. 1986. *Olduvai Gorge Vol. 4. Homo habilis: Skulls, endocasts and teeth.* Cambridge: Cambridge Univ. Press (in press).

Tobias, P.V., K.Copley & C.K.Brain 1977. South Africa In K.P.Oakley, B.G.Campbell & T.I.Molleson (eds.), *Catalogue of fossil hominids. Part I: Africa* (2nd ed.): 95-151. London: British Museum (Natural History).

Vrba, E.S. 1981. The Kromdraai australopithecine site revisited in 1980; recent investigations and results. *Ann. Transv. Mus.* 33: 17-60.

White, T.D. 1977. New fossil hominids from Laetoli, Tanzania. *Am. J. Phys. Anthropol.* 46: 197-230.

White, T.D. 1980. Additional fossil hominids from Laetoli, Tanzania: 1976-1979 specimens. *Am. J. Phys. Anthropol.* 53: 487-504.

ARCHAEOLOGY

SOME REFLECTIONS ON PREHISTORIC FIELD WEEDS

W.VAN ZEIST
Biologisch-Archaeologisch Instituut, State University of Groningen, The Netherlands

SUMMARY

In addition to cultivated plants, various wild plant species, among which field weeds, are represented in the archaeobotanical record. For the Near East the recognition of field weeds in archaeological sites poses some extra problems. Various possible field weeds could equally well have formed part of the natural steppe vegetation. Particularly informative on ancient field weeds are the charred remains of crop-plant supplies. Weed seeds in these supplies are very likely from weeds which grew in the fields concerned and which were harvested together with the crop. Most of the field weeds of Neolithic Europe were not introduced from the Near East but must have been of more local origin.

INTRODUCTION

It need not be expounded here that in the history of Old World agriculture, the Near East has played a decisive role. It was in this area that food production was first practised, viz. animal breeding and plant cultivation. From the Near East, farming spread over Europe, North Africa and eastwards over large parts of Asia.

Numerous studies of plant remains preserved in ancient settlement sites have resulted in a fairly detailed knowledge of the crop-plant assortment of prehistoric farmers in the Near East and Europe, to which areas the discussion of this chapter will be confined. So far, the oldest indications of domestic plants in SW Asia stem from occupation deposits dated shortly after 10 000 BP (in uncalibrated radiocarbon years). There is at present much discussion on possible proto-agricultural practices before 10 000 BP. However, whatever the kind of plant manipulation by Late-Palaeolithic/Mesolithic man may have been, it did not lead to the emergence of domestic plants, that is to say, of plants which had acquired recognizable morphological features which distinguish them from the ancestral wild forms. It is not until after

10 000 BP (c. 8 800 BC in calendar years) that crop-plant remains which show the traits indicative of the domestic forms appear in the archaeological record.

Early farmers in the Near East took various plants into cultivation, as a result of which a fairly large crop-plant assortment was available by 7 000 BC. The cultivars included various wheats and barleys, leguminous crops, such as pea, lentil and bitter vetch, and linseed. After farming had been well established in the Near East, agriculture expanded to Europe (and other regions) around 6 800 BC. With the technique of farming, the crop plants migrated to Europe. The plants grown by the earliest farming communities in SE Europe were all of SW Asian origin. At a later stage, species that had not come from the Near East were added to the crop-plant assortment of European farmers, e.g. opium poppy and broomcorn millet.

Ancient crop plants will not be the subject of the present discussion. In this chapter some special attention will be paid to the – usually unwanted – companions of plant cultivation, viz. the field weeds. Following Zohary (1973: 634) field weeds are conveniently defined here as plants which occur in cultivated habitats among crops. The category of field weeds is also indicated by the term segetal flora. The appreciation of field weeds varies greatly. For the farmer field weeds are a nuisance because they compete with the crop, and great efforts are made to get rid of the weeds as much as possible, by weeding and other measures. For the botanist field weeds are an interesting object of study, e.g. the adaptations of weeds that enable them to maintain themselves in the particular conditions of cultivated fields. For the nature lover an 'old-fashioned' grain field is a feast for the eyes. For an interesting discussion of the phenomenon 'field weeds', the reader is referred to Harlan (1975, chapter 4, pp. 83-103: What is a weed?).

To the archaeobotanist field weeds demonstrated for ancient sites may provide information on the conditions of the fields. For instance, are there indications of irrigation agriculture or of salinization of the fields? Was the crop autumn sown (winter grain) or spring sown (summer grain)? In the present chapter these aspects will not be considered, but the discussion will be focussed on the basic question of how to recognize field weeds in the archaeobotanical record, particularly in that of Near Eastern sites. Below it will be discussed along which lines information on the weed flora of the fields of ancient farmers in the Near East can be obtained. Thereupon, the question of the origin of the weeds in the fields of early (Neolithic) farmers in Europe will briefly be dealt with. Did the segetal flora migrate from the Near East together with the crop plants or were the weeds in the fields of the European farmers local adaptations?

THE ARCHAEOBOTANICAL EVIDENCE

Information on field weeds in ancient times must come exclusively from the archaeological plant record. The present-day segetal flora of a particular area does not tell us much about that of thousands of years ago. Thus, many of the present-day Near Eastern field weeds (leaving aside the species introduced from America, Australia and other far-away areas during the last 400 to 500 years) have not (yet) been recorded archaeobotanically. This may in part be due to the deficient archaeological plant record (see below), but in particular from archaeobotanical studies in Central and West Europe it has become clear that in the course of time the field-weed flora changed quite markedly. The introduction of new crop plants and particularly shifts to other agricultural techniques must have entailed changes in the segetal flora.

The present discussion on ancient Near Eastern field weeds is based upon carbonized plant remains. Charred vegetable material is not the only source of information on prehistoric and early-historical plant husbandry in SW Asia. Imprints of seeds in pottery and burnt clay have provided information on crop plants and on wild plant species. For some Near Eastern archaeological sites desiccated plant remains have been reported. Occasionally non-carbonized vegetable remains could be recovered from water-logged archaeological deposits. However, for information on field weeds in the ancient Near East we depend largely upon charred material. In evaluating the archaeobotanical evidence it should be borne in mind that there are some major shortcomings inherent to the material.

In the first place, plants of which the remains are found in occupation deposits had to be brought to the site. For cultivated plants and for wild species collected by the inhabitants of the site for one purpose or another, it is self-evident that they were carried to the living quarters. However, field weeds and many other plants were usually not transported to the site intentionally. The method of harvesting the crop will greatly have determined the representation of the field-weed flora in the archaeological record. In the case of cereals, if the plants were cut low on the straw and if subsequently the sheaves were brought to the settlement for further processing, the crop-cleaning waste fractions may have contained many weed seeds. On the other hand, if the grain was reaped by uprooting the plants or by plucking the ears, only a few weeds will have been among the harvested crop. Thus, in theory fields could have been really weed-infested, whereas field weeds landed up in the site only occasionally.

Carbonization implies that the plant remains have been in contact with fire. Irrespective of the cause and the nature of these fires, carbonization automatically entails a selection. Of the more sturdy seed types, substantial numbers may become carbonized in a fire, whereas other, more fragile types are usually burnt to ashes. On the other hand, the procurement of fuel for heating purposes was very likely a source of charred seeds in archaeological deposits. In areas with a scarcity of wood, dung is generally used for heating. It has been demonstrated that various seeds pass through the digestive tract of domestic (and undoubtedly also of wild) animals undamaged. The burning of dung and dung cakes may result in the charring of seeds incorporated in the animal droppings. As a consequence, the presence of at least some of the

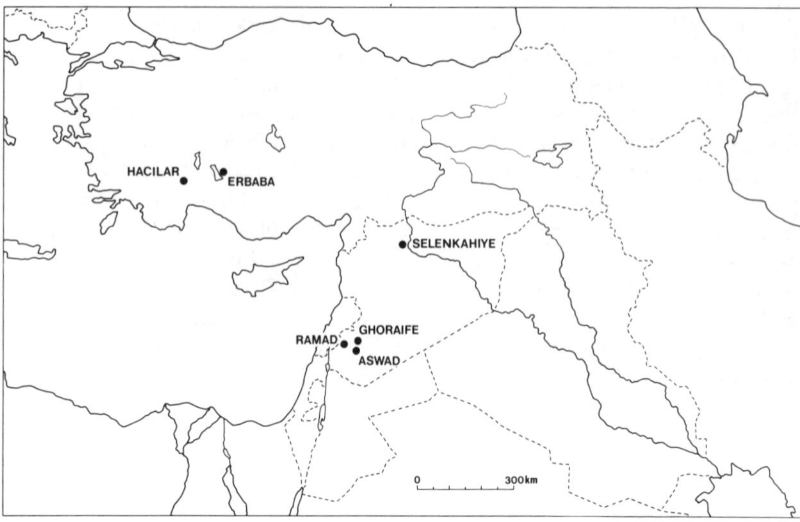

Figure 1. Near Eastern sites referred to in this chapter.

charred seeds in archaeological sites may be due to this practice (cf. Miller 1982, Miller & Smart 1984). Moreover, it is likely that the domestic animals grazed on the fields after the harvest, and this may have had an interesting effect on the archaeological plant record. Field weeds which on account of the harvesting method practised hardly had a chance of being brought to the site, may through the intermediacy of animal dung nevertheless be represented among the charred plant remains.

From the above it will be clear that we are left with many uncertainties with regard to the representation of the ancient segetal flora in the archaeological record. Another complication results from the fact that the seeds of wild plants could usually not be identified to the species level. Various seed types include species from segetal as well as from natural habitats. How can, in spite of the complicating factors mentioned above, information on ancient field weeds be gleaned from the archaeological plant record? One may gain little from theoretical considerations on what may be expected under certain premises. A pragmatic approach may be more successful. In this chapter we will start from the plant record and we will see to which conclusions on the segetal flora the plant remains give occasion. For such an approach sites which yielded sufficiently great numbers of seeds of possible field weeds are most appropriate. For that reason the discussion of Near Eastern field weeds will be focussed on two sites of which the economy was based upon food production.

Information on the ecology of Near Eastern plant taxa is derived from

Zohary (1962, 1973), Edgecombe (1970) and the flora works of Post & Dinsmore (1932), Mouterde (1966-1983) and Zohary & Feinbrun-Dothan (1966-1986). Field-weed seeds and fruits from Near Eastern sites are depicted in Figures 2-4.

THE SITES

One of the sites to be discussed here is Tell Ramad, c. 20 km SW of Damascus, at an elevation of c. 900 m. The natural vegetation of the Ramad area, with a mean annual precipitation of about 250 mm, is a pistachio-almond forest steppe, that is, a steppe with scattered tree growth. Excavations at Ramad have been carried out by H.de Contenson and W.J.van Liere in 1963 and 1965 and by H.de Contenson between 1966 and 1973. Almost all occupation at Ramad belongs archaeologically to the aceramic Neolithic (PPNB) and is radiocarbon dated to 8 200-7 900 BP, corresponding to approximately 7 100-6 700 BC in calendar years. Two aceramic occupation phases are distinguished. The vegetable remains recovered from this site demonstrate convincingly that plant cultivation was practised. The crop plants grown by the Ramad farmers include emmer wheat, free-threshing wheat, hulled and naked barley, pea, lentil and linseed (Van Zeist & Bakker-Heeres 1982 (1985)).

The Bronze Age site of Selenkahiye was founded on the right bank of the Euphrates, some 80 km E of Aleppo, at an elevation of c. 300 m. At present, a large section of the North Syrian Euphrates river valley is under water (Lake Assad) as a result of the construction of a dam at Tabqa. Annual precipitation in the Selenkahiye area is about 200 mm on average. The uplands are naturally covered by steppe vegetation, whereas poplar forest constitutes the natural vegetation of the annually flooded river-valley bottom. Excavations at the site were carried out under the direction of M.N.van Loon in 1967 and 1972-1975. The main occupation at Selenkahiye is dated from about 2 400 to 1 900 BC. Hulled barley must have been an important crop. Other plants cultivated in the Selenkahiye area include free-threshing wheat, emmer wheat, lentil, pea, chick-pea, grass pea and (probably) grape (Van Zeist & Bakker-Heeres, in press).

Almost all the samples taken for palaeobotanical examination were from occupational deposits which in general showed only scattered plant remains. In the field, the flora remains were retrieved from the soil samples by means of manual water flotation. Reasonably pure charred grain samples (only Selenkahiye) were not subjected to water flotation.

The results of the examination of some groups of samples from the two sites are summarized in Table 1. In this table mean percentages are presented. These frequencies were obtained as follows. First for each of the samples the percentages of the wild plant species were calculated based upon a sum of all the seeds and fruits, except those of domestic and of potentially wild food plants (e.g. *Pistacia, Capparis, Crataegus*). Thereupon the mean percentages

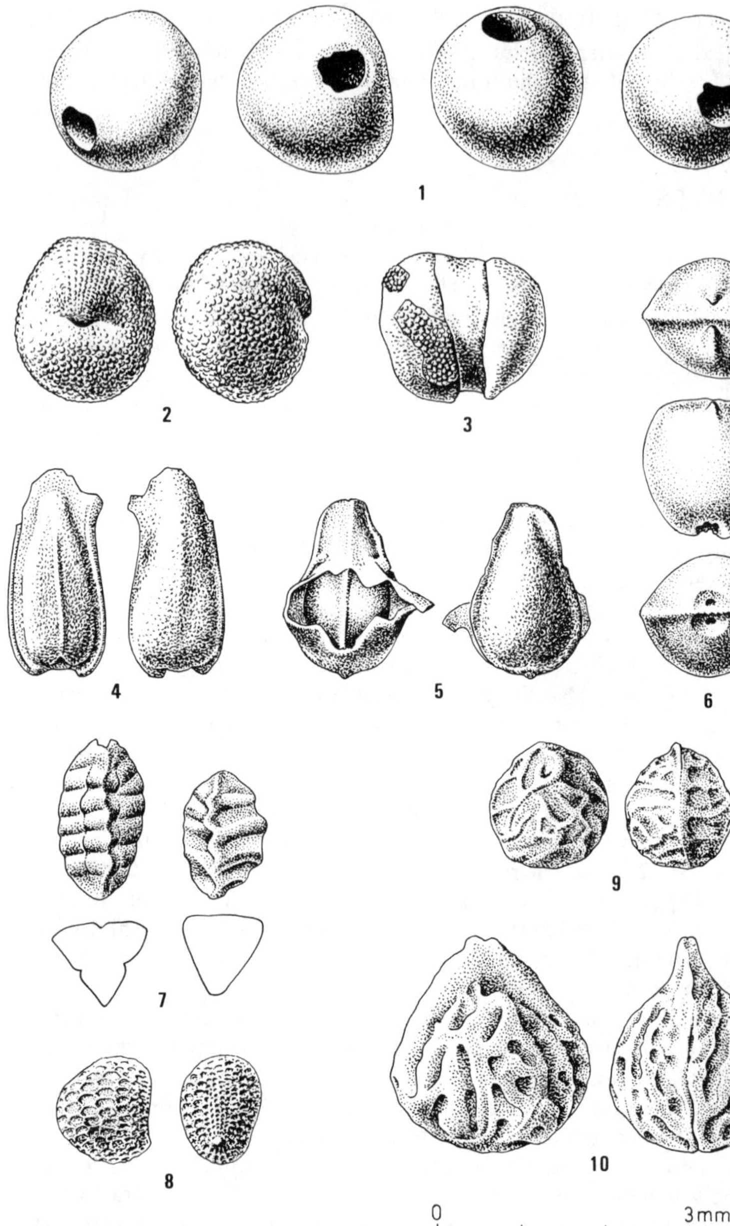

Figure 2. Archaeological seeds and fruits (Figs. 2-4). 1. *Bellevalia*; 2, 3. *Vaccaria pyramidata*; 4, 5. *Valerianella vesicaria*-type; 6. *Fumaria*; 7. *Androsace maxima*; 8. *Glaucium*; 9, 10. *Adonis*.

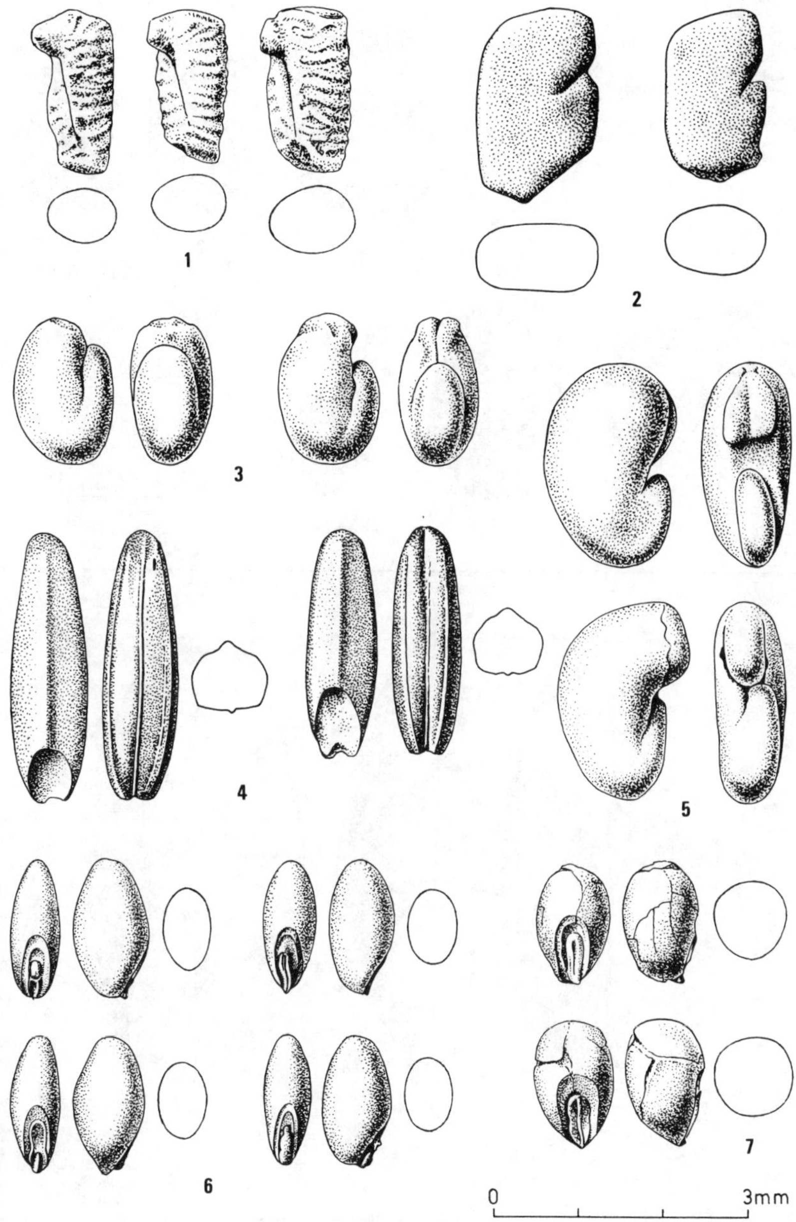

Figure 3. 1. *Trigonella astroites*-type; 2. *Astragalus*; 3. *Melilotus*; 4. *Eremopyrum*; 5. *Medicago*; 6. *Phalaris*; 7. *Echinaria.*

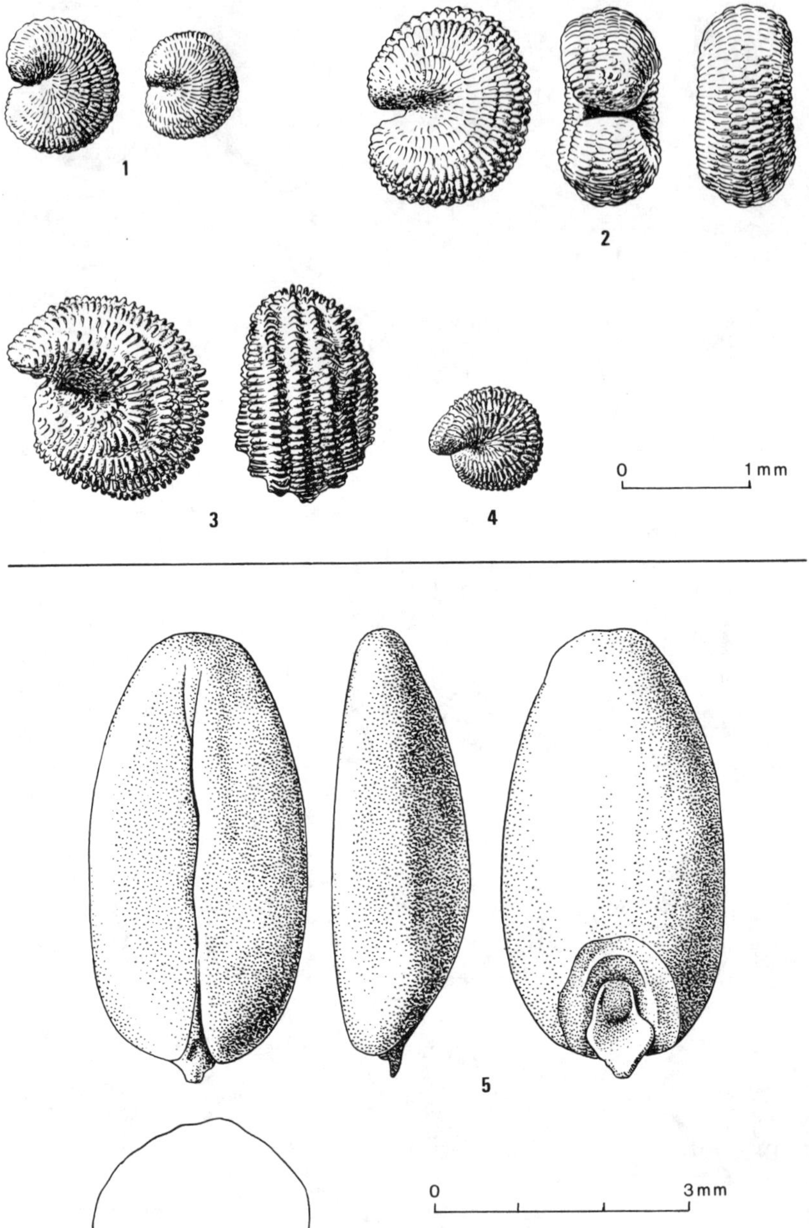

Figure 4. 1, 2. *Silene*; 3, 4. *Gypsophila*; 5. *Aegilops*.

for each of the series of samples were determined. Cultivated species and (potential) wild food plants are not shown in Table 1.

Table 2 needs no explanation. This table shows the (calculated) numbers of seeds of wild plant taxa per 200 g of barley (c. 16 000 grains).

Ramad

The Ramad samples are from two areas indicated in Table 1 as C 8 and M 4. The fill in the C 8 square consisted mainly of ashy refuse deposits.

As appears from Table 1, quite a variety of seed types of wild plant species could be established for Ramad; some types are represented by great numbers of seeds in more than one sample. As will be discussed below, various of these seed types could very well have belonged to field weeds, but the material itself does not provide conclusive evidence for this assumption. The plant remains in the Ramad samples are very likely of mixed origin. The many glume wheat spikelet remains in most of the samples indicate that at least part of the contents is made up of the waste of crop processing activities. However, this is no proof that the weed seeds or at least part of them also originate from crop cleaning residues and that as a consequence they are of grain-field weeds. None of the Ramad samples clearly consists of the remains of a grain supply in some stage of crop processing. Speculations on weeds in the fields of the Ramad farmers must necessarily be based upon the comparison with the present-day segetal flora. The 'actualistic principle' is often applied in interpreting archaeobotanical data. In connection with ancient field weeds the following may be remarked.

In Europe, in areas which are naturally forest-covered, wild plant species represented in the archaeobotanical record are confidently assigned to ecological groups or vegetation types in which they are found at present (or were found until recently). Thus, irrespective of the archaeological context from which the sample was taken and of the composition of the botanical content of the sample, certain species are listed as field weeds because they still occur in segetal habitats or are known to have been common among the crop in the recent past. This procedure appears to be justified because if at the time present-day segetal species already formed part of the natural vegetation, they were probably not particularly numerous there, and chances that seeds of specimens from natural habitats would end up in archaeological deposits are slight. However, for a site like Ramad and for other early farming sites in a steppe or forest-steppe environment, the distinction between field weeds and plants from natural vegetation types entails some extra difficulties.

It is generally assumed that Near Eastern field weeds originate from steppe vegetation. Various steppe plants would have intruded into the fields of ancient farmers and would have adapted themselves to the particular conditions in cultivated fields by natural selection of forms most suitable to segetal habitats. Admittedly, in this respect some reserve is due. Zohary (1973: 648) mentions that for a great number of segetal plants of supposedly East Mediter-

Table 1. Mean percentages of weed-seed types in cultural fill samples from Selenkahiye and Ramad (for explanation, see text). Seed types identified to the family level and unidentified seeds are not shown in this table, in consequence of which the total percentages in each of the groups of samples is less than 100

	SLK 1967	SLK 1974	Ramad C 8	Ramad M 4	Ramad M 4
Phase			II	II	I
Number of samples	8	9	20	6	5
Alkanna	0.06	–	–	–	–
Arnebia decumbens	11.25	32.56	–	–	–
Arnebia linearifolia	0.06	–	–	–	–
Heliotropium	0.50	0.10	0.05	0.08	–
Lithospermum arvense	0.10	–	0.31	2.84	6.98
Lithospermum tenuiflorum	2.06	6.24	0.85	1.08	3.10
Cleome	0.03	–	–	–	–
Silene	6.49	2.28	0.51	4.82	0.17
Gypsophila	2.54	0.46	–	0.12	–
Vaccaria pyramidata	0.34	0.11	0.20	0.20	1.10
Chenopodium	0.18	0.10	–	–	–
Suaeda	10.78	1.30	–	–	–
Helianthemum cf. *ledifolium*	–	–	8.82	8.06	6.85
Artemisia	0.39	–	–	–	–
Centaurea	0.29	0.29	0.01	–	–
Onopordon	0.08	0.28	–	–	–
Convolvulus-type	0.01	0.17	0.04	–	–
Alyssum-type	0.04	–	–	0.06	–
Euclidium	–	0.03	–	–	–
Lepidium-type	–	0.03	–	–	–
(cf.) *Carex*	0.05	0.20	0.04	0.26	–
Scirpus maritimus	–	0.25	0.46	0.10	–
Cephalaria syriaca	0.01	0.10	0.005	–	–
Aegilops	2.49	3.26	0.03	0.08	–
Avena	0.10	–	0.66	1.36	–
Bromus	1.90	0.30	0.42	0.16	0.33
Echinaria	0.08	–	1.45	4.10	0.45
Eremopyrum	9.38	9.56	0.50	1.14	–
Hordeum (wild)	1.71	1.05	0.02	0.08	–
Lolium perenne/rigidum	1.49	0.94	10.74	2.08	24.17
Lolium remotum-type	–	–	3.23	10.40	9.80
Lolium temulentum-type	0.58	–	–	–	–
Phalaris	1.44	0.44	0.15	0.18	0.25
Gramineae type A	–	–	0.66	3.72	2.02
Gramineae type B	2.21	1.34	–	–	–
Gramineae type C	4.08	0.60	–	–	–
Ajuga	0.14	–	–	–	–
Teuchrium	0.11	–	0.13	0.10	–
Ziziphora	–	–	0.005	0.02	–
Astragalus	4.89	3.35	28.60	14.96	28.03
Coronilla	1.25	0.17	0.25	–	–
Lotus-type	0.05	0.03	–	–	–
Medicago radiata	0.10	0.26	0.65	0.40	0.45
Medicago spec.	0.41	0.18	3.32	0.90	0.016
Melilotus	9.78	2.49	2.15	0.24	2.37

Table 1. (continued)

Phase / Number of samples	SLK 1967	SLK 1974	Ramad C 8 II	Ramad M 4 II	Ramad M 4 I
Number of samples	8	9	20	6	5
Onobrychis	0.10	–	0.10	0.10	–
Prosopis	0.64	0.88	–	–	–
Scorpiurus	0.03	–	–	–	–
Trifolium-type	0.04	–	–	–	–
Trigonella astroites-type	5.53	0.86	9.18	3.26	0.68
Vicia spec.	0.34	3.18	2.92	5.60	1.68
Bellevalia	1.10	4.04	3.25	0.30	1.10
'Ornithogalum'	0.10	0.03	12.02	19.34	6.60
Linum spec.	0.01	–	–	–	–
Malva	2.46	2.07	0.005	–	–
Fumaria	0.29	0.62	0.005	–	–
Glaucium	0.80	1.27	0.005	–	–
Papaver	0.01	–	–	–	–
Plantago	0.10	0.32	0.11	0.18	–
Androsace maxima	2.73	0.17	2.36	1.28	0.95
Polygonum (venantianum)	1.56	3.48	0.09	–	–
Rumex (pulcher)	0.21	1.24	0.13	0.72	0.45
Adonis	1.04	2.14	0.43	0.20	0.90
Ranunculus	–	0.10	–	–	–
Crucianella	0.04	0.03	–	0.02	–
Galium	2.58	3.73	1.47	0.10	–
Scrophularia	0.01	0.08	–	–	–
Veronica persica-type	0.01	–	–	–	–
Hyoscyamus	0.13	–	–	–	–
Thymelaea	0.05	0.80	–	–	–
Ammi majus	0.06	–	0.12	0.46	0.10
Bifora	–	–	0.15	–	–
Bupleurum (subovatum-type)	0.19	0.12	0.005	–	–
Torilis nodosa-type	–	–	0.13	0.48	–
Valerianella	0.14	0.39	–	–	–
Verbena	–	–	–	–	0.02
*Peganum harmala**	++	++	–	–	–

* Not included in sum upon which percentages are based.

ranean or Southwest Asian origin no natural habitats are known, and that they are found exclusively in disturbed environments. Leaving this question aside, one must consider the possibility that various plant taxa attested for settlement sites in a steppe-like environment formed part of the natural vegetation but at the same time occurred in the fields of ancient farmers. The problem of distinguishing between the two groups of species is further aggravated by the fact that a great number of seed types could not be identified to the species level.

Table 2. Selenkahiye. Numbers of weed seeds calculated per 200 g of barley (c. 16 000 grains)

		W 12 room 10			W 12 room 9							
	1975-SB 79	80	83	84	85	81	82	1967-S 219	S 44	1972-S 625	S 629	S 891
Volume of sample or part of of it examined (in grams)	140	210	260	155	208	194	181	240	229	50	50	45
Arnebia decumbens	1.4	–	–	–	–	–	–	8.3	–	4	–	4.4
Lithospermum tenuiflorum	–	–	–	–	–	–	–	–	–	–	–	4.4
Gypsophila	5.7	–	2.3	–	2.9	–	–	–	–	–	–	–
Silene	5.7	0.9	3.8	3.9	6.7	3.1	3.3	–	–	–	–	–
Vaccaria	2.8	1.9	3.8	–	–	4.1	1.1	–	–	–	–	–
Unident. Cruciferae	–	–	0.8	–	–	–	–	–	–	–	–	–
Aegilops	1.4	–	1.2	1.3	1.0	1.0	–	2.5	3.1	–	–	17.8
Bromus sterilis-type	0.7	–	–	–	–	–	–	–	–	–	–	–
Eremopyrum	–	0.9	0.8	1.3	–	0.5	–	0.8	–	–	–	13.3
Hordeum (wild)	–	–	–	–	–	–	1.1	0.8	–	–	–	–
Gramineae type B	15.7	–	9.2	–	–	3.1	2.2	–	–	–	–	4.4
Gramineae type C	34.2	0.9	20.7	–	1.9	10.3	11.0	–	–	–	–	–
Unident. Gramineae	1.4	0.9	1.5	0.6	–	–	–	2.5	–	2.5	–	–
Astragalus	4.3	–	3.8	1.3	–	1.0	–	–	–	–	–	–
Melilotus	6.4	4.7	40.7	180.8	57.6	3.6	1.1	2.5	3.9	–	–	4.4
Pisum/Vicia/Lens	0.7	–	1.5	5.2	4.3	–	1.1	27.5	–	–	4	–
Prosopis	–	–	–	–	–	–	–	0.4	–	–	–	–
Trigonella astroites-type	4.3	–	2.3	–	1.0	1.5	1.7	0.8	–	–	–	–
Bellevalia	–	–	0.8	–	1.9	–	1.1	9.2	–	–	–	4.4
Malva	1.4	–	1.5	3.9	1.9	1.0	–	–	–	–	–	–
Fumaria	–	–	0.8	–	–	–	–	–	–	–	–	–
Glaucium	1.4	–	–	–	–	–	–	–	–	–	–	–
Androsace maxima	1.4	–	–	–	–	–	–	–	–	–	–	–
Adonis	1.4	–	–	–	–	–	–	–	–	–	–	–
Galium	–	–	–	–	–	–	–	0.8	–	–	–	–
Thumelaea	1.4	–	–	–	–	–	–	–	–	–	–	–
Valerianella coronata-type	–	–	–	1.3	–	–	–	–	–	–	–	–
Unidentified	25.6	6.6	2.3	2.6	4.8	3.1	–	X	2.3	4	–	–

Undeniable field weeds at Ramad are *Vaccaria pyramidata* (cowherb) and *Cephalaria syriaca* (Syrian scabious). The latter is a noxious weed in present-day Near Eastern grain fields, but judging from the scarcity of its seeds in Ramad and other sites this species probably did not play an important role in the field-weed flora of ancient times. *Bellevalia* was probably a common weed in the fields of the Ramad farmers; *B.trifoliata* is a constituent of present-day segetal communities. The identity of the seed type indicated as '*Ornithogalum*' is still problematical. The good representation of *Androsace maxima* at Ramad suggests that this species occurred frequently in the fields. It is reported for natural as well as for segetal vegetation types.

Various grasses attested for Ramad could very well have occurred in fields. Thus, *Avena barbata* and *A.sterilis* are grain-field weeds. *Eremopyrum bonaepartis* is a natural constituent of steppe vegetation but is also reported for fields. *Lolium (rigidum)* and *Phalaris (paradoxa)* can both occur in fields. Rye grass must have been widespread, most likely not confined to fields. If the identification of the *Lolium remotum* caryopses is correct, it is tempting to assume that this species was a weed in linseed fields. The rather great numbers of *Echinaria (capitata)* seeds indicate that this grass was common in the vicinity of the site, probably in fields.

Leguminous taxa may have played a rather prominent role in the field-weed flora. *Vicia* (vetch), *Melilotus* (melilot) and *Medicago* (medick) species are common field weeds at present. Various trigonel species (*Trigonella astroites*-type) occur on cultivated land as well as in steppe vegetation. The interpretation of the great numbers of *Astragalus* (milk vetch) seeds presents a problem. *Astragalus* species are reported for fields, but representatives of this genus are particularly found in steppe and desert-steppe vegetation types. One wonders whether the large quantities of milk vetch seeds in the Ramad deposits rather point to the intensive collecting of shrubby *Astragalus* species for fuel.

Other possible field weeds of the Ramad farmers include *Lithospermum arvense* (corn-gromwell), *Silene* (catchfly), *Adonis* (pheasant's eye), *Galium* (cleavers), *Ammi majus* (common bishop's weed) and *Bifora testiculata* (small coriander).

The archaeobotanical record suggests that a whole array of weeds was present in the fields of the Ramad farmers. However, inferences on the Ramad field weeds are based exclusively upon present-day analogues. As has been stated above, this may involve a danger in areas with a steppe-like vegetation. Circumstantial evidence could provide firmer information on ancient field weeds. In this respect Bronze Age Selenkahiye, in a steppe environment, offers interesting perspectives.

Selenkahiye

During the excavations at Selenkahiye some charred grain deposits came to light. These grain concentrations were the remains of barley that had been fully processed, that is to say, threshed and cleaned of straw fragments, chaff and weed seeds. The cleaning of the barley crop had not resulted in the complete removal of weed seeds. The results of the examination for remaining weed seeds are shown in Table 2. As the volumes of the samples or parts thereof examined vary, the numbers of weed seeds have been converted to values per 200 g of barley. The contamination with weed seeds varies considerably. In some samples only a few types are present in small numbers of seeds, whereas in other samples the variation in seed types as well as the seed frequencies are much higher. This must be due in part to differences in the degree of accuracy in cleaning the crop of contaminants and in part to differences in the composition of the weed flora in the grain fields concerned.

Samples SLK 75 - SB 79 to 85 are from a thick layer of charred grains on the floor of two adjacent rooms. The differences in weed-seed contents indicate that the barley that had been stored there originated from more than one grain supply. The barley of the various supplies must have been grown in different fields with a different weed flora and have been processed and stored separately. It is likely that during and after the fire which occasioned the carbonization some mixing of the grain had taken place. Each sample does not necessarily represent an originally separate supply, but the weed seeds (and the barley) in some of the samples may have been the result of the mixing of two (or more?) different stocks.

It is assumed here that by far the majority of the seeds in the barley samples are from weeds which occurred in the fields and which had been harvested together with the cereal crop. Consequently, these charred grain samples provide direct information on the field-weed flora. Some seeds may have become mixed with the grain after harvesting, either before or after carbonization. Of the plant taxa listed in Table 2, *Arnebia decumbens, Lithospermum tenuiflorum* and *Thymelaea* are not mentioned in flora works as grain-field weeds.

Various caryophyllaceous species were found in the Selenkahiye grain fields: *Vaccaria pyramidata, Gypsophila* and *Silene. Gypsophila pilosa, G. viscosa, Silene conoidea, S.aegyptica* and *S.palaestina* are reported for fields. The identity of the gramineous fruit types B and C could not be established, but it is clear that these grasses must have been quite common field weeds. Various *Aegilops* (goat's-face grass) are found in fields, such as *Ae. crassa, Ae. triuncialis* and *Ae. cylindrica.* It is likely that also *Eremopyrum bonaepartis* was found in the Selenkahiye grain fields.

Melilotus (melilot) is predominant among the weed seeds in some of the charred grain samples, suggesting that a common grain-field weed is concerned here. *Melilotus indicus* and *M.sulcatus* are constituents of segetal plant communities. The fairly good representation of *Astragalus* indicates that milk-vetch species were indeed among the field weeds at Selenkahiye (see the discussion of Ramad). *Astragalus aleppicus, A.graecus, A.suberosus* and *A.hamosus* are reported for fields, to mention a few. The *Pisum/Vicia/Lens*-type is enigmatic in that the identity of the seeds could not always be determined satisfactorily. Distinguishing between *Vicia* (vetch) seeds and underdeveloped, atypical peas and lentils proved problematical. Various *Vicia* species occur in fields, e.g. *V.angustifolia, V.villosa* and *V.tetrasperma.*

As for the other taxa represented in the barley samples, constituents of the present-day segetal flora are *Adonis dentata, Bellevalia trifoliata, Fumaria parviflora, Glaucium aleppicum, Malva aegyptica* and *Valerianella coronata.* The composition of the seed types in the various samples does not point to a rather uniform field-weed flora. The composition of the weeds must have differed quite markedly between one field and another. The array of field weeds was probably much greater than is indicated in Table 2. Only those weed species that were in seed at harvesting time had a chance of being represented in the charred grain samples. That in fact more field weeds may have been present is suggested by the data presented in Table 1.

In Table 1, mean percentages of seeds of wild plants in two series of cultural fill samples from Selenkahiye are shown. It need not be emphasized that not all wild seed types in the Selenkahiye cultural fill samples are from field weeds. Thus, *Arnebia* and *Lithospermum tenuiflorum* very probably did not grow in fields, but the significance of boraginaceous seeds in archaeological deposits forms a problem anyhow. *Suaeda* probably occurred only in the steppe vegetation and some other taxa, such as *Plantago, Hyoscyamus* and *Peganum harmala,* may predominantly have been found in disturbed habitats other than fields.

All the weed-seed types established for the charred grain samples are well represented in the cultural fill samples. To mention a few: *Silene, Gypsophila, Aegilops* and other Gramineae, *Melilotus, Astragalus, Trigonella, Bellevalia, Malva.* On the other hand, many more potential field weeds are listed in Table 1 than have been found in the charred grain samples. Some of them may, indeed, not have been a constituent of the Selenkahiye field-weed flora, but the absence of others in the samples from Table 2 must be due to other factors. Thus, *Cephalaria syriaca* is a typical grain-field weed not represented in the charred grain samples. *Phalaris, Coronilla, Medicago* and *Bupleurum* may be expected to have occurred in fields. One may assume that with a greater number of charred grain concentrations from the site, a larger variety of field weeds would have been demonstrated.

In conclusion, the information on the field-weed flora provided by the charred grain samples is far from complete, but at least one may safely assume that, with a single exception, field weeds are concerned here. In theory, crop-processing waste could be even more informative, but remains of 'pure' crop-cleaning residues may not be found because this material landed up on the refuse dump where it got mixed with other vegetable debris.

From a comparison of the Ramad and Selenkahiye seed records it appears that both sites have many (possible) segetal plant taxa in common. Of course, there are differences which may be due in part to different climatological and pedological conditions and in part to the time difference of more than 4 000 years. The fair resemblance between both sites may be adduced in support of the assumption that (most of) the potential field-weed taxa attested for Ramad did indeed form part of the segetal flora of the latter site. Consequently, we conclude, albeit with some reserve, that the plant taxa concerned occurred as weeds in the fields of the Ramad farmers.

THE WEED FLORA OF NEOLITHIC EUROPE

In this final section attention will be paid to one particular aspect of the weed flora of Neolithic Europe. An attempt will be made to ascertain to what extent field weeds of the early European farmers could have been of Near Eastern origin as were (most of) the crop plants. It is not unreasonable to hypothesize that weeds in the fields of early Neolithic farmers in Europe came predominantly from SW Asia. The crop plants were introduced from that area and together with these the associated field weeds could have spread through Europe. Moreover, at the time Europe was forest-covered and in forested areas fields constitute a steppe-like environment created by man. As herbs from the forest ground vegetation only rarely developed into field weeds, the introduced weeds of Near Eastern stock would not have met serious competition. Some of the weed species would have dropped out on the way to Central Europe because the climatic conditions became too unfavourable, but others would have adapted themselves to environmental conditions quite different from those in their homeland, as did the crop plants. That various weeds did indeed migrate through Europe in the footprints of Neolithic man is demonstrated by palynological research. A well-known example is *Plantago lanceolata.* However, this attractive looking hypothesis on the origin of European weeds is not supported by archaeobotanical data.

In Table 3 it is shown which wild plant taxa have been attested for Neolithic sites in a few areas. The choice of the sites (areas) to be included was determined by the availability of archaeobotanical information. Thus, Greece is not represented because only a few occasional weeds are reported for Greek Neolithic sites. Wild fruit trees and shrubs, such as *Pyrus malus, Rubus fruticosus* and *Corylus avellana,* are not shown in Table 3 and neither are other woody perennials. Not all the taxa listed represent weeds, but some of them were probably only found in natural habitats. The latter are indicated with an asterisk in Table 3. The sites or areas which will come up for discussion are shown in Figures 1 and 5.

The earliest farmers' culture in Central Europe was the Bandkeramik culture (5 300-4 900 BC) which was found in a large territory embracing Czechoslovakia, Hungary, Austria, Poland, Moldavia (USSR), East and West Germany, SE Holland (South Limburg) and France. The distribution of Bandkeramik settlement sites largely coincides with the Central European loess area. From a large number of Bandkeramik sites plant remains have been examined (Willerding 1980, 1983), as a result of which considerable botanical data are

Table 3. Presence of weed-seed types in various Neolithic sites. Seeds present in sites from the Damascus basin but not in one or more of the other sites are not included in this table [sp: the species identity could not be ascertained. Non-field weeds are indicated with an asterisk.]

	Bandkeramik	Gomolava	Chevdar/Kazanluk	Hacilar	Erbaba	Damascus basin
**Agropyron caninum*	+	-	-	-	-	-
Agropyron repens	+	-	-	-	-	-
Bromus arvensis	+	-	-	-	-	-
**Bromus hordaceus*	+	-	-	-	-	sp
Bromus tectorum	+	-	-	-	-	-
**Carex muricata*	+	-	-	-	-	-
Digitaria ischaemum	+	-	-	-	-	-
Echinochloa crus-galli	+	-	-	-	-	-
Echium vulgare	+	-	-	-	-	-
Galium spurium	+	-	-	-	-	-
Lapsana communis	+	-	-	-	-	-
Lathyrus tuberosus	+	-	-	-	-	-
Leontodon autumnalis	+	-	-	-	-	-
Malva sylvestris	+	-	-	-	-	-
**Phleum pratense*	+	-	-	-	-	-
Plantago lanceolata	+	-	-	-	-	-
**Poa nemoralis*	+	-	-	-	-	-
**Poa trivialis*	+	-	-	-	-	-
**Polygonum amphibium*	+	-	-	-	-	-
Polygonum lapathifolium	+	-	-	-	-	-
Polygonum persicaria	+	-	-	-	-	-
Ranunculus repens	+	-	-	-	-	-
**Rumex sanguineus*	+	-	-	-	-	-
Setaria viridis	+	-	-	-	-	-
Silene dioica	+	-	-	-	-	-
**Silene nutans*	+	-	-	-	-	-
Silene vulgaris	+	-	-	-	-	-
Sinapis arvensis	+	-	-	-	-	-
**Stachys sylvatica*	+	-	-	-	-	-
Stellaria graminea	+	-	-	-	-	-
Stellaria media	+	-	-	-	-	-
Trifolium arvense	+	-	-	-	-	-
Trifolium repens	+	-	-	-	-	-
Veronica arvensis	+	-	-	-	-	-
Vicia hirsuta	+	-	-	-	-	-
Vicia tetrasperma	+	-	-	-	-	-
Agrostemma githago	+	+	-	-	-	-
Bromus secalinus	+	+	-	-	-	-
Bromus sterilis	+	+	-	-	-	+
Chenopodium hybridum	+	+	-	-	-	-
Chenopodium polyspermum	+	+	-	-	-	-

Table 3. (continued)

	Bandkeramik	Gomolova	Chevdar/Kazanluk	Hacilar	Erbaba	Damascus basin
Polygonum convolvulus	+	+	-	-	-	-
Avena (fatua)	+	+	+	-	-	sp
Chenopodium album	+	+	+	+	+	-
Atriplex patula	+	-	+	-	-	sp
Galium aparine	+	-	+	-	-	-
Rumex acetosella	+	-	+	-	-	-
Rumex crispus	+	-	+	-	-	-
Solanum (nigrum)	+	-	+	-	-	-
Polygonum aviculare	+	-	+	+	-	-
Verbena officinalis	+	-	-	+	-	+
**Atropa belladonna*	-	+	-	-	-	-
Euphorbia helioscopia	-	+	-	-	-	-
Malva spec.	-	+	+	-	-	+
Sambucus ebulus	-	+	+	-	-	-
Rumex spec.	-	+	+	-	+	+
Galium spec.	-	+	-	-	+	+
Silene spec.	-	+	-	-	+	+
Plantago spec.	-	+	-	-	+	+
Vicia spec.	-	+	-	-	+	+
Arenaria	-	-	+	-	-	-
Astragalus spec.	-	-	+	-	-	+
Brassica spec.	-	-	+	-	-	-
Galium mollugo	-	-	+	-	-	-
Linum cf. *bienne*	-	-	+	-	-	-
Polygonum spec.	-	-	+	-	-	-
Potentilla spec.	-	-	+	-	-	-
Verbascum spec.	-	-	+	-	-	-
Chenopodium spec.	-	-	+	+	-	+
Medicago spec.	-	-	+	-	+	+
Aegilops umbellulata	-	-	-	+	-	sp
Ajuga chia	-	-	-	+	-	-
Fumaria (officinalis)	-	-	-	+	-	+
Heliotropium	-	-	-	+	-	+
Malva nicaeensis	-	-	-	+	-	-
Polygonum bellardi	-	-	-	+	-	-
Silene dichotoma	-	-	-	+	-	-
Taeniatherum crinitum	-	-	-	+	-	-
Trigonella noeana	-	-	-	+	-	-
Vaccaria pyramidata	-	-	-	+	-	+
Vicia noeana	-	-	-	+	-	-
Adonis	-	-	-	+	+	+
Convolvulus (arvensis)	-	-	-	+	+	+
Lithospermum arvense	-	-	-	+	+	+
**Scirpus maritimus*	-	-	-	+	+	+

Table 3. (continued)

	Bandkeramik	Gomolava	Chevdar/Kazanluk	Hacilar	Erbaba	Damascus basin
Triticum boeoticum	-	-	-	+	+	-
Ajuga spec.	-	-	-	-	+	-
Alyssum-type	-	-	-	-	+	-
Anagallis (arvensis)	-	-	-	-	+	-
Anchusa officinalis	-	-	-	-	+	-
**Eleocharis*	-	-	-	-	+	-
Hyoscyamus	-	-	-	-	+	-
Saponaria-type	-	-	-	-	+	-
Stellaria spec.	-	-	-	-	+	-
Teucrium botrys-type	-	-	-	-	+	sp
**Thymelaea*	-	-	-	-	+	-
Ziziphora	-	-	-	-	+	+

available for this archaeological culture. In addition to cultivated plants, many wild plant taxa are reported for the Bandkeramik. Of the latter the herbs are listed in Table 3. It will be evident that not all these species occurred in the whole of the Bandkeramik territory, but the weed flora will have shown local and regional differences. Some of the weed-seed types were found in many sites, quite often in rather great numbers, whereas others have only rarely been reported for Bandkeramik sites, e.g. *Agrostemma githago*.

The Vinča-Pločnik (Vinča C/D) culture of the basal layers at Gomolava, dated to 4 800-4 400 BC, is by no means the oldest farming culture in Yugoslavia. However, no information on weeds in sites of the preceding Vinča-Tordos and Starčevo cultures is available at present. For that reason the weeds from Neolithic Gomolava (unpublished data of the author) are shown in Table 3.

Much research on archaeological plant remains from Bulgaria has been carried out, but usually this concerns mainly cultivated plants. From the Neolithic sites of Chevdar and Kazanluk, of the Karanovo I culture, dated to 6 000-5 600 BC, a fair number of wild plant taxa have been identified by Dennell (1978).

The two late-Neolithic sites from Turkey included in Table 3 were situated in a woodland environment. In that respect they form, as it were, the transition from the steppe and forest steppe region to the forest-covered European territory. Information on vegetable remains from Neolithic sites in NW Anatolia is still lacking. Plant remains from Hacilar have been examined by

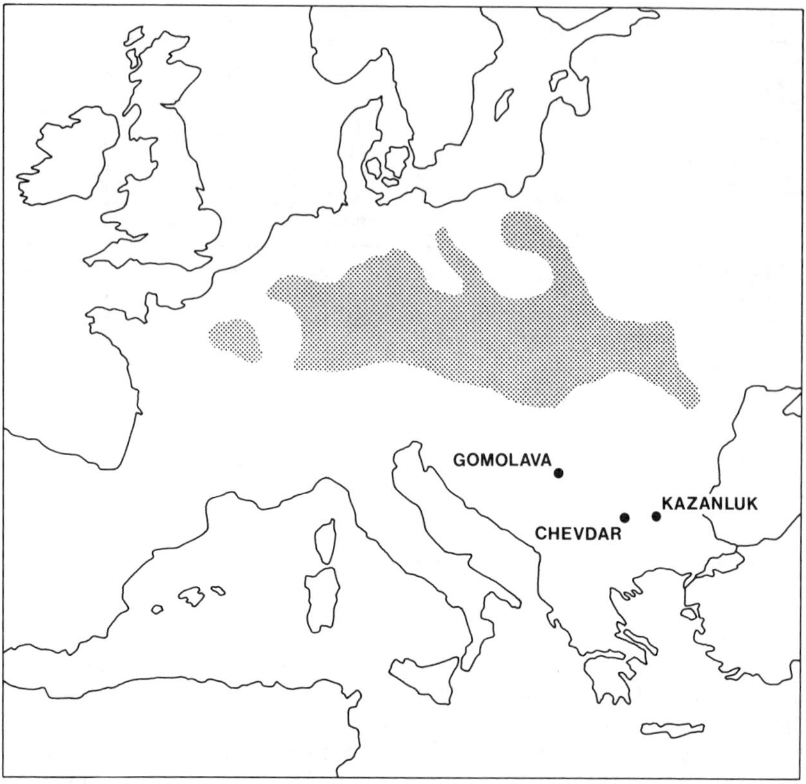

Figure 5. European sites of which vegetable remains are presented in Table 3. The shaded area is the distribution area of the Bandkeramik culture.

Helbaek (1970); those from Erbaba by Van Zeist & Buitenhuis (1983).

Finally it is indicated which of the plant taxa listed in Table 3 are represented in Neolithic sites in the Damascus basin: Ramad, Ghoraifé, Aswad (Van Zeist & Bakker-Heeres 1982 (1985)). The occupation of the latter two sites largely precedes that of Ramad. Species attested for the Damascus basin sites but not reported for one or more of the other sites under discusssion are not shown in Table 3.

It is evident from Table 3 that the Bandkeramik sites and those from the Damascus basin have only few weeds in common. One could argue that the climatic conditions in the latter area are rather extreme (more arid), as compared to those of Central Europe, and that this could account for most of the differences in the weed floras. However, if Hacilar and Erbaba are drawn into the comparison, the number of Bandkeramik weed types that is also attested for Neolithic SW Asian sites remains low. Species demonstrated so far for both the Central European Bandkeramik and the Neolithic Near East are *Chenopodium album, Polygonum aviculare, Bromus sterilis* and *Verbena officinalis.* The *Bromus* spec. fruit type from Ramad may include one or more of the Bandkeramik *Bromus* species other than *B.sterilis,* but certainly not *B.secalinus,* of which species quite large numbers of caryopses are reported for Bandkeramik sites. Whether or not the *Avena* and *Atriplex* seeds in the Damascus basin sites could be of *Avena fatua* and *Atriplex patula,* respectively, cannot be ascertained. Common Bandkeramik (field) weeds, such as *Galium aparine, Polygonum lapathifolium, Polygonum persicaria* and *Polygonum convolvulus (Fallopia convolvulus),* are conspicuously absent in the SW Asian sites. These species are equally not represented in Near Eastern sites of younger date, e.g. Selenkahiye. On the other hand, species which must have been common field weeds in the Neolithic Near East, e.g. *Trigonella, Adrosace maxima, Bellevalia, Melilotus* and *Vaccaria pyramidata,* re not reported for the Bandkeramik and other Neolithic cultures in Europe. Some reservation is due here. The archaeological plant record is (still) very fragmentary, and in Neolithic times Central Europe may have had more weed species in common with the Near East than is suggested by Table 3. Be this as it may, great differences will remain between the Neolithic weed floras of both areas.

The conclusion is unavoidable that few field weeds and other species of disturbed habitats migrated in the wake of the expanding Neolithic agriculture from the Near East right into Central Europe. It looks as if many of the Near Eastern weeds were not able to adapt themselves to the environmental conditions in the greater part of Europe. If on the way to Central Europe more and more weeds of Near Eastern origin had successively dropped out, one may speculate that weed floras of areas closer to the Near East show more resemblance to that of the latter area than the Bandkeramik weed flora does. The weeds attested for Kazanluk and Chevdar do not provide any firm support for this hypothesis. The Bulgarian sites have more species in common with the Bandkeramik sites (*Avena (fatua)* to *Polygonum aviculare* in Table 3) than with the Near Eastern sites. Moreover, all the taxa in common with the Near East are identified to the genus level (*Malva* spec., etc.) implying that it is uncertain whether the same species are concerned. Similarly, the Gomolava taxa in common with the Near East but not with the Bandkeramik are all at the genus level.

The archaeobotanical evidence available at present rather indicates that most of the Near Eastern weeds did not reach Europe in Neolithic times. Many of the weeds of the early Neolithic farmers of Syria, SE and SW Turkey may not even have made it to NW Anatolia, for which area no information on Neolithic weeds has been published so far. One must perforce assume that the majority of the European field weeds originated from the native flora. Potential field weeds were probably found in open vegetation types, that is to say, in vegetation not dominated by trees. Such vegetation types may have occurred on exposed, dry rocky sites, in stream valleys that were periodically flooded, on steep slopes and in places along rivers where organic material drifted ashore, thus creating natural nitrate-rich habitats. In addition, some of the weeds of the Bandkeramik farmers may have originated from the steppe of Central Asia which was supposedly the homeland of broomcorn millet (*Panicum miliaceum*), a crop that was cultivated in part of the Bandkeramik territory.

It looks as if various species of probably Near Eastern origin migrated to and through Europe only during a later period. Thus, *Lithospermum arvense, Convolvulus arvensis, Anagallis (arvensis)* and *Fumaria (officinalis)* listed for Near Eastern sites in Table 3, have so far not been reported for Neolithic Europe. However, these weeds do occur in European archaeological sites of later periods, suggesting that the introduction of the species concerned took place thousands of years after the first farmers had settled. The migration of weeds from the Near East and other areas to Europe must have been a continuous phenomenon. Conversely, weeds of European origin may have spread to other regions, including the Near East.

In conclusion, the archaeological plant record indicates that the majority of the weeds in the fields of early Neolithic farmers in Europe were not introduced from the Near East but that they must have been of European origin.

REFERENCES

Dennell, R.W. 1978. *Early farming in South Bulgaria from the VI to the III Millennia B.C.* BAR International Series (Supplementary) 45, 304 pp.

Edgecombe, W.S. 1970. *Weeds of Lebanon.* Beirut: American University of Beirut. 457 pp.

Harlan, J.R. 1975. *Crops and man.* Madison, Wisconsin: American Society of Agronomy and Crop Science Society of America. 295 pp.

Helbaek, H. 1970. The plant husbandry of Hacilar. In J.Mellaart (ed.), *Excavations at Hacilar*: 189-244. Edinburgh: Edinburgh University Press.

Miller, N.F. 1982. Economy and environment of Malyan, a third millennium B.C. urban center in southern Iran. Ph.D. dissertation (Anthrop.), The University of Michigan, Ann Arbor. Univ. Microfilms, Ann Arbor. 440 pp.

Miller, N.F. & T.L.Smart 1984. Intentional burning of dung as fuel: a mechanism for the incorporation of charred seeds into the archaeological record. *J. Ethnobiol.* 2: 15-28.

Mouterde, P. 1966-1983. *Nouvelle flore du Liban et de la Syrie* (3 vols.). Beyrouth: Editions de l'Imprimerie Catholique.

Post, G.E. & J.E.Dinsmore 1932. *Flora of Syria, Palestine and Sinai* (2 vols.). Beirut: American Press.

Van Zeist, W. & J.A.H.Bakker-Heeres 1982 (1985). Archaeobotanical studies in the Levant. 1. Neolithic sites in the Damascus basin: Aswad, Ghoraifé, Ramad. *Palaeohistoria* 24: 165-256.

Van Zeist, W. & J.A.H.Bakker-Heeres (in press). Archaeobotanical studies in the Levant. 4. Bronze Age sites on the North Syrian Euphrates. *Palaeohistoria* 27.

Van Zeist, W. & H.Buitenhuis 1983. A palaeobotanical study of Neolithic Erbaba, Turkey. *Anatolica* 10: 47-89.
Willerding, U. 1980. Zum Ackerbau der Bandkeramiker. *Beitr. zur Archäologie Nordwestdeutschlands und Mitteleuropas* 16: 421-456.
Willerding, U. 1983. Zum ältesten Ackerbau in Niedersachsen. *Archäologische Mitt. aus Nordwestdeutschland,* Beiheft 1: 179-219.
Zohary, M. 1962. *Plant life of Palestine, Israel and Jordan.* New York: Ronald Press. 262 pp.
Zohary, M. 1973. *Geobotanical foundations of the Middle East.* Stuttgart/Amsterdam: Gustav Fischer Verlag/Swets & Zeitlinger. 739 pp.
Zohary, M. & N.Feinbrun-Dothan 1966-1986. *Flora Palaestina* (4 vols.). Jerusalem: The Israel Academy of Sciences and Humanities.

FISHING, FOWLING AND HUNTING IN LATE PALAEOLITHIC TIMES IN THE NILE VALLEY IN UPPER EGYPT*

A.GAUTIER
Laboratorium voor Paleontologie, Rijksuniversiteit Gent, Belgium

INTRODUCTION

This review chapter is the outcome of archaeozoological work carried out in the Egyptian Nile Valley, which started with the international archaeological salvage campaign in the region of what was to become Lake Nasser (Wendorf 1968, Wendorf & Schild 1976, Wendorf et al. 1979, 1980; Vermeersch 1978). In the course of the two decades setting in with the mentioned campaign, our views on prehistoric man in the Egyptian Nile Valley changed considerably and we tend to view the late persistence of hunter-gatherers in this valley no longer as an indication that it was indeed a hunters' paradise. Especially the recent extensive research program dealing with the Late Palaeolithic in Wadi Kubbaniya (Wendorf et al., in preparation) following the discovery of barley there (Wendorf et al. 1979) and the fact that this barley became itself a 'non-event' (Wendorf et al. 1984) prompted a re-evaluation of our previous ideas. The work on the Wadi Kubbaniya fish remains (Gautier & Van Neer, in press; see also Van Neer 1984) has been most vital in this re-appraisal, as well as the comparison with archaeozoological data from the Nile Valley near Khartoum (Gautier 1983, 1984, in press).

We are dealing in the following with man-animal relations in the Late Palaeolithic between 20 000 and 8 000 BP mainly in Upper Egypt, since evidence for Lower Egypt is almost non-existent. Moreover, no fundamental changes are visible in the faunal record for the period considered. Therefore, in this chapter we assume that the evidence available can be used to characterize very generally the period considered; anyhow, the evidence is too restricted to permit any speculations on shorter intervals within this general time period. The bibliography is kept minimal. Interested readers are referred to the sources already listed, which we will not repeat in most cases, and to those listed in the following.

*A first version of this text was prepared for the session World Population at 18 000 of the World Archaeological Congress 1986. The exclusion of South African and Namibian scientists from the congress forced us to withdraw our contribution.

As Tables 1, 2 and 3 show, the spectrum of animal resources exploited by Upper Palaeolithic man in Egypt is quite impressive. It includes freshwater molluscs, fish, reptiles, birds and mammals. These various groups are reviewed in the next sections.

MOLLUSCS

The only mollusc apparently regularly on the menu is a unionid known as *Unio abyssinicus.* Its valves occur in great quantity in certain sites but unfortunately no systematic study has been made of its exact archaeozoological repartition, nor are we well informed about the exact habits of the animal. It is a Palaearctic form now confined to the Ethiopian highland and hence an exotic form in prehistoric Egypt, indicative of colder water. As to its habitat preferences, we can assume that it will have been essentially an inhabitant of open shore environments of the main channels of the Nile. Consequently people could collect this mussel most effectively when the Nile was at its lowest, during the non-flood season.

FISH

The Late Palaeolithic fish fauna known until now from the Egyptian Nile Valley (Table 1) is definitely dominated by the catfish *Clarias gariepinus.* Most of the fish listed in Table 1 migrate each year with the early Nile flood in the alluvial plain, for spawning. When, where and how these fish can be caught can be inferred from the foregoing, ethnographical data, the exact composition of the ichthyofauna per site and size analyses of the fish present. Such detailed research has been carried out for the sites in Wadi Kubbaniya near Aswan by Dr W. van Neer (Gautier & Van Neer, in press). The sites there can be divided into two groups. The first group contains many sites with mainly large and few small catfish; these indicate catfish harvesting during the early flood phase perhaps followed by some collecting later on. The second class contains a few sites with quite appreciable numbers of tilapia and only small catfish, indicating fishing in residual pools as are formed when the flood is receding. The fishing techniques may have included handpicking, the use of wounding and striking gear (clubs, spears), cover pots, lines with fish gorges (many possible bone gorges were collected), dams, fences, traps; trampling and deoxigenation of the water; poisoning with ichthyotoxic plants. Probably the techniques used changed according to the availability of the fish (early flood bonanza of catfish) and the situation (inundated marginal floodplain; residual pools). All members of the groups gathered for fishing may have been involved in the early flood harvest; something comparable is still the case in the southern Sudan with the riverine Nilotic tribes. Thus catfish could have been gathered in great numbers and part of this annual harvest

Table 1. Fish collected from Late Palaeolithic sites in the Egyptian Nile Valley*

Eel (*Anguilla anguilla*)	R
Mormyridae	RR
Cyprinidae including *Barbus bynni, Labeo* sp.	RR
Bagridae including *Bagrus* sp., *Chrysichthys* sp.	RR
Clarias gariepinus	FF
Clariidae including *Heterobranchus* sp.	RR
Mochokidae including *Synodontis* sp.	RR
Centropomidae including Nile perch (*Lates niloticus*)	RR
Chiclidae including tilapia (*Tilapia* sp.)	R

*R = rare, RR = very rare, FF = very frequent. Based on the data in Greenwood (1968), Churcher (1972), Greenwood & Todd (1976), Gautier et al. (1980) and Gautier & Van Neer (Wadi Kubbaniya; in press).

was maybe prepared for later consumption by drying or smoking, but no evidence for such activities has been detected yet.

TURTLE

The only reptile on the menu appears to have been the soft-shelled turtle, *Trionyx unguis.* This freshwater turtle grows to sizes of 90 cm (antero-posterior length of the carapace) and weighs up to 50 kg. It predates on fish and other aquatic creatures and is not an easy prey itself, for the animals bite dangerously when cornered (Villiers 1958). How prehistoric man caught *Trionyx* is not clear, but it was not a very regular item on the menu. Perhaps people went occasionally for stray animals in the floodplain into which the turtles may have been lured by the spawning fishes.

BIRDS

The avifauna in Table 2 forms quite an impressive list, which would probably still be longer if all material was identified to species. At Wadi Kubbaniya rallids, no doubt mostly referrable to the black coot, form about 52% of the total identified avifaunal sample. Anatids, again most referrable to the species listed (white-fronted goose to smew), account for about 38%, and the little grebe for about 5%. We do not know to what degree Wadi Kubbaniya is representative for the whole Nile Valley, but we probably may accept that fowling was focusing on aquatic or near shore birds. Moreover, it appears that most of the species found until now are winter visitors, which nowadays come to Egypt from September to April. Thus fowling would have been essentially a winter activity. Various devices may have been used for it, including nets, traps, and bow and arrow.

Table 2. Birds known from Late Palaeolithic sites in the Egyptian Nile Valley*

Little grebe (*Tachybaptus ruficollis*)	F
Great crested grebe (*Podiceps cristatus*)	RR
Red-necked grebe? (*P. grisegena*)	RR
Great white pelican? (*Pelecanus onocrotalus*)	RR
Cormorant (*Phalacocorax carbo*)	R
Grey heron (*Ardea cinerea*)	RR
(Black) stork (*Ciconia nigra*?)	RR
Wood ibis (*Mycteria ibis*)	RR
Spoonbill (*Platalea leucorodia*)	RR
Greater flamingo (*Phoenicopterus antiquorum*)	RR
Spur-winged goose (*Plectropterus gambensis*)	RR
White-fronted goose (*Anser albifrons*)	R
Bean goose (*Anser fabalis*)	RR
Brent or red-breasted goose (*Branta bernicla/B. ruficollis*)	RR
Shelduck (*Tadorna tadorna*)	RR
Ruddy shelduck (*Tadorna ferruginea*)	RR
Mallard (*Anas platyrhynchos*)	R
Teal (*Anas crecca*)	RR
Widgeon (*Anas penelope*)	RR
Pintail (*Anas acuta*)	R
Garganey (*Anas guerguedula*)	RR
Pochard (*Athya ferina*)	RR
Trifted duck? (*Athya fuligula*)	RR
Goosander (*Mergus merganser*)	RR
Red-breasted goosander (*Mergus serrata*)	RR
Smew (*Mergus albellus*)	RR
Black kite (*Milvus migrans*	RR
Osprey (*Pandion haliaetus*)	RR
Golden eagle (*Aquila chrysaëtos*)	RR
Peregrine falcon (*Falco peregrinus*)	RR
Quail (*Coturnix coturnix*)	RR
Black coot (*Fulica atra*)	F
Crane (*Grus grus*)	RR
Demoiselle crane (*Anthropoides virgo*)	RR
Curlew? (*Numenius arquatus*)	RR
Gulls and terns (Laridae)	RR

*R = rare (less than 20), RR = very rare (less than 10), F = frequent (more than 35). Based on Tchernov (1968), Churcher (1972), Ballmann (1980), Gautier et al. (1980) and Gautier & Van Neer (in press; preliminary identifications provided by Mrs D.Matthiesen, Gainesville). Where possible, species are recorded; in one case (Laridae) only the family could be indicated.

MAMMALS

The mammals hunted by the prehistoric inhabitants of the Egyptian Nile Valley are listed in Table 3. Most mammalian assemblages are composed essentially of hartebeest, wild cattle and dorcas gazelle. The frequencies of this trio in the samples collected from the various sites and regions (Fayum,

Table 3. Mammals used by Late Palaeolithic man in the Egyptian Nile valley*

Hare (*Lepus capensis*)	RR
Bandicoot rat (*Nesokia indica*)	RR
North African porcupine (*Hystrix cristata*)	RR
Red fox (*Vulpes vulpes*)	RR
Golden jackal/wolf (*Canis aureus/C. lupus*)**	RR
Wild cat (*Felis silvestris*)	RR
Otter (*Lutra* sp.)	RR
Spotted hyena (*Crocuta crocuta*)	RR
Hyena (*Hyaena hyaena*)	RR
Rock hyrax (*Procavia capensis*	RR
Hippopotamus (*Hippopotamus amphibius*)	RR
Warthog (*Phacochoerus aethiopicus*)	RR
Rhinoceros (*Diceros diceros, Ceratotherium simum*)	RR
Wild camel (*Camelus thomasi*)	RR
Giraffe (*Giraffa camelopardalis*)	RR
Wild ass (*Equus africanus*)	RR
Wild cattle (*Bos primigenius*)	FF
Hartebeest (*Alcelaphus buselaphus*)	FF
Kob (*Kobus kob*)	RR
Barbary sheep (*Ammotragus lervia*)	RR
Dorcas gazelle (*Gazella dorcas*)	F
Dama (gazelle) (*Gazella dama*)	RR

*R = rare, RR = very rare, FF = very frequent. Based on Gautier (1968, 1976a, 1976b, 1978, 1984), Churcher (1972), Gautier et al. (1980), Gautier & Van Neer (in press).
**See Ferguson (1981).

Isna and Idfu, Kom-Ombo, Elkab, Wadi Kubbaniya, Wadi Halfa) are approximately 36%, 53% and 11%. The frequencies calculated for each region separately indicate clearly that dorcas gazelle is essentially a visitor from the desert and that hartebeest is better adapted to dry conditions than wild cattle. As a whole this mammalian game spectrum is decidedly narrow, especially when compared with the one from mid-Holocene sites in the region around Khartoum in the Central Sudan (Gautier 1983, 1984, in press). There we find some 45 species, among which there are at least ten antelopes, African buffalo and other large herbivores. North of the Batn-el-Haggar we counted six ruminants in a total fauna of some 23 species. We think that essentially a combination of climatic factors (e.g. aridity) and azonal ecological (annual flooding) factors is responsible for the restricted spectrum of the Egyptian Nile Valley.

As to how the game was killed, little or no direct evidence has yet been found. One generally accepts that the techniques of Late Palaeolithic hunters were fairly sophisticated. Their equipment may have included traps and snares for smaller mammals, while bow and arrow, spears and other piercing gear may have been used for bigger game. Perhaps the Nilotic hunters often drove the large herbivores into shallow water, where their movements were impeded, before attacking them. Snares on the paths used by dorcas gazelles coming from the valley fringes to the water were perhaps quite effective killers. Hence the absolute frequencies calculated for this small herbivore may exaggerate its availability with respect to that of the duo hartebeest-aurochs. This duo will be discussed separately.

NATURAL VEGETATION OF THE EGYPTIAN NILE VALLEY

For several millennia tree cutting for building and for making fuel, overgrazing by livestock, irrigation practices and the introduction of various exotic plants have totally altered the vegetation within the boundaries of the Egyptian Nile Valley. Especially the effects of the increasingly elaborate and continuous irrigation may distort the way we imagine the natural landscape of the Egyptian Nile Valley, Surely, it was appreciably less green and productive than the huge oasis now stretching from Nubia to the delta.

Various scholars have tried to give landscape reconstructions of the Egyptian Nile Valley (Butzer 1959, Butzer & Hansen 1968, Hassan 1974). In his detailed description of the Egyptian alluvial flats in Holocene times before the advent of artificial irrigation, Butzer (1976) stresses the fact that these flats form basically a floodplain of the convex variety. It is marked by natural levees a few meters (1-3 m) above the seasonally inundated alluvial flats, which have their lowest areas distant from the river, often near the outer valley margins. The levees, both active and abandoned, divide the alluvial flats into discrete flood basins. The Nile would ideally rise to bankful stage in southern Egypt by mid-August and then spread out through overflow channels and breaches across the levees into these basins. Depending on the height of flood and local topographical differences which would not exceed 2 m, more or fewer floodbasins would be inundated. After several weeks or months, depending on the importance of the flood and the relative elevation of the flood basin, the alluvial flats would emerge in response to the falling riverhead, dropping groundwater, infiltration, evaporation and drainage back to the main channel. The first basins in Upper Egypt are normally dry by early October. By late November all but the lowest basin hollows in the northernmost valley would be drained, but marshes would persist in isolated valley-margin backswamps and other topographic lows. In his description of the prehistoric Nile Valley of Egypt Butzer (1976) refers specifically to the Omo River in the Omo lowlands (Ethiopia). In his view, in this region restricted rains contribute to plant growth, but the dominant factor would be the seasonal inundations. Most obvious in the Omo lowlands is the basic division between evergreen gallery forest along the channels and tributaries and the grassland or acacia-shrub savanna on the seasonally inundated flats (Butzer 1971). The importance of floods is also emphasized by Delany & Happold (1979), who note that the mesic conditions of the river Nile in northern Sudan do not extend more than a few meters beyond the river banks.

Springuel (1981) found that the natural vegetation on higher ground on the islands of the First Cataract near Aswan would be an 'edaphic climax' consisting of acacia-shrub and woodland. Riverine woodland with acacia was also described on seasonally inundated land along the Blue Nile (Mubarak et al. 1982). We may expect a comparable vegetation on the Egyptian floodplain provided that we make some amendments for the absence of any appreciable rain along the Egyptian Nile, at least in the south and in the period considered here, and perhaps for the fact that the flood plain may be generally 'drier' than the islands mentioned.

Dr Wendorf (Dallas) and Dr Schild (Warszawa) in their review of this chapter drew my attention to the fact that all identifiable charcoal remains from Wadi Kubbaniya are tamarisk and that acacia seedlings are killed by the flood. Acacia may therefore have been restricted to land that was not or no longer regularly flooded.

THE DUO HARTEBEEST-WILD CATTLE

As said, the duo hartebeest-wild cattle formed the main mammalian game resource in the Egyptian Nile Valley, in number and certainly in weight, dorcas gazelle attaining but about 0.1 and 0.03 of the respective weights of hartebeest and wild cattle. In East Africa hartebeest appear to be inhabitants of woodlands, moist savannas and somewhat drier grassland, but not of marshes, swamps and valley grasslands, very dry savannas or very dry woodland (Kingdon 1982a: 21, chart). This antelope is a typical grazer of coarse, medium to long grasses. Once a prolific and often dominant ruminant in most of subsaharan Africa, hartebeest have probably suffered the greatest contraction in range of all African ruminants. Self-advertisement and marked sedentarism make the animals an easy prey for hunters. However, the major cause for their decline in numbers appears to be the expansion of domestic cattle in Africa. Both hartebeest and domestic cattle are bulk-roughage feeders living essentially on a grass diet; thus they are not very selective of the kind of parts of grass they eat during the drier seasons. Several authors have also stressed the fact that hartebeests prefer to live in ecotones, usually between scrub or woodland and open grassland. In East Africa they prefer *Acacia* savanna and hardpans fringing humid, cotton soil valleys (*mbugas*) within woodlands. Hartebeest would live near and in the latter at the end of the dry season, spreading in the woodland with the arrival of the rains (Kingdon 1982b: 505-523).

Aurochs or wild cattle was a very widely distributed Palaearctic bovid; this suggests an appreciable ecological tolerance. In Western Europe the animal is generally not found in faunal assemblages believed to reflect colder Pleistocene phases; in these cases it is replaced by the steppe bison or wisent (*Bison priscus*) (Kurtén 1968). Toepfer (1963: 37, Fig. 9) considers wild cattle as a typical inhabitant of lush oceanic biotopes in Pleistocene Europe. The present day descendant of the aurochs, our domestic cattle, ranges from (semi-)arid to wooded regions, but no doubt special selection unconsciously induced or consciously applied by man has broadened the tolerance of the domestic form appreciably. Generally speaking, aurochs appear to be fairly tolerant herd animals adapted to good grassland without or with wooded vegetation. We do not know, however, to what extent they were migrant, responding to seasonal changes in their various habitats. As its domestic descendants, the African wild form may well have occupied habitats overlapping with that of hartebeest.

The already mentioned attempts to reconstruct the prehistoric landscape of the Egyptian Nile Valley can be combined with the ecological data on our bovid duo. We can describe this valley as consisting of, perhaps very restricted, gallery woodland near the main channels, and a mosaic of acacia-shrub savanna and restricted less dry acacia-woodland around permanent or semi-permanent marshy depressions mostly situated towards the valley margins. Hartebeest found acceptable grazing in the shrubland in the earlier post-flood phase, and in the wetter areas with woodland when conditions became drier. Here the comparison with the situation of hartebeest in East Africa ends. With the flood this 'beest' and aurochs had to move away, each according to its particular needs, to not submerged higher ground and into the wadis, from which they returned after the flood. Presumably hartebeest started moving earlier than wild cattle.

AVAILABLE UNGULATE BIOMASS

We can try to estimate the available ungulate biomass in the alluvial flats of the Egyptian Nile. This exercise is fraught with difficulties because of the lack of precise concepts and data. The maximum ungulate biomass in 'dry savanna' given in the literature available to us, amounts to about 5 000 kg/km^2; estimates for dry, Sahelian environments do not exceed 200 kg (Bourlière 1963; Delany & Happold 1979: 311, Table 11.4). Our estimate for the Egyptian Nile Valley should take into account the animals' annual movements to not inundated heights and into the wadis. Moreover, not all of the ungulate biomass pertains to hartebeest and aurochs. Hippopotamus adds its quite appreciable weight (1 400 kg). Indeed, although passing most of its life in the water, the hippo makes regular incursions into the land (0.5-3 km) to graze and it may consume up to 60 kg of grasses and herbs daily (Verheyen 1954). Densities of hippopotamus along the prehistoric Nile are difficult to guess. This pachyderm is a rare item in the bag of our prehistoric hunters, but perhaps they did not care much for this bulky and irascible creature. Bourlière (1963) gives estimates of 7.2 and 8 animals per km^2 in Queen Elisabeth Park, Uganda and in the Rwindi-Rutshuru Plain, Zaire, but Haltenorth & Diller (1979) refer to as much as 100 animals per km on the shores of the Great Lakes.

The outcome of the restrictions formulated are that the available biomass of our bovid duo may have been rather low, perhaps with animal densities per km^2 of a few animals only. Thus mammalian game was perhaps rather hard to come by for our Later Palaeolithic hunters. We do however not want to calculate theoretical yields of animal protein etc. as Hassan (1974; 1981: 22) has done. In our opinion such computations should be based on secondary productivity data and not on biomass census data; unfortunately, such productivity data are lacking.

RESOURCE SCHEDULING

Resource scheduling models for ancient man in the Egyptial Nile Valley (Hassan 1974, 1984; Butzer 1976) go all back to Niemeyer (1936) who described the annual cycle of rural life in the Mamluk period in relation to the Nile's behaviour. No doubt the Late Palaeolithic annual round was also connected with the ebb and flood of the Nile, but its precise sequence and timing may be difficult to establish. First, sites may have been reoccupied at different seasons but their faunal samples cannot be separated in the subsequent seasonal components. Second, sites occupied during the low flood near the main Nile are not well known, being hidden or destroyed by later alluvial and agricultural processes. Be it as it may, our model should no doubt include an annual catfish harvesting season in the earlier flood period (August-September?). People may have lingered on near their fishing sites until early winter, combining then some fishing, big game hunting and fowling. As the flood receded, the game until now concentrated in the drier valley parts and the wadis migrated to the floodplain where good grazing was shooting up; many birds no doubt followed and occupied the Nile shores. People went after these creatures and dispersed over the floodplain, creating maybe many but mostly ephemeral and easily obliterated sites. Now they were hunting and fowling, and fishing in residual ponds. When the birds left, people went on hunting and fishing, but may have combined these activities of necessity with the collecting of molluscs. Then the flood came back and the cycle could start again.

IMPORTANCE OF THE MAIN ANIMAL RESOURCES

The importance of these resources (fish, birds, mammals) throughout the year cannot be assessed, since the complete resource scheduling cycle is not represented in the archaeological record. Second, if we take for example the evidence from Wadi Kubbaniya, where fish is decidedly dominant (some 95% of the number of remains of the main resources; birds 4%; mammals 1%), how can be translate the quantitative data obtained into available food? Differential preservation, different food/refuse ratios, the so-called 'Schlepp effect' (sensu Perkins 1969) etc., bias the excavated record and, in our opinion, most attempts to compute quantities of food are at best heuristic models, which unfortunately we mistake rather easily for 'exact' science! Anyhow, we believe on the strength of the total evidence, archaeological and other, and because of the sheer number of fish remains that the latter resource was indeed primordial at Wadi Kubbaniya. As already pointed out, surplus catfish was perhaps prepared for later consumption. Such fish do however last but a few months and would not have carried our fishermen over into the next floodseason. More important perhaps is the fact that catfish was a predictable resource, which was easily exploited with much success. The migrating birds were also predictable, but no doubt less easily captured; and how did their total available biomass compare with that of the fish? As to the mammalian resources, we have already pointed out that these were perhaps the least reliable, especially when the food was chasing the game to higher terrain in the valley and into the wadis. The importance of plants is beyond the scope of this chapter, but the available evidence (see for example Wendorf & Schild 1976b) and general considerations on the relative importance of plant resources with recent hunter-gatherers suggest that these may have been another very predictable resource. Plants can have played a major role in the floodplain during the low water season, replacing as it were the fish that people found at Wadi Kubbaniya and related sites in the highwater season.

CONCLUSIONS

As later people, Late Palaeolithic man along the Egyptian Nile probably exploited his environment according to an annual schedule in response to the ebb and flood of the river, combining fish harvesting, fowling, big game hunting, plant harvesting and shell collecting according to the season. We feel that until now we have overestimated unconsciously the carrying capacity of the primeval Egyptian Nile Valley and its big game biomass available to people (hartebeest, wild cattle). Catfish appears to have been a very important flood resource because of its easy harvest in appreciable quantities, the possible conservation of surplus, but most of all its predictability. Plants may have been a resource of comparable importance in the low water season. Fish (and plants) can thus have played a prominent role in making the Egyptian Nile Valley an acceptable place for Late Palaeolithic hunter-gatherers, and in the persistence of their way of life there. The archaeological evidence for the suggested annual round is very incomplete. Because of the cut-and-fill history of the Nile and intense agricultural activities near the river during several millennia, this evidence may remain incomplete for ever, precluding any attempts at palaeodemographic and other dangerous exercises.

REFERENCES

Ballmann, P. 1980. Report on the avian remains from sites in Egyptian Nubia, Upper Egypt and the Fayum. In A.E.Close (ed.), *Loaves and fishes: the prehistory of Wadi Kubbaniya*: 307-310. Dallas: Southern Methodist Univ. Press.

Bourlière, F. 1963. Observations on the ecology of some large African mammals. In F.C.Howell & F.Bourlière (eds.), *African ecology and human evolution.* Viking Fund Publ. in Anthropology 36: 43-64.

Butzer, K. 1959. Die Naturlandschaft Agyptens wáhrend der Vorgeschichte und der Dynastische Zeit. *Akad. Wiss. Lit. Mainz, Math.-Naturw. Kl.* 1959 (2): 43-122.

Butzer, K. 1971. *Recent history of an Ethiopian delta.* Chicago: Univ. Chicago Dept. Geogr. Research Paper 136.

Butzer, K. 1976. *Early hydraulic civilisation in Egypt. A study in cultural ecology.* Chicago: Univ. Chicago Press.

Butzer, K.W. & C.L.Hansen 1968. *Desert and river in Nubia.* Chicago: Univ. Chicago Press, 562 pp.

Churcher, C.S. 1972. Late Pleistocene vertebrates from archaeological sites in the plain of Kom Ombo, Upper Egypt. *Royal Ontario Museum* 82.

Delany, M.J. & D.C.D.Happold 1979. *Ecology of African mammals.* London/New York: Longman.

Ferguson, W.W. 1981. The systematic position of *Canis aureus lupaster* (Carnivora: Canidae) and the occurrence of *Canis lupus* in North Africa, Egypt and Sinai. *Mammalia* 45: 459-468.

Gautier, A. 1968. Mammalian remains of the northern Sudan and southern Egypt. In F.Wendorf (ed.), *The prehistory of Nubia*: 80-99. Dallas: Southern Methodist Univ. Press.

Gautier, A. 1976a. Freshwater molluscs and mammals from Upper Palaeolithic sites near Idfu and Isna. In F.Wendorf & R.Schild (eds.), *Prehistory of the Nile valley*: 349-364. New York: Academic Press.

Gautier, A. 1976b. Appendix II. Animal remains from archaeological sites of Terminal Palaeolithic to Old Kingdom age in the Fayum. In F.Wendorf & R.Schild (eds.), *Prehistory of the Nile valley*: 369-381. New York: Academic Press.

Gautier, A. 1978. La faune de vertébrés des sites épipaléolithiques d'Elkab. In P.Vermeersch (ed.), *Elkab II. L'Elkabien, Epipaléolithique de la vallée du Nil Egyptien*: 103-114. Brussels: Leuven Univ. Press.

Gautier, A. 1983. Animal life along the prehistoric Nile: the evidence from Saggai 1 and Geili (Sudan). *Origini* 12: 50-115.

Gautier, A. 1984. Quaternary mammals and archaeozoology of Egypt and the Sudan: a survey. In L.Krzyzaniak & M.Kobusiewicz (eds.), *Origin and early development of food-producing cultures in north-eastern Africa*: 43-56. Poznan: Poznan Archaeol. Museum.

Gautier, A. (in press). A general review of the known prehistoric faunas of the central Sudanese Nile valley. In L.Krzyzaniak & M.Kobusiewicz (eds.), *Late Prehistory of the Nile basin and the Sahara.* Poznan: Poznan Archaeol. Museum.

Gautier, A., P.Ballmann & W.Van Neer 1980. Appendix 1. Molluscs, fish, birds and mammals from the Late Paleolithic sites in Wadi Kubbaniya. In A.E.Close (ed.), *Loaves and fishes: the prehistory of Wadi Kubbaniya*: 281-295. Dallas: Southern Methodist Univ. Press.

Gautier, A. & W.van Neer (in press). *The animal remains from the Late Palaeolithic sequence in Wadi Kubbaniya.*

Greenwood, P.H. 1968. Fish remains in the prehistory of Nubia. In F.Wendorf (ed.), *The prehistory of Nubia*: 100-109. Dallas: Southern Methodist Univ. Press.

Greenwood, P.H. & E.Todd 1976. Fish remains from Palaeolithic sites near Idfu and Isna. In F.Wendorf & R.Schild (eds.), *Prehistory of the Nile Valley:* 383-388. New York: Academic Press.

Haltenorth, Th. & H.Diller 1979. *Elseviers Gids van de Afrikaanse Zoogdieren.* Amsterdam/Brussels: Elsevier.

Hassan, F. 1974. *The archaeology of the Dishwa plain, Egypt: a study of a Late Palaeolithic settlement.* Geol. Survey of Egypt Paper 59.

Hassan, F.A. 1981. *Demographic archaeology.* New York: Academic Press.

Hassan, F.A. 1984. Environment and subsistence in predynastic Egypt. In J.D.Clark & S.A.Brandt (eds.), *From hunters to farmers*: 57-64. Berkeley: Univ. California Press.

Kingdon, J. 1982a. *East African Mammals* (III C. Bovids). London/New York/San Francisco: Academic Press.

Kingdon, J. 1982b. *East African Mammals* (III D. Bovids). London/New York/San Francisco: Academic Press.

Kurtén, B. 1968. *Pleistocene mammals of Europe.* London: Weidenfeld & Nicholson.

Mubarak, M.Q., E.A.Bari, G.E.Wickens & M.A.J.Williams 1982. The vegetation of the Sudan. In M.A.J.Williams & D.A.Adamson (eds.), *A land between two rivers*: 143-164. Rotterdam: Balkema.

Niemeyer, W. 1936. *Aegypten zur Zeit der Mamluken.* Berlin: Reiner.

Perkins, D. 1969. Fauna of Catal Hüyük: evidence for early cattle domestication in Anatolia. *Science* 164: 177-179.

Springuel, I.V. 1981. Studies on the natural vegetation of the islands of the First Cataract at Aswan, Egypt. Ph.D. dissertation, Botany Dept. Aswan Fac. Science, Assiut Univ.

Tchernov, E. 1968. Peregrine falcon and purple gallinule of Late Pleistocene age in Sudanese Aswan reservoir area. *The Auk* 85 (1): 133.

Toepfer, V. 1963. *Tierwelt des Eiszeitalters.* Leipzig: Akademische Verlagsgesellschaft.

Van Neer, W. 1984. The use of fish remains in African archaeozoology. In N.Desse-Berset (ed.), *Proc. of 2nd Fish Osteoarchaeology Meeting*: 155-167. Paris: Centre National de Recherches Scientifiques.

Verheyen, R. 1954. *Monographie éthologique de l'hippopotame.* Brussels: Inst. Parcs Nat. du Congo Belge.

Vermeersch, P. 1978. *Elkab II. L'Elkabien, Epipaléolithique de la vallée du Nil Egyptien.* Brussels: Leuven Univ. Press.

Villiers, A. 1958. *Tortues et crocodiles de l'Afrique noire française.* Initiations Africaines 15. Dakar: Inst. Français d'Afrique Noire.

Wendorf, F. (ed.) 1968. *The prehistory of Nubia* (I & II). Dallas: Southern Methodist Univ. Press.

Wendorf, F. & R.Schild (eds.) 1976a. *Prehistory of the Nile valley.* New York: Academic Press.

Wendorf, F. & R.Schild 1976b. The use of ground Grain during the Late Palaeolithic of the lower Nile valley, Egypt. In J.Harlan (ed.), *Origins of African plant domestication*: 269-288. The Hague: Mouton.

Wendorf, F., R.Schild, N.Hadidi, A.Close, M.Kobusiewicz, H.Wiekowska, B.Issawai & B.Haas 1979. Use of barley in the Egyptian Late Palaeolithic. *Science* 205: 1341-1347.

Wendorf, F., R.Schild & A.Close 1980. *Loaves and fishes.* Dallas: Southern Methodist Univ. Press.

Wendorf, F., R.Schild, A.E.Close, D.J.Donahue, A.J.T.Jull, T.H.Zabel, H.Wiekowska, M.Kobusiewicz, B.Issawi & N.el Hadidi 1984. New radiocarbon dates on the cereals from Wadi Kubbaniya. *Science* 225: 645-646.

EARLY DOMESTIC CATTLE IN THE EASTERN SAHARA

F.WENDORF/A.E.CLOSE
Department of Anthropology, Southern Methodist University, Dallas, Texas, USA
R.SCHILD
Polska Akademia Nauk, Instytut Historii Kultury Materialnej, Warsaw, Poland

This chapter is a response to Smith's review article on 'Prehistory of the Eastern Sahara' (Wendorf & Schild 1980), published in volume 16 of *Palaeoecology of Africa* (Smith 1984). In that article, Smith deals with two aspects of the Eastern Sahara publication: the claim for very early domesticated cereals in Wadi Kubbaniya some 18 000 years ago, and the claim for domesticated cattle in the Eastern Sahara over 9 000 years ago.

Smith's discussion of the Kubbaniya material contains a number of errors, particularly concerning the reconstruction of the Kubbaniyan environment 18 500-17 000 years ago. The environment at that time was, indeed, colder than that of today, as Smith correctly noted, but it was certainly not wetter. All the evidence indicates that this was a period of extreme aridity (Wendorf & Schild 1980: 234-236; Wendorf et al. 1980: 11-47), and not at all comparable to the modern environment in the wadis west of the Delta, which receive measurable amounts of winter rainfall. The improbability that cereals would grow naturally in such an environment is, however, moot, since we now believe that the cereal grains found in the Kubbaniyan sites were intrusive (Wendorf et al. 1984a).

In contrast to Kubbaniya, further research in the Eastern Sahara has significantly strengthened the case for the presence of early domestic cattle there (Wendorf et al. 1984b). The new data are derived from a series of Holocene sites in the area of Bir Kiseiba (22°34′N, 29°53′E), some 250 km west of the Nile; there has also been a restudy specifically of the 'large bovids' from the sites at Nabta originally published in 1980. This new information has not led to any significant alteration of our views, but rather to clarification, and, in the light of Smith's comments on the Nabta material, we feel it is worthwhile to restate our arguments and make more explicit the way in which we reached our conclusions.

Our method is that of 'strong inference' (Platt 1964), involving the formulation of alternative hypotheses, the testing of these to exclude one or more of them and the adoption of what remains. This is based on the principle that 'any conclusion that is not an exclusion is insecure' (Platt 1964: 347), or, to paraphrase Popper, that since another, later explanation may be as good or better, there is no such thing as proof and science advances only by disproofs.

The data with which we have to deal are derived from 23 early Holocene sites in the Eastern Sahara, which have yielded 29 cultural assemblages. These are grouped into Early Neolithic, Middle Neolithic and Late Neolithic, on the basis of dating and coincidental similarities in artifact types. The Early Neolithic is further subdivided into four consecutive stages, of which the first three were formerly called Terminal Palaeolithic (Wendorf & Schild 1980). The groups are not intended to correlate, as Smith implies, with Gabriel's (1977) Initial, Main and Final Neolithic. Details of these groupings and of their numerous radiocarbon dates have been published elsewhere (Wendorf & Schild 1980, Wendorf et al. 1984b) and need not be repeated here. Table 1 summarises the groups, their chronological ranges and the frequency of large bovid bones associated with each cultural assemblage. Site E-77-1, which yielded some of the large bovid bones discussed by Smith, is omitted from the table, since it is a mixed Early and Middle Neolithic site. It is obvious from Table 1 that although large bovid bones are very rare, they occur in *every* faunal assemblage with a significant number of identifiable mammal bones. (The collection from the upper layers of E-79-4, with 41 identifiable mammal bones, is the largest not to contain large bovids.)

Since large bovid bones are so rare, it can be hypothesised that they might be intrusive into the sites. There are three taphonomic classes of intrusive faunal remains: geological, 'pene-contemporaneous' and recent (Gautier 1984). The bovid bones are not fossilised and so are not geological intrusives. There are no large bovids in the Eastern Sahara today and have not been for several millennia, so they cannot be recent intrusives. If intrusive, the bovid bones must be pene-contemporaneous with the sites. This would imply that there were large bovids in the Eastern Sahara during the early Holocene, dying at random across the landscape, and that we have found the remains of those which chanced to die in archaeological sites. As far as surface remains are concerned, the most cursory visit to the Eastern Sahara shows that there is not a random scatter of large bovid bones across the surface of the desert; they occur only in archaeological sites. It is more difficult to determine the overall subsurface distribution of large bovid bones, but the hundreds of square metres of stratigraphic trenches excavated away from the sites have never yielded such bones, suggesting that the subsurface large bovid bones also occur only in archaeological sites. The correlation between sites and large bovid bones is almost perfect (imperfect only because not all sites yielded bovid bones) and we may reject the hypothesis of intrusion. The large bovid bones are in true *archaeological* association with the sites.

Large bovids occurring in Holocene contexts in northeastern Africa may be giant buffalo (*Pelorovis antiquus,* formerly *Homoioceras antiquus*), African buffalo (*Syncerus caffer*), wild cattle (*Bos primigenius*) or domestic

Table 1. Frequencies of large bovid bones and of all identifiable mammal bones in Eastern Saharan Neolithic sites

Region	Site		Large bovid bones	Identifiable mammal bones
Early Neolithic of El Adam type: 9 500-9 000 BP				
Nabta	E-77-7		2	6
Kiseiba	E-79-8		9	387
Kiseiba	E-80-4		9	50
Early Neolithic of El Kortein type: 8 800-8 500 BP				
Nabta	E-75-6,	lower	1	138
Nabta	E-77-3		1	29
Kiseiba	E-80-1,	Area A	0	0
Kiseiba	E-80-2		0	0
Early Neolithic of El Ghorab type: 8 500-8 200 BP				
Kharga	E-76-6		0	0
Dyke area	E-72-5		0	0
Nabta	E-77-6		0	0
Kiseiba	E-79-1		0	15
Kiseiba	E-79-3		0	7
Kiseiba	E-79-4,	lower	1	130
Kiseiba	E-79-8,	Area X	0	2
Early Neolithic of El Nabta type: 8 200-7 900 BP				
Nabta	E-75-6,	upper	1	410
Kiseiba	E-79-4,	upper	0	41
Kiseiba	E-79-5		0	24
Kiseiba	E-80-1,	Areas C & D	2	60
Kiseiba	E-80-3		0	0
Middle Neolithic: 7 700-6 200 BP				
Nabta	E-75-8,	lower	39	603
Nabta	E-77-5		1	63
Nabta	E-77-5A		0	0
Kiseiba	E-79-2		1	459
Kiseiba	E-79-6		3	100
Kiseiba	E-79-7		0	22
Late Neolithic: 6 200-5 000 BP				
Nabta	E-75-8,	upper	20	143
Kiseiba	E-79-4,	'Late Neol'	0	0
Kiseiba	E-79-5B		3	3
Kiseiba	E-79-9		0	0

cattle (*B.primigenius* f. taurus). The identification with giant buffalo may be rejected on both osteometric and morphological grounds (Gautier 1984). African buffalo falls within the same size-range as cattle and Smith (1984: 319) implies that the Eastern Saharan bovids might well be African buffalo The entire collection has been carefully re-examined to solve precisely this problem and, morphologically, the bones are very unlikely to belong to *Syncerus* (Gautier 1984). Metrically and morphologically, the large bovid bones can be assigned to cattle.

We therefore have cattle bones in true association with archaeological sites in the Eastern Sahara from about 9 500 to about 5 000 BP. The crucial question is whether these are the bones of wild cattle or of domestic cattle. Smith seeks to show that the earlier cattle, at least, were wild on the basis of size, although size alone is a very treacherous criterion to use as an indicator of domestication. Change in size may result not only from genetic isolation, but also from environmental change or stress. European wild cattle were smaller in the Holocene than in the Pleistocene (Bökönyi 1974: 99; Lasota-Moskalewska 1980: 121) without being any whit less wild, and there are several well-documented instances of both reduction and increase in size of domestic cattle during and after the European Neolithic (opera cit.). It is, in any case, very unlikely that there was a complete barrier to interbreeding between wild and domestic cattle during Neolithic times anywhere, and interbreeding seems actually to have been encouraged in, for example, the Globular Amphorae culture, where domestic cattle were fully as large as their wild relatives (Lasota-Moskalewska 1977). Size, then, is ambiguous. The samples from the Neolithic of the Eastern Sahara fall overall within the size-ranges of larger domestic cattle and of smaller wild cattle (Gautier 1984). This could actually be taken to support their domestic status, since the environmental conditions were so difficult (see below) that any wild cattle would have undergone a significant reduction in size, purely as a response to stress. In the end, however, the samples are so limited as to be inconclusive.

The critical argument, and one which Smith completely ignores, is the ecological one. All of the faunal assemblages are dominated by small dorcas gazelle (*Gazella dorcas*) and hare (*Lepus capensis*), with, usually, minor quantities of the larger dama gazelle (*G.dama*). In the absence of cattle, this would indicate an environment with a very low carrying capacity, restricting the ruminants essentially to gazelles. This is in accord with the sedimentological and pedological studies, which also show an arid environment. Except for two fragments of oryx or addax from the Middle Neolithic of E-75-8, there are *no* remains of animals intermediate in size between the gazelles and the cattle. Large bovids are known from other parts of the Sahara at this and earlier periods, but in every case are associated with a full spectrum of animals of other sizes: the giant buffalo in the Middle Palaeolithic of Bir Tarfawi, for example, occurs with white rhinoceros, wild camel, warthog and large and medium-sized antelopes, as well as gazelles (Gautier 1980). What is known of the ecology of modern arid and semi-arid environments suggests that a fauna composed of gazelles, hare and cattle, with nothing in between, would be surprising at the very least. The carrying capacity of the Eastern Sahara in the early Holocene was simply unsuitable for herds of wild cattle.

There is also the problem of water. Gazelle can survive for extended periods on only the moisture they obtain from green vegetation. Hare are not quite so desert-adapted, but can obtain sufficient moisture from dew. Cattle need actually to drink water at least every other day. There was no permanent, standing water in the Holocene Sahara (although there had been during the period of the Middle Palaeolithic), only ephemeral playa lakes. These filled during the rainy season but stood dry for much of the year, at which time the only way to obtain water was to dig wells for it, as we know the human occupants of the area did. We may therefore reject the hypothesis that cattle were an integral part of the natural, wild fauna of the Eastern Sahara in the early Holocene.

We know that wild cattle were numerous in the lush environment of the Nile Valley at this time, and it can be hypothesised that each year the Nilotic cattle ranged westward into the Desert of their own accord, to graze on the new grass after the summer rains, and returned to the Valley before the dry season. This might be possible for Nabta (100 km west of the Nile), but is extremely unlikely for Bir Kiseiba (250 km west of the Nile), and, again, makes very little ecological sense. In the Valley, the wild cattle are accompanied, as expected, by the full range of animals of other sizes (Gautier 1976, 1978). If the cattle went far out into the Desert, why did the medium-sized animals not do likewise? This is a particularly pertinent question for the hartebeest, which was quite common in the Valley, which is much better adapted to aridity than are cattle, but which was absent from the Eastern Sahara. We may, again, reject the hypothesis that the animals in the Desert were wild cattle coming from the Nile Valley.

We are left with the hypothesis that the cattle bones, found in true archaeological association with Eastern Saharan sites of the tenth to sixth millennia BP, are those of animals that were to some extent domesticated. The process of biological domestication may still have been at an early stage, which would accord with the rather large size of the animals, but there seems to have been very effective human control of their movements. This is simply a working hypothesis and has not been 'proved'. Nothing is proved in the natural sciences; the theory of gravity has been a useful and reliable predictor of future events for several centuries, but it remains unproven. If another, later hypothesis should be as good or better, then we may reject the hypothesis of domestic cattle. For the moment, however, it seems to be our closest approximation to what was actually happening in the Eastern Sahara during the early Holocene.

Like other useful hypotheses, this not only answers questions but also raises them. If the cattle were in the Desert as domestic animals, where were they domesticated? Further, since our evidence suggests that the desert was not occupied year-round until about 8 100 BP, where did the early cattle-herders take their herds during the dry season? Smith (1984: 323) suggests that the cattle were domesticated from local wild stock. Since there were no local wild stock we may discount this suggestion. The process of domestication must at least have begun in an area where wild cattle occurred naturally; that is, an area where water was more readily and permanently available. Such an area could also have served as a refuge during the dry season in the desert. In the early Holocene, there were two such regions adjacent to the Eastern

Sahara: western Sudan to the South and the Nile Valley to the East.

If, as seems most likely, the early Holocene wet periods in the Eastern Sahara resulted from a northward shift of the summer monsoon belt, then northern Sudan received more rainfall than did southern Egypt, and wild cattle may have occurred naturally across the modern Sahel. It can be hypothesised that they were domesticated there, and that in the Eastern Sahara we have the northern section of a North-South pattern of transhumance. The prehistory of western Sudan remains largely unknown. However, preliminary survey work in the area of Merga, in norther Darfur, has shown that while there are early Holocene sites, they differ markedly from contemporary sites in Egypt in both lithics and ceramics, and their faunas do not include cattle. This lack may be due to the lower quality of forage in the higher-rainfall zone. Under modern Sahelian conditions and particularly in sandy soils, vegetation which grows where water, rather than nitrogen, is the limiting factor (less than 300 mm of rain per annum) is richer in protein, leading to the curious situation that at the end of the dry season there is more good forage in the northern Sahel than in the southern Sahel or Savannah (Breman & De Wit 1983). This might explain why there was early Neolithic cattle-herding in southern Egypt, which corresponded to the modern northern Sahel, but apparently not in northern Sudan, which was more like the modern southern Sahel. In any case, it seems unlikely that the early Holocene pastoralists of the Eastern Sahara either came from, or returned to, the South.

The method of strong inference leads us to suppose that the first cattle-pastoralists came from the Nile Valley. The latitudes of Bir Kiseiba and Nabta suggest, specifically, the area between the First and Second Cataracts, which, unfortunately, now lies beneath Lake Nasser. A few early Holocene sites are known from this part of the Valley (Schild et al. 1968), and resemble the Desert occurrences in lithic technology and typology, and in a shared preference for Egyptian flint; they also yielded many cattle bones. However, recent re-examination of the Nilotic cattle bones (Gautier 1984) has shown that the cattle were larger than those in the Desert and fall within the size-range of wild cattle, *Bos primigenius.* This apparently contradictory situation is not, in fact, at variance with a Nilotic origin for the Saharan domestic cattle. Cattle bones are extremely rare in the Saharan sites (Table 1), indicating that they were not kept primarily for meat, but rather for products such as milk and blood, as are the cattle of many East African pastoralists to this day. Such a restricted use of the domestic animals would be maintained during the dry seasons in the Nile Valley – it would make no sense to slaughter them – while the Nilotic wild cattle could be hunted for meat. This leads to the curious situation where the camp-sites of cattle-pastoralists apparently contain the bones only of wild cattle. The bones of occasional domestic animals that died (or were killed) in the Valley are statistically lost among the much larger wild population. This implies that the domestic cattle are archaeologically invisible not only in the dry season sites of established pastoralists, but also in those sites occupied when domestication was first beginning. We may recognise the early Saharan cattle as domestic on ecological grounds, but are quite unable to recognise the first domestic cattle in the Nile Valley.

In short, and despite Smith's efforts to the contrary, our data indicate the presence of domestic cattle in the Eastern Sahara by the tenth millennium BP. We suppose that these were cattle first domesticated at an earlier, but unguessable, date in the Nile Valley. It is unfortunate that Smith chose to devote almost all of his review article to what was a rather minor point in a much farther reaching prehistory. It is doubly unfortunate that Smith did not know that further research and considerably more data (Wendorf et al. 1984b) have only confirmed those initial tentative suggestions that he regards as so misleading.

REFERENCES

Bökönyi, S. 1974. *History of domestic animals in Central and Eastern Europe.* Budapest: Akademíai Kiado.

Breman, H. & C.T.de Wit 1983. Rangeland productivity and exploitation in the Sahel. *Science* 221: 1341-1347.

Gabriel, B. 1977. Early and mid-Holocene climate in the eastern central Sahara. In D.Dalby, R.J.Harrison-Church & F.Bezzaz (eds.), *Drought in Africa.* London: International Africa Institute.

Gautier, A. 1976. Freshwater mollusks and mammals from Upper Palaeolithic sites near Idfu and Isna. In F.Wendorf & R.Schild, *Prehistory of the Nile Valley.* New York: Academic Press.

Gautier, A. 1978. La faune de vertébrés des sites épipaléolithiques d'Elkab. In P.M.Vermeersch, *Elkab II. L'Elkabien, Epipaléolithique de la Vallée de Nil Egyptien.* Leuven: Fondation Egyptienne Reine Elisabeth & Universitaire Pers.

Gautier, A. 1980. Contributions to the archaeozoology of Egypt. In F.Wendorf & R. Schild, *Prehistory of the Eastern Sahara.* New York: Academic Press.

Gautier, A. 1984. Archaeozoology of the Bir Kiseiba region, Eastern Sahara. In F.Wendorf, R. Schild (assemblers) & A.E.Close (ed.), *Cattle-keepers of the Eastern Sahara: the Neolithic of Bir Kiseiba.* Dallas: Southern Methodist University, Department of Anthropology.

Lasota-Moskalewska, A. 1977. Ocena morfologiczna kości bydlęcych z 'Gajowizny'. In J.Kowalczyk (ed.), *Cmentarzysko Kultury Amfor Kulistych w. Złotej Sandomierskiej.* Wrocław: Ossolineum.

Lasota-Moskalewska, A. 1980. Morphological changes of domestic cattle skeletons from Neolithic age to the beginning of the Iron Age. *Wiadomości Archeologiczne* 45: 119-163.

Platt, J.R. 1964. Strong inference. *Science* 146: 347-353.

Schild, R., M.Chmielewska & H.Więckowska 1968. The Arkinian and Shamarkian industries. In F.Wendorf (ed.), *Prehistory of Nubia* Vol. II. Dallas: Fort Burgwin Research Center/SMU Press.

Smith, A.B. 1984. The origins of food production in northeast Africa. In J.A.Coetzee & E.M.van Zinderen Bakker Sir., *Palaeoecology of Africa and the surrounding islands* Vol. 16. Rotterdam: Balkema.

Wendorf, F. & R.Schild 1980. *Prehistory of the Eastern Sahara.* New York: Academic Press.

Wendorf, F., R.Schild (assemblers) & A.E.Close (ed.) 1980. *Loaves and fishes: the prehistory of Wadi Kubbaniya.* Dallas: Southern Methodist University, Department of Anthropology.

Wendorf, F. et al. 1984a. New radiocarbon dates on the cereals from Wadi Kubbaniya. *Science* 225: 645-646.

Wendorf, F., R.Schild (assemblers) & A.E.Close (ed.) 1984b. *Cattle-keepers of the Eastern Sahara: the Neolithic of Bir Kiseiba.* Dallas: Southern Methodist University, Department of Anthropology.

ARCHAEOLOGICAL SEDIMENTS FROM LEOPARD'S HILL CAVE, ZAMBIA

R.L.BURGESS
University of Chicago, Illinois, USA

SUMMARY

Matrix attached to the excavated fauna from Leopard's Hill Cave (Zambia) was systematically removed to provide preliminary data on the important archaeo-sedimentary sequence at this site. The analytical results show remarkable overall continuity in textural composition, but indicate a prominent role for human and animal activities in deposition. Evidence for more rapid chemical disaggregation of the schistose roof rock after 16 500 BP may indicate a slightly moister environment than during the previous nine millennia.

METHODS AND RESULTS

Leopard's Hill Cave, in south-central Zambia (15°25′S; 28°45′E), is located just south of the Muchinga Escarpment, on the plateau, at more than 1 220 m above sea-level. The area receives about 815 mm of rainfall annually, mostly during the summer, and the surrounding vegetation formerly was a *Brachystegia* woodland, now cleared for farming. The cave, one of a series at the base of a dolomitic ridge, is significant for its unusually continuous record of cultural and environmental changes over the past 25 000 years. Excavations by L.Hodges and J.D.Clark in 1958, as well as an extensive lithic analysis, were the topic of a dissertation by Miller (1969). A preliminary analysis of the faunal material by Kaufulu et al. (1978) has recently been superseded by the detailed study of Klein (1984).

Although the site represents a key Later Stone Age (Nachikufan) sequence, resting on a non-descript, macrolithic 'Proto-LSA', no geological study was carried out during the original excavations and subsequent reinvestigation (Miller 1969). The faunal collections reported on by Klein (1984) contained variable, but often considerable amounts of attached matrix. In order to provide at least preliminary data for this important archaeo-sedimentological sequence, they were systematically detached. Eighteen of these samples, of

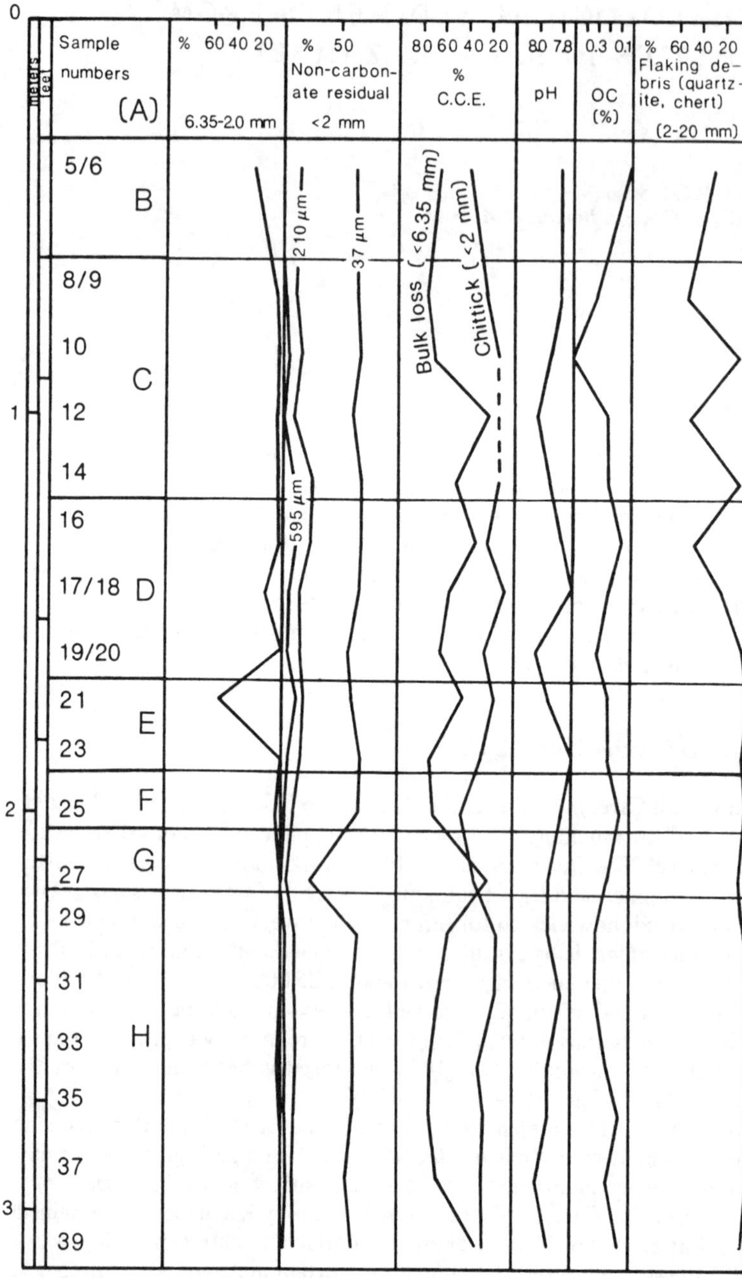

Figure 1. Results of analytical program

which four represented a combination of two superposed levels, were sufficiently large (mean 26.7 g) for an exploratory study. Publication appears warranted because no Quaternary cave sediments have yet been studied from Zambia, and only one such site is available from Zimbabwe (Redcliff, see Klein 1978).

The analytical program involved decalcification in dilute hydrochloric acid, wet-sieving with a set of standard mesh sieves, and removal of flaking micro-debris (quartzite and chert) from the fraction coarser than 2 mm during microscopic scanning. These basic data are presented in Figure 1, with textures as described by standard pedological nomenclature in Table 1. Statistical parameters are not shown: because of the unorthodox nature of the samples, in part cemented to bone, they would lend an unwarranted impression of precision. However, the granulometric results are so uniform that the Folk median and sorting indices (and their 1 σ values) show minimal variation (4.58 ± 0.66 and 2.30 ± 0.18 respectively for the fraction under 2 mm). Munsell colours (dry) and calcium carbonate equivalent (CCE) were determined for whole samples by both bulk loss in acid (fraction under 6.4 mm) and the Chittick gasometric method (fraction under 2 mm). Finally pH, organic carbon (OC), and phosphorus (P) were analyzed by Archaeological Services, Inc. There was no 'available' P in any of the samples due to the high degree of calcification.

The results can be discussed briefly, with reference to the stratigraphic column presented in Table 1 (based in part on Miller (1969)). The cumbersome descriptive terminology has been arbitrarily replaced by units A through H, as shown.

The non-carbonate columns in Table 1 exhibit remarkably little change through time. Of interest are a higher fine silt/clay fraction in unit G, somewhat coarser sands in parts of C, D and E, and higher grit levels in B, D and, particularly, E. Interpretation of these grain sizes and their variability is impeded by a lack of comparative surface samples from above and below the cave entrance, and especially by a lack of appropriate lithological information. Although described as a dolomite cave, all the primary minerals – quartz, feldspar, and mica – are derived from a light gray, metamorphosed, sandy shale, suggesting that the cave formed by solution under a non-carbonate interbed of some sort. Derived hematite concretions are present, and common in units E and H. The carbonates represented are secondary, including stalagtite fragments (units E and H); calcified concretions, often with mesocrystalline calcite (units C, B and H); and a variable degree of matrix saturation by calcite, often in combination with phosphates. In fact, most of the difference between the Chittick and bulk-soluble CCE values is due to mobilization of phosphates and other 'organic' compounds by the latter procedure. There also appears to be a stalagmite growth in the cave, comformable with units C and D (Miller 1969: Fig. 12.1).

Although the cave strata dip downwards toward the back, the processes of sediment accumulation can only be determined in the field; however, the

Table 1. Summary of stratigraphic information

Stratigraphic unit	Cultural designation	^{14}C dates	Munsell colour (dry)	Texture	Excavation data
A. 'Upper dark earth' 0-30 cm	Iron Age	1 415 + 125 (SR 126)	Not analyzed	Not analyzed	
B. 'Lower dark earth' 30-61 cm	Nachikufan III	2 865 + 95 (SR 141) 3 500 + 80 (UCLA 1 429 C)	7.5 yr brown to dark brown	Very calcareous, gritty sandy loam	Much large rubble thin black crust; living floor at base
C. 'Red/brown earth' 61-122 cm	Nachikufan II A/B	9 700 + 85 (UCLA 1 290)	7.5 yr brown to dark brown	Calcareous sandy loam	Numerous $CaCO_3$ lenses; calcified concretions in basal sample
D. 'Dark brown earth' 122-168 cm	Nachikufan I		7.5 yr brown to dark brown to 10 yr brown to 5 yr red brown	Calcareous sandy loam; in part gritty	Coarse river sand, fine pebbles, and rolled schist and bone in top 8 cm; calcified concretions in samples to 152 cm; unconformable basal contact
E. 'Brown earth' 122-191 cm	Nachikufan I	16 715 + 95 (SR 138)	7.5 yr brown to dark brown	Calcareous sandy loam to gritty sandy loam	
F. 'Breccia' 191-206 cm			7.5 yr brown	Very calcareous sandy loam	
G. 'Gray brown earth' 206-221 cm	Macrolithic	16 400 + 265 (UCLA 1 291) 21 550 + 950 (GXO 957)	7.5 yr brown	Very calcareous sandy loam	Much snail shell noted
H. 'Red/brown earth' 221-312 cm	Macrolithic	22 600 + 510 (UCLA 1 429 B) 23 600 + 360 (UCLA 1 429 A)	7.5 yr brown	Calcareous sandy loam	$CaCO_3$ lenses present local stalactite fragments

degree of cave enclosure and the similarity of detached roof rock with the disaggregated residual sediment suggest a principally autochthonous origin. Under magnification, all the debris coarser than 2 mm show a similar, marked degree of chemical weathering, implying that the type of breakdown has remained basically similar. However, the variable amounts of such debris in units B, D and E, as well as the peak of finer material in G, seem to suggest that rates of sedimentation and/or chemical weathering have varied through time; this is demonstrated by an 84.4% coefficient of variability for the amount of sediment coarser than 6.4 mm among the eighteen samples analyzed.

Plotting the average ^{14}C for each unit (see Miller 1969) in relation to mean sediment accumulation (Table 1) produces uniform sedimentation rates (27 cm per 1 000 years) throughout, except for a hiatus between B and C. If on the other hand, both dates cited for unit G are correct and in sequence, accumulation of level G may have been unusually slow (6 cm per 1 000 years), while that of H would have been more rapid than average (42 cm per 1 000 years), with units E and F deposited almost instantaneously. The high percentage of fine silt/clay from level G supports the latter interpretation, although it is at least equally probable that the ^{14}C dates are, in part, misleading. In any event, the accumulation rates do not allow for more rapid deposition to explain the grit peaks in units B, D and E. They can therefore best be attributed to more effective or rapid chemical disaggregation of the roof rock. This would suggest more persistent or more rapid water percolation from above since 16 500 BP, i.e. a slightly moister environment than during deposition of the earlier units.

Microscopic scanning does not bear out the presence of coarse river sand and fine pebbles in unit D, as claimed in the original excavation notes (Miller 1969: 347). These notes also indicate that unit H rested on a 'breccia', for which there would appear to be no samples (Miller 1969: Fig. 12.1). However, the depth of samples 37 and 39, according to other field notes (Miller 1969: 306), puts them within this breccia; the material in the coarser-than-6.4 mm fraction consists of fragments of edge-corroded schist, some large and dense hematite concentrations, and pieces of quartzite. This appears to be the 'breccia', but it is unclear whether this rubble was related to more rapid, chemical roof disaggregation or to eluviation of the finer matrix, e.g. by running water.

A second possible perspective on environmental change is provided by the carbonates, specifically the presence of broken stalagtites in samples 21, 31, 33, 35, 37 and 39, and of cementation horizons in samples 14, 16, 18, 33 and 37. Interpretation is ambiguous without the benefit of appropriate field evaluation, and the background Chittick CCE values for these same strata are all average or even well below. In default of analog data, these precipitates could equally well reflect increased carbonate solution and mobilization upslope, or reduced carbonate mobilization and more rapid precipitation within the cave.

It must be emphasized that the background physical composition of the Leopard's Hill strata is primarily homogeneous, implying relatively subdued changes in the controlling environmental parameters. Most of the variability evident reflects on cultural and biological factors. The flaking micro-debris prominent in levels B, C and D records prominent artifact-making within the cave. Hearths and ash horizons are present throughout the column, particuarly in units B to E (see Miller 1969: Fig. 12.1). The designation 'red' earth, for level C, is not borne out in the Munsell colour record, but the sands from samples 5/6, 7, 10/12, 15, 31 and 34 (units B-D, H) frequently show ferruginous films or discolouration, due to oxidation in hearth zones. The abundance of micro-mammalian bone in samples 33 and 37 and the high micro-bone weights in levels 3, 11 and 28 (Kaufulu et al. 1978: Fig. 5b) argue that owl pellets were a significant component in at least unit H. Micro-bone again peaks in level 16, but was largely identified as bat (Miller 1969: 305). Altogether, the strong organic carbon readings in upper C, lower D and upper H, combined with the high non-carbonate solubility in upper C, lower D, in E and in most of H, suggest plant and animal residues – due either to human activity or animal predators in the cave.

In conclusion, Leopard's Hill Cave does not suggest major environmental changes during the last 25 000 years or so, which is in keeping with the faunal interpretation of Klein (1984). More rapid chemical disaggregation of the schistose roof rock since 16 500 BP may possibly indicate an increase in moisture, but the ambiguous record of carbonate precipitation cautions against any climatic inferences without proper field investigation. By contrast, the record of human and animal activities within the cave is prominent, if not pre-eminent.

ACKNOWLEDGEMENTS

Professors J.Desmond Clark, Sheryl E.Miller and Richard G.Klein facilitated study of the Leopard's Hill Cave sediments, while Professor Karl W.Butzer provided suggestions for analytical procedures and interpretation.

REFERENCES

Kaufulu, Z., K.Schick & J.Sept 1978. Preliminary analysis of the faunal assemblage from Leopard's Hill Cave, Zambia. Unpublished manuscript.

Klein, R.G. 1978. Preliminary analysis of the mammalian fauna from Redcliff Stone Age cave site, Rhodesia. *Occasional Papers, Nat. Mus. Rhodesia* A 4 (2): 74-80.

Klein, R.G. 1984. Later Stone Age faunal samples from Heuningsneskrans Shelter (Transvaal) and Leopard's Hill Cave (Zambia). *S. Afr. Archaeological Bull.* 39: 109-116.

Miller, S.E. 1969. The Nachikufan industries of the Late Stone Age in Zambia. Unpublished Ph.D. dissertation, Univ. of California, Berkeley.

BOOK REVIEWS AND ANNOUNCEMENTS

Palaeoenvironments of holocene fossil Man in Morocco. Contribution of palynology. By A.Ballouche

Holocene palaeoenvironments of fossil Man in Morocco are described. Palynological data are combined with data obtained by other methods.

In the first part, the methodology is explained, especially the definition and the meaning of the principal palynological taxa selected. The pollen morphology of the three species of the genus *Pistacia* L. in Morocco (*P. atlantica, P.lentiscus, P.terebinthus*) has been studied. Their specific determinations is particularly important for palaeoecological reconstructions.

The second part presents the palynological study of holocene deposits from sites located in low relief areas of Morocco, in two different geographic regions.

Atlantic Morocco: On the basis of pollen analysis of lagoonal and estuarine deposits and coastal peats (Oualidia, Wad Tahadart, Larache) the palaeoenvironments could be reconstructed. From the history of these palaeoenvironments, we can deduce the palaeogeographical and climatic evolution of the sites. The human impact, whose modalities are specified, is progressively more important from the middle Holocene onward. Anthropogenic events are responsible for the recent vegetational landscapes.

Eastern Morocco: Holocene lacustrine and travertine formations of Ksabi basin (Middle Moulouya) have been analysed. An important change of palaeoenvironments related to a major climatic occurrence is observed around 8 500 years BP.

Pollen analysis of archaeological sites (Caves of El-Khil and Kaf Taht el-Ghar) shows some palaeoethnological aspects of the neolithic occupation of North Morocco: cultivation, animal husbandry and selective food gathering.

The results are viewed in the regional context of North Africa and the North-western Sahara.

Key-words: Morocco; Holocene; palynology; pollen; *Pistacia*; palaeoenvironments; palaeogeography; palaeoclimate; action of Man.

Paléoenvironnements de l'Homme fossile holocène au Maroc. Résumé de Thèse de l'Université de Bordeaux I. Par A.Ballouche

Les paléoenvironnements holocènes de l'Homme fossile au Maroc sont précisés, les résultats obetnus par la palynologie étant intégrés à ceux provenant d'autres méthodes.

Dans la première partie, la méthodologie est exposée, avec la définition et la signification des principaux taxons palynologiques retenus. La morphologie pollinique des trois espèces marocaines du genre *Pistacia* L. (*P.atlantica, P. lentiscus, P.terebinthus*), dont la distinction est particulièrement importante pour les reconstitutions paléoécologiques, est étudiée.

La seconde partie est consacrée à l'étude palynologique de sédiments holocènes en zones non montagneuses, dans deux domaines géographiques différents:

Maroc atlantique: l'Analyse palynologique de sédiments lagunaires, estuariens et de tourbes littorales (Oualidia, Tahadart, Région de Larache) permet la reconstitution des paléoenvironnements végétaux. De l'histoire de ces paléoenvironnements est déduite l'évolution paléogéographique et climatique des sites. L'action de l'homme, dont les modalités sont précisées, est de plus en plus forte depuis l'holocène moyen. Elle est à l'origine des paysages végétaux actuels.

Maroc oriental: l'Etude a porté sur des dépôts lacustres et travertineux de l'Holocène inférieur et moyen du Bassin de Ksabi (Moyenne Moulouya). Un important changement dans les paléoenvironnements, attribué à un événement climatique majeur, se manifeste vers 8 500 ans BP.

Certains aspects paléoethnologiques du Néolithique du Maroc septentrional ont pu être mis en évidence par l'analyse pollinique de sites archéologiques en grotte (El-Khil, Kaf taht el-Ghar): cultures, élevages, collecte sélective de certaines espèces végétales.

L'ensemble des résultats est élargi à l'échelle de l'Afrique du Nord et du Sahara Nord-occidental.

Mots-clés: Maroc; Holocène; palynologie; pollen; *Pistacia*; paléoenvironnements; paléogéographie; paléoclimat; action de l'Homme.

Changements globaux en Afrique durant le Quaternaire. Passé-présent-futur. INQUA-ASEQUA Symposium International 1986, 527 pp. Edited by H. Faure, L.Faure and E.S.Diop. Obtainable from: ORSTOM. Librairie Diffusion, 70 route d'Aulnay, F93140 Bondy, France; 300 FF + postage.

This impressive volume contains the extended summaries of the 130 papers which were submitted at the INQUA-ASEQUA symposium held at Dakar, Senegal, from 21-28 April 1986. These summaries, which cover up to six or even eight pages, are of the nature of concise articles as most of them are provided with tables, diagrams and extended bibliographies.

The meeting at Dakar was organised by INQUA in co-operation with ASEQUA (Association Sénégalaise pour l'Etude du Quaternaire de l'ouest Africain). It was supported and patronized by an impressive host of international scientific unions and received substantial financial support from the French research organisations CNRS and ORSTOM.

An analysis of the countries from which the 130 summaries originate gives the following picture: 54 come from France, 30 from African states, while 12 are the result of the co-operation between France and an African state. Similarly four contributions come from the co-operation of France with a non-African state and 29 summaries originate from other non-African states. Germany is represented by nine papers and England by one. The figures illustrate the considerable scientific activity of France and its former colonies in Quaternary research of Africa. A total of 90 summaries is written in French and 40 in English.

The symposium was concentrated on the Sahara and surrounding countries and oceans. Outside this region six papers deal with Kenya, one with Angola and two with South Africa. Regions that receive much attention are the Sahara (14 papers), Senegal (14 papers), and Niger (5 papers). Subjects of high priority which are discussed are: lacustrine and lake deposits (14 papers), sea levels (9), mangroves (6), archaeology (7), dating (7) and pollen (13). A great number of general aspects receive much attention, such as: the astronomical theory, palaeomagnetism, geodesy, deep sea cores, albedo and long climatic sequences.

It is not possible to discuss the wealth of information presented in this volume. Surveying the content of this report it is realised that Africa consists of a very large French speaking part and a smaller English speaking section. The counterpart of the Dakar meeting was the SASQUA symposium organised at Mbabane (Swaziland) in 1983 on *Late Cainozoic Palaeoclimates of the Southern Hemisphere.* In the proceedings of the Mbabane meeting (Vogel 1984) East and southern Africa are treated in 24 articles, all written in English. Both these meetings were very successful and the two volumes which resulted from them present a very valuable survey of the state of the art.

The division of Africa in two linguistic parts is important when considering the future development of the natural sciences in the continent. The organising societies concerned with Quaternary research in Africa, INQUA, ASEQUA and SASQUA, could pay more attention to this situation. *Palaeoecology of Africa* can also assist here to bridge the gap between the two parts of Africa by publishing extended reviews and summaries of research in the 'other' language. The Dakar volume gives a good example as a small number of summaries is presented in both French and English! A continuation of this system would greatly assist in our understanding of the Quaternary problems of Africa, which does not consist of two halves but stretches so well balanced over the equator from 35°N to 35°S.

The Dakar volume is a store-house of information for every one interested in the Quaternary history of Africa. It testifies to the enormous growth of our knowledge during the last twenty years, when ASEQUA started its activities. The front cover of the volume deserves special attention as it gives an impressive coloured 'Météostat' scanning satellite picture of the thermic conditions of Africa between 15-21 November 1984.

The organisers of the symposium and especially Professor and Madame Faure and Dr Diop deserve deep gratitude from all Quaternarists for arranging this meeting and for publishing this excellent book.

Reference: Vogel, J.C. (ed.) 1984. *Late Cainozoic Palaeoclimates of the Southern Hemisphere.* Balkema, Rotterdam, 520 pp.

E.M. v.Z.B.

ADDRESSES

BALLOUCHE, A., INSAP, Ave John Kennedy, Cashier Postal, Rabat-Souissi, Morocco.
BAUMHAUER, R., Geographisches Institut der Universität Würzburg, Am Hubland, D 8700-Würzburg, W.Germany.
BESLER, H., Geographisches Institut, Universität zu Köln, Albertus-Magnus Platz, 5000 Köln, W.Germany.
BURGESS, R.L., University of Chicago, Illinois 60637, USA.
BURNEY, D.A., Department of Zoology, Duke University, Durham, NC 27706, USA.
CLARK, J.DESMOND, Department of Anthropology, University of California, Berkeley, Ca 94720, USA.
CLOSE, A.E., Department of Anthropology, Southern Methodist University, Dallas, Texas 75275, USA.
COETZEE, J.A., Department of Botany, University of the Orange Free State, Bloemfontein 9300, South Africa.
COHEN, A.S., Department of Geosciences, University of Arizona, Tucson, AZ 85721, USA.
DELIBRIAS, G., Centre des Faibles Radioactivités, Domaines du CNRS, 91190, Gif s/Yvette, France.
DESHAYES, M., Deputy Director, Institut Français de Pondichéry, 10 St Louis Str. BP 33, Pondicherry 605001, India.
DIESTER-HAASS, L., Geographisches Institut der Universität Mannheim, Abt. für Geologie, Postfach 2428, 6800 Mannheim, W.Germany.
FAURE, H., Laboratoire de Géologie du Quaternaire, CNRS, Luminy, 13288 Marseille, Cedex 2, France.
FLOHN, H., Meteorologisches Institut, Universität Bonn, Auf dem Hügel 20, 53 Bonn-Endenich, W.Germany.
GAUTIER, A., Laboratorium voor Paleontologie, Rijksuniversiteit Gent, Krijgslaan 281/S 8, B 9000 Gent, Belgium.
HALL, K.J., Geography Department, University of Natal, Pietermaritzburg 3201, South Africa.
HAMILTON, A., Department of Environmental Studies, Coleraine Co., Londonderry, UK.
HEINE, K., Institut für Geographie, Universität Regensburg, POB 397, D-8400, Regensburg, W.Germany.
HOOGHIEMSTRA, H., Institute of Palynology and Quaternary Sciences, University of Göttingen, Wilhelm-Weber-Strasse 2, D-3400 Göttingen, W.Germany. New address: Hugo de Vries-Laboratorium, Kruislaan 318, 1098 SM Amsterdam, the Netherlands.
KRÖPELIN, S., Institut für Physische Geographie, Freie Universität Berlin, Grünewaldstr. 35, D-1000 Berlin (West) 41, W.Germany.
KUKLA, G.J., Lamont-Doherty Geological Observatory of Columbia University, Palisades, NY 10964, USA.

LANCASTER, N., Department of Geology, Arizona State University, Tempe, AZ 85287, USA.
MAHANEY, W.C., Department of Geography, Atkinson College, York University, 4700 Keele Street, North York, Ontario M3J 1P3, Canada.
MALEY, J., ORSTOM, UR 103 & CNRS, UA 327, Institut des Sciences de l'Evolution, Laboratoire de Palynologie, Université des Sciences et Techniques du Languedoc, Montpellier 34060 F, France.
MAUD, R.R., 68 Ridge Road, Durban 4001, South Africa.
MECO, J., Escuela Universitaria del Profesorado de EGB, Universidad de la Laguna, Las Palmas 34004, España.
MERCER, J.D., Ohio State University, Institute of Polar Studies, 103 Mendenhall, 125 South Oval Mall, Columbus, Ohio 43210, USA.
NEUMANN, K., Botanisches Institut der J.W.Goethe Universität, Frankfurt am Main, W.Germany.
PARTRIDGE, T.C., School of Earth Sciences and Palaeoanthropology Research Group, University of the Witwatersrand, 1 Jan Smuts Avenue, Johannesburg 2001 South Africa.
PETIT-MAIRE, N., Directeur de Recherches, Laboratoire de Géologie du Quaternaire, CNRS, Case 907, 13288 Marseille, Cedex 9, France.
POMEL, S., Institut de Géographie, Faculté des Lettres, Bd.M.Schuman, 13000 Aix-en-Provence, France.
RISER, J., Laboratoire de Géologie du Quaternaire, CNRS, Luminy, Case 907, 13288 Marseille, Cedex 9, France.
ROGNON, P., Géodynamique des milieux continentaux et UA 722 (CNRS) UER des Sciences de la Terre, Université Pierre et Marie Curie, 4 Place Jussieu, 75252 Paris, Cedex 05, France.
ROSSO, J.C., Laboratoire de Géologie du Quaternaire, CNRS, Luminy, Case 907, 13288 Marseille, Cedex 9, France.
SCHILD, R., Polska Akademia Nauk, Instytut Historii Kultury Materialnej, ul. Swierczewskiego 105, 00-140 Warsaw, Poland.
SCHULZ, E., Geographisches Institut der Universität Würzburg, Am Hubland, D-8700 Würzburg, W.Germany.
SKOWRONEK, A., Geographisches Institut der Universität Würzburg, Am Hubland, D-8700 Würzburg, W.Germany.
TOBIAS, P.V., Palaeo-anthropology Research Unit, Department of Anatomy, University of the Witwatersrand, Medical School, Parktown, Johannesburg 2193, South Africa.
VAN ZEIST, W., Biologisch-Archaeologisch Instituut, State University of Groningen, Poststraat 6, 9712 ER Groningen, the Netherlands.
VINCENS, A., Laboratoire de Géologie du Quaternaire, CNRS, Case 907, Luminy, 13288 Marseille, Cedex 9, France.
WENDORF, F., Department of Anthropology, Southern Methodist University, Dallas, Texas 75275, USA.